Lecture Notes in Mathematics

Edited by A. Dold and B. Eckmann

1046

Algebraic K-Theory, Number Theory, Geometry and Analysis

Proceedings of the International Conference held at Bielefeld, Federal Republic of Germany, July 26–30, 1982

Edited by A. Bak

Springer-Verlag
Berlin Heidelberg New York Tokyo 1984

Editor

Anthony Bak
Fakultät für Mathematik, Universität Bielefeld
4800 Bielefeld, Federal Republic of Germany

AMS Subject Classifications (1980): 12A35, 12A60, 12A62, 12A70, 13D15, 14C25, 14C35, 14F15, 14F99, 16A54, 18F25, 18F30, 18G30, 20C99, 46-02, 46M20, 46M99, 57R52, 57R67

ISBN 3-540-12891-3 Springer-Verlag Berlin Heidelberg New York Tokyo
ISBN 0-387-12891-3 Springer-Verlag New York Heidelberg Berlin Tokyo

Printing and binding: Beltz Offsetdruck, Hemsbach/Bergstr.
2146/3140-543210

PREFACE

A conference on algebraic K-theory and its occurrences in algebraic number theory, algebraic topology, algebraic geometry, and analysis was held in Bielefeld from July 26 to July 30, 1982. The conference was supported jointly by the Deutsche Forschungsgemeinschaft, the office of the Rektor Prof. K. P. Grotemeyer, and the research contracts of Professors H. Abels, U. Rehmann, F. Waldhausen, and the editor of this volume. The volume contains papers submitted by scientists invited to the conference. Roughly half the articles published here record details of an address delivered at the conference.

It is well known that algebraic K-theory came into being as an attempt to systemize in algebraic terms certain developments in algebraic geometry and algebraic topology and conversely, to provide algebraic tools to apply to the latter. Since its inception, the subject has continually undergone refinement and extensions, due in part to its internal dynamics, but more importantly to its contact with other branches of mathematics. The processes above have made a precise definition of the subject impossible and have at the same time contributed to breaking down traditional boundaries between established branches of mathematics. Today, algebraic K-theory is a loosely defined collection of concepts and methodologies, no longer purely algebraic, which are providing a common basis for a wide and ever growing variety of specialisms in mathematics. The papers of the current volume are part of an ongoing record of the above.

It is a pleasure for me to give thanks to Günter Habdank, Wilhelm Koberg, Ulf Rehmann, and Siegfried Splitthoff for their untiring and alert assistance during the conference. I would like also to thank our secretaries Frau Gudrun Dräger and Frau Elsa Oberschelp-Asensio for their help in making these proceedings available.

Anthony Bak

Bielefeld, October, 1983

LIST OF PARTICIPANTS AND AUTHORS

Dr. Anthony Bak
Fakultät für Mathematik
Universität Bielefeld
Universitätsstr. 1
4800 Bielefeld 1
Federal Republic Germany

Dr. Jon Berrick
Department of Mathematics
National University of Singapore
Kent Ridge
Singapore 0511

Dr. J. Brinkhuis
Department of Mathematics
King's College
Strand
London WC 2R,
England

Dr. Jerzy Browkin
Instytut Matematyki UW
Palac Kultury i Nauki
PL-00-901 Warszawa
Poland

Dr. John L. Bryant
Department of Mathematics
Florida State University
Tallahassee, Fl. 32306
USA

Dr. D. Burghelea
Ohio State University
Columbus, Ohio 43210
USA

Dr. Ph. Cassou-Noguès
U.E.R. de Mathématiques
Université de Bordeaux I,
Laboratoire associé au C.N.R.S.N°226,
33405 Talence Cedex
France

Dr. Ruth Charney
I.H.E.S.
35, Route de Chartres
91440 Bures-sur-Yvette
France

Dr. Jean-Louis Colliot-Thélène
81 av. du general Leclerc
75014 Paris
France

Dr. Joachim Cuntz
Department of Mathematics
University of Pennsylvania
Philadelphia, PA 19104
USA

Dr. Z. Fiedorowicz
Ohio State University
Columbus, Ohio 43210
USA

Dr. Henri Gillet
Department of Mathematics
Princeton University
Fine Hall, Box 37
Princeton, NJ 08544
USA

Dr. Thomas Goodwillie
Department of Mathematics
Harvard University
Cambridge, MA 02138, USA

Dr. Daniel R. Grayson
Department of Mathematics
University of Illinois
Urbana, Ill.61801
USA

Dr. Daniel Guin
Département de Mathématiques
Université Louis Pasteur
6700 Strasbourg
France

Dr. Jürgen Hurrelbrink
Department of Mathematics
Louisiana State University
Baton Rouge, Louisiana 70803
USA

Dr. Kiyoshi Igusa
Department of Mathematics
Brandeis University
Waltham, Mass. 02254
USA

Dr. W. van der Kallen
Department of Mathematics
Postbus 80.010
508 TA Utrecht
The Netherlands

Dr. Christian Kassel
Départment de Mathématiques
Université de Strasbourg
7 rue Déscartes
67084 Strasbourg
France

Dr. M.E. Keating
Mathematics Department
National University of Singapore
Kent Ridge
Singapore 0511

Dr. M. Kervaire
Institute de Mathématique
2 rue du Lievre
Genève 1211
Switzerland

Dr. Frans Keune
Mathematisch Instituut
Katholike Universiteit
Toernooiveld
6525 ED Nijmegen
The Netherlands

Dr. Manfred Kolster
Mathematisches Institut
Westf. Wilhelms-Universität
Einsteinstr. 62
4400 Münster
Federal Republic Germany

Dr. F. Kouwenhoven
Mathematisch Instituut
Rijksuniversiteit Utrecht
Budapestlaan 6
3508 TA Utrecht
The Netherlands

Dr. Adoremi Kuku
Department of Mathematics
University of Ibadan
Ibadan, 21051
Nigeria

Dr. Jean-Louis Loday
Départment de Mathématique
Université Louis Pasteur
7, rue René Descartes
67084 Strasbourg Cedex
France

Dr. Bruce A. Magurn
Department of Mathematics
Cornell University
Ithaca NY 14850
USA

Dr. Hans J. Munkholm
Matematisk Institut
Odense Universitet
Campusvej 55
5230 Odense
Denmark

Dr. R. Oliver
Matematisk Institut
Ny Munkegade
Aarhus Universitet
8000 Aarhus C
Denmark

Dr. William Pardon
Department of Mathematics
Duke University
Durham, NC 27706
USA

Dr. Erik K. Pedersen
Department of Mathematics
Odense Universitet
Campusvej 55
5230 Odense M
Denmark

Dr. Ulf Rehmann
Fakultät für Mathematik
Universität Bielefeld
Universitätsstr. 1
4800 Bielefeld 1
Federal Republic Germany

Dr. J.-J. Sansuc
Ministère des Universités
Ecole Normale Supérieure
45, rue D'Ulm
75230 Paris Cedex 05
France

Dr. John R. Silvester
Department of Mathematics
King's College London, Strand
London, WC2R 2LS
England

Dr. Victor P. Snaith
Department of Mathematics
University of Western Ontario
London, Ontario N6A 5B7
Canada

Dr. Christophe Soulé
U.E.R. de Mathématiques
Université de Paris VII
75003 Paris Cedex
France

Dr. Michael R. Stein
Department of Mathematics
Northwestern University
Lunt Hall
Evanston, Illinois 60201
USA

Dr. Jan Stienstra
Mathematisches Institut
der Rijksuniversiteit te Leiden
Wassenaarseweg 80
2300 RA Leiden
The Netherlands

Dr. Jan Strooker
Mathematisches Institut
der Rijksuniversiteit
Budapestlaan, De Uithof
Utrecht
The Netherlands

Dr. Ulrich Stuhler
Fachbereich 7, Mathematik
Gesamtschule Wuppertal
Gaußstr. 20
5600 Wuppertal
Federal Republic Germany

Dr. A.A. Suslin
Department of Mathematics
Fontanka 27
Leningrad 191011
U.S.S.R.

Dr. Martin Taylor
Department of Mathematics
Trinity College
Cambridge
England

Dr. Pierre Vogel
Institut de Mathématiques
Université de Nantes
2, rue de la Houssinière
44072 Nantes Cedex
France

Dr. Wolrad Vogell
Fakultät für Mathematik
Universität Bielefeld
Universitätsstr. 1
4800 Bielefeld 1
Federal Republic Germany

Dr. Ton Vorst
Department of Mathematics
Econometric Institute
Erasmus University
P.O. Box 1738
3000 DR Rotterdam
The Netherlands

Dr. Jack Wagoner
Department of Mathematics
University of California
Berkeley, California 94704
USA

Dr. F. Waldhausen
Fakultät für Mathematik
Universität Bielefeld
Universitätsstr. 1
4800 Bielefeld 1
Federal Republic Germany

Dr. Charles Weibel
Department of Mathematics
Rutgers University
New Brunswick, N.J. 08903
USA

Dr. Bruce Williams
Department of Mathematics
University of Notre Dame
Notre Dame, Indiana 46556
USA

PROGRAM OF CONFERENCE

Monday, 26. 7.:

9:30 Opening ceremonies

10:00 V. Snaith: A survey of Bott periodic algebraic K-Theory

11:30 B. Williams: Surgery and higher K-Theory

14.15 R. Oliver: Class groups and the Artin cokernel

16:00 E.I. Pedersen: On the $K_{-i}(R)$-functors

Tuesday, 27. 7.:

9:30 J. Cuntz: C*-Algebras and K-Theory

11:00 M. Kolster: Generalized Dennis-Stein symbols and K_2

14:15 M. Taylor: Galois module structure of rings of integers

16:00 J. Wagoner: The global to p-adic problem in algebraic K-Theory

17:15 J. Stienstra: Formal group laws, O-cycles and K_2

Wednesday, 28. 7.:

9:30 C. Soulé: On the rank of K-groups of varieties over $\mathbb{Q}$

11:00 A. Berrick: The plus-construction on a fibration

14:15 C. Kassel: Central extension of Lie algebras

16:00 J.-L. Loday: Connes' homology, Lie algebras and K-Theory

17:15 F. Keune: K_2 of a 1-fold stable ring

17:15 R. Charney: K-Theory of ideals

Thursday, 29. 7.:

9:30 V. Snaith: Some cases of the Lichtenbaum conjecture

11:00 H. Munkholm: Transfer maps on K- and L-Theory coming from group extensions $\mathbb{Z}/n\mathbb{Z} \to \pi \to \varphi$

14:15 W. v.d. Kallen: The (Merkurjev-)Suslin preprints

16:00 J.-L. Colliot-Thélène: Chow groups of rational surfaces

17:15 W. Pardon: Zero cycles and quadratic forms

17:15 A. Kuku: Equivariant K-Theory for profinite groups and Galois cohomology

20:00 J. Brinkhuis: Galois modules and embeddings of number fields

Friday, 30. 7.:

9:30 D. Grayson: Lattices in Euclidean space

11:00 J. Hurrelbrink: On the Birch-Tate conjecture and generation by symbols

14:15 H. Gillet: K-Theory and cohomology of simplicial schemes with applications to regular maps and intersection theory on module problems

16:00 C. Weibel: Double relative K_1

17:15 A. Vorst: On the K_n-regularity index

TABLE OF CONTENTS

A.J. BERRICK, Group epimorphisms preserving perfect radicals, and the plus-construction . . . 1

J. BRINKHUIS, K_2 and Galois extensions of fields . . . 13

J. BROWKIN and J. HURRELBRINK, On the generation of $K_2(\sigma)$ by symbols . . . 29

D. BURGHELEA and Z. FIEDOROWICZ, Hermitian algebraic K-theory of topological spaces . . . 32

R.M. CHARNEY, A note on excision in K-theory . . . 47

J. CUNTZ, K-theory and C^*-algebras . . . 55

H. GILLET, Homological descent for the K-theory of coherent sheaves . . . 80

K. IGUSA, What happens to Hatcher and Wagoner's formula for $\pi_0 C(M)$ when the first Postnikov invariant of M is nontrival? . . . 104

W. v.d. KALLEN, A note on excision for K_2 . . . 173

M.E. KEATING, The K-theory of triangular rings and orders . . . 178

F. KEUNE, The K_2 of a 1-fold stable ring . . . 193

M. KOLSTER, On the Birch-Tate conjecture for maximal real subfields of cyclotomic fields . . . 229

A. KUKU, Equivariant K-theory and the cohomology of profinite groups . . . 235

H.J. MUNKHOLM, Transfer maps on K_i, $i \leq 1$, associated with certain periodic resolutions . . . 245

R. OLIVER, An exact sequence involving $K_1(\hat{\mathbb{Z}}_p\pi)$ and $K_2(\hat{\mathbb{Z}}_p\pi)$ . . . 255

W. PARDON, A relation between Witt groups and zero-cycles in a regular ring . . . 261

V. SNAITH, An upper bound for algebraic K-theory . . . 329

V. SNAITH, Towards the Lichtenbaum-Quillen conjecture concerning the algebraic K-theory of schemes . . . 349

A.A. SUSLIN, Homology of GL_n, characteristic classes and Milnor K-theory . . . 357

M.R. STEIN, Generating the cokernel of $K_3\mathbb{Z} \to K_3\mathbb{F}_7$ . . . 376

P. VOGEL, Une nouvelle famille de groupes en L-theorie algebrique . . . 385

T. VORST, A survey on the K-theory of polynomial extensions . . . 422

C.A. WEIBEL, Algebraic K-theory and the Adams e-invariant . . . 442

C.A. WEIBEL, Mennicke-type symbols for relative K_2 . . . 451

GROUP EPIMORPHISMS PRESERVING PERFECT RADICALS,

AND THE PLUS-CONSTRUCTION

A. J. Berrick

If one uses the plus-construction approach to algebraic K-theory, then various key problems (for example, obstruction to excision) have natural topological formulations (such as a fibre sequence being plus-constructive). Here we swing the pendulum back, to obtain an algebraic setting for discussing these matters. Central to this viewpoint is a certain class of group homomorphisms.

1. Definition

To define the class, recall that a group P is perfect if equal to its commutator subgroup [P,P]. Because the homomorphic image of a commutator is a commutator, so

1.1 *the homomorphic image of a perfect group is again perfect.*

The class of perfect subgroups of a given group G is therefore closed under automorphisms of G. It is also evidently closed under group union, because if each subgroup H_α's generated by its commutators, then so is the subgroup the H_α's all generate. Thus the class admits a maximum element, the *perfect radical* $\mathcal{P}G$ of G, which must be a fully invariant subgroup. The construction is functorial because, from 1.1,

1.2 *if* $\phi : G \to H$ *is a homomorphism, then* $\phi(\mathcal{P}G) \leq \mathcal{P}(\phi G) \leq \mathcal{P}H$.

An alternative approach here is to view $\mathcal{P}G$ as the intersection of the transfinite derived series of G.

We enquire under what conditions equality holds in (1.2). Since it is the image ϕG to which we wish to restrict consideration, suppose ϕ to be an epimorphism. We seek hypotheses to ensure that $\phi(\mathcal{P}G) = \mathcal{P}(\phi G)$, in other words, that ϕ is EP^2R - an Epimorphism (or, in context, Extension) Preserving Perfect Radicals.

For an extreme example of an epimorphism which is not EP^2R, let the extension $R \rightarrowtail F \overset{\phi}{\twoheadrightarrow} P$ correspond to a free presentation of a perfect group P. Thus F, being free, has only free non-trivial subgroups, and no free group can be perfect. So although $\mathcal{P}F = 1$, making $\phi\mathcal{P}F = 1$, we have $\mathcal{P}P = P$. On the other hand, there are no examples from finite group theory : we shall see that any surjection with finite kernel is EP^2R.

Note that solubility of G forces triviality of $\mathcal{P}G$ (that is, G is *hypoabelian*) since among n-th derived groups we must have $(\mathcal{P}G)^{(n)} \leq G^{(n)}$. In fact,

1.3 *The following three conditions are equivalent :-*

(i) G *is soluble;*

(ii) $\mathcal{P}G = 1$ *and for some* i $G^{(i)}$ *is finite;*

(iii) $\mathcal{P}G = 1$ *and for some* j $G^{(j)}/Z(G^{(j)})$ *is finite.*

The proof is an easy exercise, save for a lemma of Schur to the effect that $G^{(i)}/Z(G^{(i)})$ finite implies $G^{(i+1)}$ finite.

2. Algebraic results

Our main purpose here is to establish (in (2.3) below) conditions on a group G and normal subgroup N = Ker ϕ sufficient to ensure that an epimorphism $\phi : G \twoheadrightarrow Q$ is EP^2R. Several such conditions are collated as Proposition 2.3. We find it useful (for Lemma 3.7) to relate one of these to the following question. Since evidently the composition of EP^2R maps is EP^2R, what can be said about the converse? Again, it is easy to see that if $\psi : G \to H$, $\phi: H \to Q$ have composite $\phi\circ\psi : G \to Q$ EP^2R then ϕ must also be EP^2R; so, under what conditions will ψ be EP^2R as well?

LEMMA 2.1 : *Let* $\psi : G \to H, \phi : H \to Q$ *be epimorphisms such that, for some finite* n,

$$(\text{Ker } \phi)^{(n)} \leq \psi\mathcal{P}G.$$

Then

(a) ϕ *is* EP^2R; *and*

(b) ψ *is* EP^2R *if and only if* $\phi\circ\psi$ *is* EP^2R.

Proof. Let $J \trianglelefteq H$ denote the inverse image $\phi^{-1}(PQ)$. From the lemma (2.2) below, $J^{(n)} = J^{(n+1)}(\mathrm{Ker}\ \phi)^{(n)}$. However it is known that $(\mathrm{Ker}\ \phi)^{(n)} \leq \psi PG \leq PH = (PH)^{(n+1)} \leq J^{(n+1)}$ (after (1.2)). So $J^{(n)} \leq J^{(n+1)}$, whence $J^{(n)}$, being perfect, lies in - and is therefore equal to - PH. Thus PH has the same image as $J^{(n)}$, namely $PQ^{(n)} = PQ$. This establishes (a), and thence one implication of (b), when $\phi\circ\psi$ is the composite of EP^2R maps.

Conversely, if $\phi\circ\psi$ is EP^2R, then PH and ψPG both have image PQ, from which one deduces that $PH = (\psi PG).\mathrm{Ker}\ \phi$. Then

$$PH = PH^{(n)} = ((\psi PG).\mathrm{Ker}\ \phi)^{(n)} = \psi PG.(\mathrm{Ker}\ \phi)^{(n)}$$

because $\psi PG \trianglelefteq H$. By assumption this last subgroup lies in ψPG, leaving $\psi PG = PH$ as required.

I am grateful to Prof. B. Hartley for suggesting an argument which led me to the following lemma.

LEMMA 2.2 : *Let* $K \hookrightarrow J \twoheadrightarrow P$ *be an extension where* P *is a perfect group. Then for all finite* $m \geq n \geq 0$,

$$J^{(n)} = J^{(m)}K^{(n)}.$$

Proof. For each fixed m, argue by induction on n. The case $n = 0$ results from the fact that $P = P^{(m)}$, with the latter the image of $J^{(m)}$. On the other hand, whenever $J^{(n-1)} = J^{(m)}K^{(n-1)}$ it follows that

$$J^{(n)} = [J^{(m)}K^{(n-1)}, J^{(m)}K^{(n-1)}] = J^{(m)}K^{(n)}.$$

An immediate application of Lemma 2.1(a) (or (b) with ψ as the identity map) is (iii) of the following result.

PROPOSITION 2.3 : *An extension* $N \hookrightarrow G \overset{\phi}{\twoheadrightarrow} Q$ *is* EP^2R *provided either*

(i) *it is split;*

(ii) $G^{(m)} \leq N.PG$ *for some finite* m;

(iii) $N^{(n)} \leq PG$ *for some finite* n; *or*

(iv) *the homomorphism* $G \to \mathrm{Aut}(N/PN)$, *induced by conjugation, has hypoabelian image in* $\mathrm{Out}(N/PN) = \mathrm{Aut}(N/PN)/\mathrm{Inn}(N/PN)$.

REMARKS. Of the above hypotheses, (iii) is perhaps the most useful, in its application to central and perfect extensions, or, more generally, to extensions whose kernel has finite derived length. The finite ordinals occurring in (ii), (iii) may not be replaced by infinite ones, as the example of a free presentation of a perfect group reveals. Finally, the study of the outer automorphisms induced by an extension is pursued further in [4].

Proof. (i) is just a two-fold application of (1.2), since if $\psi : Q \to G$ is right-inverse to ϕ, then $\phi PG \geq \phi\psi PQ = PQ$. For (ii) we also borrow from (1.2) the fact that $\phi PG \leq PQ$. Now $PQ \leq Q^{(m)}$ for all finite m, so that $PQ \leq \phi(G^{(m)}.N) \leq \phi(PG.N) = \phi PG$.

Thus only (iv) demands further discussion, in the course of which we shall make use of the following immediate consequence of (iii).

COROLLARY 2.4 : *If* $P \trianglelefteq G$ *is perfect, then* $P(G/P) = PG/P$.

By a little extra work, we shall in fact establish the following strengthening of (iv), whose converse will be proved by topological means in §3 below. That its automorphism homomorphism condition is weaker than (iv) may be seen from the commuting diagram

$$\begin{array}{ccccc} G & \to & \mathrm{Aut}(N) & \to & \mathrm{Aut}(N/PN) \\ \downarrow & & \downarrow & & \downarrow \\ Q & \to & \mathrm{Out}(N) & \to & \mathrm{Out}(N/PN) \end{array} .$$

PROPOSITION 2.5 : *An extension* $N \hookrightarrow G \overset{\phi}{\twoheadrightarrow} Q$ *induces the trivial homomorphism* $PQ \to \mathrm{Out}(N/PN)$ *if and only if both*

a) ϕ *is* EP^2R; *and*

b) $[PG,N] = PN$.

In fact, it is only necessary to prove this result when $PN = 1$, since by (2.4) ϕ is EP^2R if and only if $N/PN \hookrightarrow G/PN \twoheadrightarrow Q$ is. (2.4) also implies the equivalence of $[PG,N] = PN$ and $[P(G/PN), N/PN] = 1$. We shall therefore demonstrate this notationally simpler, special case...

LEMMA 2.6 : *An extension* $N \hookrightarrow G \overset{\phi}{\twoheadrightarrow} Q$ *with* $\mathcal{P}N = 1$ *induces the trivial map* $\mathcal{P}Q \to \mathrm{Out}(N)$ *if and only if both*

a) ϕ *is* EP^2R; *and*

b) $[\mathcal{P}G, N] = 1$.

To obtain (a) first, we restrict to the induced extension $N \hookrightarrow H \twoheadrightarrow \mathcal{P}Q$ over $\mathcal{P}Q$. Proceeding as in [4], we observe that the homomorphism $H \to \mathrm{Out}(N)$ has kernel $N.C_H(N)$ and so embeds $H/N.C_H(N)$ in $\mathrm{Out}(N)$. This makes $H/N.C_H(N)$ precisely the image of $\mathcal{P}Q \cong H/N$ in $\mathrm{Out}(N)$ and therefore trivial. Consequently $H = N.C_H(N)$ and there is an extension $N \cap C_H(N) = Z(N) \hookrightarrow C_H(N) \twoheadrightarrow \mathcal{P}Q$. Since the centre of N has derived length at most 1, condition (2.3)iii) is satisfied, whence $C_H(N) \twoheadrightarrow \mathcal{P}Q$ is EP^2R. However, $C_H(N) \leq G$, so that $\phi : G \twoheadrightarrow Q$ is EP^2R too.

From the fact that $\phi(\mathcal{P}C_H(N)) = \mathcal{P}Q$ we have that $\mathcal{P}G \leq N.\mathcal{P}C_H(N)$, giving $\mathcal{P}G = \mathcal{P}(N.\mathcal{P}C_H(N))$. So (2.6)(b) is an immediate application of the next lemma.

LEMMA 2.7 : *Suppose* $A = BC$ *where* B, C *are commuting normal subgroups of* A. *Then*

a) $[\mathcal{P}A, B] = \mathcal{P}B$, $[\mathcal{P}A, C] = \mathcal{P}C$; *and*

b) $\mathcal{P}A = \mathcal{P}B.\mathcal{P}C$.

REMARK 2.8 : *Conditions* (2.7)(a) *and* (b) *are equivalent.*

Note that by applying (2.7)(b) instead of (2.7)(a) we obtain the following. (From (2.6)(a) we may identify H above as $N.\mathcal{P}G$.)

COROLLARY 2.9 : *If the extension* $N \hookrightarrow G \twoheadrightarrow Q$ *induces the trivial map* $\mathcal{P}Q \to \mathrm{Out}(N)$, *then* $\mathcal{P}G = \mathcal{P}N.\mathcal{P}C_{N.\mathcal{P}G}(N)$.

To prove the lemma we verify the assertions P_α, Q_α below, for each ordinal α.

$$P_\alpha : \mathcal{P}A \leq B^{(\alpha)}.C .$$

$$Q_\alpha : [\mathcal{P}A, B] \leq B^{(\alpha)} .$$

Evidently P_0 is true. If α is a limit ordinal, then the truth of Q_β for all $\beta < \alpha$ clearly forces that of Q_α; otherwise, the truth of $P_{\alpha-1}$ gives that of P_α, since

$$\mathcal{P}A = [\mathcal{P}A, \mathcal{P}A] \leq [B^{(\alpha-1)}.C,\ B^{(\alpha-1)}.C] \leq B^{(\alpha)}.C .$$

On the other hand, because when Q_α holds

$$PA = [PA, B.C] \leq [PA, B].C \leq B^{(\alpha)}.C ,$$

we have Q_α implying P_α; the converse is even easier. So transfinite induction clinches all P_α, Q_α. From the latter,

$$[PA, B] \leq \bigcap_\alpha B^{(\alpha)} = PB = [PB, B] \leq [PA, B] .$$

Since the hypotheses of (2.7) require $B \cap C$ to be central in A, that lemma has an interesting generalization whose proof uses Proposition 2.3 (and thereby Lemma 2.7).

PROPOSITION 2.10 : *Suppose* $A = B.C$ *for normal subgroups* B, C *of* A. *If* $B \cap C$ *has either its derived series of finite length or* $\mathrm{Out}(B \cap C)$ *or* $\mathrm{Out}(B \cap C / P(B \cap C))$ *hypoabelian, then*

$$PA.(B \cap C) = PB.PC.(B \cap C) .$$

Let us first check that (2.10) does indeed imply (2.7). For,

$$\begin{aligned} PA &= [PA.(B \cap C),\ PA.(B \cap C)] \\ &= [PB.PC.(B \cap C),\ PB.PC.(B \cap C)] \\ &= [PB.PC,\ PB.PC] = PB.PC. \end{aligned}$$

Now the the proof of (2.10). From the decomposition

$$A/B \cap C \cong B/B \cap C \times C/B \cap C ,$$

we have (after, e.g., (2.3)i))

$$P(A/B \cap C) \cong P(B/B \cap C) \times P(C/B \cap C) .$$

We identify these perfect radicals by appealing to (2.3)iii), iv). Thus any extension $B \cap C \hookrightarrow D \twoheadrightarrow D/B \cap C$ is EP^2R. In particular, we deduce that

$$PA.(B \cap C)/B \cap C \cong PB.(B \cap C)/B \cap C \times PC.(B \cap C)/B \cap C ;$$

whence the result.

3. Connections with the plus-construction

In algebraic K-theory, the starting-point for the definition of K_1A is the Whitehead lemma on the general linear group over the ring A. This identifies $\mathcal{P}GLA$ as EA, the subgroup generated by elementary matrices. Moreover, EA = [GLA,GLA], making GLA a member of the class of groups of finite derived length, which we have seen is well-suited to a discussion of epimorphisms preserving perfect radicals. It was of course the classical fact that $K_1A = GLA/\mathcal{P}GLA$ which led Quillen to the plus-construction and thence the definition $K_iA = \pi_i(BGLA^+)$. Since then there has been much further use of the plus-construction, in topology and geometry. The main motivation for the study of EP^2R maps (at the present time) lies in their close relation to natural questions concerning the plus-construction. Here are two examples (respectively [1 (5.11)] and [1 (6.8)]).

(3.1) *Suppose* $f : X \to Y$ *has connected fibre. Then the commuting diagram*

$$\begin{array}{ccc} X & \xrightarrow{q_X} & X^+ \\ {\scriptstyle f}\downarrow & & \downarrow{\scriptstyle f^+} \\ Y & \xrightarrow{q_Y} & Y^+ \end{array}$$

is co-Cartesian if and only if $\pi_1(f)$ *is* EP^2R.

(3.2) *If the fibre sequence* $F \to E \overset{p}{\to} B$ *(with* F,E,B *connected) is plus-constructive (that is,* F^+ *is also the homotopy fibre of* p^+*), then* $\pi_1(p)$ *is* EP^2R.

The proof of (3.2) in [1] involves a diagram-chasing argument whose evident irreversibility casts considerable doubt on the converse. This suspicion can be validated in two ways.

First, it is possible to state precisely what further condition is needed in order to characterise those fibre sequences (of connected spaces) which are plus-constructive. Thus [3] reveals the following.

(3.3) *A fibre sequence* $F \to E \overset{p}{\to} B$ *is plus-constructive if and only if* $\pi_1(p)$ *is* EP^2R *and the induced action of* $\mathcal{P}\pi_1(E)$ *on* $\pi_*(F^+)$ *is trivial.*

In the special case where F,E,B are classifying spaces of groups with $\pi_1(F)$ hypoabelian, this simplifies greatly.

(3.4) *Let* $N \hookrightarrow G \overset{\phi}{\twoheadrightarrow} Q$ *be a group extension, with* $PN = 1$. *Then the fibre sequence* $BN \to BG \to BQ$ *is plus-constructive if and only if both* ϕ *is* EP^2R *and* $[PG,N] = 1$.

This comes very close to implying the as-yet unproved part of (2.5) above. The gap is filled by the following corollary to the main theorem of [3].

PROPOSITION 3.5 : *An extension* $N \hookrightarrow G \twoheadrightarrow Q$ *with* $PN = 1$ *induces the trivial homomorphism* $PQ \to \mathrm{Out}(N)$ *if and only if the fibre sequence* $BN \to BG \to BQ$ *is plus-constructive.*

The proof of (3.5) (given [3]) simply consists in identifying the group $\mathrm{AUT}(BN^+)$ of (free) homotopy classes of self-homotopy equivalences of BN^+. Since $PN = 1$, this is just $\mathrm{AUT}(BN) = \pi_0(\mathrm{Aut}(BN))$ which is indeed $\mathrm{Out}(N)$ [5 p.42].

This argument does not require that the total space and base space be classifying spaces. There is therefore the following corollary, in some sense dual to the result [2] that every fibre sequence whose base has hypoabelian fundamental group is plus-constructive.

COROLLARY 3.6 : *Suppose both* N *and* Out(N) *are hypoabelian. Then every fibre sequence with fibre* BN *is plus-constructive.*

Algebraic K-theory provides important examples concerning (2.5) and (3.4); it demonstrates the irredundancy of the condition on the subgroup $[PG,N]$. We take for G the general linear group GLUT (after [1 p.27]) on the ring $\begin{pmatrix} A & A \\ 0 & A \end{pmatrix}$ of upper triangular 2×2 - matrices over a given ring A. The split ring epimorphism

$$\begin{pmatrix} A & A \\ 0 & A \end{pmatrix} \longrightarrow \begin{pmatrix} A & 0 \\ 0 & A \end{pmatrix}$$

induces a split epimorphism $\phi : \mathrm{GLUT} \to GL(A \oplus A)$ of general linear groups. By (2.3)i) above, ϕ is certainly EP^2R. Now $N = \mathrm{Ker}\ \phi$ intersects with $GL_n UT$ as matrices $\begin{pmatrix} I_n & M \\ 0 & I_n \end{pmatrix}$ where M is an arbitrary $n \times n$ - matrix over A. So multiplication in N corresponds to addition of finite matrices, making N an abelian subgroup. However, N fails to commute with $PGLUT = EUT$. For example, in GL_2UT the matrices

$$\left[\begin{array}{c|c} I_2 & \begin{matrix} 0 & 0 \\ 1 & 0 \end{matrix} \\ \hline 0 & I_2 \end{array}\right] \quad \varepsilon \ N_\cap GL_2UT$$

and $e^1_{12} \oplus I_2 \; \varepsilon \; E_2UT$ have non-trivial commutator. Hence (by (2.5)) the image of $E(A \oplus A)$ in $Out(N)$ is non-trivial, and from (3.4) the fibre sequence $BN \to BGLUT \xrightarrow{B\phi} BGL(A \oplus A)$ is not plus-constructive. This fact allows the existence of the homotopy equivalence $B\phi^+ : BGLUT^+ \to BGL(A \oplus A)^+$ (since $BN^+ = BN$ is patently not contractible), whose significance embraces the non-extension of lower K-theory Mayer-Vietoris sequences [1 ch.3] and the additivity of the natural map from the A-representation ring of a group G to the algebraic K-group $KA^0(BG)$ [1 (12.3)], [6]. The latter enables one to define further structure on algebraic K-theory [1 chs 12,13].

More generally, for any ring epimorphism $A \twoheadrightarrow A_1$ the induced map $\psi : EA \to EA_1$ is evidently EP^2R (EA is after all perfect). However, since EA has trivial centre, $B\psi$ is only plus-constructive if ψ is an isomorphism. Again, this is the fact which makes so difficult the description of the fibre of $BEA^+ \to BEA_1^+$, and thereby the relative terms of the long exact sequence linking K_*A to K_*A_1.

The second discussion concerning a converse to (3.2) above consists in a characterisation of fibre sequences $F \to E \overset{p}{\to} B$ in terms of the relation between the (homotopy) fibres $F = F_p$ and F_{p^+}. Recall that, to have $F_{p^+} = F^+$, one requires of the map $F \to F_{p^+}$ that it induce an epimorphism on fundamental groups (with kernel $P\pi_1(F)$) and an isomorphism on homology (with arbitrary coefficients). It is thus to be expected that these two data should arise as distinct aspects of the original fibre sequence. This expectation is fulfilled in (3.8), (3.9) below. First, a lemma based on (2.1).

LEMMA 3.7 : *Suppose* $A \overset{f}{\to} B \overset{g}{\to} C$ *is a fibre sequence with* $P\pi_1(C) = 1$, *and that a map* $p : E \to B$ *(with connected homotopy fibre) induces* $p' : A\times_B E \to A$. *Then* $\pi_1(p)$ *is* EP^2R *if and only if* $\pi_1(p')$ *is.*

To prove (3.7), we consider the following commuting diagram, which is exact only in respect of the first pairs of horizontal maps.

$$\begin{array}{ccccccccc}
\pi_2(C) \to \pi_1(A\times_B E) & \xrightarrow{\pi_1(f')} & \operatorname{Im}\pi_1(f') & \hookrightarrow & \pi_1(E) & \xrightarrow{\pi_1(g\circ p)} & \pi_1(C) \\
\Big\| \qquad \downarrow \pi_1(p') & & \downarrow \pi_1(p)| & & \downarrow \pi_1(p) & & \Big\| \\
\pi_2(C) \to \pi_1(A) & \xrightarrow{\pi_1(f)} & \operatorname{Ker}\pi_1(g) & \hookrightarrow & \pi_1(B) & \xrightarrow{\pi_1(g)} & \pi_1(C)
\end{array}$$

Since $\pi_2(C)$ is abelian, Lemma 2.1 yields (a) that $\pi_1(f')$ is EP^2R, and (b) that $\pi_1(p')$ is EP^2R precisely when $\pi_1(p\circ f') = \pi_1(f\circ p')$ is, which by (a) occurs precisely when the restriction $\pi_1(p)|$ is EP^2R. However, the hypothesis that $P\pi_1(C) = 1$ forces $P\pi_1(B)$ to lie inside $\operatorname{Ker}\pi_1(g)$, making $\pi_1(p)|$ EP^2R if and only if $\pi_1(p)$ is.

PROPOSITION 3.8 : *For any fibre sequence* $F \to E \overset{p}{\to} B$ *with* F *connected, the induced fundamental group homomorphism* $\pi_1(F) \to \pi_1(F_{p^+})$ *is surjective if and only if* $\pi_1(p)$ *is* EP^2R.

Proof. By [2 Lemma 2.1] and the lemma above, neither assertion is affected by pulling-back over the fibre sequence $AB \to B \to B^+$ (since of course $P\pi_1(B^+) = 1$). Therefore assume B is already an acyclic space. Then B^+ is contractible, so that the fibre sequence $F_{p^+} \to E^+ \to B^+$ pulls back over $B \to B^+$ to induce a map of fibre sequences over B:

$$\begin{array}{ccc}
F & \xrightarrow{Fg} & F_{p^+} \simeq E^+ \\
& \searrow \qquad \swarrow & \\
& E \overset{g}{\to} E^+\times B & \\
& p \searrow \quad \swarrow & \\
& B &
\end{array}$$

As the maps of total spaces and of fibres share a common homotopy fibre [1 p.35], it follows that $\pi_1(Fg)$ maps $\pi_1(F)$ onto $\pi_1(F_{p^+})$ if and only if

$$\pi_1(g) : \pi_1(E) \to \pi_1(E^+\times B) = \pi_1(B) \times \pi_1(E)/P\pi_1(E)$$

is surjective. Now write π for $\pi_1(E)$, P for $P\pi_1(E)$ and K for $\operatorname{Ker}\pi_1(p)$, so that $\pi_1(B) \cong \pi/K$. There is a map of group extensions

$$\begin{array}{ccccc} KP & \rightarrowtail & \pi & \twoheadrightarrow & \pi/KP \\ \downarrow & & \downarrow \pi_1(g) & & \downarrow \Delta \\ KP/K \times KP/P & \rightarrowtail & \pi/K \times \pi/P & \twoheadrightarrow & \pi/KP \times \pi/KP \end{array} ,$$

where the left-hand surjection is the canonical projection

$$KP \twoheadrightarrow KP/K\cap P \cong P/K\cap P \times K/K\cap P \cong KP/K \times KP/P \ .$$

Since the diagonal inclusion Δ is an isomorphism precisely when its domain is trivial, the diagram shows that $\pi_1(g)$ is onto if and only if $\pi = KP$, which is to say, $\pi_1(E) = \text{Ker}\ \pi_1(p).P\pi_1(E)$. Now the image $\pi_1(B)$ of $\pi_1(p)$ is perfect; thus the last equality is equivalent to $\pi_1(p)$ being EP^2R, as required.

An alternative, and potentially faster, proof (kindly suggested by the referee) uses a generalisation of the Snake Lemma to the case of non-abelian groups. This is applied to the map of the exact sequences of fundamental groups corresponding to p and p^+. The obvious difficulty, of defining the relevant cokernel (to $\pi_1(Fg)$), is circumvented, because the assumption that B is acyclic ensures that $\pi_1(Fg)$ has its image normal in $\pi_1(F_{p^+})$.

Of course (3.2) is an immediate corollary to this proposition. By means of one of the techniques introduced in the proof above, we can also characterise those fibre sequences for which the induced map $F \to F_{p^+}$ is an integral homology equivalence.

PROPOSITION 3.9 : *For any fibre sequence* $F \to E \xrightarrow{p} B$ *with* F *connected, the induced homology homomorphism* $H_*(F) \to H_*(F_{p^+})$ *is an isomorphism if and only if* $P\pi_1(B)$ *acts trivially on* $H_*(F)$.

The sufficiency of the $P\pi_1(B)$-action condition is just [2 Lemma 4.2]; here we demonstrate its necessity. We do this by again taking the induced fibre sequences over AB

$$\begin{array}{ccc} F \to F_{p^+} \simeq \overline{E}^+ \\ \searrow \quad \swarrow \\ \overline{E} \to \overline{E}^+ \times AB \\ \searrow \quad \swarrow \\ AB \end{array}$$

(This is the same diagram as previously. The notation differs here because we shall have to refer to the original fibre sequence.) Since the right-hand fibre sequence is a product fibration, its action of $\pi_1(AB)$ on $H_*(F_{p^+})$ is certainly trivial. The given homology isomorphism forces the action on $H_*(F)$ to be trivial too. However this action is induced from the original action of $\pi_1(B)$ on $H_*(F)$ via the pull-back over $AB \to B$ of the original fibre sequence. Hence the result follows from the fact that $P\pi_1(B)$ is the image of the homomorphism $\pi_1(AB) \to \pi_1(B)$.

References

1. A. J. Berrick, *An Approach to Algebraic K-Theory,* Research Notes in Math. 56, Pitman (London, 1982).

2. ——— , The plus-construction and fibrations, *Quart. J. Math. Oxford* (2) 33 (1982), 149-157.

3. ——— , Characterisation of plus-constructive fibrations, *Advances in Math.,* to appear.

4. ——— , Group extensions and their trivialisation, to appear.

5. P. E. Conner & F. Raymond, Deforming homotopy equivalences to homeomorphisms in aspherical manifolds, *Bull. Amer. Math. Soc.* 83 (1977), 36-85.

6. D. Quillen, Characteristic classes of representations, in *Algebraic K-Theory Evanston* 1976, Lecture Notes in Math. 551 (Springer, 1976), 189-216.

Department of Mathematics
National University of Singapore
Kent Ridge, Singapore 0511
Republic of Singapore

K_2 AND GALOIS EXTENSIONS OF FIELDS*

Jan Brinkhuis

By the main theorem of Galois theory, one knows that for each finite Galois extension of fields F/E with Galois group Γ, the fixed field F^Γ is equal to E. Equivalently, in terms of algebraic K-theory, the map α_i from K_iE to $(K_iF)^\Gamma$, induced by the inclusion of E into F, is an isomorphism if $i = 1$. The situation is different for $i = 2$: both kern α_2 and cokern α_2 can be non-trivial. At present all one knows in general about these groups is that they are $|\Gamma|$ - torsion. In the hope that they are interesting invariants, I have tried to compute them in a few special cases.

Let k be a field containing a primitive n-th root of unity ξ, let X be a transcendental over k and consider the Kummer extension $k(X)/k(X^n)$. The map α_2 has then the same kernel and cokernel as the map "raising to the n-th power" from the multiplicative group k^* to the subgroup of k^* of those elements x for which the Matsumoto symbol $\{x, \xi\}$ is trivial in K_2k. The cokernel of this last map has been computed by Tate in the case that n is a prime number and that k is a number field. It is then the product of $r_2 + 1$ cyclic groups of order n, where r_2 is the number of complex embeddings of k (see Th. 6.3 of [T]).

Let k be a field of characteristic $p > o$ and let X be a transcendental over k. We consider the Artin-Schreier extension $k(X)/k(X^p-X)$. The map α_2 is then an isomorphism. In fact the injectivity follows a fortiori from Suslin's result that the K_2 of a field of characteristic p has no p-torsion (see Th. 1.10 of [S]).

Let Γ be a finite group of rotations of the unit sphere. To this one associates classically a faithful action of Γ on the field $F = \mathbb{C}(X)$ of rational functions of one variable X over the complex numbers. We consider the Galois extension F/F^Γ. We will show how to calculate kern α_2 and cokern α_2 from the given presentation of Γ as rotation group of the sphere.

Finally we will consider Galois extensions F/E of number fields. If $E = \mathbb{Q}$ and F is totally real abelian, we will derive via Theorem (6.6) of [T] an explicit formula for kern α_2. Moreover for arbitrary cyclic extensions F/E of number fields it is proved that one has $|\text{cokern } \alpha_2| = |\text{kern } \alpha_2|$ if the transfer map tr : $K_2F \to K_2E$ is onto : this last condition is for example satisfied if $E = \mathbb{Q}$ and if moreover F is totally real.

* I would like to thank Prof. J.R. Strooker for introducing me to algebraic K-theory. Prof. U. Rehmann kindly communicated the results in section 4 to me.

Part one : extensions of type $k(X)/k(X)^{\Gamma}$.

1. An exact sequence.

(1.1) Let k be a field. Let the field F be a purely transcendental extension of k of degree one. Let Γ be a finite group of automorphisms of F over k. Our aim is to compute the kernel and cokernel of the map α_2 from $K_2(F^{\Gamma})$ to $(K_2F)^{\Gamma}$ which is induced by the inclusion of F^{Γ} into F.

(1.2) With valuation on F- or on F^{Γ}- we will always mean a discrete valuation which is trivial on k. For each valuation w on F, we let e_w denote the ramification index of w over F^{Γ} and we let Γ_w denote the decomposition group of w, that is $\Gamma_w = \{\gamma \in \Gamma \mid \gamma(w) = w\}$. We recall that e_w depends only on v, the valuation on F^{Γ} which lies under w. So we can write e_v instead of e_w. We let S denote the finite set of valuations w on F with $e_w \neq 1$ and we let R denote the finite set of valuations v on F^{Γ} with $e_v \neq 1$. We set $n = |\Gamma|$.

For each $r \in N$ we consider the maps

$$(1.3)\quad k^* \xrightarrow{d_1} k^* \xrightarrow{d_2} K_2k$$

where $d_1 = d_1(k,r)$ is defined by $x \mapsto x^r$ and $d_2 = d_2(k,r)$ by $y \mapsto \{y, \xi_r\}$ where ξ_r is a chosen generator of ker d_1 (Warning: notice that ξ_r need not be a primitive r-th root of unity). Clearly (1.3) is a complex, that is $d_2 d_1 = 0$. Let $H_0(k,r)$ and $H_1(k,r)$ denote the homology groups of this complex, that is

$$(1.4)\quad H_0(k,r) = \ker d_1$$

$$H_1(k,r) = \frac{\ker d_2}{\operatorname{im} d_1}$$

(1.5) Theorem. We make the following assumptions.

(1.6) (i) All $w \in S$ have residue field k.

(ii) The groups Γ_w $(w \in S)$ generate Γ.

Then there exists a canonical exact sequence

$$(1.7)\qquad 1 \to \operatorname{kern} \alpha_2 \to \bigoplus_{v\in R} H_0(k,e_v) \xrightarrow{t} H_0(k,n) \to \operatorname{cokern} \alpha_2 \to \bigoplus_{v\in R} H_1(k,e_v) \xrightarrow{u} H_1(k,n)$$

where t is given by $(x_v)_v \to \prod_v x_v$ and u by $(y_v)_v \to \prod_v y_v^{\,n/e_v}$

(1.8) Proof. The main problem is to give, for each $\tilde{w}\in S$, a good criterion for an element of K_2F to be $\Gamma_{\tilde{w}}$-fixed.

For each valuation w on F we let k(w) denote its residue field. Recall that one has a homomorphism from K_2F to $k(w)^*$ given by $x \to x_w$ where, for a Matsumoto symbol $x = \{f, g\}$, x_w is defined to be the

class of $(-1)^{w(f)w(g)} f^{w(g)} g^{-w(f)}$.

(1.9) Lemma. For each $\tilde{w}\in S$, the following conditions on an element $x\in K_2F$ are equivalent

(i) x is $\Gamma_{\tilde{w}}$-fixed

(ii) $(x_w)_w$ is $\Gamma_{\tilde{w}}$-fixed and $x_{\tilde{w}}\in \ker d_2(k,e_{\tilde{w}})$

(1.10) Proof of the Lemma. By assumption (1.6)(i) $\tilde{w}$ has residue field k. Let X^{-1} be a local parameter of $\tilde{w}$, that is $X\in F$, $F = k(X)$ and $\tilde{w}(X^{-1}) = 1$. Thus $\tilde{w} = w_\infty$ is the valuation "at $X = \infty$", given by $w_\infty(f) = -\deg_X(f)$. Notice that, for each $\gamma \in \Gamma_{w_\infty}$, we have

$$(1.11)\qquad \gamma(X) = a_\gamma X + b_\gamma \text{ with } a_\gamma\in k^*,\ b_\gamma\in k.$$

We need the standard split exact sequence (see [M] p. 106)

$$(1.12)\qquad 1 \to K_2k \overset{c}{\underset{\leftarrow}{\to}} K_2F \to \bigoplus_{w\neq w_\infty} k(w)^* \to 1$$

where the last map is given by $x \to (x_w)_w$ and where the splitting homomorphism c is defined by $\{f,g\} \to \{\text{lead. coef. } f, \text{ lead. coef. } g\}$. Notice that the homomorphisms of this sequence, except c, preserve Γ_{w_∞}-action. Now we come to a crucial formula

$$(1.13)\qquad c({}^{\gamma}x) = c(x)\ \{x_{w_\infty}, a_\gamma\} \text{ for all } \gamma\in\Gamma_{w_\infty} \text{ and all } x \in K_2F$$

It is enough to prove (1.13) for $x = \{f,g\}$ with $f, g \in k[X]$, $f = a X^r + \ldots\ldots, g = b X^s + \ldots\ldots$

(1.14) $$c({}^{\gamma}x) = c(\{a\, a_\gamma^r\, X^r + \ldots\ldots,\ b\, a_\gamma^s\, X^s + \ldots\}) =$$

$$\{a\, a_\gamma^r,\ b a_\gamma^s\} = \{a,\ b\}\ \{(-1)^{rs}\, a^s b^{-r},\ a_\gamma\} = c(x)\ \{x_{W_\infty},\ a_\gamma\}$$

By (1.12) and (1.13) the following two conditions on an element $x \in K_2F$ are equivalent

(i) x is Γ_{W_∞} - fixed

(ii)' $(x_W)_W$ is Γ_{W_∞} - fixed and $\{x_{W_\infty},\ a_\gamma\} = 1$ for all $\gamma \in \Gamma_{W_\infty}$.

Now we observe that, in order to complete the proof of Lemma (1.9), it suffices to prove the following statement.

(1.15) The sequence

$$\Gamma_{W_\infty} \xrightarrow{\ m\ } k^* \xrightarrow{\ d_2(k,e_{W_\infty})\ } k^*$$

is exact with d_2 as in (1.3) and with m defined by $\gamma \to a_\gamma$ (see (1.11)).

(1.16) Notice first that the group kern m can be embedded into the additive group of the field k by $\gamma \to b_\gamma$ (with b_γ as in (1.11)).

(1.17) Assume that char $k = 0$. Then, by (1.16) and the finiteness of Γ, it follows that kern $m = 0$, that is, m is injective, so $|m(\Gamma_{W_\infty})| = |\Gamma_{W_\infty}|$ which is, by (1.6)(ii), equal to e_{W_∞}. The rest of the proof of (1.15) is easy.

(1.18) Assume now that char $k = p > 0$. Then kern m has, by (1.16), exponent p, so $|\text{kern } m|$ is a power of p. As a field of char p has no non-trivial p-th roots of unity, we have $(|\text{im } m|, p) = 1$. As $e_{W_\infty} = |\Gamma_{W_\infty}| = |\text{kern } m| \cdot |\text{im } m|$ we conclude that $|\text{im } m|$ is the non p-part of e_{W_∞}. The rest of the proof of (1.15) is again easy. This finishes the proof of Lemma (1.9). □

(1.19) Remark. In the proof of (1.15) we have seen that assumption (1.6)(i) implies the presence of certain roots of unity in k : for each v with $e_v \neq 1$, the polynomial $X^{e_v} - 1$ splits completely into linear factors over k.

Now we continue the proof of Theorem (1.5). We need the following facts.

(1.20) $\left(\bigoplus_{w|v} k(w)^*\right)^\Gamma = k(v)^*$ for each valuation v on F^Γ

(1.21) F, and so, by Lüroth's theorem, F^Γ, is a purely transcendental extension of k of degree one and for such extensions L/k one has the following exact sequence (which we have used already in a slightly different form : see (1.12)):

(1.22) $1 \to K_2k \to K_2L \to \bigoplus_w k(w)^* \to k^* \to 1$

Here the first map is induced by inclusion $k \hookrightarrow L$, the second map is defined by $x \to (x_w)_w$ (as in 1.12) and the last map is defined by

$$(x_w)_w \to \prod_w N_{k(w)/k}\, x_w,$$

where N = norm.

Now we are going to construct a commutative diagram

$$\begin{array}{ccccccccc} 1 \to & K_2k & \to & K_2(F^\Gamma) & \to & \bigoplus_v k(v)^* & \longrightarrow & k^* & \longrightarrow 1 \\ & \downarrow \mathrm{id} & & \downarrow \alpha_2 & & \downarrow (d_1(k(v),e_v))_v & & \downarrow d_1(k,n) & \\ 1 \to & K_2k & \to & (K_2F)^\Gamma & \to & \bigoplus_v \mathrm{kern}\ d_2(k(v),e_v) & \to & \mathrm{kern}\ d_2(k,n) & \end{array} \tag{1.23}$$

with exact horizontal sequences.

(1.24) Remark. If $e_v = 1$, then

$$\ker d_2\ (k(v),\ e_v) = k(v)^*$$

and

$$d_1\ (k(v),\ e_v) = id_{k(v)^*}.$$

As top sequence in (1.23) we take the exact sequence (1.22) for $L = F^\Gamma$.

One readily verifies using (1.9), (1.20) and (1.24) that there is a unique commutative diagram with exact horizontal sequences

$$(1.25)\quad \begin{array}{ccccccccccc} 1 & \to & K_2k & \to & (K_2F)^\Gamma & \to & \bigoplus_v \operatorname{kern} d_2(k(v),e_v) & \to & \operatorname{kern} d_2(k,n) & & \\ & & \downarrow{\scriptstyle id} & & \downarrow & & \downarrow & & \downarrow{\scriptstyle *} & & \\ 1 & \to & K_2k & \to & K_2F & \to & \bigoplus_w k(w)^* & \longrightarrow & k^* & \longrightarrow & 1 \end{array}$$

such that the bottom sequence is (1.22) for $L = F$ and such that the vertical maps are the obvious ones, e.g. the third one is defined via (1.20).

As bottom sequence in (1.23) we now take the top sequence in (1.25). The entire diagram (1.23) is now constructed. Its commutativity is easily verified.

Finally one produces the desired exact sequence (1.7) by a standard lemma - of - the - snake - type argument in diagram (1.23), using definition (1.4). □

2. Special cases.

Recall that the map from $PGl_2(k)$ to $Aut(k(X)/k)$ defined by

$$\overline{\begin{pmatrix} a & b \\ c & d \end{pmatrix}} \longrightarrow \left(X \to \left(\frac{aX + b}{cX + d}\right)\right)$$

is an isomorphism of groups. Therefore every finite subgroup Γ of $PGl_2(k)$ gives us an example of a Galois extension of fields $k(X)/k(X)^\Gamma$. Often the conditions (1.6)(i) and (ii) are satisfied and then Theorem (1.5) can be applied.

We will consider in this section the following special cases [always assuming (1.6)(i) and (ii)].

I Γ is a cyclic group

II k is a finite field

III $k = \mathbb{C}$.

(2.1) Case I. We have to distinguish two cases

(2.2) I (a) char $k = n$, so n is a prime p

I (b) char $k \neq n$

Let σ be a generator of Γ. By choosing a suitable generator X of F over k one can always achieve [using (1.6)(i) and (ii)] the following in the respective cases

(2.3) I (a) $\sigma(X) = X + 1$

I (b) $\sigma(X) = \xi X$

for some primitive n-th root of unity ξ in k.

(2.4) I(a) We are considering the Artin-Schreier extension $F = k(X)/E = k(X^p - X)$. One readily sees that the valuation w_∞ on F has ramification index $p = [F : E]$ over E. We recall that the valuations on $F = k(X)$ other than w_∞ correspond to the maximal ideals of $k[X]$. For each conjugate $\sigma^i(X)$ of X over E which is $\neq X$, the difference $\sigma^i(X) - X = i$ [by (2.2) I (a)] is clearly a unit in the ring $k[X]$. By ramification theory it follows now that no valuation on F other than w_∞ is ramified over E. We find therefore using Theorem (1.5) that kern $\alpha_2 = 1$ and cokern $\alpha_2 = 1$, that is, α_2 is an isomorphism.

(2.5) I (b). Now we are considering the Kummer extension $F = k(X)/E = k(X^n)$. With similar arguments as above one finds that w_∞ and w_0 (the valuation "at $X = 0$") are the only valuations on F which ramify over E. They have both ramification index $n = |\Gamma|$. This gives via Theorem (1.5)

(2.6) kern $\alpha_2 \simeq \mu_n$, the group of n-th roots of unity in k

cokern $\alpha_2 = \{x \in k^* \mid \{x, \xi\} = 1 \text{ in } K_2k\}/(k^*)^n$

(2.7) Examples [in case I (a)]

1) k is finite : as the K_2 of a finite field is trivial, we find cokern α_2 $\approx k^*/(k^*)^n$, so cokern α_2 is cyclic of order n.

2) k is algebraically closed : then $k^* = (k^*)^n$, so cokern $\alpha_2 = 1$.

3) k is an algebraic number field and n is a prime number : combining (2.6) with Th (6.3) of [T] we find that cokern α_2 is a product of $r_2 + 1$ cyclic groups of order n (where r_2 is the number of complex embeddings of k).

(2.8) Case II. k is a finite field $\mathbb{F}_q$. Here we will only consider the example that Γ is the group of all automorphisms of $F = k(X)$ over k of the form $X \to aX + b$ with $a \in k^*$ and $b \in k$. One easily finds that there are two valuations of $E = F^\Gamma$ which are ramified in F, one of them has $e = q - 1$, $f = 1$, $g = q$, the other one has $e = q\,(q - 1)$, $f = g = 1$. Using moreover that $K_2\mathbb{F}_q = 1$, it follows from the exact sequence (1.7) that both kern α_2 and cokern α_2 are cyclic of order $q - 1$.

(2.9) Case III $k = \mathbb{C}$. Now Theorem (1.5) boils down to the following.

(2.10) The map α_2 has the same kernel and cokernel as the map from $\bigoplus_{v \in R} \mu_{e_v}$ to μ_n given by $(x_v)_v \to \prod_v x_v$ (where μ_r denotes the group of r-th roots of unity in $\mathbb{C}$).

(2.11) Now we recall some wellknown facts (see e.g. [V]). Place the complex plan $\mathbb{C}$ in three dimensional Euclidean space E_3 and place also a two dimensional unit sphere S_2 in E_3 such that its centre coindices with the origin 0 of the complex plane. Let T (= top) be one of the two points on the sphere of which the perpendicular projection on the complex plane is 0. Projecting <u>from centre T</u> gives a bijection between C and S_2-T. Furthermore we let T correspond to "infinity" ∞. Thus a bijection between the projective line $\mathbb{C} \cup \{\infty\}$ and S_2 is established. It is a classical theorem that each finite subgroup of $PGl_2(\mathbb{C})$ is conjugate to one which corresponds to a group of rotations of S_2. These are the following ones

(i) C_n, the cyclic group of order n.
(ii) D_n, the dihedral group of order n (n even).
(iii) T, the tetrahedral group.
(iv) O, the octahedral group.
(v) I, the icosahedral group.

(2.12) One easily computes the following table.

Γ		kern α_2	cokern α_2
C_n		(n)	0
D_n	$n \equiv 0 \pmod 4$	(2,2)	(2)
	$n \equiv 2 \pmod 4$	(2)	0
T		(3)	(2)
O		(2)	(2)
I		0	(2)

(2.13) For example, take the case Γ = T and consider the action of T on the sphere. One sees readily that there are precisely three T-orbits with less than $|T| = 12$ elements : the set 0_1 of vertices (of the tetrahedron), the set 0_2 of midpoints of the edges and the set 0_3 of barycentres of the triangles. We observe that the points of the sphere correspond to the valuations on F. Let v_i be the valuation on E $= F^\Gamma$ corresponding to 0_i, then $e_{v_i} = |T| \cdot |0_i|^{-1}$ $(i = 1,2,3)$.

We conclude now, using (2.10), that α_2 has the same kernel and cokernel as the map from $\mu_2 \oplus \mu_3 \oplus \mu_2$ to μ_{12} given by $(x,y,z) \to x\,y\,z$. So kern $\alpha_2 \simeq (3)$ and cokern $\alpha_2 \simeq (2)$.

(2.14) Remark. The same approach as the one given above (in sections 1. and 2.) should also work in the case of higher K's. In particular, in the case I (b), that is $E = k(X^n)$ and $F = k(X)$, one should also get that the map $\alpha_i : K_iE \to (K_iF)^\Gamma$ has the same kernel and cokernel as the map from $K_{i-1}k$ to

$$\{x \in K_{i-1}k \mid x \circ \xi = 1 \text{ in } K_ik\}/n(K_{i-1}k)$$

given by $x \to nx$. Here o denotes the K-theoretic cupproduct from $K_{i-1}k \times K_1k$ to K_ik.

Part two : some extensions of number fields.

3. Consequences of a theorem of Tate.

Firstly we recall the definition of the $\Omega_{\mathbb{Q}}$ - module $(\mathbb{Q}/\mathbb{Z})(2)$ from [T] [where $\Omega_{\mathbb{Q}} = \mathrm{Gal}(\bar{\mathbb{Q}}/\mathbb{Q})$].

(3.1) $$(\mathbb{Q}/\mathbb{Z})(2) = \bigoplus_{\ell} (\mathbb{Q}_\ell/\mathbb{Z}_\ell)(2)$$

(sum over all prime numbers)

$$(\mathbb{Q}_\ell/\mathbb{Z}_\ell)(2) = \mathbb{Q}_\ell/\mathbb{Z}_\ell \otimes_{\mathbb{Z}_\ell} \varprojlim_r \mu_{\ell^r} \otimes \varprojlim_r \mu_{\ell^r}$$

Here the inverse limits are taken with respect to the maps $\mu_{\ell^{r+1}} \to \mu_{\ell^r}$ given by $x \to x^\ell$; the action of $\Omega_{\mathbb{Q}}$ is defined as follows : trivially on $\mathbb{Q}_\ell/\mathbb{Z}_\ell$ and via the Galois action on the other factors. Now we recall Theorem (6.6) and its Corollary from [T] :

Theorem (i). For each totally real number field F, there is a canonical isomorphism

$$g_F : H^1(\Omega_F, (\mathbb{Q}/\mathbb{Z})(2)) \xrightarrow{\sim} K_2F$$

(ii) Moreover for each subfield E of F the following diagram is commutative

$$\begin{array}{ccc} H^1(\Omega_E, (\mathbb{Q}/\mathbb{Z})(2)) & \xrightarrow{g_E} & K_2E \\ \downarrow & & \downarrow \\ H^1(\Omega_F, (\mathbb{Q}/\mathbb{Z})(2)) & \xrightarrow{g_F} & K_2F \end{array}$$

where the vertical maps are the canonical ones.

(3.2) Corollary. Let F/E be a Galois extension of totally real number fields with Galois group Γ. Then there are canonical isomorphisms

$$\operatorname{kern} \alpha_2 \simeq H^1(\Gamma, (\mathbb{Q}/\mathbb{Z})(2)^{\Omega_F})$$

$$\operatorname{cokern} \alpha_2 \simeq \operatorname{kern}\,[H^2(\Gamma, (\mathbb{Q}/\mathbb{Z})(2)^{\Omega_F}) \overset{\inf}{\to} H^2(\Omega_E, (\mathbb{Q}/\mathbb{Z})(2))]$$

Proof : immediate from the Theorem by using the exact sequence of Hochschild-Serre.□

(3.3) In the rest of this section we restrict ourself to the following case : $E = \mathbb{Q}$ and F is a totally real abelian number field.

(3.4) We define, for each prime number ℓ, F_ℓ to be the smallest number field for which Ω_{F_ℓ} acts trivially on $[(\mathbb{Q}_\ell/\mathbb{Z}_\ell)(2)]^{\Omega_F}$ (it is of course a subfield of F).

We need a Lemma.

(3.5) Lemma. $H^i(G(F_\ell/\mathbb{Q}), (\mathbb{Q}_\ell/\mathbb{Z}_\ell)(2)^{\Omega_F}) = 0$ for all $i \geq 1$.

Outline of the proof. By standard arguments, using the exact sequence of Hochschild-Serre, it is seen that it suffices to prove the following :

let $\Delta_1 \subset \Delta_2$ be open subgroups of $(1 + \ell\,\mathbb{Z}_\ell)^2$ then

$H^i(\Delta_2/\Delta_1, (\mu_{\ell^\infty})^{\Delta_1}) = 0$ for all $i \geq 1$.

($\mathbb{Z}_\ell^*$ acts on the group μ_{ℓ^∞} of ℓ-th-power roots of unity by power raising).

As Δ_2/Δ_1 is necessarily cyclic and as $(\mu_{\ell^\infty})^{\Delta_1}$ is finite, it suffices to prove that

$$\hat{H}^0\left[\Delta_2/\Delta_1, (\mu_{\ell^\infty})^{\Delta_1}\right] \text{ is trivial}$$

(reduced cohomology)

Now this cohomology group is, by definition, equal to the quotient A/B where

$$A = \left((\mu_{\ell^\infty})^{\Delta_1}\right)^{\Delta_2/\Delta_1}$$

$$B = N_{\mathbb{Q}(\mu_{\ell^s})/\mathbb{Q}(\mu_{\ell^r})}(\mu_{\ell^s}) \quad (N = \text{norm})$$

where we have written $\Delta_2 = 1 + \ell^r\mathbb{Z}_\ell$ and $\Delta_1 = 1 + \ell^s\mathbb{Z}_\ell$

One readily checks that A and B are both equal to μ_{ℓ^r}, so A/B is the trivial group, as desired. □

(3.6) Proposition. Let F be a totally real abelian extension of $E = \mathbb{Q}$. Then kern α_2 is isomorphic to

$$\text{Hom}\,(G(F/F_2), \mathbb{Z}/8\mathbb{Z}) \oplus \text{Hom}\,(G(F/F_3), \mathbb{Z}/3\mathbb{Z})$$

with F_ℓ as in (3.4).

Proof. By Cor (3.2) we have

$$\text{kern}\ \alpha_2 \cong H^1(G(F/\mathbb{Q}), (\mathbb{Q}/\mathbb{Z})(2)^{\Omega_F})$$

By Lemma (3.5) this is isomorphic to the direct sum over all prime numbers ℓ of

$$H^1(G(F/F_\ell),\ (\mathbb{Q}_\ell/\mathbb{Z}_\ell)(2)^{\Omega_F})^{G(F_\ell/\mathbb{Q})}$$

This is, as $G(F/F_\ell)$ acts, by definition (3.4), trivially on $(\mathbb{Q}_\ell/\mathbb{Z}_\ell)(2)^{\Omega_\mathbb{Q}}$ the same as

$$\mathrm{Hom}_{G(F_\ell/\mathbb{Q})}(G(F/F_\ell),\ (\mathbb{Q}_\ell/\mathbb{Z}_\ell)(2)^{\Omega_{F_\ell}})$$

This is, as $G(F_\ell/\mathbb{Q})$ acts trivially on $G(F/F_\ell)$ ($G(F/\mathbb{Q})$ is abelian), equal to

$$\mathrm{Hom}\,(G(F/F_\ell),\ (\mathbb{Q}_\ell/\mathbb{Z}_\ell)(2)^{\Omega_\mathbb{Q}})$$

The following obvious table finishes the proof

ℓ	$(\mathbb{Q}_\ell/\mathbb{Z}_\ell)(2)^{\Omega_\mathbb{Q}}$
2	$\frac{1}{8}\mathbb{Z}_8/\mathbb{Z}_8$
3	$\frac{1}{3}\mathbb{Z}_3/\mathbb{Z}_3$
$\neq 2,3$	(0)

□

(3.7) Examples.

1) Let $F = \mathbb{Q}\sqrt{d}$, d a square free natural number. By Proposition (3.6) one gets

kern $\alpha_2 = \mathbb{Z}/2\mathbb{Z}$ if $d \neq 2$; in fact kern $\alpha_2 = \langle\{-1,\ d\}\rangle$

kern $\alpha_2 = 0$ if $d = 2$

2) Let $F = \mathbb{Q}(\cos\frac{2\pi}{7})$. It is readily checked that the three elements $\{2, 7\}^i$ $(0 \leq i \leq 2)$ in $K_2\mathbb{Q}$ lie in the kernel of α_2. On the other hand, by Proposition (3.6), one finds kern $\alpha_2 \approx \mathbb{Z}/3\mathbb{Z}$. So we have got hold of the whole kernel.

4. On the cokernel of α_2.

One easily gets information on cokern α_2 as follows. For an arbitrary finite Galois field extension F/E with group Γ the following diagram commutes:

$$\begin{array}{ccc} K_2F & \xrightarrow{N_\Gamma} & (K_2F)^\Gamma \\ & {\scriptstyle tr}\searrow \quad \nearrow{\scriptstyle \alpha_2} & \\ & K_2E & \end{array}$$

tr = transfer F/E

$$N_\Gamma = \sum_{\sigma\in\Gamma} \sigma$$

Hence, if $\Gamma = \langle\sigma\rangle$ is <u>cyclic</u>, then

$$(K_2F)^{1-\sigma} \subset \text{kern tr} \subset \text{kern } N_\Gamma$$

Now, in the cyclic case, if F is global, then

$$|\hat{H}^{-1}(\Gamma, K_2F)| = |\hat{H}^0(\Gamma, K_2F)|$$

Proof. By Moore-Bass-Tate etc. there is an exact sequence of Γ-modules

$$0 \to W(F) \to K_2F \to \bigoplus_v (\bigoplus_{w|v} \mu_w) \to \mu_F \to 1$$

where v are non complex primes of E, w such of F, μ_w = roots of unity of F_w, μ_F = roots of unity of F. The Γ-modules W(F), $(\bigoplus_{w|v} \mu_w)$, μ_F are all <u>finite</u>. A calculation of Herbrand-quotients will now give the assertion. □

Assume now $tr : K_2F \to K_2E$ is onto. Then

$$\text{kern } N_\Gamma \twoheadrightarrow \text{kern } \alpha_2, \text{ hence } \hat{H}^{-1}(\Gamma, K_2F) \twoheadrightarrow \text{kern } \alpha_2$$

and

$$\text{cokern } \alpha_2 = \text{cokern } N_\Gamma = \hat{H}^{o}(\Gamma, K_2F).$$

This implies in the global case :

$$|\text{kern } (\hat{H}^{-1}(\Gamma, K_2F) \twoheadrightarrow \text{kern } \alpha_2)| \cdot |\text{kern } \alpha_2| = |\text{cokern } \alpha_2|.$$

But, by [S], Thm (1.3), kern tr = $(K_2F)^{1-\sigma}$, hence $\hat{H}^{-1}(\Gamma, K_2F) \simeq$ kern α_2, and we get for arbitrary cyclic Galois extensions F/E with $tr : K_2F \to K_2E$ surjective the relation $|\text{kern } \alpha_2| = |\text{cokern } \alpha_2|$. But, if $E = \mathbb{Q}$ and F is totally real then always $tr : K_2F \to K_2E$ is onto, since this is true locally everywhere (cf [M], Appendix) and since $W(\mathbb{Q}) = \{0\}$.

It is not too difficult to describe the general situation of cyclic extensions of global fields along the given arguments.

References.

[1] Dennis, R.K. and Stein, M.R. The functor K_2 : A survey of computations and problems, in Algebraic K-theory II, Lecture Notes in Math., 342, Springer, Berlin, 1973.

[2] Milnor, J. Introduction to Algebraic K-theory, Annals of Math. Studies, 72, Princeton, U.P., Princeton, 1971.

[3] Suslin. Torsion in K_2 of fields. Preprint.

[4] Tate, J. Relations between K_2 and Galois cohomology, Invent. Math., 36 (1976), 257-274.

[5] Val, Patrick Du, Homographies, Quaternions and Rotations, Oxford Mathematical Monographs, Oxford, Clarendon Press, 1964.

Faculty of Economics
Department of Mathematics H7 - 11
Erasmus University
P.O. Box 1738
3000 DR ROTTERDAM

ON THE GENERATION OF $K_2(\mathcal{O})$ BY SYMBOLS

by

Jerzy Browkin and Jürgen Hurrelbrink

Instytut Matematyki UW
Pałac Kultury i Nauki
PL-00-901 Warszawa
Poland

Department of Mathematics
Louisiana State University
Baton Rouge, La. 70803
USA

In this note we determine all real quadratic fields, for which the tame kernel $K_2(\mathcal{O})$ is generated by Steinberg symbols. It turns out that there are only three fields of this kind.

Let F be a real quadratic number field; denote by $\mathcal{O}$ the ring of integers of F, by d_F the discriminant of F.

<u>THEOREM:</u> $K_2(\mathcal{O})$ is generated by Steinberg symbols

if and only if $d_F = 5$, 8 or 13.

Proof: For $d_F = 5$, 8 and 13 in [5] presentations are given for the groups $SL_2(\mathcal{O})$, which imply that $K_2(\mathcal{O})$ consists of four Steinberg symbols.

Conversely: if $K_2(\mathcal{O})$ is generated by Steinberg symbols, it is proved in [1] that $d_F = 8$, p or $8p$ with a prime $p \equiv 5(8)$. Now, if $K_2(\mathcal{O})$ is generated by Steinberg symbols, it consists of such symbols, since in the real quadratic case there are at most four ones, namely 1, $\{-1,-1\}$, $\{-1,\varepsilon\}$, $\{-1,-\varepsilon\}$; here ε denotes the fundamental unit. So, the order of $K_2(\mathcal{O})$ is a divisor of four. It also has to be a multiple of four as we see from the determination of the 2-Sylow subgroup of $K_2(\mathcal{O})$ in [1]. Hence we conclude: $\#K_2(\mathcal{O}) = 4$.

On the other hand: the ℓ-part of $\#K_2(\mathcal{O})$ is equal to the ℓ-part of the rational integer $w_2(F)\zeta_F(-1)$ for all odd primes ℓ. Here ζ_F denotes Dedekind's zeta-function of F, and $w_2(F)$ is the largest integer N such that the Galois group of the cyclotomic extension $F(\mu_N)$ over F is annihilated by 2. This has been proved for all totally real number fields F, which are abelian, [6]. Moreover, for $d_F = 8$, p or 8p with a prime $p \equiv 5(8)$ it also holds $w_2(F)\zeta_F(-1) \equiv 4(8)$ by [2], [7]. This implies $\#K_2(\mathcal{O}) = w_2(F)\zeta_F(-1) = 4$. So, $w_2(F)d_F^{3/2}(4\pi^4)^{-1}\zeta_F(2) = 4$, by the functional equation of ζ_F. Because of $w_2(F) \geq 24$ and $\zeta_F(2) > 1$ this yields $d_F^3 < 4^2 \cdot \frac{\pi^8}{36}$ i.e. $d_F < 17$. Thus, we arrive at $d_F = 5$, 8 or 13.

Also in the following, F is assumed to be real quadratic. If $K_2(\mathcal{O})$ is generated by Steinberg symbols, we observed above that $K_2(\mathcal{O})$ is as small as possible, namely $\#K_2(\mathcal{O}) = 4$. Similarly, one obtains

<u>THEOREM:</u> The order of $K_2(\mathcal{O})$ is equal to 4 if and only if $d_F = 5$, 8, 12 or 13.

Moreover, the wild kernel $N(\mathcal{O})$ is a subgroup of the tame kernel $K_2(\mathcal{O})$ of index at least 4. So, if $K_2(\mathcal{O})$ is generated by Steinberg symbols, $N(\mathcal{O})$ is trivial. In [3] all real quadratic fields with this property have been determined:

<u>THEOREM:</u> The wild kernel $N(\mathcal{O})$ is trivial if and only if $d_F = 5$, 8, 12, 13, 17, 24 or 33.

The order of the subgroup $\mathrm{Symb}(\mathcal{O})$ of $K_2(\mathcal{O})$ generated by Steinberg symbols is a divisor of four. Hence $\#\mathrm{Symb}(\mathcal{O}) = 2$ or 4, since F has a real embedding and consequently $\{-1,-1\}$ is non trivial. Again, if $K_2(\mathcal{O})$ is generated by Steinberg symbols, we observed above that $\#\mathrm{Symb}(\mathcal{O}) = 4$. It has been proved in [4]

<u>THEOREM:</u> The order of $\mathrm{Symb}(\mathcal{O})$ is equal to 4 if and only if $d_F = 8$ or $d_F \neq 4(a^2 \pm 2b^2)$ for $a,b \in \mathbb{Z}$.

It is a consequence of the first theorem that for almost all real quadratic fields F and for all $n \geq 2$ the groups $SL_n(\mathcal{O})$ - which in fact are generated by elementary matrices - cannot be presented by just adding to the usual Steinberg relations the ones which correspond to Steinberg symbols. What about adding the relations which correspond to Dennis-Stein symbols?

REFERENCES

[1] J. Browkin, A. Schinzel, On Sylow 2-subgroups of $K_2\mathcal{O}_F$ for quadratic number fields F, J. reine angew. Math. 331 (1981), 104-113.

[2] J. Hurrelbrink, On the size of certain K-groups, Comm. in Alg. 10 (17) (1982), 1873-1889.

[3] J. Hurrelbrink, On the wild kernel, to appear in Archiv Math.

[4] B. Hoffman, K.F. Hettling, J. Browkin, On the group generated by symbols in $K_2\mathcal{O}_F$ for real quadratic fields F, to appear.

[5] F. Kirchheimer, J. Wolfart, Explizite Präsentation gewisser Hilbertscher Modulgruppen durch Erzeugende und Relationen, J. reine angew. Math. 315 (1980), 139-173.

[6] B. Mazur, A. Wiles, to appear.

[7] J. Urbanowicz, On the 2-primary part of a conjecture of Birch and Tate, to appear in Acta Arithmetica.

Hermitian algebraic K-theory of topological spaces

by

D. Burgheleat and Z. Fiedorowicz
The Ohio State University
Columbus, Ohio 43210

1. Introduction. In [W] Waldhausen extended Quillen's higher algebraic K-theory to topological rings, in order to analyze the homotopy type of the stable concordance spaces. In this paper we briefly describe an analogous extension of Karoubi's Hermitian algebraic K-theory and indicate some applications to groups of self-homeomorphisms of manifolds.

Let R_* be a simplicial ring with an involution such that 0 and 1 are in different path components. Let $\varepsilon \in R_0$ be a central element such that $\varepsilon\bar{\varepsilon} = 1$. Then $\pi_0 R_*$ is a discrete ring with involution and there is an evident projection $\rho: R_* \to \pi_0 R_*$. By abuse we denote $\varepsilon = \rho(\varepsilon)$.

Definition 1.1. Let ${}_\varepsilon O_n(R_*)$ denote the simplicial category whose objects are n x n matrices A such that $A^* = \bar{\varepsilon}A$ and such that the projection of A via ρ yields a nonsingular matrix over $\pi_0 R_*$. Here A^* denotes the conjugation transpose of A. A morphism $\lambda: A \to B$ is an n x n matrix over R_* whose projection over $\pi_0 R_*$ is nonsingular and such that $A = \lambda^* B \lambda$. We denote by ${}_\varepsilon O(R_*)$ the simplicial category $\coprod_{n>0} {}_\varepsilon O_n(R_*)$. Then the classifying space $B_\varepsilon O(R_*)$ may be group completed to form a space which we shall denote $B_\varepsilon O(R_*)^+$ (cf [M]). We denote ${}_\varepsilon\mathcal{L}(R_*) = B_\varepsilon O(R_*)^+$.

---

†Supported by NSF grant.

Remark 1.2. It is useful to note that the classifying space $B_{\varepsilon}O_n(R_*)$ has an alternative description as the two sided bar construction $B(\mathrm{Sym}_n^{\varepsilon}(R_*)\ ,\ \widetilde{GL}_n(R_*)\ ,\ *)$. Here $\widetilde{GL}_n(R_*)$ is the simplicial monoid of $n \times n$ matrices over R_* which project to nonsingular matrices over $\pi_0 R_*$ and $\mathrm{Sym}_n^{\varepsilon}(R_*)$ is the simplicial set of all $n \times n$ matrices $A \in \widetilde{GL}_n(R_*)$ such that $A^* = \varepsilon A$. The action of $\widetilde{GL_n}(R_*)$ on $\mathrm{Sym}_n^{\varepsilon}(R_*)$ is via $(A\ ,\ \lambda) \to \lambda^* A \lambda$.

We illustrate the plausibility of this definition with two examples where these notions reduce to the familiar ones of Karoubi's Hermitian algebraic K-theory and topological K-theory respectively.

Example 1.3. Let $R_* = R$ be a discrete ring. Then ${}_{\varepsilon}O(R)$ is also discrete. An object $A \in {}_{\varepsilon}O_n(R)$ corresponds to a nondegenerate ε-Hermitian form on R^n . A morphism $\lambda : A \to B$ corresponds to an isometry from $(R^n\ ,\ A)$ to $(R^n\ ,\ B)$. Thus ${}_{\varepsilon}O(R)$ is precisely the category of ε-Hermitian forms on finitely generated free modules over R . Thus ${}_{\varepsilon}\widetilde{L}(R)$ corresponds to Karoubi's definition of Hermitian algebraic K-theory of R except that Karoubi uses finitely generated projective modules instead of free modules, which results in $\pi_0({}_{\varepsilon}\widetilde{L}(R))$ differing from his definition. We shall show below how this defect can be remedied.

Example 1.4. Let $\mathbb{C}$ denote the topological ring of complex numbers with complex conjugation as the involution. Define a topological category. ${}_1O_n(\mathbb{C})$ as in 1.1, except requiring that all matrices involved be themselves nonsingular rather than their projections over $\pi_0\mathbb{C} = 0$. Then as in Remark 1.2

$$B_1 O_n(\mathbb{C}) = B(\mathrm{Sym}_n^1(\mathbb{C})\ ,\ GL_n(\mathbb{C}),\ *) = \mathrm{Sym}_n^1(\mathbb{C}) \times_{GL_n(\mathbb{C})} B(GL_n(\mathbb{C})\ ,\ GL_n(\mathbb{C})\ ,\ *)$$

$$= \mathrm{Sym}_n^1(\mathbb{C}) \times_{GL_n(\mathbb{C})} EGL_n(\mathbb{C})$$

is the Borel construction on the space $\mathrm{Sym}_n^1(\mathbb{C})$ of $n \times n$ nondegenerate Hermitian matrices. Now $\mathrm{Sym}_n^1(\mathbb{C}) = \coprod_{p+q=n} \mathrm{Sym}_{p,q}^1(\mathbb{C})$ where $\mathrm{Sym}_{p,q}^1(\mathbb{C})$

denotes the space of $n \times n$ Hermitian matrices of type (p, q). Since $GL_n(\mathbb{C})$ acts transitively on $Sym^1_{p,q}(\mathbb{C})$ we have $Sym^1_{p,q}(\mathbb{C}) \simeq U_{p,q}(\mathbb{C})\backslash GL_n(\mathbb{C})$ where $U_{p,q}(\mathbb{C})$ is the unitary group of type (p, q). Hence

$$B_1O_n(\mathbb{C}) \cong \coprod_{p+q=n} U_{p,q}(\mathbb{C})\backslash GL_n(\mathbb{C}) \times_{GL_n(\mathbb{C})} EGL_n(\mathbb{C}) = \coprod_{p+q=n} U_{p,q}(\mathbb{C})\backslash EGL_n(\mathbb{C})$$

$$\simeq \coprod_{p+q=n} BU_{p,q}(\mathbb{C}) \simeq \coprod_{p+q=n} BU(p) \times BU(q)$$

It follows that $B_1O(\mathbb{C})^+ \simeq BU \times BU \times \mathbb{Z} \times \mathbb{Z}$.

2. Basic properties of Hermitian algebraic K-theory simplicial rings We summarize the main properties of our construction in the following result

Theorem 2.1. (a) If $\mathcal{C}$ is a Serre class which contains all 2-torsion groups and $f : R_* \to T_*$ is a homomorphism of simplicial rings with involution such that $f(\varepsilon) = \varepsilon$ and which induces an isomorphism on π_0 and with $\pi_i f = 0 \bmod \mathcal{C}$ for $i \leq n$, then $\pi_i({}_\varepsilon\tilde{L}(f)) = 0 \bmod \mathcal{C}$ for $i \leq n + 1$.

(b) There are natural pairings ${}_\varepsilon\tilde{L}(R_*) \wedge {}_\delta\tilde{L}(T_*) \to {}_{\varepsilon\delta}\tilde{L}(R_* \otimes_{\mathbb{Z}} T_*)$ which are associative, commutative and unital up to homotopy.

(c) $\pi_1({}_\varepsilon\tilde{L}(SR_*)) \otimes \mathbb{Z}[1/2] \cong {}_\varepsilon L_0(\pi_0 R_*) \otimes \mathbb{Z}[1/2]$

(d) If $\eta \in \pi_1({}_1\tilde{L}(S\mathbb{Z})) \otimes \mathbb{Z}[1/2]$ denotes the generator, then the adjoint of

$$S^1 \wedge {}_\varepsilon\tilde{L}(R_*) \xrightarrow{\eta \wedge 1} {}_1\tilde{L}(S\mathbb{Z}) \wedge {}_\varepsilon\tilde{L}(R_*) \to {}_\varepsilon\tilde{L}(SR_*)$$

induces an equivalence on basepoint components away from 2.

In (c) and (d) above we denoted $S\mathbb{Z}$ the suspension of the ring $\mathbb{Z}$: the quotient of the ring of $\infty \times \infty$ integral matrices, which are row and column finite, by the ideal of finite matrices. We denoted $SR_* = S\mathbb{Z} \otimes_{\mathbb{Z}} R_*$ and ${}_\varepsilon L_0(\pi_0 R_*)$ the Grothendieck group of all nondegenerate ε-Hermitian forms on finitely generated projective modules over $\pi_0 R_*$.

In view of 2.1(c) and (d) we can associate to ${}_\varepsilon\tilde{L}(R_*)$ a (nonconnective) Ω-spectrum and simultaneously remedy the discrepancy with Karoubi's theory which we mentioned in 1.3 above.

Definition 2.2. Define a 1/2 local Ω-spectrum $\underset{\varepsilon}{\underline{L}}(R_*) = \{\Omega^{n+1}{}_\varepsilon\tilde{L}(S^{n+1}R_*)\}$ with structure maps as in 2.1(d). We obtain a functor

$\underset{\varepsilon}{\underline{L}}$: simplicial rings with involution $\to$ Ω-spectra.

We have the following consequential properties.

Theorem 2.3. (a) If R_* is discrete then $\underset{\varepsilon}{\underline{L}}(R_*)$ agrees with Karoubi's definition (away from 2).

(b) If $f : R_* \to T_*$ induces an isomorphism on π_0 and $\pi_i f = 0$ modulo a Serre class $\mathcal{C}$ for $i \leq n$ then $\pi_i(\underset{\varepsilon}{\underline{L}}(f)) = 0$ mod $\mathcal{C}$ for $-\infty < i \leq n + 1$

(c) There are natural pairings of spectra $\underset{\varepsilon}{\underline{L}}(R_*) \wedge {}_\delta\underline{L}(T_*) \to {}_{\varepsilon\delta}L(R_* \otimes_{\mathbb{Z}} T_*)$ which are associative, commutative and unital up to homotopy.

(d) Multiplication by $\langle -1 \rangle \in {}_1L_0(\mathbb{Z}) \otimes \mathbb{Z}[1/2] = \pi_0({}_1\underline{L}(\mathbb{Z}))$ defines a natural involution on $\underset{\varepsilon}{\underline{L}}(R_*)$ inducing a splitting $\underset{\varepsilon}{\underline{L}}(R_*) \simeq \underset{\varepsilon}{\underline{L}}^s(R_*) \times \underset{\varepsilon}{\underline{L}}^a(R_*)$ into symmetric and antisymmetric parts. This splitting is compatible with pairings.

(e) There is a Karoubi type periodicity $\Omega^2\underset{\varepsilon}{\underline{L}}^a(R_*) \simeq \underset{-\varepsilon}{\underline{L}}^a(R_*)$ and $\rho : R_* \to \pi_0R_*$ induces an equivalence $\underset{\varepsilon}{\underline{L}}^a(R_*) \simeq \underset{\varepsilon}{\underline{L}}^a(\pi_0R_*)$.

Part (e) follows immediately from parts (b), (c) and Karoubi's proof of periodicity for discrete rings with involution. Specifically, Karoubi constructs elements $u \in \pi_{-2}({}_{-1}\underline{L}^a(\mathbb{Z}))$ and $\tilde{v} \in \pi_2({}_{-1}\underline{L}^a(\mathbb{Z}))$ whose cup product is ± 32 in $\pi_0({}_1\underline{L}^a(\mathbb{Z}))$. It follows that cup product by u and $\tilde{v}$ induce the first equivalence of part (e) . (Karoubi defines u and $\tilde{v}$ as elements in the Witt groups of $\mathbb{Z}$. However, it follows from Theorem 3.2 below that Karoubi's Witt groups are the same as the homotopy groups of $\underset{\varepsilon}{\underline{L}}^a(R)$ away from 2 for any discrete ring R .) By part (b) the map $R_* \to \pi_0R_*$ induces an isomorphism on the negative homotopy groups of $\underset{\varepsilon}{\underline{L}}^a(\ )$. Periodicity then implies the second equivalence of part (e).

3. The hyperbolic map.

Recall that Waldhausen has defined the algebraic K-theory of a simplicial ring R_* to be $K(R_*) = B\widetilde{GL}(R_*)^+ \times K_0(\pi_0 R_*)$ (cf. 1.3 above) which is then equipped with the structure of an Ω-spectrum $\underline{\underline{K}}(R_*)$ in much the same way as in 2.2 above.

When we try to define a hyperbolic map $H : \underline{\underline{K}}(R_*) \to {}_\varepsilon\underline{\underline{L}}(R_*)$ we face certain difficulties, the foremost of which is that the usual formula

$$A \to \begin{pmatrix} A & 0 \\ 0 & (A^*)^{-1} \end{pmatrix}$$

does not make sense, since matrices in $\widetilde{GL}_n(R_*)$ are not invertible. The following construction which is due to D.M. Kan enables us to overcome this problem.

Construction 3.1. Let G be a group, let G^{op} denote G with the opposite multiplication and let $\iota : G \to G^{op}$ denote the inverse map. Then $B\iota : BG \to BG^{op}$ is canonically homotopic to the map κ given by

$$\kappa[(g_1,g_2,\dots,g_n) \; ; \; (t_0,\dots,t_n)] = [(g_n,g_{n-1},\dots,g_1) \; ; \; (t_n,t_{n-1},\dots,t_0)]$$

The virtue of this construction is that the map κ makes perfect sense when G is replaced by a monoid such as $\widetilde{GL}_n(R_*)$.

We use the description of $B_\varepsilon O_{2n}(R_*)$ as the two sided bar construction $B(\mathrm{Sym}^\varepsilon_{2n}(R_*) \, , \, \widetilde{GL}_{2n}(R_*) \, , \, *)$ (cf. Remark 1.2) and consider the resulting fibration sequence

$$(*) \qquad \mathrm{Sym}^\varepsilon_{2n}(R_*) \to B_\varepsilon O_{2n}(R_*) \to B\widetilde{GL}_{2n}(R_*)$$

When we pull back the fibration along the chain of maps

$$B\widetilde{GL}_n(R_*) \times B\widetilde{GL}_n(R_*)^{op} \xrightarrow{1\times T} B\widetilde{GL}_n(R_*) \times B\widetilde{GL}_n(R_*) \to B\widetilde{GL}_{2n}(R_*)$$

where T is the map induced by $\lambda \to \lambda^*$, the resulting fibration breaks up as a fiberwise disjoint union, one piece of which has the form

$$(**) \qquad \mathrm{Sym}^{\varepsilon}_{n,n}(R_*) \to E \to B\widetilde{GL}_n(R_*) \times B\widetilde{GL}_n(R_*)^{op}$$

where

$$\mathrm{Sym}^{\varepsilon}_{n,n}(R_*) = \left\{ \begin{pmatrix} A & B \\ \varepsilon B^* & C \end{pmatrix} \;\middle|\; \begin{array}{l} A^* = -\varepsilon A \,,\ C^* = -\varepsilon C \,,\ A \,,\ C \ \text{project to} \\ \text{the zero matrix over } \ \pi_0 R_* \,,\ B \in \widetilde{GL}_n(R_*) \end{array} \right\}$$

The fibration (**) decomposes as the Whitney sum of three fibrations corresponding to the blocks A, B and C. The fibrations corresponding to the blocks A and C have evident zero sections. The pullback of the fibration corresponding to the block B along the map

$$B\widetilde{GL}_n(R_*) \xrightarrow{(1,\kappa)} B\widetilde{GL}_n(R_*) \times B\widetilde{GL}_n(R_*)^{op} ,$$

where κ is the aforementioned Kan map, has a cross-section up to canonical homotopy. The net result of this is a cross-section up to homotopy of the pullback of (*) along $1 \oplus \tau$, where $\tau = T \cdot \kappa$ is an involution on $B\widetilde{GL}_{2n}(R_*)$. This defines a hyperbolic map $B\widetilde{GL}_n(R_*) \to B_{\varepsilon}O_{2n}(R_*)$ which induces a map of spectra $H : \underline{\underline{K}}(R_*) \to {}_{\varepsilon}\underline{\underline{L}}(R_*)$ whose properties are summarized below.

<u>Theorem 3.2.</u> Let $H : \underline{\underline{K}}(R_*) \to {}_{\varepsilon}\underline{\underline{L}}(R_*)$ denote the hyperbolic map and let $F : {}_{\varepsilon}\underline{\underline{L}}(R_*) \to \underline{\underline{K}}(R_*)$ denote the forgetful map. Then the folowing diagrams commute up to homotopy

$$\begin{array}{ccccc} \underline{\underline{K}}(R_*) & \overset{H}{\to} & {}_{\varepsilon}\underline{\underline{L}}(R_*) & \overset{F}{\to} & \underline{\underline{K}}(R_*) \\ \| & & \downarrow{\scriptstyle\sigma} & & \| \\ \underline{\underline{K}}(R_*) & \overset{H}{\to} & {}_{\varepsilon}\underline{\underline{L}}(R_*) & \overset{F}{\to} & \underline{\underline{K}}(R_*) \end{array} \qquad \begin{array}{ccccc} {}_{\varepsilon}\underline{\underline{L}}(R_*) & \overset{F}{\to} & \underline{\underline{K}}(R_*) & \overset{H}{\to} & {}_{\varepsilon}\underline{\underline{L}}(R_*) \\ \| & & \downarrow{\scriptstyle\tau} & & \| \\ {}_{\varepsilon}\underline{\underline{L}}(R_*) & \overset{F}{\to} & \underline{\underline{K}}(R_*) & \overset{H}{\to} & {}_{\varepsilon}\underline{\underline{L}}(R_*) \end{array}$$

$$\underbrace{\underline{\underline{K}}(R_*) \longrightarrow \underline{\underline{K}}(R_*)}_{1+\tau} \qquad\qquad \underbrace{{}_{\varepsilon}\underline{\underline{L}}(R_*) \longrightarrow {}_{\varepsilon}\underline{\underline{L}}(R_*)}_{1+\sigma}$$

(Here τ is the involution on $\underline{\underline{K}}(R_*)$ described in the preceding paragraph and σ is the involution on ${}_\varepsilon\underline{\underline{L}}(R_*)$ described in Theorem 2.3.) Away from 2 there results an equivalence of the symmetric spectra $\underline{\underline{K}}^s(R_*) \simeq {}_\varepsilon\underline{\underline{L}}^s(R_*)$. In particular ${}_\varepsilon\underline{\underline{L}}^s(R_*)$ is independent of ε.

4. The Hermitian algebraic K-theory of topological spaces.

We consider connected topological spaces X with orientations, by which we mean a homomorphism $\omega : \Omega_*X \to \mathbb{Z}/2 = \{\pm 1\} \subseteq \mathbb{Z}$, where we understand Ω_*X to denote any suitable simplicial loop group of X, such as the Kan loop group. In applications X will be a manifold and ω will be induced by the first Stieffel-Whitney class of the tangent bundle of X. To the pair (X, ω) we can associate the simplicial group ring $\mathbb{Z}[\Omega_*X]$ with involution given by $\overline{(\Sigma m_i g_i)} = \Sigma m_i \omega(g_i) g_i^{-1}$, and we let $\varepsilon = \pm 1$. For the sake of conciseness we will denote

${}_\varepsilon\underline{\underline{L}}(X) \equiv {}_\varepsilon\underline{\underline{L}}(\mathbb{Z}[\Omega_*X])$, $\underline{\underline{K}}(X) = \underline{\underline{K}}(\mathbb{Z}[\Omega_*X])$, ${}_\varepsilon\underline{\underline{L}}^s(X) \equiv {}_\varepsilon\underline{\underline{L}}^s(\mathbb{Z}[\Omega_*X])$, etc. In this notation we can restate the results of the preceding two sections as

__Theorem 4.2.__ (a) If $f : X \to Y$ is map of spaces with orientations such that f induces an isomorphism on π_1 and such that $\pi_i f = 0 \mod \mathcal{C}$ for $i \leq n$ and for $\mathcal{C}$ a Serre class, then $\pi_i({}_\varepsilon\underline{\underline{L}}(f)) = 0 (\mathcal{C})$ for $-\infty < i \leq n$.

(b) There are natural pairings of spectra ${}_\varepsilon\underline{\underline{L}}(X) \wedge {}_\delta\underline{\underline{L}}(Y) \to {}_{\varepsilon\delta}\underline{\underline{L}}(X \times Y)$ $\underline{\underline{K}}(X) \wedge \underline{\underline{K}}(Y) \to \underline{\underline{K}}(X \times Y)$ which are associative, commutative and unital up to homotopy.

(c) There are natural involutions on ${}_\varepsilon\underline{\underline{L}}(X)$ and $\underline{\underline{K}}(X)$ inducing splittings ${}_\varepsilon\underline{\underline{L}}(X) \simeq {}_\varepsilon\underline{\underline{L}}^s(X) \times {}_\varepsilon\underline{\underline{L}}^a(X)$, $\underline{\underline{K}}(X) \simeq \underline{\underline{K}}^s(X) \times \underline{\underline{K}}^a(X)$ away from 2, into symmetric and antisymmetric parts. These splittings are compatible with pairings.

(d) The natural map $X \to K(\pi_1 X, 1)$ induces an equivalence ${}_\varepsilon\underline{\underline{L}}^a(X) \simeq {}_\varepsilon\mathbb{L}(\pi_1 X, \omega)$, away from 2, where ${}_\varepsilon\mathbb{L}(\pi_1 X, \omega)$ is the 4-periodic spectrum whose homotopy groups are the Wall surgery groups of $(\pi_1 X, \omega)$.

In particular ${}_1\underline{\underline{L}}^a(*) \simeq \underline{\underline{BO}}[1/2]$ the periodic spectrum of real topological K-theory.

(e) The forgetful map $F : {}_\varepsilon\underline{\underline{L}}(X) \to \underline{\underline{K}}(X)$ induces an equivalence ${}_\varepsilon\underline{\underline{L}}^s(X) \simeq \underline{\underline{K}}^s(X)$ away from 2. Hence ${}_{-1}\underline{\underline{L}}^s(X) \simeq {}_{+1}\underline{\underline{L}}^s(X)$.

In order to connect the algebraic K-theory $K(X)$ with spaces of concordances, Loday and Waldhausen constructed an assembly map

$$\underline{\underline{K}}(*) \wedge X_+ \to \underline{\underline{K}}(X)$$

(Here X_+ denotes X with a disjoint basepoint adjoined.) Waldhausen then defined a Whitehead spectrum $\underline{\underline{Wh}}^K(X)$ by means of the fibration sequence

$$\underline{\underline{K}}(*) \wedge X_+ \to \underline{\underline{K}}(X) \to \underline{\underline{Wh}}^K(X)$$

We denote by $\underline{Wh}^K(X)$ the zeroth space of the Ω-spectrum $\underline{\underline{Wh}}^K(X)$. There is a Hermitian analogue of these constructions which we sketch below.

Construction 4.3. Let $\tilde{X} \to X$ be the covering space associated with ω regarded as a principal $\mathbb{Z}/2$ bundle. If A is a $\mathbb{Z}/2$-space with basepoint (fixed under the $\mathbb{Z}/2$ action), the associated bundle $A \times_\omega \tilde{X} \to X$ has a cross section $s : X \to A \times_\omega \tilde{X}$ and we consider $A \wedge_\omega \tilde{X} = A \times_\omega \tilde{X}/s(X)$. It is easy to see that if $\underline{\underline{A}} = \{A_i\}$ is a spectrum with an involution, $\{A_i \wedge \tilde{X}\}$ defines a spectrum $\underline{\underline{A}} \wedge_\omega \tilde{X}_+$ which is a quotient spectrum of $\underline{\underline{A}} \wedge \tilde{X}_+$. Regarding $\tilde{X}$ as the two-sided bar construction $B(\mathbb{Z}/2 , \Omega_* X , *)$ we can define a $\mathbb{Z}/2$ equivariant map $\tilde{X}_+ \to {}_1L(X)$. Smashing this map with ${}_\varepsilon L(*)$ and applying pairings, we obtain a map of spectra.

$${}_\varepsilon\underline{\underline{L}}(*) \wedge \tilde{X}_+ \to {}_\varepsilon\underline{\underline{L}}(*) \wedge {}_1\underline{\underline{L}}(X) \to {}_\varepsilon\underline{\underline{L}}(X)$$

which factors through the quotient to define an assembly map

$${}_\varepsilon\underline{\underline{L}}(*) \wedge_\omega \tilde{X}_+ \to {}_\varepsilon\underline{\underline{L}}(X)$$

We define the Hermitian Whitehead spectrum $\underline{\underline{Wh}}^{\varepsilon L}(X)$ by means of the fibration sequence

$$ {}_{\varepsilon}\underline{\underline{L}} \wedge_{\omega} \tilde{X}_{+} \to {}_{\varepsilon}\underline{\underline{L}}(X) \to \underline{\underline{Wh}}^{\varepsilon L}(X) $$

We denote by $Wh^{\varepsilon L}(X)$ the zeroth space of the Ω-spectrum $\underline{\underline{Wh}}^{\varepsilon L}(X)$.

We summarize below the main properties of the assembly map and of the Whitehead space. The proofs are straightforward applications of the definitions and of Theorem 4.2.

Theorem 4.4. (a) The following diagram commutes up to canonical homotopies

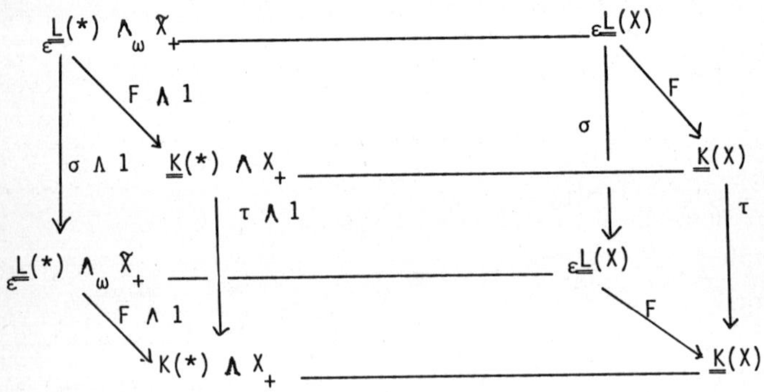

where τ and σ are the respective involutions on $\underline{\underline{K}}(\ )$ and ${}_{\varepsilon}\underline{\underline{L}}(\ )$.

(b) The involution σ induces an involution of the fibration sequence

$$ {}_{\varepsilon}\underline{\underline{L}}(*) \wedge_{\omega} \tilde{X}_{+} \to {}_{\varepsilon}\underline{\underline{L}}(X) \to \underline{\underline{Wh}}^{\varepsilon L}(X) , $$

splitting it (away from 2) into a product of symmetric parts

$$ {}_{\varepsilon}\underline{\underline{L}}^{s}(*) \wedge X_{+} \to {}_{\varepsilon}\underline{\underline{L}}^{s}(X) \to \underline{\underline{Wh}}^{\varepsilon L,s}(X) , $$

and antisymmetric parts

$$ {}_{\varepsilon}\underline{\underline{L}}^{a}(*) \wedge_{\omega} \tilde{X}_{+} \to {}_{\varepsilon}\underline{\underline{L}}^{a}(X) \to \underline{\underline{Wh}}^{\varepsilon L,a}(X) $$

In particular we have a splitting

$$Wh^{\varepsilon L}(X) \simeq Wh^{\varepsilon L,s}(X) \times Wh^{\varepsilon L,a}(X)$$

There is a similar splitting of the fibration sequence

$$K(*) \wedge X_+ \to \underline{\underline{K}}(X) \to \underline{\underline{Wh}}^K(X)$$

(c) The forgetful map $F : {}_{\varepsilon}\underline{\underline{L}}(X) \to \underline{\underline{K}}(X)$ induces an equivalence of fibration sequences

$$\begin{array}{ccccc} {}_{\varepsilon}L^s(*) \wedge X_+ & \to & {}_{\varepsilon}\underline{\underline{L}}^s(X) & \to & \underline{\underline{Wh}}^{\varepsilon L,s}(X) \\ \downarrow \wr & & \downarrow \wr & & \downarrow \wr \\ K^s(*) \wedge X_+ & \to & \underline{\underline{K}}^s(X) & \to & \underline{\underline{Wh}}^{K,s}(X) \end{array}$$

In particular $Wh^{\varepsilon L,s}(X) \simeq Wh^{K,s}(X)$.

(d) There is an equivalence of fibration sequences

$$\begin{array}{ccccc} {}_{\varepsilon}\underline{\underline{L}}^a(*) \wedge_\omega \tilde{X}_+ & \to & {}_{\varepsilon}\underline{\underline{L}}^a(X) & \to & \underline{\underline{Wh}}^{\varepsilon L,a}(X) \\ \downarrow \wr & & \downarrow \wr & & \downarrow \wr \\ {}_{\varepsilon}\underline{\underline{L}}^a(0)[\frac{1}{2}] \wedge_\omega \tilde{X}_+ & \to & {}_{\varepsilon}\underline{\underline{L}}(\pi_1 X,\omega)\ [\frac{1}{2}] & \to & \underline{\underline{Wh}}^{\varepsilon L}(X)\ [\frac{1}{2}] \end{array}$$

where ${}_{\varepsilon}\underline{\underline{L}}(\pi_1 X , \omega)$ is the 4-periodic Wall surgery spectrum.

5. Geometric applications.

By combining the results of the preceding sections with the work of Hatcher and Waldhausen relating algebraic K-theory to spaces of concordances one obtains an analogous result relating Hermitian algebraic K-theory to spaces of self-homeomorphisms of manifolds. At the π_0-level this was anticipated by Hsiang and Sharpe [HS].

Notation 5.1. Let M^n be a compact manifold possibly with a boundary. We denote by $H(M, \partial M)$, resp. $\mathrm{Homeo}(M, \partial M)$, the topological monoid, resp. group, of self-homotopy equivalences, resp. self-homeomorphisms which restrict to the identity on ∂M. We denote by $H(M, \partial M)/\mathrm{Homeo}(M, \partial M)$ the homotopy fiber of $B\mathrm{Homeo}(M, \partial M) \to BH(M, \partial M)$. We will denote by $\phi : N \to N$ a function of the form $\phi(n) = [\frac{n - \alpha}{\omega}]$ where α, ω are positive integers and [ ] denotes the greatest integer function.

Theorem 5.1. There is a $\phi : N \to N$ as above, such that for any compact triangulable manifold M^{2n}, $n \geq 3$ the basepoint components of the spaces $H(M^{2n}, \partial M^{2n})/\mathrm{Homeo}(M^{2n}, \partial M^{2n})$ and $\Omega \mathrm{Wh}^{\varepsilon L}(M, \omega)$ are $\phi(2n)$-rationally equivalent where $\varepsilon = (-1)^n$ and where $\omega = w_1 \in H^1(M; /2)$ is the first Stiefel-Whitney class of M.

Remark 5.2. Theorem 5.1 is not far from the result of [BL]. However the present formulation puts these results in a different perspective. The integers ω and α mentioned above come from Hatcher' stability theorem for concordances. Igusa's announced proof of the stability theorem gives $\omega = 7$ with a potential improvement to $\omega = 4$.

Remark 5.3. There is a version of Theorem 5.1 for odd dimensional manifolds M^{2n+1} as well. In that version one replaces the functor ${}_{\varepsilon}\underline{L}(X)$ by the functor ${}_{\varepsilon}\underline{U}(X)$, the fiber of the hyperbolic map $H : \underline{K}(X) \to {}_{\varepsilon}\underline{L}(X)$.

Remark 5.4. We strongly believe that the present rational version of Theorem 5.1 can be improved to a $\frac{1}{2}$ local version if our present Hermitian algebraic K-theory $_{\varepsilon}L(X) = {}_{\varepsilon}\underline{L}(\mathbb{Z}[\Omega_{*}X])$ of simplicial rings is replaced by a suitably defined Hermitian algebraic K-theory $_{\varepsilon}\underline{LA}(X) = {}_{\varepsilon}\underline{L}(\Omega^{\infty}S^{\infty}[\Omega_{*}X])$ of rings up to homotopy. In such a theory the role of orientation $\omega : \Omega_{*}X \to \mathbb{Z}/2 = \{\underline{+}1\} \subseteq \mathbb{Z}$ will be supplanted by a stable spherical fibration ξ yielding a homorphism $\Omega\xi : \Omega \to \Omega BF \simeq F \subseteq \Omega^{\infty}S^{\infty}$. It is perhaps even possible that there exists a global version of Theorem 5.1 but one can not say anything definite at present because both Hermitian algebraic K-theory and the spaces $H(M , \partial M)/\mathrm{Homeo}(M , \partial M)$ are poorly understood at the prime 2.

Theorem 5.1 follows directly from the following three results and Theorem 4.4(b).

Proposition 5.5. If M^n is a compact triangulable manifold and $\widetilde{\mathrm{Homeo}}(M , \partial M)$ denotes the topological group of block homeomorphisms which restrict to the identify on ∂M, then the spaces $H(M , \partial M)/\mathrm{Homeo}(M , \partial M)$ and $H(M , \partial M)/\widetilde{\mathrm{Homeo}}(M , \partial M) \times \widetilde{\mathrm{Homeo}}(M , \partial M)/\mathrm{Homeo}(M , \partial M)$ are $\phi(n)$-homotopy equivalent away from 2 .

This is essentially Corollary C9 of [BL] slightly improved. Here is a sketch of its proof. If M is a compact topological manifold there exists a homotopy commutative diagram

$$\begin{array}{lccccc}
(*) & & B\mathcal{P}(M) & \longrightarrow & B\mathcal{F}(M) & \longrightarrow & BH(M) \\
 & & \uparrow\iota & & \uparrow & & \uparrow \\
(**) & \widetilde{\mathrm{Homeo}}(M , \partial M)/_{\mathrm{Homeo}(M , \partial N)} & & \to & B\mathrm{Homeo}(M , \partial N) & \to & B\widetilde{\mathrm{Homeo}}(M , \partial M)
\end{array}$$

where $\mathcal{P}(M) = \lim\limits_{n\to\infty} C(M \times I^n)$ and (*) and (**) are fibrations[†].
The fibration (*) can be defined for any compact ANR , X , and it is a simple-homotopy invariant. With any stable microbundle ξ over X one can

---

[†] C(M) denotes the group of concordances of the manifold M , i.e. $\mathrm{Homeo}(M \times I , \partial M \times I - M \times o)$. For a definition of (*) see [B].

construct a geometric involution on $B\mathcal{P}(X)$ which depends only on the spherical class of ξ see [BL]. This involution decomposes $B\mathcal{P}(M)_{odd}$ as $B\mathcal{P}(M)^s \times B\mathcal{P}(M)^a$ and if M is a compact triangulable manifold it is shown in [BL] that $(\widetilde{Homeo}\ (M\ ,\ \partial M)/Homeo\ (M\ ,\ \partial M))_{odd}$ is $\phi(n)$-homotopy equivalent to $B\mathcal{P}(M)^s$, resp. $B\mathcal{P}(M)^a$, whenever $n \equiv 0 \bmod 2$, resp. $n \equiv 1 \bmod 2$ and ι identifies to the appropriate inclusion.

This diagram implies the existence of the commutative diagram

$$\begin{array}{lccccc} (***) & B\mathcal{P}(M) & \longrightarrow & E & \longrightarrow & H(M,\partial M)/\widetilde{Homeo}(M,\partial M) \\ & \uparrow\iota & & \uparrow & & \uparrow \\ (****) & \widetilde{Homeo}(M,\partial M)/Homeo(M,\partial M) & \to & H(M,\partial M)/Homeo(M,\partial M) & \to & H(M,\partial M)/\widetilde{Homeo}(M,\partial M) \end{array}$$

where (***) is the pullback of (*) by the composition $H(M\ ,\ \partial M)/\widetilde{Homeo}\ (M\ ,\ \partial M) \to B\widetilde{Homeo}\ (M\ ,\ \partial M) \to BH(M)$, hence is trivial. Then after $\frac{1}{2}$-localization (****) has to be trivial. Q.E.D.

<u>Proposition 5.6.</u> Let M^{2n} be a compact manifold, $n \geq 3$. Then there is a homotopy equivalence on basepoint components

$$(H(M\ ,\ \partial M)/\widetilde{Homeo}\ (M\ ,\ \partial M))\ [\tfrac{1}{2}] \simeq \Omega Wh^{\varepsilon^{L,a}}(M)$$

where $\varepsilon = (-1)^n$.

<u>Proof.</u> By Theorem 4.4(d) we have an equivalence of fibrations

$$\Omega\ (\underline{\underline{L}}_{\varepsilon}^{a}(*) \wedge_{\omega} M_{+}) \to \Omega\ \underline{\underline{L}}_{\varepsilon}^{a}(M) \to \underline{\underline{\Omega Wh}}^{\varepsilon^{L,a}}(M)$$

$$\Omega\ (\underline{\underline{L}}_{\varepsilon}(0)[\tfrac{1}{2}] \wedge_{\omega} M_{+}) \to \Omega_{\varepsilon}L(\pi_1\ M\ ,\ \omega_1)\ [\tfrac{1}{2}] \to \underline{\underline{Wh}}^{\varepsilon^{L}}(M)[\tfrac{1}{2}]$$

By periodicity the later fibration is in turn equivalent to

$$\Omega^{2n+1}({}_1\underline{L}(0)[\tfrac{1}{2}] \wedge_\omega \tilde{M}_+) \to \Omega^{2n+1}{}_1L(\pi_1 M , \omega_1) [\tfrac{1}{2}] \to \Omega^{2n+1} \mathrm{Wh}^{1^L}(M)[\tfrac{1}{2}]$$

Noting that ${}_1L(0) [\frac{1}{2}] \simeq \underline{BO} [\frac{1}{2}]$ and noting that the pair $(M , \partial M)$ has Poincare duality of formal dimension $2n$ with respect to the spectrum $\underline{BO} [\frac{1}{2}]$ we obtain the equivalent fibration

$$\Omega\mathrm{Map} (M , \underline{BO} [\tfrac{1}{2}]) \to \Omega^{2n+1} {}_1\underline{L} (\pi_1 M , \omega_1) [\tfrac{1}{2}] \to \Omega^{2n+1} \underline{\mathrm{Wh}}^{1^L}(M)$$

Passing to the zeroth spaces of these spectra one notices that the resulting fibration can be none other than the Sullivan-Wall surgery fibration sequence

$$\mathrm{Map} (SM , BO [\tfrac{1}{2}]) \to \Omega^{2n+1}{}_1L(\pi_1 M , \omega_1) [\tfrac{1}{2}] \to H(M , \partial M)/\widetilde{\mathrm{Homeo}} (M , \partial M)$$

whence the result is proved.

<u>Proposition 5.7.</u> Let M^{2n} be a compact triangulable manifold, $n \geq 3$. Then the basepoint components of the spaces $\widetilde{\mathrm{Homeo}}(M , \partial M)/\mathrm{Homeo}(M , \partial M)$ and $\Omega\mathrm{Wh}^{\epsilon^L,s}(M)$ are $\phi(2n)$ rationally equivalent.

<u>Proof.</u> Waldhausen has constructed a rational equivalence between $\Omega\mathrm{Wh}^K(M)$ and the classifying space $B\mathcal{P}(M)$. Subsequently Hsiang and Jahren and Vogell [V] have shown that under the equivalence the geometric involution on $B\mathcal{P}(M)$ corresponds to the algebraic involution on $\Omega\mathrm{Wh}^K(M)$ defined in the preceding section. Consequently there is a rational equivalence of the symmetric parts $\Omega\mathrm{Wh}^{K,s}(M)$ and $B\mathcal{P}(M)^s$. As we have mentioned in the proof of Proposition 5.5 $\widetilde{\mathrm{Homeo}} (M^{2n} , \partial M^{2n})/\mathrm{Homeo} (M^{2n} , \partial M^{2n})_{odd}$ is $\phi(2n)$-homotopy equivalent to $B\mathcal{P}(M)^s$ and thus it is $\phi(2n)$-rationally equivalent to $\Omega\mathrm{Wh}^{K,s}(M)$. By Theorem 4.4 c) $\Omega\mathrm{Wh}^{K,s}(M)$ and $\Omega\mathrm{Wh}^{\epsilon^L,s}(M)$ are homotopy equivalent. These observations complete the proof of Proposition 5.7 and of Theorem 5.1.

References

[B] D. Burghelea - Compact ANR fibrations and stability for concordances I (preprint).

[BL] D. Burghelea and R. Lashof - Geometric transfer and the homotopy type of the automorphism groups of manifolds, T.A.M.S. 1982, pp. 1-39.

[HS] W.C. Hsiang and R.W. Sharpe - Parametrized surgery and isotopy, Pacific Jour. of Math. 67 (1976) No. 2, pp. 401-459.

[K] M. Karoubi - Theorie de Quillen et homologie du groupe orthogonal, Le theoreme fundamental de la K-theorie Hermitienne, Ann of Math 112 (1980), pp. 207-238, 238-282.

[I] K. Igusa - Stability theorem for smooth concordances (an outline) preprint ORSAY 1981.

[M] J.P. May - E_∞ spaces, groups completions, and permutative categories, London Math. Soc. Lecture Note Series 11(1976), pp. 61-94.

[V] W. Vogell - The canonical involution on the algebraic K-theory of spaces, to appear in Proc. Conf. Alg. Top. Aarhus (1982) Lecture Notes in Math.

[W] F. Waldhausen - Algebraic K-theory of topological spaces I , Proc. Symposia in Pure Math 32 (1978), 35-62.

A NOTE ON EXCISION IN K-THEORY

by

Ruth M. CHARNEY

Let R be a ring with unit and $J \subset R$ a 2-sided ideal. Define $F_{R,J}$ to be the homotopy fiber:

$$F_{R,J} \to BGL(R)^+ \to B\overline{GL}(R/J)^+$$

where $\overline{GL}(R/J)$ = image $GL(R) \to GL(R/J)$. It is well known that this fiber depends not only on the structure of J as a ring without unit, but on the ring R as well. This is known as the "failure of excision" in algebraic K-theory, and it is for this reason that we have, in general, neither Mayer-Vietoris sequences in K-theory, nor long exact sequences of the form

$$\to K_i(J) \to K_i(R) \to K_i(R/J) \to K_{i-1}(J) \to$$

(with $K_i(J)$ defined independently of R). It has been shown, however, that under certain conditions, excision does hold, (eg. Charney [3], and Weibel [7], [8]). In this note, we prove a more general theorem to that effect. I would like to thank A. Suslin and W. van der Kallen for their helpful comments and I.H.E.S. for its hospitality.

The independence of the fibers $F_{R,J}$ from R is most easily formulated by comparing them with a fixed space F_J. For this purpose, define

$$\Gamma_n(J) = \text{Kernel } (GL_n(J \oplus \mathbb{Z}) \to GL_n(\mathbb{Z}))$$

$$\Gamma(J) = \varinjlim \Gamma_n(J).$$

Let

$$\Sigma_n = n \times n \text{ permutation matrices}$$

$$\cong n^{th} \text{ symmetric group}$$

$$\Sigma = \varinjlim \Sigma_n$$

Then Σ acts on $\Gamma(J)$ by conjugation so we can form the semi-direct product $\tilde{\Gamma}(J) = \Gamma(J) \rtimes \Sigma$. This may be thought of as the group of matrices in $GL(J \oplus \mathbb{Z})$ whose image in $GL(\mathbb{Z})$ is a permutation matrix. Consider the fibration

$$B\Gamma(J) \to B\tilde{\Gamma}(J) \to B\Sigma .$$

The base and total spaces of this fibration have associated +-construction which are

∞-loop spaces. We define F_J to be the homotopy fiber

$$F_J \to B\tilde{\Gamma}(J)^+ \to B\Sigma^+ .$$

Then for any ring R containing J as a 2-sided ideal, we have a natural map of fibrations

$$\begin{array}{ccc} F_J & \xrightarrow{f_{R,J}} & F_{R,J} \\ \downarrow & & \downarrow \\ B\tilde{\Gamma}(J)^+ & \longrightarrow & BGL(R)^+ \\ \downarrow & & \downarrow \\ B\Sigma^+ & \longrightarrow & B\overline{GL}(R/J)^+ \end{array}$$

Definition: Let J be a ring without unit, $M \subseteq \mathbb{Z}$ a multiplicative set. We say J is an M-excision ideal if for any ring with unit R containing J as a 2-sided ideal, $f_{R,J}$ induces isomorphisms $\pi_*(F_J) \otimes M^{-1}\mathbb{Z} \cong \pi_*(F_{R,J}) \otimes M^{-1}\mathbb{Z}$.

Remarks: 1. For a space X, let $M^{-1}X$ denote the $M^{-1}\mathbb{Z}$-completion of X in the sense of Bousfield-Kan, [1]. Then the definition states that J is an M-excision ideal if $f_{R,J}$ induces a homotopy equivalence $M^{-1}F_J \simeq M^{-1}F_{R,J}$.

2. In [3], the space F_J is denoted by $B\Gamma(J)^+$. This notation is a bit misleading as F_J is not, in the usual sense, a +-construction on $B\Gamma(J)$. However, the theorem which follows shows that this notation is justified in certain cases. Also in [3], an alternate description of F_J is given as the geometric realization of the category $\langle P,S \rangle$ where P and S are the monoidal categories

$$P = \coprod_{n\geq 0} \underline{\underline{\Sigma_n}} \qquad S = \coprod_{n\geq 0} \underline{\underline{\tilde{\Gamma}_n(J)}}$$

and $\langle\ ,\ \rangle$ is the construction defined by D. Grayson (after D. Quillen) in [4].

We now state and prove our main theorem.

Theorem: Let J be a ring without unit and $M \subseteq \mathbb{Z}$ a multiplicative set such that $J \otimes M^{-1}\mathbb{Z} = 0$ or $J \otimes M^{-1}\mathbb{Z}$ has a unit. Then J is an M-excision ideal and $H_*(F_J; M^{-1}\mathbb{Z}) \cong H_*(\Gamma(J); M^{-1}\mathbb{Z})$.

Note, in particular, that if J is a 2-sided ideal in a ring R (with unit) such that R/J is annihilated by $m \in \mathbb{Z}$, then the hypothesis of the theorem is satisfied by M = the multiplicative set generated by m. The main step in the proof of the theorem is the following lemma.

Lemma: Under the hypothesis of the theorem, Σ acts trivially on $H_*(\Gamma(J);M^{-1}\mathbb{Z})$.

<u>Proof of Lemma</u>: It suffices to show that for all $n, s > 0$, Σ_n acts trivially on the image of $H_*(\Gamma_s(J);M^{-1}\mathbb{Z})$. Choose $k > n,s$. Let $R = J \oplus \mathbb{Z}$ and let $E_k^{\pm}(R)$ be the subgroup of $GL_k(R)$ generated by the elementary matrices e_{ik}^r, e_{kj}^r, $r \in R$, and the matrix $\tau = \begin{bmatrix} 1 & & 0 \\ & \ddots & \\ 0 & & 1_{-1} \end{bmatrix}$. The elementary matrices e_{ik}^r, e_{kj}^r generate all of the elementary matrix group $E_k(R)$. Since $E_k(R)$ contains all $k \times k$ permutation matrices up to sign, we have $\Sigma_n \subset \Sigma_k \subset E_k^{\pm}(R)$. It suffices, therefore, to show that these generators act trivially.

Clearly, since $k > s$, τ acts trivially on $\Gamma_s(J)$ and hence on the image of its homology. For e_{ik}^r, we factor the inclusion $\Gamma_s(J) \subset \Gamma(J)$ into

$$\Gamma_s(J) \subset \left[\begin{array}{c|c} \Gamma_{k-1}(J) & * \\ \hline 0 & 1 \end{array}\right] = \Gamma_{k-1}(J) \ltimes J^{k-1} \subset \Gamma(J)$$

and show that e_{ik}^r acts trivially on $H_*(\Gamma_{k-1}(J) \ltimes J^{k-1};\ M^{-1}\mathbb{Z})$.

By hypothesis, $J \otimes M^{-1}\mathbb{Z}$ contains a unit $\alpha = a \otimes \frac{1}{m}$, $m \in M$, (with $\alpha = 0$, if $J \otimes M^{-1}\mathbb{Z} = 0$). It follows that right multiplication by a,

$$r_a : J \to J$$

induces isomorphisms

$$r_{a_*} : H_*(J^{k-1};\ M^{-1}\mathbb{Z}) \circlearrowleft \cong$$

By the Lyndon-Hochschild-Serre spectral sequence, we therefore have isomorphisms

$$(\mathrm{id} \ltimes r_a)_* : H_*(\Gamma_{k-1}(J) \ltimes J^{k-1};\ M^{-1}\mathbb{Z}) \circlearrowleft \cong$$

Since J is a 2-sided ideal of R, right multiplication by a may also be viewed as a map

$$r_a : R \to J.$$

Thus, the above isomorphism factors through

$$\Gamma_{k-1}(J) \ltimes J^{k-1} \xrightarrow{\text{inc}} \Gamma_{k-1}(J) \ltimes R^{k-1} \xrightarrow{\text{id} \ltimes r_a} \Gamma_{k-1}(J) \ltimes J^{k-1}$$

$$\Gamma_{k-1}(J) \ltimes R^{k-1} = \left[\begin{array}{c|c} \Gamma_{k-1}(J) & * \\ \hline 0 & 1 \end{array}\right] \quad (* \in R^{k-1})$$

Now $e^r_{ik} \in \Gamma_{k-1}(J) \ltimes R^{k-1}$ hence it acts trivially on the homology of this group. We conclude that e^r_{ik} acts trivially on $H_*(\Gamma_{k-1}(J) \ltimes J^{k-1}; M^{-1}\mathbb{Z})$.

The proof for e^r_{kj} is the same using $\left[\begin{array}{c|c} \Gamma_{k-1}(J) & 0 \\ \hline * & 1 \end{array}\right]$ and left multiplication by a. □

<u>Proof of theorem</u>: Consider the map of fibrations

$$\begin{array}{ccc} B\Gamma(J) & \xrightarrow{J} & F_J \\ \downarrow & & \downarrow \\ B\tilde{\Gamma}(J) & \longrightarrow & B\tilde{\Gamma}(J)^+ \\ \downarrow & & \downarrow \\ B\Sigma & \longrightarrow & B\Sigma^+ \end{array}$$

The right-hand sequence is a fibration of loop-spaces, hence π_1 of the base acts trivially on the homology of the fiber (with $M^{-1}\mathbb{Z}$-coefficients). By the lemma, the same holds for the left-hand fibration. Hence we can apply the spectral sequence comparison theorem to conclude that φ_J induces isomorphisms $H_*(B\Gamma(J); M^{-1}\mathbb{Z}) \cong H_*(F_J; M^{-1}\mathbb{Z})$.

Now we apply the same argument to the fibrations

$$\begin{array}{ccc} B\Gamma(J) & \xrightarrow{\varphi_{R,J}} & F_{R,J} \\ \downarrow & & \downarrow \\ BGL(R) & \longrightarrow & BGL(R)^+ \\ \downarrow & & \downarrow \\ B\overline{GL}(R/J) & \longrightarrow & B\overline{GL}(R/J)^+ \end{array}$$

To do so, we need only verify that $\overline{GL}(R/J)$ acts trivially on $H_*(\Gamma(J); M^{-1}Z)$. But any $g \in \overline{GL}(R/J)$ can be written as $g = \sigma g_1 \sigma^{-1}$ where $\sigma \in \Sigma$ and

$g_1 \in \left[\begin{array}{c|c} I_n & 0 \ldots \\ \hline 0 & \overline{GL}(R/J) \\ \vdots & \end{array}\right] \subset \overline{GL}(R/J)$. Hence, by the lemma, g acts trivially on the image of $H_*(\Gamma_n(J); M^{-1}\mathbb{Z})$. Since this is true for n arbitrarily large, the action is trivial on $H_*(\Gamma(J); M^{-1}\mathbb{Z})$. As above, therefore, we conclude that $\varphi_{R,J}$

induces isomorphisms on $H_*(\ ;M^{-1}\mathbb{Z})$.

Finally, mapping the first diagram to the second, we obtain a homotopy commutative triangle

$$\begin{array}{ccc} F_J & \xrightarrow{f_{R,J}} & F_{R,J} \\ & \nwarrow^{\varphi_J} \quad \nearrow_{\varphi_{R,J}} & \\ & B\Gamma & \end{array}$$

Hence $f_{R,J}$ also induces isomorphisms on $H_*(\ ;M^{-1}\mathbb{Z})$. Since $f_{R,J}$ is an H-space map, it follows that it induces isomorphisms (cf. [1]),

$$\pi_*(F_J) \otimes M^{-1}\mathbb{Z} \cong \pi_*(F_{R,J}) \otimes M^{-1}\mathbb{Z}\,. \qquad \square$$

We remark that these same methods may be used to give a simple proof of a theorem of L. Vaserstein ([6], §17):

<u>Theorem</u>: Let J be a ring without unit. Suppose that for any finite subset $x_1,\dots,x_n \in J$, there exist $a,b \in J$ such that $x_i a = x_i$ and $bx_i = x_i$ for $i = 1,\dots,n$. Then J is an excision ideal (i.e. a $\{1\}$-excision ideal).

<u>Remark</u>: Vaserstein's theorem is actually stronger than this as he requires only the existence of a <u>right</u> "identity".

<u>Proof</u>: Let J' be the subring generated by $x_1,\dots,x_n$. Show that Σ acts trivially on the image of $H_*(\Gamma_s(J')) \to H_*(\Gamma(J))$ using the factorizations

$$\Gamma_s(J') \hookrightarrow \Gamma_{k-1}(J') \ltimes (J')^{k-1} \hookrightarrow \Gamma_{k-1}(J') \ltimes R^{k-1} \underset{\mathrm{id}\ltimes \ell_b}{\overset{\mathrm{id}\ltimes r_a}{\rightrightarrows}} \Gamma_{k-1}(J') \ltimes J^{k-1} \hookrightarrow \Gamma(J).$$

We leave the details to the reader. $\square$

<u>Applications</u>: For R a ring with unit and $m \in \mathbb{Z}^+$, let $K_i(R)_{1/m} \equiv K_i(R) \otimes \mathbb{Z}\,[\frac{1}{m}]$

1. The cartesian square of rings

$$\begin{array}{ccc} \mathbb{Z}[\mathbb{Z}/p\mathbb{Z}] & \longrightarrow & \mathbb{Z}[\xi_p] \\ \downarrow & & \downarrow \\ \mathbb{Z} & \longrightarrow & \mathbb{Z}/p\mathbb{Z} \end{array} \qquad \xi_p = e^{2\pi i/p}$$

has an associated Mayer-Vietoris sequence

$$\cdots \to K_i(\mathbb{Z}[\mathbb{Z}/p\mathbb{Z}])_{1/p} \to K_i(\mathbb{Z}[\xi_p])_{1/p} \oplus K_i(\mathbb{Z})_{1/p} \to K_i(\mathbb{Z}/p\mathbb{Z})_{1/p} \to$$
$$K_{i-1}(\mathbb{Z}[\mathbb{Z}/p\mathbb{Z}])_{1/p} \to \cdots$$

2. Let $\mathcal{O}$ = ring of integers in a number field
M = an order in $\mathcal{O}$
J = conductor ideal
m = smallest integer > 0 in $J \cap \mathbb{Z}$

Then the square

$$\begin{array}{ccc} M & \hookrightarrow & \mathcal{O} \\ \downarrow & & \downarrow \\ M/J & \hookrightarrow & \mathcal{O}/J \end{array}$$

has an associated $K_*(\ )_{1/m}$ Mayer-Vietoris sequence. Moreover, M/J and $\mathcal{O}/J$ are finite rings annihilated by m, hence their $K_*(\ )_{1/m}$-theories are the same as those of the reduced ring $(M/J)_{red}$ and $(\mathcal{O}/J)_{red}$, (Weibel [7]). The reduced rings are products of finite fields hence their K-groups are completely computable and $K_*(M/J_{red}) \to K_*(\mathcal{O}/J_{red})$ is injective, (Quillen [5]). Thus, the Mayer-Vietoris sequence may be reduced to a long exact sequence relating the $K_*(\ )_{1/m}$-theories of M and $\mathcal{O}$.

For example, suppose M is the smallest order containing J, i.e. $M = \mathbb{Z} + J$. Write J as a product of prime ideals

$$J = \mathfrak{p}_1^{e_1} \cdots \mathfrak{p}_k^{e_k}$$

and suppose, for simplicity, that the $\mathfrak{p}_i$'s have distinct residue characteristics r_i. Then

$$(M/J)_{red} = (\mathbb{Z}/m\mathbb{Z})_{red} = \mathbb{F}_{r_1} \times \cdots \times \mathbb{F}_{r_k}$$
$$(\mathcal{O}/J)_{red} = \mathcal{O}/\mathfrak{p}_1 \times\cdots\times \mathcal{O}/\mathfrak{p}_k = \mathbb{F}_{r_1^{f_1}} \times \cdots \times \mathbb{F}_{r_k^{f_k}} .$$

So $K_{2j}(M/J_{red}) = K_{2j}(\mathcal{O}/J_{red}) = 0$ and

$$0 \to K_{2j-1}(M/J_{red}) \to K_{2j-1}(\mathcal{O}/J_{red}) \to \prod_i \mathbb{Z}/n_i\mathbb{Z} \to 0$$

where

$$n_i = (r_i^{f_i j}-1)/(r_i^j-1) = 1 + r_i^j + r_i^{2j} + \cdots + r_i^{(f_i-1)j}$$

The Mayer-Vietoris sequence, therefore, reduces to exact sequences

$$0 \to K_{2j-1}(M)_{1/m} \to K_{2j-1}(\mathcal{O})_{1/m} \to (\prod_i \mathbb{Z}/n_i\mathbb{Z}) \otimes \mathbb{Z}[\tfrac{1}{m}] \to K_{2j-2}(M)_{1/m} \to K_{2j-2}(\mathcal{O})_{1/m} \to 0$$

For the particular case of a real quadratic number field, see [3].

We conclude with some remarks about excision in K-theory with coefficients. Let $n \in \mathbb{Z}^+$ and recall (or see [2]) that for a space X, one can define $\pi_*(X;\mathbb{Z}/n)$ so that there is a long exact sequence

$$\longrightarrow \pi_i(X) \xrightarrow{\times n} \pi_i(X) \longrightarrow \pi_i(X;\mathbb{Z}/n) \longrightarrow \pi_{i-1}(X) \xrightarrow{\times n} \pi_{i-1}(X) \longrightarrow$$

(In general, $\pi_i(X;\mathbb{Z}/n)$ is defined only for $i \geq 2$ and has a group structure only for $i \geq 3$. However, if X is an ∞-loop space, then by delooping, $\pi_i(X;\mathbb{Z}/n)$ can be defined for all $i \geq 1$ so that it is an abelian group and fits in the above long exact sequence). For R a ring with unit, one defines K-theory with $\mathbb{Z}/n$ - coefficients by

$$K_i(R;\mathbb{Z}/n) \equiv \pi_i(BGL(R)^+;\mathbb{Z}/n) \quad i \geq 1.$$

Now suppose $M = \{m \in Z \mid (m,n) = 1\}$. Then it is easily seen that multiplication by $m \in M$ is invertible on $\pi_*(X,\mathbb{Z}/n)$. Thus for an M-excision ideal $J \subset R$,

$$\pi_*(F_{R,J}) \otimes M^{-1}\mathbb{Z} \cong \pi_*(F_J) \otimes M^{-1}\mathbb{Z}$$

implies

$$\pi_*(F_{R,J};\mathbb{Z}/n) \cong \pi_*(F_J;\mathbb{Z}/n)$$

Letting $\mathbb{Z}_{(n)}$ denote $M^{-1}\mathbb{Z}$ in this case, we obtain:

<u>Theorem</u>: K-theory with $\mathbb{Z}/n$-coefficients satisfies excision on the class of ideals J such that $J \otimes \mathbb{Z}_{(n)} = 0$ or $J \otimes \mathbb{Z}_{(n)}$ has a unit.

REFERENCES:

[1] A. Bousfield and D. Kan, Homotopy Limits, Completions and Localizations, Lecture Notes in Math. 304, Springer-Verlag (1972).

[2] W. Browder, Algebraic K-theory with coefficients $\mathbb{Z}/p$, Lecture Notes in Math. 657 (1978), 40-85.

[3] R. Charney, K-theory of ideals, London, Ontario (1981), AMS Current Trends in Algebraic Topology, CMS Conference Proceedings, Vol. 2, Part 1, 3-18.

[4] D. Grayson, after D. Quillen, Higher algebraic K-theory II, Lecture Notes in Math. 551, Springer-Verlag (1976), 217-240.

[5] D. Quillen, On the cohomology and K-theory of the general linear group over a finite field, Ann. of Math. 96 (1972), 552-586.

[6] L. Vaserstein, Foundations of algebraic K-theory, Russian Math. Surveys, vol. 31, n°4 (1976), 89-156.

[7] C. Weibel, Mayer-Vietoris sequences in mod p K-theory, Olberwolfach (1980), Lecture Notes in Math., to appear.

[8] C. Weibel, Mayer-Vietoris sequences and module structures on NK_*, Lecture Notes in Math. 854 (1981), 466-493.

K-theory and C*-algebras

Joachim Cuntz*
Department of Mathematics
University of Pennsylvania
Philadelphia, PA 19104

In this article, we want to give an introduction to the basic ideas of C*-algebraic K- and KK-theory, which is not too technical and does not assume too much familiarity with C*-algebras on the part of the reader. Given the diversity and the rapid growth of the subject, we necessarily have to make a choice on the topics to be covered. This choice of course reflects the author's preferences and present interests. We decided to concentrate on K-theory proper rather than on its applications in index theory, etc. and to leave out comments on the (rather young) history of the subject. Along the way, we prove a number of new results.

It has become quite clear recently that K-theory for C*-algebras is best viewed as a particular case of Kasparov's KK-theory, which at the same time generalizes the Ext-theory of Brown-Douglas-Fillmore. Therefore, in section 2, we only give a brief sketch of the "classical" K-theory for C*-algebras based on projective modules over C*-algebras or on the homotopy groups of groups of units. The main part of the article really only begins in section 4 where we give an easy and rather general proof of Bott periodicity for arbitrary homotopy-invariant, half-exact and stable functors defined on the category of C*-algebras. This proof is based on a generalized "homotopy-equivalence" between the "Toeplitz algebra" $\mathcal{T}$ and the algebra $\mathbb{C}$ of complex numbers. In section 5 we introduce KK-theory following the approach in [20] . The group $KK(A,B)$ appears here as the set of homotopy classes of "quasi-homomorphisms" between two C*-algebras A and B . We discuss Kasparov's product $KK(A_0,A_1) \times KK(A_1,A_2) \to KK(A_0,A_2)$ which generalizes the composition of two homomorphisms. An invertible element in $KK(A,B)$ gives what we called above a "generalized homotopy equivalence" between A and B . We exhibit a number of new invertible elements. We then define a pairing $E(A) \times KK(A,B) \to E(B)$ for every homotopy-invariant, half-exact and stable functor E from C*-algebras to abelian groups. This pairing is compatible (associative) with the Kasparov product, so that the existence of an invertible element in $KK(A,B)$ implies that any such functor E takes the same values on A and B . Finally, in section 6 we use the previous results to give a characterization of the K-functor on (a large class of) C*-algebras by five simple axioms.

For the convenience of the reader, we include a rather long list of references, which, however, is by no means complete.

---

*partially supported by NSF

1. C*-algebras.

Let H be a Hilbert space and $L(H)$ the algebra of bounded operators on H. A C*-algebra is, by definition, a subalgebra of $L(H)$, for some H, which is complete in the operator norm and closed under taking the adjoint x^* of an operator x. Such an algebra can be characterized abstractly as a complete normed algebra A over $\mathbb{C}$ with an antilinear involution $*: A \to A$ such that $\|x^*x\| = \|x\|^2$ for all $x \in A$. Every commutative C*-algebra is isomorphic to the algebra $C_0(X)$ of complex-valued continuous functions, vanishing at ∞, on some locally compact space X, with sup-norm and complex conjugation of functions. In particular, every commutative, closed and *-invariant subalgebra of an arbitrary C*-algebra is of this form. Thus, a C*-algebra is the non-commutative version of $C_0(X)$.

We list a few typical examples of C*-algebras.

1.1. The algebra $M_n(\mathbb{C})$ of $n \times n$ complex matrices with operator norm from the action on $\mathbb{C}^n$.

1.2. Let H be a Hilbert space. Then $L(H)$, the algebra of bounded operators on H, and $K(H)$, the algebra of compact operators on H are C*-algebras. $K(H)$ is an ideal in $L(H)$, and the quotient $Q(H) = L(H)/K(H)$ is a C*-algebra called the Calkin algebra. If $H = \mathbb{C}^n$, then $K(\mathbb{C}^n) = L(\mathbb{C}^n) \cong M_n(\mathbb{C})$. The algebra of compact operators on a Hilbert space of countably infinite dimension is usually denoted simply by K.

1.3 Let M be a compact manifold. Every pseudodifferential operator of order σ gives a bounded operator on $L^2(M)$. Let ψ_0, ψ_{-1} denote the closure, in the operator norm, of the algebras of pseudodifferential operators of order ≤ 0 and ≤ -1, respectively. Then $\Psi_{-1} = K(L^2(M))$ and the symbol map σ gives an exact sequence

$$0 \to K(L^2(M)) \to \psi_0 \overset{\sigma}{\to} C(X) \to 0$$

where X is a certain compact space (the cotangent sphere bundle on M).

1.4. If in 1.3 one considers pseudodifferential operators on a manifold M with a group action or foliation, differentiating in leaf direction, then one obtains much more complicated C*-algebras for Ψ_{-1}.

1.5 Let G be a discrete or, more generally, a locally compact group. The full group C*-algebra $C^*(G)$ and the reduced C*-algebra $C^*_{red}(G)$ are obtained as the completion, of the complex group ring of G (the L^1-algebra of G in the non-discrete case), with respect to the largest possible C*-norm and the norm induced by the left regular representation on $L^2(G)$, respectively.

1.6. Let $\mathcal{T}$ be the C*-algebra generated by a non-unitary isometry v (i.e. $v^*v = 1$, $vv^* \neq 1$) on some Hilbert space. If v' is another such isometry then the map $v \to v'$ extends to an isomorphism $\mathcal{T} \to \mathcal{T}'$ of the C*-algebras generated by v and v'. $\mathcal{T}$ is called the Toeplitz-algebra since it can also be described as the C*-algebra of all Toeplitz operators on the unit disc.

1.7. Let $v_1,\ldots v_n$ be isometries on a Hilbert space such that $\sum_{i=1}^{n} v_i v_i^* = 1$. For the C*-algebra generated by these isometries a uniqueness theorem similar to the one for the Toeplitz algebra under 1.6, holds. This C*-algebra is denoted by $\mathcal{O}_n$. It is simple, i.e. it does not contain any non-trivial closed ideal.

1.8. On the Hilbert space $L^2(S^1)$, the C*-algebra $C(S^1)$ acts by multiplication $f(\xi) = f\xi$, $f \in C(S^1)$, $\xi \in L^2(S^1)$. For $\theta \in [0,1]$ let A_θ denote the C*-algebra generated by these multiplication operators and the unitary u_θ given by rotation by $2\pi\theta$ on $L^2(S^1)$, $(u_\theta\xi)(s) = \xi(s-2\pi\theta)$ $\xi \in L^2(S^1)$. For irrational θ, A_θ is simple.

1.9. If $A_1 \overset{\phi_1}{\to} A_2 \overset{\phi_2}{\to} A_3 \to \ldots$ is a sequence of homomorphisms, we can form the C*-algebraic inductive limit $\varinjlim_{\phi_i} A_i$ as the completion, with respect to the natural C*-norm, of the algebraic inductive limit of this sequence. If $\mathbb{C} = M_1(\mathbb{C}) \to M_2(\mathbb{C}) \to M_3(\mathbb{C}) \to \ldots$ are the natural embeddings in the upper left corner, one obtains as inductive limit the algebra K of compact operators on a separable infinite-dimensional Hilbert space ($\ell^2(\mathbb{N}) = \varinjlim \mathbb{C}^n$). In general, a C*-algebraic inductive limit of a sequence of finite-dimensional C*-algebras is called an AF-algebra.

2. K-theory.

If A and B are C*-algebras one can form their tensor product $A \otimes B$ (we are talking here about the "minimal" tensor product) which is again a C*-algebra. Special cases are $M_n(\mathbb{C}) \otimes A \cong M_n(A)$, the $n \times n$-matrices over A, $C_0(X) \otimes A \cong C_0(X,A)$, the continuous A-valued functions on X, and $C_0(X_1) \otimes C_0(X_2) \cong C_0(X_1 \times X_2)$, X, X_1, X_2 locally compact spaces.

An element e of a C*-algebra is called a projection if $e^2 = e = e^*$. Two projections are called homotopic if they can be connected by a continuous path of projections. If A is a C*-algebra with unit one defines an abelian group $K(A)$ as the enveloping group (Grothendieck-group) of the semigroup of homotopy classes of projections in $K \otimes A = \varinjlim M_n(A)$. Here the sum of two homotopy classes $[e]$ and $[f]$ is defined as $[e'+f']$ where $e' \in [e]$, $f' \in [f]$ and $e'f' = 0$.

For a C*-algebra A without unit one can define $K(A)$ as the kernel of the homomorphism $K(\tilde{A}) \to K(\mathbb{C})$ induced by the quotient map $\tilde{A} \to \mathbb{C}$, where $\tilde{A}$ is the "one-point compactification" of A, i.e. A with a unit adjoined. K is a covariant functor from C*-algebras to abelian groups which among other things satisfies the three following basic axioms.

(E1) K is homotopy invariant: If $\phi_0, \phi_1 : A \to B$ are homomorphisms which are homotopic (in the topology of pointwise convergence), then ϕ_0 and ϕ_1 induce the same homomorphisms ϕ_{0*}, ϕ_{1*} from $K(A)$ to $K(B)$.

(E2) K is half-exact: If $0 \to J \xrightarrow{i} A \xrightarrow{q} B \to 0$ is an exact sequence of C*-algebras, then $K(J) \xrightarrow{i_*} K(A) \xrightarrow{q_*} K(B)$ is exact.

(E3) K is stable: The natural inclusion $\kappa : A \to K \otimes A = \varinjlim M_n(A)$ induces an isomorphism $\kappa_* : K(A) \to K(K \otimes A)$.

The stability axiom (E3) is quite natural for functors defined on the category of C*-algebras, since $K \otimes A \cong K \otimes B$ for two C*-algebras A and B, if and only if A and B are strongly Morita equivalent, cf. [7].

Starting from $K(A)$ one can, as usual, define higher functors $K_{-i}(A)$ for $i \in \mathbb{N}$ by $K_{-i}(A) = K(C_0(\mathbb{R}^i) \otimes A)$. The Bott periodicity theorem (for complex C*-algebras) states that there is a natural isomorphism $K_{-i}(A) \cong K_{-i-2}(A)$. Thus we are left with only two functors denoted K_0 and K_1 which form a generalized cohomology theory. This means essentially that an exact sequence $0 \to J \xrightarrow{i} A \xrightarrow{q} B \to 0$ of C*-algebras induces a (periodic) long exact sequence

$$\ldots \to K_i(B) \overset{\delta}{\to} K_{i+1}(J) \overset{i_*}{\to} K_{i+1}(A) \overset{q_*}{\to} K_{i+1}(B) \overset{\delta}{\to} \ldots \qquad (i\in \mathbb{Z}/2)$$

It is easy to see that $K_1(A) \cong \pi_0(GL^\infty(A)) = GL^\infty(A)/GL_0^\infty(A)$ where $GL^\infty(A)$ is the group of invertible elements in $K\otimes A^\sim$ and $GL_0^\infty(A)$ its connected component of the identity. More generally one then obtains $K_{-i}(A) \cong \pi_{i-1}(GL^\infty(A))$. The long exact sequence together with the usual machinery of topology gives effective means to compute the K-groups for commutative C*-algebras (locally compact spaces). For non-commutative C*-algebras the situation is different. It follows trivially from the definitions, that

$$K_0(M_n(\mathbb{C})) = \mathbb{Z}\ , \ K_1(M_n(\mathbb{C})) = 0$$

and for an infinite dimensional Hilbert space H: $K_0(K(H)) = \mathbb{Z}$, $K_1(K(H)) = 0$ $K_0(L(H)) = K_1(L(H)) = 0$. However, to go beyond this and trivial conclusions therefrom one needs new methods.

Typically, one tries to construct a "generalized homotopy equivalence" (an invertible element in $KK(A,B)$, see below) between a C*-algebra A, whose K-groups one wants to know and a C*-algebra B whose K-groups one knows. The K-groups of A will then be naturally isomorphic to those of B. The first non-trivial computations of the K-groups of simple C*-algebras were for O_n and A_θ [15],[45] using such a method. O_n and A_θ can be represented as quotients

$$0 \to K \to E_n \to O_n \to 0$$

$$0 \to K\otimes C(S^1) \to T_\theta \to A_\theta \to 0$$

and there are equivalences $K \sim \mathbb{C}$, $E_n \sim \mathbb{C}$, $K\otimes C(S^1) \sim C(S^1)$, $T_\theta \sim C(S^1)$.* From this one immediately obtains exact sequences giving the K-groups of O_n and A_θ: $K_0(O_n) = \mathbb{Z}/(n-1)$, $K_1(O_n) = 0$, $K_0(A_\theta) = K_1(A_\theta) = \mathbb{Z}^2$. A similar method will work to determine the K-homology groups. K-homology is the homology theory dual to K-theory developed by Brown-Douglas-Fillmore, and by Kasparov who started from ideas by Atiyah. We will give a formal definition of the K-homology groups below in the frame of Kasparov's KK-theory. Sometimes, one can see immediately what one of these groups is. For instance the K-homology group $K^1(O_n)$ was shown in [42] and [44] to be $\mathbb{Z}/(n-1)$ before the K-groups or the group $K^0(O_n)$ were known.

Most of the time it is very advantageous to consider K-theory and K-homology simultaneously. This leads to KK-theory.

Another approach to the computation of the K-groups of C*-algebras like (O_n or) A_θ appears in [12], [28], [34]. Here the idea essen-

* In [15],[45] only the isomorphism of the K-groups was proved. We will exhibit the concrete equivalences below.

tially is to set up an equivalence directly between A_θ and $C(S^1 \times S^1) \otimes C_0(\mathbb{R})$. A. Connes was interested in determining $K_0(A_\theta)$, since $K \otimes A_\theta$ can also be described as the C*-algebra of pseudodifferential operators of order ≤ -1 on the 2-torus foliated by lines of irrational slope θ . The index of an elliptic pseudodifferential operator on this foliated manifold takes values in $K_0(A_\theta)$.

Remark. Restricting the functors K_i described above to commutative C*-algebras, one obtains, of course, the ordinary Atiyah-Hirzebruch K-theory.

3. Traces and K_0 .

A genuine C*-algebra concept which enters the picture here is the one of a trace. A finite trace τ on a C*-algebra A is a positive $(\tau(x^*x) \geq 0)$ functional from A to $\mathbb{C}$ such that $\tau(xy) = \tau(yx)$ for all $x, y \in A$. Every such finite (or more generally densely defined) trace τ induces a homomorphism $\tau_*: K_0(A) \to \mathbb{R}$ by $\tau_*([e]) = \mathrm{Tr} \otimes \tau\,(e)$ for a projection e in $M_n(\mathbb{C}) \otimes A$ and Tr the ordinary trace on $M_n(\mathbb{C})$. It sometimes happens that a non-commutative C*-algebra A has a unique trace ("invariant measure").

This is for instance true for the C*-algebras A_θ . Here $K_0(A_\theta) \cong K_1(A_\theta) = \mathbb{Z}^2$ for all θ , so the A_θ 's are not distinguished by their K-groups. However each A_θ for irrational θ has a unique trace mapping one of the generators of $K_0(A_\theta)$ to 1 , the other to θ . Thus the range of the trace homomorphism τ_* on $K_0(A_\theta)$ is $\mathbb{Z} + \mathbb{Z}\theta$, showing that A_θ is isomorphic to A_θ , only if $\theta' = 1 - \theta$, (in which case they are in fact isomorphic). If P is an elliptic operator on the 2-torus differentiating along the leaves of the foliation by lines of irrational slope, then ind P is an element of $K_0(A_\theta)$. By taking $\tau_*(\text{ind } P)$ we get a real number in $\mathbb{Z} + \mathbb{Z}\theta$ which is the real valued index of P .

4. More general functors on C*-algebras.

Of course, the K- and K-homology groups are not the only topological invariants associated with a C*-algebra. For a rather thorough discussion of other possible functors see [84]. We will see, however, that a functor E which is defined on the class of all C*-algebras (or a suitable subclass like all nuclear or all separable C*-algebras) and is half-exact, stable and homotopy-invariant, i.e. satisfies axioms (E1)-(E3) from above cannot be too far from K-theory. In this section we will use a very simple argument to show that any such functor satisfies Bott periodicity, $E(A\otimes C_0(\mathbb{R}^2)) \cong E(A)$ for all A. In section 5, we will more generally construct a pairing between E and KK which will imply many more restrictions on E. An example of a covariant functor satisfying (E1) - (E3) is $E(A) = K_0(A\otimes B)$ or more generally $E(A) = KK(B_1, A\otimes B_2)$ for fixed B, B_1, B_2. It could well be that these are the only possibilities for E.

Let us first gather some easy general properties of stable half-exact homotopy functors (with values in abelian groups). To be specific we only consider covariant functors. The contravariant case is exactly analogous.

4.1 Proposition.

Let E satisfy (E1) - (E3).

(a) Let $E_{-i}(A) = E(A\otimes C_0(\mathbb{R}^i))$. If $0 \to J \overset{i}{\to} A \overset{q}{\to} B \to 0$ is an exact sequence of C*-algebras, then there is a long exact sequence

$$\to E_{-i-1}(B) \overset{\delta}{\to} E_{-i}(J) \xrightarrow{E_{-i}(i)} E_{-i}(A) \xrightarrow{E_{-i}(q)} E_{-i}(B) \overset{\delta}{\to} \dots$$

$$\dots \overset{\delta}{\to} E(J) \xrightarrow{E(i)} E(A) \xrightarrow{E(q)} E(B)$$

(b) If $0 \to J \overset{i}{\to} A \overset{q}{\to} B \to 0$ is split exact then there is a split exact sequence

$$0 \to E(J) \xrightarrow{E(i)} E(A) \xrightarrow{E(q)} E(B) \to 0$$

(c) E is additive: if $\phi_1, \phi_2: A \to B$ are homomorphisms such that $\phi_1(A)\phi_2(A) = \{0\}$ then $E(\phi_1+\phi_2) = E(\phi_1) + E(\phi_2)$ (note that $\phi_1 + \phi_2$ is a homomorphism).

Proof. (a) This is proved, using a general procedure from topology in [33, §7 Lemma 5].

(b) follows immediately from (a).

(c) Let $A_1 = \phi_1(A)$, $A_2 = \phi_2(A)$ and $j: A_1\oplus A_2 \to B$ the inclusion map. By half-exactness we have $E(A_1\oplus A_2) = E(A_1) \oplus E(A_2)$. Writing ϕ_i' for

ϕ_i considered as a homomorphism from A to A_i, $i = 1,2$, we moreover have $E(\phi_1+\phi_2) = E(j)(E(\pi_1)+E(\pi_2))E(\phi_1'+\phi_2') =$ $E(j)E(\pi_1(\phi_1'+\phi_2') + E(j)E(\pi_2(\phi_1'+\phi_2')) = E(\phi_1) + E(\phi_2)$ with $\pi_i\colon A_1 \oplus A_2 \to A_1 \oplus A_2$ the projection onto $A_i \subset A_1 \oplus A_2$ (clearly $E(\pi_1) + E(\pi_2) = \mathrm{id}$). q.e.d.

Consider the Toeplitz algebra T generated by the isometry v, see 1.6 above. T contains an ideal isomorphic to K and there is an exact sequence

$$0 \to K \to T \overset{q}{\to} C(S^1) \to 0$$

Set $e = 1 - vv^*$. Then e is a one-dimensional projection in the ideal K.

We embed $K \otimes T$ as an ideal in a larger C*-algebra $\hat{T}$ as follows. Let $\hat{T}$ be the C*-subalgebra of $T \otimes T$ generated by $K \otimes T$ and $T \otimes 1$. Write $\hat{v} = v \otimes 1$, $\hat{e} = e \otimes 1 = 1 - \hat{v}\hat{v}^*$, $v' = e \otimes v$, $e' = e \otimes e$. The projection $\hat{e}$ is the unit for $C^*(v') \cong T$.

4.2 Lemma. There is a continuous path ϕ_t, $t \in [-1,1]$ of homomorphisms from T to $\hat{T}$ such that

$$\phi_{-1}(v) = \hat{v}(1-\hat{e}) + v'$$
$$\phi_0(v) = \hat{v}$$
$$\phi_1(v) = \hat{v}(1-\hat{e}) + \hat{e}$$

Proof. (cf. [16]).

It is easy to check that the expressions

$$u_{-1} = \hat{v}(1-\hat{e})\hat{v}^* + v'\hat{v}^* + \hat{v}v'^* + e'$$
$$u_1 = \hat{v}(1-\hat{e})\hat{v} + \hat{e}\hat{v}^* + \hat{v}\hat{e}$$

define self-adjoint unitaries in $\hat{T}$ such that $u_{-1}\hat{v} = \hat{v}(1-\hat{e}) + v'$ and $u_1\hat{v} = \hat{v}(1-\hat{e}) + \hat{e}$.

Let u_t, $t \in [-1,1]$ be a continuous path of unitaries in $\hat{T}$ connecting $u_{-1}, u_0 = 1$ and u_1. By the uniqueness theorem for T, the map $v \to u_t\hat{v}$ extends, for each t, to a homomorphism $\phi_t\colon T \to \hat{T}$. q.e.d.

Let E now be a functor satisfying (E1)-(E3) and $p\colon T \to \mathbb{C}$ the composition of $q\colon T \to C(S^1)$ and evaluation at $1\colon C(S^1) \to \mathbb{C}$.

4.3 Proposition. (a) The homomorphisms $E(j)\colon E(\mathbb{C}) \to E(T)$ and $E(p)\colon E(T) \to E(\mathbb{C})$ are inverse to each other, thus isomorphisms.
(b) $E(\mathrm{id}_A \otimes j)$ and $E(\mathrm{id}_A \otimes p)$ give isomorphisms between $E(A)$ and $E(A \otimes T)$ for arbitrary A.

Proof. Since $pj = id_{\mathbb{C}}$ it is clear that $E(p)E(j) = 1$. To show that $E(j)E(p) = 1$ let $\overline{T}$ be the subalgebra of $\hat{T} \oplus T$ defined by $\overline{T} = \{(x,y) \mid \hat{q}(x) = q(y)\}$ where $\hat{q}: \hat{T} \to \hat{T}/K \otimes T$ is the quotient map. With $J = K \otimes T \oplus \{0\} \subset \overline{T}$ we have a split exact sequence.

$$0 \to J \to \overline{T} \overset{\pi}{\to} T \to 0$$

Let $\alpha, \alpha_0, \alpha_1$ be the homomorphisms from T to $\overline{T}$ defined by $\alpha(v) = \hat{v}(1-\hat{e}) \oplus v$ $\quad \alpha_0(v) = \phi_0(v) \oplus v$, $\alpha_1(v) = \phi_1(v) \oplus v$. By the lemma above α_0 is homotopic to α_1. Let moreover κ be the canonical embedding of T in $K \otimes T \cong J$. Then

$$\alpha_0 = \alpha + \kappa id_T$$

$$\alpha_1 = \alpha + \kappa(jp)$$

Since $\pi\alpha = \pi\alpha_0 = \pi\alpha_1$ the image of $E(\alpha_0) - E(\alpha)$ and $E(\alpha_1) - E(\alpha)$ lies in $E(J)$ in the splitting $E(\overline{T}) = E(J) \oplus E(T)$. By the identities above, $E(\alpha_0) - E(\alpha) = E(\kappa id_T)$ and $E(\alpha_1) - E(\alpha) = E(\kappa(jp))$ (using additivity of E). Since by homotopy invariance $E(\alpha_0) = E(\alpha_1)$ and by stability $E(\kappa)$ is an isomorphism, the assertion follows.

(b) follows upon tensoring in the proof of (a) all algebras by A and all homomorphisms by id_A. q.e.d.

4.4 Theorem. Let E be a functor satisfying (E1)-(E3). Then there is a natural isomorphism $E(A) \cong E_{-1}(A \otimes C_0(\mathbb{R}))$ for every A.

Proof. Let $T_0 = \operatorname{Ker} p \subset T$. The sequence $0 \to T_0 \overset{\rho}{\to} T \overset{p}{\to} \mathbb{C} \to 0$ is split exact thus $E(\rho)$ is injective. On the other hand, by Proposition 4.3, $E(\rho) = E(jp)E(\rho) = E(jp\rho) = 0$. Thus $E(T_0) = 0$. The exact sequence of C*-algebras $0 \to A \otimes K \to A \otimes T_0 \to A \otimes C_0(\mathbb{R}) \to 0$ therefore gives rise to the exact sequence

$$0 \to E_{-1}(A \otimes C_0(\mathbb{R})) \to E(A \otimes K) \to 0$$

whence the desired isomorphism. q.e.d.

Remark. The argument given above also works for real C*-algebras. One then has to work with the real C*-algebras $T_{\mathbb{R}}$ generated by an isometry v for which there is an exact sequence

$$0 \to K_{\mathbb{R}} \to T_{\mathbb{R}} \to C_0^{\mathbb{R}}(i\mathbb{R}) \to 0$$

where $C_0^{\mathbb{R}}(i\mathbb{R})$ stands for the algebra of fixed points in $C_0(\mathbb{R})$ for the complex conjugation $c(f)(s) = \overline{f}(-s)$. One then obtains $E(A) \cong E(A \otimes C_0^{\mathbb{R}}(\mathbb{R}) \otimes C_0^{\mathbb{R}}(i\mathbb{R}))$ for all A.

5. KK-theory.

The full power of C*-algebraic K-theory comes in with the KK-theory developed by Kasparov. It contains the ordinary K-theory and K-homology at the same time and is the right frame for nearly all K-theoretic computations. In the complex case KK-theory associates with a pair of C*-algebras A and B two groups $KK_0(A,B)$ and $KK_1(A,B)$. $KK_0(A,B)$ has a natural interpretation as "K-homomorphisms" from A to B, while $KK_1(A,B)$ represents equivalence classes of extensions of A by B (this generalizes the ideas of Brown-Douglas-Fillmore).

To motivate the definition recall first that, for a C*-algebra B with unit, $K_0(B)$ was defined as the group generated by homotopy classes of projections in $K\otimes B$. Now a projection e in $K\otimes B$ can be viewed as a homomorphism $\mathbb{C} \to K\otimes B$ mapping 1 to e. Conversely, every homomorphism $\phi: \mathbb{C} \to K\otimes B$ is of this form with $e = \phi(1)$. Thus $K_0(B)$ may also be described as the group generated by homotopy classes of homomorphisms from $\mathbb{C}$ to $K\otimes B$. To define $K_0(B)$ for non-unital B, one adjoins a unit to B to obtain $\tilde{B}$, and effectively considers homotopy classes of pairs of projections $(e,\bar{e})$ in $K\otimes\tilde{B}$ whose difference lies in $K\otimes B$. The group consisting of such homotopy classes, divided by the classes of trivial pairs of the form (e,e) gives $K_0(B)$. Again one may reformulate this as the group generated by classes of pairs of homomorphisms $\phi,\bar{\phi}:\mathbb{C} \to K\otimes\tilde{B}$ such that $\phi(x) - \bar{\phi}(x) \in K\otimes B$ for all $x \in \mathbb{C}$, divided by the classes of trivial pairs.

Let now A and B be two arbitrary C*-algebras. A <u>quasihomomorphism</u> from A to B is a pair $(\phi,\bar{\phi})$ of homomorphisms from A to E, where E is a C*-algebra containing a subalgebra J of B as an ideal, such that $\phi(x) - \bar{\phi}(x) \in J$ for all $x \in A$. We write $(\phi,\bar{\phi}): A \to E \triangleright J \subset B$. Usually we assume that E is generated, as a C*-algebra, by $\phi(A)$ and $\bar{\phi}(A)$. With $(\phi,\bar{\phi})$ we associate a linear map $D:A \to B$ and a bilinear map $Q: A \times A \to B$ by $D = \phi - \bar{\phi}$ and $Q(x,y) = \bar{\phi}(x)D(y)$, $x,y \in A$. Two quasihomomorphisms $(\phi_0,\bar{\phi}_0)$ and $(\phi_1,\bar{\phi}_1)$ are called homotopic if there are quasihomomorphisms $(\psi_t,\bar{\psi}_t)$, $t \in [0,1]$ such that $(\psi_i,\bar{\psi}_i) = (\phi_i,\bar{\phi}_i)$, $i = 0,1$, and such that the maps $t \to D_t(x) = \psi_t(x) - \bar{\psi}_t(x)$ and $t \to Q_t(x,y) = \bar{\psi}_t(x)D_t(y)$ are continuous for all fixed x and y in A (in particular $(\phi_0,\bar{\phi}_0)$ and $(\phi_1,\bar{\phi}_1)$ are "homotopic" if they define the same D and Q).

We write $KK(A,B)$ for the set of all homotopy classes of quasihomomorphisms from A to $K\otimes B$. Then $KK(A,B)$ is an abelian group with the following operation of orthogonal addition. For $[\phi_i,\bar{\phi}_i] \in KK(A,B)$, $(\phi_i,\bar{\phi}_i): A \to E_i \triangleright J_i \subset K\otimes B$, $i = 1,2$, we define

$[\phi_1,\bar\phi_1] + [\phi_2,\bar\phi_2] = [\phi_1\oplus\phi_2,\bar\phi_1\oplus\bar\phi_2]$ where

$$(\phi_1\oplus\phi_2,\bar\phi_1\oplus\bar\phi_2): A \to E_1\oplus E_2 \rhd J_1\oplus J_2 \subset M_2(K\otimes B) \cong K\otimes B$$

The inverse for $[\phi,\bar\phi]$ is of course $[\bar\phi,\phi]$.

The definition of $KK(A,B)$ given here is different from Kasparov's definition in [33]. For justification and a detailed exposition see [20]. Note that, what we call a quasihomomorphism here, is called a pre-quasihomomorphism in [20].

A basic ingredient of Kasparov's theory is the product $KK(A_0,A_1) \times KK(A_1,A_2) \to KK(A_0,A_2)$ which generalizes the composition of homomorphisms from A_0 to A_1 and A_1 to A_2. This product is used to prove most properties of KK , like Bott periodicity, exact sequences in both variables, the isomorphism between $KK_1(A,B)$ and the group $Ext(A,B)$ of extensions of A by B , etc. ... . To define the product, we have to assume that A_0 is separable. Let then $(\phi,\bar\phi): A_0 \to E_1 \rhd J_1 \subset K\otimes A_1$ and $(\psi,\bar\psi): A_1 \to E_2 \rhd J_2 \subset K\otimes A_2$ be quasihomomorphisms. We replace $\psi,\bar\psi$ by $id\otimes\psi$, $id\otimes\bar\psi$ to get homomorphisms, still denoted $\psi,\bar\psi$, from $K\otimes A_1$ to $K\otimes E_2$. We may assume, cf. [20] , that the restrictions of $\psi,\bar\psi$ to J_1 extend to homomorphisms (still denoted $\psi,\bar\psi$) from E_1 to an algebra E , that contains $(K\otimes)E_2$ as an ideal (such that $\psi(e)x = \lim_\lambda \psi(e)\psi(u_\lambda)x$, where $e \in E_1$, $x \in E_2$ and u_λ is an approximate unit for J_1).

Let A be the smallest C*-subalgebra of $M_2(E)$ which contains $\alpha(A_0)$, $\bar\alpha(A_0)$ with

$$\alpha = \begin{pmatrix} \psi\phi & 0 \\ 0 & \bar\psi\bar\phi \end{pmatrix} \qquad \bar\alpha = \begin{pmatrix} \bar\psi\phi & 0 \\ 0 & \psi\bar\phi \end{pmatrix}$$

and is invariant under conjugation by $\begin{pmatrix} 0 & 1 \\ 1 & 0 \end{pmatrix}$. It is easy to see (using a technical theorem of Kasparov or another theorem by G.K. Pedersen) that there exists an automorphism τ of A , of the form $\tau = e^\delta$, δ a bounded *-derivation of A , such that $\alpha(x) - \tau\bar\alpha(x) \in M_2(J_2)$ for all $x \in A_0$. It is easy to see that the homotopy class of the quasihomomorphism $(\alpha,\tau\bar\alpha)$ from A_0 to $K\otimes A_2$ does not depend on the choice of τ and is also independent of the choice of $(\phi,\bar\phi)$, $(\psi,\bar\psi)$ in their homotopy classes.

Thus we can define the product $[\phi,\bar\phi][\psi,\bar\psi]$ as $[\alpha,\tau\bar\alpha]$. This product is bilinear and, most important, associative. (The proof of existence and associativity of the product depends very heavily on C*-algebra techniques.) In particular, $KK(A,A)$ becomes, for every separable C*-algebra A , a ring with unit 1_A given by the class of the identity automorphism of A .

Let us collect a few more of the simplest properties of KK; [33]:

1. There is a tensor product $KK(A_1,B_1) \times KK(A_2,B_2) \to KK(A_1 \otimes A_2, B_1 \otimes B_2)$ which is associative and commutative (this can be obtained from the product described above).

2. Set $KK_0(A,B) = KK(A,B)$ and
$KK_1(A,B) = KK(C_0(\mathbb{R}) \otimes A, B) \cong KK(A, C_0(\mathbb{R}) \otimes B)$.
Then $KK_*(\mathbb{C},B) = K_*(B)$ and $KK_*(A,\mathbb{C}) = K^*(A)$ (K-homology). In particular, $KK_1(A,\mathbb{C}) = \text{Ext } A$, the Brown-Douglas-Fillmore group, if A is "nuclear", and $KK_1(A,\mathbb{C})$ = group of invertible elements in Ext A , for general separable A .

3. One has Bott periodicity
$KK(A,B) \cong KK(C_0(\mathbb{R}^2) \otimes A, B) \cong KK(C_0(\mathbb{R}) \otimes A, C_0(\mathbb{R}) \otimes B) \cong KK(A, C_0(\mathbb{R}^2) \otimes B)$.

4. For each fixed separable nuclear A there is a long exact sequence for the functors $KK_n(A,\cdot)$. For fixed B , the functors $KK_n(\cdot,B)$ satisfy a long exact sequence on the category of separable nuclear C*-algebras.

Let now E be a (covariant) functor from C*-algebras to, say, abelian groups which satisfies axioms (E1),(E2),(E3). We construct a pairing $E(A) \times KK(A,B) \to E(B)$ as follows. Let $(\phi,\bar{\phi}): A \to E \triangleright J \subset K \otimes B$ be a quasihomomorphism from A to $K \otimes B$. Define $\phi',\bar{\phi}': A \to A \oplus E$ by $\phi' = \text{id} \oplus \phi$, $\bar{\phi}' = \text{id} \oplus \bar{\phi}$ and let E' denote the C*-algebra generated by $\phi'(A)$ and $\bar{\phi}'(A)$. If J' denotes the ideal $E' \cap (\{0\} \oplus J)$, then $(\phi',\bar{\phi}'): A \to E' \triangleright J' \subset K \otimes B$ is homotopic to $(\phi,\bar{\phi})$ and there is a split exact sequence $0 \to J' \to E' \overset{\pi}{\to} A \to 0$ (with lift ϕ'). For $e \in E(A)$ we now write $e(\phi',\bar{\phi}) = E(j)(E(\phi')(e) - E(\bar{\phi}')(e))$ where $j: J \to K \otimes B$ is the embedding. Clearly, $E(\phi')(e) - E(\bar{\phi}')(e) \in E(J) \subset E(E')$ (since $\pi\phi' = \pi\bar{\phi}'$) and $e(\phi,\bar{\phi})$ is an element of $E(K \otimes B) \cong E(B)$.

5.1 <u>Lemma</u>. (a) If $(\phi,\bar{\phi})$ and $(\psi,\bar{\psi})$ are quasihomomorphisms from A to $K \otimes B$, which define the same D and Q , then $e(\phi,\bar{\phi}) = e(\psi,\bar{\psi})$ for all $e \in E(A)$.
(b) If $(\phi,\bar{\phi})$ is homotopic to $(\psi,\bar{\psi})$, then $e(\phi,\bar{\phi}) = e(\psi,\bar{\psi})$.

<u>Proof</u>. (a) Let $(\phi,\bar{\phi})$ be as above and $(\psi,\bar{\psi}): A \to D \triangleright I \subset K \otimes B$. The hypothesis implies that $J' \cong I'$ and that the map $\phi'(a) \to \psi'(a)$, $\bar{\phi}'(a) \to \bar{\psi}'(a)$ for $a \in A$, extends to an isomorphism $\alpha: E' \to D'$ such that $\alpha\phi' = \psi'$, $\alpha\bar{\phi}' = \bar{\psi}'$.
(b) Let $(\sigma_t,\bar{\sigma}_t): A \to E_t \triangleright J_t \subset K \otimes B$, $t \in [0,1]$, be a homotopy connecting $(\phi,\bar{\phi})$ and $(\psi,\bar{\psi})$. Set $D_t = \sigma_t - \bar{\sigma}_t$ and $Q_t(x,y) = \bar{\sigma}_t(x)D_t(y)$,

$x,y \in A$. We write $K\otimes B([0,1])$ for the C*-algebra of continuous $K\otimes B$-valued functions on $[0,1]$. Let $D: A \to K\otimes B([0,1])$ be the map which maps x to the continuous function $t \to D_t(x)$ in $K\otimes B([0,1])$ and, similarly, Q the bilinear map from $A \times A$ to $K\otimes B([0,1])$ which maps (x,y) to the function $t \to Q_t(x,y)$. Let $\hat{J}$ be the C*-subalgebra of $K\otimes B([0,1])$ generated by the image of D and Q and $\hat{E}$ the C*-subalgebra of $\bigoplus_{t\in[0,1]} E_t$ generated by $\sigma(A)$ and $\bar{\sigma}(A)$, where $\sigma(x) = \oplus\sigma_t(x)$, $\bar{\sigma}(x) = \oplus\bar{\sigma}_t(x)$. Clearly, $\hat{J}$ is an ideal in $\hat{E}$ and $(\sigma,\bar{\sigma})$ defines a quasihomomorphism from A to $K\otimes B([0,1])$ thus an element $e(\sigma,\bar{\sigma})$ in $E(B([0,1]))$. For $t \in [0,1]$, let π_t be the evaluation map from $K\otimes B([0,1])$ to $K\otimes B$. Since E is homotopy-invariant $E(\pi_t)$ is an isomorphism (ker π_t is contractible) and, moreover, $E(\pi_t) = E(\pi_{t'})$ for all $t,t' \in [0,1]$. The assertion now follows from the fact that, by (a) , $e(\phi,\bar{\phi}) = E(\pi_0)(e(\sigma,\bar{\sigma}))$ and $e(\psi,\bar{\psi}) = E(\pi_1)(e(\sigma,\bar{\sigma}))$. q.e.d.

The lemma shows that the product $(e,x) \to ex = e(\phi,\bar{\phi})$, for $x = [\phi,\bar{\phi}]$, is well defined. The usefulness of this product stems from the following theorem.

5.2 <u>Theorem</u>. Let A_0,A_1,A_2 be C*-algebras and assume that A_0 is separable. Let $e \in E(A_0)$, $x \in KK(A_0,A_1)$, $y \in KK(A_1,A_2)$. Then $(ex)y = e(xy)$ where xy is the Kasparov product.

<u>Proof</u>. The proof is not difficult, but complicated. Let $x = [\phi,\bar{\phi}]$, $y = [\psi,\bar{\psi}]$ where $(\phi,\bar{\phi}): A_0 \to E_1 \triangleright J_1 \subset K\otimes A_1$, $(\psi,\bar{\psi}): K\otimes A_1 \to E_2 \triangleright J_2 \subset K\otimes A_2$. As above set $\phi',\bar{\phi}' = \mathrm{id}\oplus\phi$, $\mathrm{id}\oplus\bar{\phi}: A \to A\oplus E_1$, and let E_1' denote the C*-algebra generated by $\phi'(A)$ and $\bar{\phi}'(A)$. As in [20,4.3], we may assume that $(\psi,\bar{\psi})|_{J_1'} = (\rho,\bar{\rho})|_{J_1'}$, where $(\rho,\bar{\rho}): E_1' \to \hat{E}_2 \triangleright \hat{J}_2 \subset K\otimes A_2$ is a quasihomomorphism from E_1' to $K\otimes A_2$. By [20,4.1] the product xy is represented by the quasihomomorphism $(\alpha,\bar{\alpha})$ with

$$\alpha = \begin{pmatrix} \rho\phi' & 0 \\ 0 & \bar{\rho}\bar{\phi}' \end{pmatrix} \qquad \bar{\alpha} = \begin{pmatrix} \bar{\rho}\phi' & 0 \\ 0 & \rho\bar{\phi}' \end{pmatrix}$$

(τ can be chosen as id here).

Write $\mathcal{D}$ for the C*-algebra generated by $\alpha(A)$ and $\bar{\alpha}(A)$. Again we apply to $\alpha,\bar{\alpha},\mathcal{D}$ the " ' " - procedure to get $\alpha',\bar{\alpha}' = \mathrm{id}\oplus\alpha,\mathrm{id}\oplus\bar{\alpha}': A \to A\oplus\mathcal{D}$, and $\mathcal{D}'$ as the C*-algebra generated by $\alpha'(A)$, $\bar{\alpha}'(A)$. Now also define $\rho',\bar{\rho}' = \mathrm{id}\oplus\rho$, $\mathrm{id}\oplus\bar{\rho}: E_1' \to E_1'\oplus\hat{E}_2$ and let R denote the C*-algebra generated by $\rho'(E_1')$ and $\bar{\rho}'(E_1')$. Let also J_2' denote the ideal $R \cap (0\oplus\hat{J}_2)$ in R . Finally let

$$\alpha'' = \begin{pmatrix} \rho'\phi' & 0 \\ 0 & \bar{\rho}'\bar{\phi}' \end{pmatrix} \quad \bar{\alpha}'' = \begin{pmatrix} \bar{\rho}'\phi' & 0 \\ 0 & \rho'\bar{\phi}' \end{pmatrix}$$

and $\mathcal{D}''$ the C*-subalgebra of $M_2(\mathcal{R})$ generated by $\alpha''(A)$, $\bar{\alpha}''(A)$. We have split exact sequences

$$0 \to J_2'' \to \mathcal{D}'' \overset{\alpha''}{\leftrightarrows} \mathcal{R} \to 0$$

$$0 \to J_2' \to \mathcal{R} \overset{\rho'}{\leftrightarrows} E_1' \to 0$$

As in the proof of 5.1(a) one has $\mathcal{D}'' \cong \mathcal{D}'$ and
$e(\alpha,\bar{\alpha}) = e(\alpha',\bar{\alpha}') = e(\alpha'',\bar{\alpha}'') = E(j_1)E(w_1)^{-1}(E(\alpha'')(e) - E(\bar{\alpha}'')(e))$, where $j_1\colon J_2'' = \mathcal{D}'' \cap M_2(J_2') \to K\otimes A_2$ and $w_1\colon J_2'' \to \mathcal{D}''$ denote the inclusions. If moreover, $j_2\colon J_2' \to K\otimes A_2$, $w_2\colon J_2' \to \mathcal{R}$, $d\colon \mathcal{D}'' \to M_2(\mathcal{R})$, $d : J_2'' \to M_2(J_2')$ denote the inclusions, we have $E(d)E(w_1) = E(w_2)E(d_0)$.
Now, since $(\rho',\bar{\rho}')_{|J_1'} \sim (\psi,\bar{\psi})_{|J_1'}$,

$$(ex)y = E(j_2)E(w_2)^{-1}[E(\rho')(E(\phi')(e) - E(\bar{\phi}')(e)) - E(\bar{\rho}')(E(\phi')(e) - E(\bar{\phi}'(e))]$$

while

$$e(xy) = E(j_1)E(w_1)^{-1}(E(\alpha'')(e) - E(\alpha'')(e)) = E(j_2)E(w_2)^{-1}E(d)(E(\alpha'')(e) - E(\alpha'')(e))$$

Since $E(d)E(\alpha'') = E(\rho'\phi') + E(\bar{\rho}'\bar{\phi}')$ and $E(d)E(\bar{\alpha}'') = E(\bar{\rho}'\phi') + E(\rho'\bar{\phi}')$, it follows that $(ex)y = e(xy)$. q.e.d.

Let us now assume that we have two C*-algebras A and B and elements $x \in KK(A,B)$, $y \in KK(B,A)$ such that $xy = 1_A$, $yx = 1_B$ (x,y are called invertible elements). Then for any C*-algebras P and Q, multiplication from the right by $1_Q\otimes x$ induces an isomorphism between $KK(P,Q\otimes A)$ and $KK(P,Q\otimes B)$. The inverse is multiplication by $1_Q\otimes y$. This follows from the associativity of the product and the fact that multiplication by $1_Q\otimes 1_B = 1_{Q\otimes B}$ clearly induces the identity. Similarly $KK(P\otimes A,Q)$ and $KK(P\otimes B,Q)$ will be naturally isomorphic. This means that A and B are isomorphic for KK-theory. Moreover, this carries over even to arbitrary stable, half-exact, homotopy functors by Theorem 5.2.

5.3 <u>Proposition</u>. Let E be a functor defined on the category of all separable C*-algebras which satisfies (E1), (E2) and (E3). Let A,B be two separable C*-algebras such that $KK(A,B)$ contains an invertible element. Then $E(P\otimes A) \cong E(P\otimes B)$ for all separable C*-algebras P .

<u>Proof</u>. If $KK(A,B)$ contains an invertible element then so does $KK(P\otimes A,P\otimes B)$. It suffices, therefore, to show that $E(A) \cong E(B)$. This, however, follows immediately from Theorem 5.2 and the fact that $e1_A = e$ for all $e \in E(A)$. q.e.d.

5.4 Proposition. For a separable C*-algebra A the following are equivalent

(1) $KK(A,A) = 0$

(2) The element 1_A in $KK(A,A)$ is zero.

(3) $KK(P\otimes A,Q)$ and $KK(P,Q\otimes A)$ are zero for all C*-algebras P and Q.

(4) There is an invertible element in $KK(A,0)$.

(5) $E(P\otimes A) = 0$ for every C*-algebra P and every functor E satisfying (E1),(E2),(E3).

Proof. Obvious.

If A satisfies these equivalent conditions we call A K-contractible.

Let us look at a number of examples of invertible elements and K-contractible C*-algebras and discuss some applications.

A. What we have done in section 4 is the following. We implicitly showed that the homomorphisms $j: \mathbb{C} \to T$ and $p: T \to \mathbb{C}$ define elements in $KK(\mathbb{C},T)$ and $KK(T,\mathbb{C})$ which are inverse to each other. Then we used the procedure of the present section to conclude that $E(T) \cong E(\mathbb{C})$ for every stable, half-exact homotopy functor E . From this Bott periodicity followed by half-exactness.

Alternatively, one can of course use the invertible element in $KK(\mathbb{C},C_0(\mathbb{R}^2))$ constructed by Kasparov in [33] in conjunction with 5.3 to prove Bott periodicity for every such E.

B. The example under A. admits the following generalization. Let α be an automorphism of a C*-algebra A . The "crossed product" $A \times_\alpha \mathbb{Z}$, of A by the action of $\mathbb{Z}$ on A given by α , is a (universal) C*-algebra generated by A and a unitary u such that $uau^* = \alpha(a)$ for all $a \in A$, see e.g. [43] . It is important to know the K-theoretic invariants for such a crossed product, since many well-known C*-algebras can be represented as a crossed product of much simpler algebras by certain automorphisms. One approach to this is similar to our proof of Bott periodicity above. The generalized Toeplitz algebra T_α , introduced by Pimsner-Voiculescu in [45], is defined as the C*-subalgebra of $(A\times_\alpha\mathbb{Z})\otimes T$ generated by $A\otimes 1$ and $u\otimes v$, v the isometry generating T (we assume here that A has a unit). One easily sees that in analogy to the exact sequence for T , one has an exact sequence

$$0 \to K\otimes A \to T_\alpha \to A\times_\alpha\mathbb{Z} \to 0$$

5.5 Proposition. Let j be the embedding of A in T_α, given by $a \to a\otimes 1$, and let $[j]$ be the corresponding element of $KK(A,T_\alpha)$. Let τ be the element $[(id, Ad(1\otimes v))]$ of $KK(T_\alpha, A)$. Then $[j]$ and τ are inverse to each other.

Proof. It is clear that $\tau[j] = 1_A$. To prove that $[j]\tau = 1_{T_\alpha}$ let $\hat{T}_\alpha \subset (A \times_\alpha \mathbb{Z}) \otimes \hat{T}$ be the C*-algebra generated by $A\otimes 1$ together with $u\otimes v'$ and $u\otimes v$ ($\hat{T}, v', \hat{v}$ as in 4.2). Let k be the homomorphism from T_α to $\hat{T}_\alpha$ which is given by the restriction of $id\otimes\phi_0$ ($\phi_0 : T \to \hat{T}$ as in 4.2) to T_α. The product $[j]\tau$ is the homotopy class of the quasihomomorphism $(k, Ad(1\otimes\hat{v})k)$ (remember that $K\otimes T_\alpha$ is an ideal in $\hat{T}_\alpha$ and note that $k - Ad(1\otimes\hat{v})k$ maps T_α into this ideal). By lemma 4.2, the homomorphism $k: T_\alpha \to \hat{T}_\alpha$ is homotopic to the restriction of $id\otimes\phi_{-1}$ to T_α which is $Ad(1\otimes\hat{v})k \oplus k'$ where $k'(a\otimes 1) = a\otimes e'$, $k'(u\otimes v) = u\otimes v'$. Thus $[j]\tau = [k'] = [id_{T_\alpha}]$. q.e.d.

5.6 Proposition. Let κ be the inclusion of A in $K\otimes A$ and s be the inclusion of $K\otimes A$ in T_α. Then the element $\tau[s\kappa]$ of $KK(A,A)$ is equal to $1_A - [\alpha^{-1}]$.

Proof. One has that $s\kappa \oplus Ad(1\otimes v) = j$ on A and $Ad(1\otimes v) = Ad\, u^{-1}\, Ad(u\otimes v) = \hat{\alpha}^{-1}\, Ad(u\otimes v)$ on T_α where $\hat{\alpha}$ is defined on T_α by $\hat{\alpha}(a\otimes 1) = \alpha(a)\otimes 1$ and $\hat{\alpha}(u\otimes v) = u\otimes v$. Since $Ad(u\otimes v)$ is inner on T_α one has $[Ad(u\otimes v)] = [id]$. Therefore $\tau[Ad(1\otimes v)] = [\alpha^{-1}]$ on A and $\tau[j] = \tau[s\kappa] + \tau[Ad(1\otimes v)] = 1_A$. q.e.d.

Let us now apply the exact sequence of K-theory to T_α. Using the isomorphism $K(A) \cong K(T_\alpha)$ given by Proposition 5.5 and taking into account Proposition 5.6 one obtains the exact sequence of Pimsner-Voiculescu [45].

$$\begin{array}{ccccc} K_0(A) & \xrightarrow{id-\alpha_*^{-1}} & K_0(A) & \xrightarrow{\ell_*} & K_0(A\times_\alpha\mathbb{Z}) \\ \uparrow & & & & \downarrow \\ K_1(A\times_\alpha\mathbb{Z}) & \xleftarrow{\ell_*} & K_1(A) & \xleftarrow{id-\alpha_*^{-1}} & K_1(A) \end{array}$$

(ℓ the inclusion of A in $A\times_\alpha\mathbb{Z}$).
Moreover, we see that an analogous exact sequence holds for every functor E statisfying (E1),(E2),(E3). The same result could be obtained from using the invertible elements in $KK(A\otimes C_0(\mathbb{R}), A\times_\beta\mathbb{R})$ exhibited in [28] and [34].

C. Let G be a locally compact group. G is called amenable if the homomorphism $t:C^*(G) \to \mathbb{C}$ induced by the trivial one-dimensional representation factors over the quotient map $\lambda:C^*(G) \to C^*_{red}(G)$. It can be shown that this is equivalent to λ being an isomorphism.

For a discrete group G one can introduce the notion of K-amenability, replacing homomorphisms with elements of KK. Thus G is K-amenable if there is an element t_{red} in $KK(C^*_{red}(G),\mathbb{C})$ such that $t_{red}\lambda = t$ in $KK(C^*(G),\mathbb{C})$. Again it can be shown that this is equivalent to invertibility of λ in $KK(C^*(G),C^*_{red}(G))$. It is shown in [19] that a rather large class of non-amenable groups is still K-amenable. This includes free groups and arbitrary subgroups of groups like $SL(2,\mathbb{Z})$. In particular every stable half-exact homotopy functor E takes the same values on $C^*(G)$ and $C^*_{red}(G)$ for such a group. For the free group F_n , it is very easy to compute the K- and K-homology groups for the full C^*-algebra $C^*(F_n)$. Since F_n is K-amenable, one obtains, at the same time, the corresponding groups for $C^*_{red}(F_n)$.

D. Assume that $\pi: A \to B$ is a surjective homomorphism between nuclear separable C^*-algebras such that $[\pi] \in KK(A,B)$ is invertible. The exact sequence for KK in the second variable then shows that $KK(\mathrm{Ker}\pi,\mathrm{Ker}\pi) = 0$, i.e. $\mathrm{Ker}\pi$ is K-contractible. For instance, the algebra $\mathcal{T}_0$ considered in 4.4 is K-contractible. Since $C^*(F_n)$ is not nuclear, it is not clear whether $\mathrm{Ker}\,\lambda \subset C^*(F_n)$ is K-contractible.

E. For the algebras $\mathcal{O}_n$ described in 1.7 one has $KK(\mathcal{O}_n,\mathcal{O}_n) = \mathbb{Z}/(n-1)$. cf. [17] . In particular, $\mathcal{O}_2$ is K-contractible and, more generally, so is $A\otimes\mathcal{O}_2$ for any A . It is difficult to distinguish K-contractible C^*-algebras from each other, if they are not different for trivial reasons. For instance, we do not know of an example of a separable nuclear simple C^*-algebra A with unit such that $A\otimes\mathcal{O}_2$ is not isomorphic to $\mathcal{O}_2$.

The algebras $\mathcal{O}_n$ are universal with the property $K_0(\mathcal{O}_n) = \mathbb{Z}/(n-1)$, $K_1(\mathcal{O}_n) = 0$. If B is another C^*-algebra with this property, then there is an element in $KK(\mathcal{O}_n,B)$ which induces an isomorphism on K-theory. These algebras are therefore the natural candidates for the definition of K-theory with coefficients in $\mathbb{Z}/(n-1)$ by $K(A,\mathbb{Z}/(n-1))$ $= K(A\otimes\mathcal{O}_n)$, cf. also [55].

6. More about homotopy functors.

We are now going to use some of the results of the previous section, to characterize K-theory by some simple axioms.

6.1 Proposition. Let E be a functor defined on a suitable class of C*-algebras satisfying (E1),(E2),(E3) such that $E(\mathbb{C}) = \mathbb{Z}$. Then there is a natural transformation $K(A) \to E(A)$.

Proof. Let e_1 be the generator of $E(\mathbb{C})$. With every $x \in K(A) = KK(\mathbb{C},A)$ we associate the product $e_1 x \in E(A)$. This is a natural transformation by 5.3 (if $\phi: A \to A'$ is a homomorphism and $[\phi]$ the element of $KK(A,A')$ determined by ϕ , then $E(\phi)(e) = e[\phi]$ and $K(\phi)(x) = x[\phi]$ for $e \in E(A)$, $x \in K(A)$) . q.e.d.

Of course, this transformation extends to a transformation $K_* \to E_*$ which is compatible with suspensions and the boundary maps in the long exact sequences for K_* and E_* .

We will add to (E1),(E2),(E3) the following axioms which are clearly satisfied by K-theory.

(E4) Continuity. Let $A = \varinjlim_{\phi_i} A_i$ be the C*-algebraic inductive limit of a system (A_i,ϕ_i) . Then $E(\varinjlim_{\phi_i} A_i) = \varinjlim_{E(\phi)_i} E(A_i)$

(E5) Normalization. $E(\mathbb{C}) = \mathbb{Z}$, $E(C_0(\mathbb{R})) = 0$.

Now let h be the smallest category of separable C*-algebras such that:

(h1) h contains $\mathbb{C}$.

(h2) h is closed under taking countable inductive limits.

(h3) If any two of the C*-algebras A_1,A_2,A_3 in an exact sequence of the form $0 \to A_1 \to A_2 \to A_3 \to 0$ belong to h , then so does the third one.

(h4) If A is in h and there is an invertible element in $KK(A,A')$, then A' belongs to h .

The crucial property of h is, of course, (h4). We restrict to separable C*-algebras, since, in (h4), invertibility in $KK(A,A')$ makes sense only for separable A and A' .

6.2 Proposition.

(a) h contains all separable commutative C*-algebras

(b) h contains all separable C*-algebras of type I .

(c) If A and B are in h , then $A \otimes B$ is in h .

(d) If the group $\mathbb{Z}$ acts on a C*-algebra A by α and if $A \in h$, then the crossed product $A \times_\alpha \mathbb{Z}$ belongs to h .

(e) If a simply connected, solvable Lie group G acts on a C*-algebra A by α and $A \in h$, then $A \times_\alpha G \in h$.

Proof. (a) Since there is an invertible element in $KK(\mathbb{C}, C_0(\mathbb{R}^2))$, one has $C_0(\mathbb{R}^2) \in h$ by (h1) and (h4) . Since $C_0(\mathbb{R})$ is a quotient of $C_0(\mathbb{R}^2)$ by an ideal of the form $C_0(\mathbb{R}^2) \oplus C_0(\mathbb{R}^2)$, also $C_0(\mathbb{R}) \in h$. More generally, it follows that $C_0(\mathbb{R}^n) \in h$ for all n . Since every compact space can be represented as an inverse limit of triangulable spaces [61,X Theorem 10.1], by (h3) and (h2) h contains $C(X)$ for every compact metrisable X.

(b) Obvious from (a) and (h4) for anybody who knows what a C*-algebra of type I is.

(c) B can be built up from $\mathbb{C}$ by finitely many of the operations in (h1) - (h4) . If we apply the same operations to the second tensor factor in $A \otimes \mathbb{C}$ we obtain $A \otimes B$.

(d) This follows from the fact that there is an invertible element in $KK(T_\alpha, A)$, cf. 5.5, and $A \rtimes_\alpha \mathbb{Z}$ is a quotient of T_α by $K \otimes A$.

(e) There is an invertible element in $KK(A \rtimes_\alpha G, A \otimes C_0(\mathbb{R}^n))$, n the dimension of G , cf [12,V Cor. 7] , [28] , [34] . q.e.d.

The author does not know of an example of a separable nuclear C*-algebra which does not belong to h . But h also contains non-nuclear C*-algebras like $C^*(F_n), C^*_{red}(F_n)$ for the free group F_n on n generators (these algebras are "homotopy equivalent" to commutative C*-algebras, cf. [19]). In fact, it seems difficult to find an example of a separable C*-algebra which is not in h .

6.3 Theorem. Let E be a covariant functor defined on h , whith values in abelian groups, satisfying (E1)-(E5). Then $E_* = K_*$ on h .

Proof. Let h' be the class of all C*-algebras A for which the transformation defined in 6.1 gives an isomorphism $K_*(A) \to E_*(A)$. By axiom (E5) and Bott-periodicity for E_* (cf. 4.4 and K_* , h' contains $\mathbb{C}$. By (E4) h' is closed under taking inductive limits. If

$A \in h'$ and there is an invertible element in $KK(A,A')$, then $A' \in h'$ by 5.3. Finally, h' is closed under taking extensions, ideals and quotients (as in $(h4)$) by an obvious application of the five lemma. This shows that $h \subset h'$. q.e.d.

This theorem is a considerable strengthening of Rosenberg's result [84, 4.3].

7. Applications of C*-algebraic K-theory.

In this final section we want to give a very brief outline of some applications of C*-algebraic K-theory.

First of all, there are many applications in C*-algebra theory itself. The K-theoretic invariants are often the only tool to distinguish C*-algebras from each other. E.g. A_θ for different θ , $\mathcal{O}_n$ for different n , $C^*_{red}(F_n)$ for different n are pairwise non-isomorphic as one sees computing their K-groups. The AF-algebras mentioned in 1.9 are classified by their ordered K_0-groups; in fact, the structure of an AF-algebra is to a very large extend determined by this ordered group [26]. Other problems in C*-algebra theory have been solved using K-theory: Brown showed that an extension of an AF-algebra by an AF-algebra is again an AF-Algebra [66], [24]; Connes used K-theory to exhibit simple C*-algebras with no non-trivial projections [12]; in fact, Kadison's conjecture that $C^*_{red}(F_n)$ is such an algebra has been verified, [46], [19]. The list can easily be extended.

An important application of KK-theory is the classification of extensions in the spirit of Brown-Douglas-Fillmore. Every exact sequence $0 \to B \to E \to A \to 0$ with nuclear separable A gives an element in $KK_1(A,B)$ which, given A and B , determines E in a rather strong way. Such extension problems often arise, for instance, for group C*-algebras or in operator theory, cf. e.g. [62], [85].

Passing from commutative to non-commutative C*-algebras, as arguments of the K-functor, does not make K-theory more complicated, but on the contrary more flexible, powerful and sometimes even easier. Basic results like Bott-periodicity become more natural in this setting. Kasparov's product and his equivariant theory for non-compact Lie groups will certainly have interesting applications.

Kasparov used his KK-theory to prove the Novikov conjecture for discrete subgroups of connected Lie groups [34]. Miscenko used an index theorem for pseudodifferential operators with coefficients in a C*-algebra [39] to prove the Novikov conjecture for a different class of groups, cf. [38], [79]. The relevant non-commutative C*-algebras coming in here are of the form $C^*(\pi)$, π the fundamental group of the manifold. Rosenberg [51] used Kasparov's and Miscenko's results and methods to obtain restrictions on manifolds admitting a Riemannian metric of positive scalar curvature, obtaining results similar to those by Gromov and Lawson [30].

There is a number of index theorems using C*-algebras, cf. e.g. [2], [10], [14], [39], [65], [35]. Typically, here the index is an element of the K_0-group of some C*-algebras rather than an element of $\mathbb{Z}$ (which is the K_0 of $\mathcal{K}$). Besides the applications already mentioned above, there are applications in harmonic analysis [9], [13], [81] (existence of discrete series representations for semi-simple and more general Lie groups) but also existence of solutions for certain difference-differential equations [11], [75] and other applications [10]. A more detailed account of this work is completely outside the scope of this report.

Topological dynamical systems give rise very naturally to (non-commutative) C*-algebras. Their topological invariants give information on the dynamical system. For topological Markov chains such C*-algebras and their invariants were studies in much detail in [36], [21], [17], [18]. For instance, reducible topological Markov chains give rise to C*-algebras which are extensions of C*-algebras corresponding to irreducible chains by each other. The elements of KK_1, obtained from these C*-algebra extensions, describe how the irreducible components of the dynamical system are glued together to give the global system, cf. also [78]. Many other more subtle invariants can be explicitly computed for topological Markov chains from the associated C*-algebras [18].

References

1. J. Anderson, A C*-algebra A for which Ext(A) is not a group, Ann. of Math. 107 (1978),455-458.

2. M.F. Atiyah, Elliptic operators, discrete groups and von Neumann algebras, Soc. Math. de France Astérisque 32/33 (1976a), 43-72.

3. M.F. Atiyah, A survey of K-theory, in "K-theory and operator algebras," ed. B. Morrel and I.M. Singer, Springer Lecture Notes in Math. 575.

4. P. Baum and A. Connes, Geometric K-theory for Lie groups and Foliations, preprint.

5. L. Brown, Stable isomorphism of hereditary subalgebras of C*-algebras, Pacific J. Math. 71 (1977), 335-348.

6. L. Brown, R. Douglas and P. Fillmore, Extensions of C*-algebras and K-homology, Ann. of Math. 105 (1977), 265-324.

7. L. Brown, P. Green and M. Rieffel, Stable isomorphism and strong Morita equivalence of C*-algebras, Pac. J. Math. 71(1977), 349-363.

8. L. Brown, Extensions and the structure of C*-algebras, Istituto Nazionale di Alta Matematica, Symp. Math. 20 (1976), 539-566.

9. M.F. Atiyah and W. Schmid, A geometric construction of the discrete series for semisimple Lie groups, Invent. math. 42(1977), 1-62.

10. A. Connes, Sur la théorie non commutative de l'intégration, in "Algèbres d'Opérateurs", ed. by P. de la Harpe, Springer Lecture Notes in Math. 725.

11. A. Connes, C*-algèbres et géometrie differéntielle, C.R. Acad. Sci. Paris, tome 290 (1980).

12. A. Connes, An analogue of the Thom isomorphism for crossed products of a C*-algebra by an action of $\mathbb{R}$, Advances in Math. 39 (1981), 31-55.

13. A. Connes and H. Moscovici, The L^2-index theorem for homogeneous spaces of Lie groups, Ann. of Math. 115 (1982), 291-330.

14. A. Connes and G. Skandalis, The longitudinal index theorem for foliations, preprint IHES.

15. J. Cuntz, K-theory for certain C*-algebras, Ann. of Math. 113 (1981), 181-197.

16. J. Cuntz, K-theory for certain C*-algebras II, J. Operator Theory 5 (1981), 101-108.

17. J. Cuntz, A class of C*-algebras and topological Markov chains II: Reducible chains and the Ext-functor for C*-algebras, Invent. math. 63(1981), 25-40.

18. J. Cuntz, On the homotopy groups of the space of endomorphisms of a C*-algebra, Proc. of the OAGR conference (Neptun, Romania, 1980), Pitman, London, to appear.

19. J. Cuntz, K-theoretic amenability for discrete groups, preprint.

20. J. Cuntz, Generalized homomorphisms between C*-algebras and KK-theory, Proc. of Mathematics-Physics conference (ZiF, Bielefeld, 1981), Springer Lecture notes in Math., to appear.

21. J. Cuntz and W. Krieger, A class of C*-algebras and topological Markov chains, Invent. math. 56(1980), 251-268.

22. R. Douglas, C*-algebra extensions and K-homology, Ann. of Math. Studies, no. 95, Princeton Univ. Press, Princeton, NJ 1980.

23. E. Effros, D. Handelman and C.L. Shen, Dimension groups and their affine representations, Amer. J. Math. 102(1980), 385-407.

24. E. Effros, Dimensions and C*-algebras, CBMS Regional Conf. Ser. in Math., no. 46, Amer. Math. Soc., 1981.

25. E.G. Effros and J. Rosenberg, C*-algebras with approximately inner flip, Pacific J. Math. 77(1978), 417-443.

26. G. Elliott, On the classification of inductive limits of sequences of semi-simple finite dimensional algebras, J. Algebra 38(1976), 29-44.

27. G. Elliott, On the K-theory of the C*-algebra generated by a projective representation of a torsion-free abelian group, Proc. of the OAGR conference (Neptun, Romania, 1980), Pitman, London, to appear.

28. T. Fack and G. Skandalis, Connes' analogue of the Thom isomorphism for the Kasparov groups, Invent. math. 64(1981), 7-14.

29. K.R. Goodearl and D. Handelman, Rank functions and K_0 of regular rings, J. Pure Appl. Algebra 7 (1976), 195-216.

30. M. Gromov and H.B. Lawson, Jr., Positive scalar curvature and the Dirac operator on complete Riemannian manifolds, preprint.

31. M. Karoubi and O. Villamayor, K-théorie algébrique et K-théorie topologique I, Math. Scand. 28 (1971), 265-307.

32. G.G. Kasparov, Topological invariants of elliptic operators, I: K-homology, Izv. Akad. Nauk SSSR, Ser. Mat. 39 (1975), 796-838.

33. G.G. Kasparov, The operator K-functor and extensions of C*-algebras, Izv. Akad. Nauk SSSR, Ser. Mat. 44 (1980), 571-636.

34. G.G. Kasparov, K-theory, group C*-algebras and higher signatures (conspectus) part 1 and 2, preprint, Chernogolovka 1981.

35. G.G. Kasparov, The index of invariant elliptic operators, K-theory and Lie group representations, preprint, Chernogolovka 1982.

36. W. Krieger, On dimension functions and topological Markov chains, Invent. Math. 56 (1980), 239-250.

37. J.A. Mingo, K-theory and multipliers of stable C*-algebras, Ph.D. Thesis, Dalhousie Univ., Halifax, N.S., 1982.

38. A.S. Miščenko, C*-algebras and K-theory, in "Algebraic Topology", Springer Lecture Notes in Math 763, 262-274.

39. A.S. Miščenko and A.T. Fomenko, The index of elliptic operators over C*-algebras, Izv. Akad. Nauk SSSR, Ser. Mat. 43(1979), 831-859.

40. A.S. Miščenko, Infinite-dimensional representations of discrete groups and higher signature, Izv. Akad. Nauk. SSSR, Ser. Mat. 38 (1974), 81-106.

41. A.S. Miščenko and J.P. Solov'ev, On infinite-dimensional representations of fundamental groups and formulas of Hirzebruch type, Dokl. Akad. Nauk SSSR 234 (1977), 761-764.

42. W. Paschke, N. Salinas, Matrix algebras over $\mathcal{O}_n$, Mich. Math. J.26, 3-12.

43. G.K. Pedersen, C*-algebras and their automorphism groups, Academic Press, London-New York-San Francisco, 1979.

44. M. Pimsner and S. Popa, The Ext-groups of some C*-algebras considered by J. Cuntz, Rev. Roum. Math. Pures et Appl. 23(1978).

45. M. Pimsner and D. Voiculescu, Exact sequences for K-groups and Ext-groups of certain cross-products C*-algebras, J. Operator Theory 4 (1980), 93-118.

46. M. Pimsner and D. Voiculescu, K-groups of reduced crossed products by free groups, J. Operator Theory 8(1982), 131-156.

47. M. Rieffel, C*-algebras associated with irrational rotations, Pacific J. Math. 93(1981), 415-429.

48. M. Rieffel, Dimension and stable rank in the K-theory of C*-algebras, preprint.

49. J.E. Roberts, New light on the mathematical structure of algebraic field theory, Proc. Symp. Pure Math. vol. 38 (ed. R.V. Kadison), part 2, 523-550.

50. J. Rosenberg, Group C*-algebras and topological invariants, Proc. of the OAGR conference (Neptun, Romania, 1980), Pitman, London, to appear.

51. J. Rosenberg, C*-algebras, positive scalar curvature and the Novikov conjecture, preprint.

52. J. Rosenberg and C. Schochet, The classification of extensions of C*-algebras, Bull. Amer. Math. Soc. 4 (1981), 105-110.

53. C. Schochet, Topological methods for C*-algebras II: geometric resolutions and the Künneth formula, Pacific J. Math. 98 (1982), 443-458.

54. C. Schochet, Topological methods for C*-algebras III: axiomatic homology, preprint.

55. C. Schochet, Topological methods for C*-algebras IV: mod p homology, preprint.

56. G. Skandalis, Some remarks on Kasparov theory, preprint, Queen's University.

57. G. Skandalis, Six term exact sequences for the KK-groups of graded C*-algebras, in preparation.

58. I.M. Singer, Some remarks on operator theory and index theory, Springer Lecture Notes in Math. 575, 129-138.

59. A.M. Torpe, K-theory for the leaf space of foliations by Reeb components, preprint, Odense.

60. J.L. Taylor, Banach algebras and topology, in "Algebras in Analysis", edited by J.H. Williamson, Academic Press (1975).

61. S. Eilenberg and N. Steenrod, Foundations of algebraic topology, Princeton Univ. Press, Princeton, 1952.

The following articles have appeared in "Operator algebras and applications (ed. R.V. Kadison), Proc. Symp. Pure Math. vol. 38 part 1, Amer. Math. Soc., 1982.

62. J. Rosenberg, Homological invariants of extensions of C*-algebras, pp. 35-76.

63. J. Cuntz, The K-groups for free products of C*-algebras, pp. 81-84.

64. J. Cuntz, The internal structure of simple C*-algebras, pp. 85-116.

65. P. Baum and R.G. Douglas, K-homology and index theory, pp. 117-173.

66. L.G. Brown, Extensions of AF-algebras: the projection lifting problem, pp. 175-176.

67. G. Elliott, On the K-theory of the C*-algebra generated by a projective representation of $\mathbb{Z}^n$, pp. 177-180.

68. W.L. Paschke, EXT-groups as K-groups, pp. 185-186.

69. D. Handelman, Positive matrices and dimension groups affiliated to topological Markov chains, pp. 191-194.

70. M. Rieffel, Morita equivalence for operator algebras, pp. 285-298.

71. M. Rieffel, Applications of strong Morita equivalence to transformation group C*-algebras, pp. 299-310.

72. P. Green, Equivariant K-theory and crossed product C*-algebras, pp. 337-338.

73. J.N. Renault, C*-algebras of groupoids and foliations, pp. 339-350.

74. A. Connes and H. Moscovici, L^2-index theory on homogeneous spaces and discrete series representations, pp. 419-434.

75. A. Connes, A survey of foliations and operator algebras, pp. 521-628.

76. T. Fack and G. Skandalis, Some properties of the C*-algebra associated with a foliation, pp. 629-632.

The following articles have appeared in "Operator algebras and K-theory", Contemporary Mathematics vol. 10 (ed. R.G. Douglas and C. Schochet), Amer. Math. Soc., 1982.

77. P. Baum and R.G. Douglas, Index theory, Cobordism, and K-homology, pp. 1-32.

78. D. Handelman, Reducible topological Markov chains via K_0-theory and Ext, pp. 41-76.

79. W.C. Hsiang and H.D. Rees, Miščenko's work on Novikov's conjecture, pp. 77-98.

80. J. Kaminker, Pseudo-differential operators and differential structures, pp. 99-128.

81. H. Moscovici, L^2-index of elliptic operators on locally symmetric spaces of finite volume, pp. 129-138.

82. W.L. Paschke, K-theory for actions of the circle group on C*-algebras, pp. 139-142.

83. M.A. Rieffel, Connes' analogue for crossed products of the Thom isomorphism, pp. 143-154.

84. J. Rosenberg, The role of K-theory in non-commutative algebraic topology, pp. 155-182.

85. N. Salinas, Some remarks on the classification of essentially n-normal operators, pp. 183-196.

Homological Descent for the K-theory of coherent sheaves

Henri Gillet

Contents:

§1 K'-theory of simplicial schemes

§2 Chow homology of simplicial schemes

§3 Envelopes and hyperenvelopes

§4 The descent theorems

§5 Applications

§6 The niveau spectral sequence

Introduction. The theory of cohomological descent has many applications and variations, including the theory of mixed Hodge structures ([D]) and Lichtenbaum's conjecture $[T_2]$. The purpose of this paper is to consider the analogous problem of homological descent, in the context of the K-theory of coherent sheaves. In order to begin discussing descent problems, we must first understand the K-theory of coherent sheaves on simplicial schemes, so in §1 I outline the K'-theory of simplicial schemes having projective face maps. §2 is a straightforward discussion of Chow homology, though the discussion of excision is new. In §3 we apply the machinery of $[SGA4]V^{bis}$ to the notion of an 'envelope' introduced in [F-G]. §4 contains the two descent theorems; one for the K-theory of coherent sheaves, one for Chow homology (as I discuss in §6, these two results are closely related). The applications described in §5 involve extending theorems about K'-theory and projective morphisms to include arbitrary morphisms. While these results have appeared elsewhere, the proof of theorem (5.1) which appears in $[G_2]$ is essentially an 'ad-hoc' proof; in fact the present paper is a result of my attempt to understand more clearly what is happening in $[G_2]$§4. Finally, in §5 I discuss the K'-theory of simplicial schemes with proper face maps; this may be viewed as a preliminary to an examination of the (vector-bundle) K-theory of arbitrary simplicial schemes which will appear in a future paper.

We will assume all schemes are noetherian, quasi-compact and separated.

§1. K'-theory of simplicial schemes

If $X_{\bullet}$ is a simplicial scheme with proper face maps, define for each $n \geq 0$ $\underline{\underline{A}}(X_n)$ to be the full exact subcategory of $\underline{\underline{M}}(X_n)$ consisting of those coherent sheaves F on X_n such that for all injective monotone maps τ: $[m] \longrightarrow [n]$, $R^pX(\tau)_*F = 0$ in $\underline{\underline{M}}(X_n)$ for all $p > 0$.

Lemma 1.1 Let τ: $[m] \longrightarrow [n]$ be an arbitrary monotone map. Then $X(\tau)_*$: $\underline{\underline{A}}(X_n) \longrightarrow \underline{\underline{M}}(X_m)$ is an exact functor.

Proof. τ factors as a composition:

$$\begin{array}{ccc} [m] & \xrightarrow{\ \alpha\ } \!\!> & [p] \\ & \searrow^{\tau} & \downarrow \beta \\ & & [n] \end{array}$$

where α is the composition of a sequence of degeneracies and β is injective , so that $X(\alpha)$: $X \longrightarrow X$ is a closed immersion and hence

$$X(\alpha)_*: \underline{\underline{M}}(X_p) \longrightarrow \underline{\underline{M}}(X_m)$$

is exact. By definition of $\underline{\underline{A}}(X_n)$,

$$X(\beta)_*: \underline{\underline{A}}(X_n) \longrightarrow \underline{\underline{M}}(X_p)$$

is exact and therefore the composition

$$X(\tau)_* = X(\alpha)_* X(\beta)_*$$

is exact too.

Lemma 1.2 Let τ be as in lemma 1. Then the image of $X(\tau)_*$ is contained in $\underline{\underline{A}}(X_m)$.

Proof We must show that for all $F \in \underline{\underline{A}}(X_n)$ and all σ: $[p] \longrightarrow [m]$, $X(\tau)_*F$ is $X(\sigma)$ acyclic. By lemma 1, F is both $X(\tau)$ and $X(\tau\sigma)$ acyclic, hence by looking at the Leray spectral sequence for the composition $X(\sigma)X(\tau)$, we see that $R^pX(\sigma)_*(X(\tau)_*F) = 0$ for $p > 0$.

Lemma 1.3 Let e: $X_{\bullet} \longrightarrow S$ be an augmented simplicial scheme with projective face maps, such that each X_i is quasi-projective over S . Then the inclusion

$$\underline{\underline{A}}(X_n) \longrightarrow \underline{\underline{M}}(X_n)$$

induces an isomorphism on K-theory for all $n \geq 0$.

Proof If L is an invertible sheaf on X_n which is very ample relative to the augmentation $e_n: X_n \longrightarrow S$, then L is also very ample relative to $X(\tau): X_n \longrightarrow X_m$ for any monotone $\tau: [m] \longrightarrow [n]$, since $e_n = e_m \cdot X(\tau)$. Hence for any coherent sheaf F on X_n, there is a $k_0 \geq 0$ such that for all $k > k_0$ and all injective monotone maps $\tau: [m] \longrightarrow [n]$, $R^p X(\tau)_* F = 0$ for $p > 0$ (observe that since there are only finitely many such τ, k_0 is constructed by taking the supremum of the k_0 that work for each individual τ). The rest of the proof proceeds as in [Q] §7, with one additional remark. It is not necessary to assume that L is ample on X_n, because all we want to be sure of is that L is generated by a finite number of its global sections; but if Y is an arbitrary scheme there is a canonical epimorphism of locally free sheaves on $\mathbb{P}^n_Y$ for any $n \geq 1$:

$$\mathcal{O}^{\oplus(n+1)}_{\mathbb{P}^n_Y} \longrightarrow \mathcal{O}_{\mathbb{P}^n_Y}(1)$$

and we are done.

Lemma 1.4 Let D be a finite diagram in the category of simplicial schemes quasi-projective over a fixed base scheme S such that for each $d \in D$, the vertex $X^d_\bullet$ has projective face maps, and for each morphism $f: d \longrightarrow e$ of D, $f: X^d_\bullet \longrightarrow X^e_\bullet$ is projective (i.e. projective in each degree). Define, for each $d \in D$ and each $n \geq 0$, $\underline{\underline{A}}(X^d_n;D)$ as before, requiring in addition that for all $f: d \longrightarrow e$ in D and all injective monotone maps $\tau: [m] \longrightarrow [n]$, that for a coherent sheaf F to belong to $\underline{\underline{A}}(X^d_n;D)$ the sheaves

$$R^p(f_n \cdot X(\tau))_* F \in \underline{\underline{M}}(X^e_m)$$

must be zero for $p > 0$. Then for all $f: d \longrightarrow e$ in D and monotone $\tau: [m] \longrightarrow [n]$ the functor

$$(f_n \cdot X(\tau))_*: \underline{\underline{A}}(X^d_n, D) \longrightarrow \underline{\underline{M}}(X^e_m)$$

is exact, and its image is contained in $\underline{\underline{A}}(X^e_m, D)$. Furthermore, for all $d \in D$ and all $n \geq 0$, the inclusion

$$\underline{\underline{A}}(X^d_n;D) \longrightarrow \underline{\underline{M}}(X^d_n)$$

induces an isomorphism on K-theory.

Proof Just as in lemmas 1 to 3.

Definition Let D be a diagram of simplicial schemes as above, and suppose $d \in D$. Define:

$$K'_*(X^d_\bullet;D) = \pi_{*+1}(|BQ\underline{\underline{A}}(X^d_\bullet;D)|)$$

where $BQ\underline{\underline{A}}(X^d;D)$ is the bisimplicial set obtained by taking the nerve of Quillen's Q construction applied to the simplicial exact category $(n \longrightarrow \underline{\underline{A}}(X^d_n;D))$.

Lemma 1.5 If D is a subdiagram of a diagram E, both satisfying the conditions of lemma 4, then there is an inclusion of simplicial exact categories:

$$\underline{\underline{A}}(X^d_\bullet;E) \longrightarrow \underline{\underline{A}}(X^d_\bullet;D)$$

inducing an isomorphism:

$$K'_*(X^d_\bullet;E) \longrightarrow K'_*(X^d_\bullet;D) .$$

Hence we may write $K'_*(X^d_\bullet)$ without confusion.

Proof The first assertion is obvious. The second comes from considering the induced map of simplicial spaces, which in degree $n \geq 0$ is:

$$|BQ\underline{\underline{A}}(X^d_n;E)| \longrightarrow |BQ\underline{\underline{A}}(X^d_n;D)|$$

This map is a weak equivalence in each degree, and hence ([B-K]XII 4.2) induces a weak equivalence

$$|BQ\underline{\underline{A}}(X^d_\bullet;E)| \longrightarrow |BQ\underline{\underline{A}}(X^d_\bullet;D)| .$$

Proposition 1.6 Let S be a scheme. The covariant function K'_* from the category of projective morphisms between noetherian schemes over S to the category of graded abelian groups extends to a functor on the category of projective morphisms between quasi-projective simplicial schemes over S with projective face maps.

Proof If $X_\bullet$ is a simplicial scheme as above, let D be the punctual diagram with single object $X_\bullet$, and define for $q \geq 0$:

$$K'_q(X_\bullet) = K'_q(X_\bullet,D) .$$

If $F: X_\bullet \longrightarrow Y_\bullet$ is a projective morphism, let D be the corresponding two object diagram. Then there is a natural map

$$f_*: K'_*(X_\bullet, D) \longrightarrow K'_*(Y_\bullet)$$

because the functors

$$f_*: \underline{\underline{A}}(X_n, D) \longrightarrow \underline{\underline{A}}(Y_n)$$

are exact for $n \geq 0$ (by lemma 1.4), so that we have a map of simplicial spaces:

$$f_*: (n \longrightarrow |BQ\underline{\underline{A}}(X_n, D)|) \longrightarrow (n \longrightarrow |BQ\underline{\underline{A}}(Y_n)|) .$$

Hence by lemma 1.5 we have a map for all $q \geq 0$:

$$f_*: K'_q(X_\bullet) \simeq K'_q(X_\bullet, D) \longrightarrow K'_q(Y_\bullet) .$$

If we have morphisms $f: X_\bullet \longrightarrow Y_\bullet$ and $g: Y_\bullet \longrightarrow Z_\bullet$, then by considering the diagram

$$D \equiv \quad \begin{array}{ccc} X_\bullet & \xrightarrow{f} & Y_\bullet \\ & {\scriptstyle g\cdot f}\searrow & \downarrow {\scriptstyle g} \\ & & Z_\bullet \end{array}$$

we see, again by lemmas 4 and 5, that $g_* f_* = (gf)_*$.

<u>Proposition 1.7</u> Let $X_\bullet$ be a simplicial scheme as above, $f: U_\bullet \longrightarrow X_\bullet$ an open subsimplicial scheme such that $X_\bullet - U_\bullet = Y_\bullet$ is a closed subsimplicial scheme of $X_\bullet$. Note that both $U_\bullet$ and $Y_\bullet$ are quasi-projective over S with projective face maps. Then there is a long exact localization sequence:

$$\longrightarrow K'_q(Y_\bullet) \xrightarrow{i_*} K'_q(X_\bullet) \xrightarrow{f^*} K'_q(U_\bullet) \xrightarrow{\partial} K'_{q-1}(Y_\bullet)$$

(where $i: Y_\bullet \longrightarrow X_\bullet$ is the natural map).

<u>Proof</u> We have a sequence of maps of simplicial spaces:

$$(j \longrightarrow |BQ\underline{\underline{A}}(Y_j)|) \xrightarrow{i_*} (j \longrightarrow |BQ\underline{\underline{A}}(X_j)|) \xrightarrow{f^*} (j \longrightarrow |BQ\underline{\underline{A}}(U_j)|)$$

which in each degree is a fibration sequence up to homotopy with connected base. (Note that i_* exists by proposition (1.6) while f^* exists because the formation of higher direct images is local in the Zariski topology). Hence as is well known (see e.g. [W]) the corresponding map on realizations is a fibration sequence up to homotopy, and we obtain the desired localization sequence by taking homotopy groups.

If $X_\bullet$ is a simplicial scheme as above, $|BQ\underline{\underline{A}}(X_\bullet)|$ has a natural infinite loop space structure, since $|BQ^n\underline{\underline{A}}(X_\bullet)|$ is a (n-1) fold delooping of $|BQ\underline{\underline{A}}(X_\bullet)|$ where $Q^n\underline{\underline{A}}(X_\bullet)$ is the iterated Q construction of Waldhausen ([W]). Given this, by [B-K]XII§5 and $[T_1]$3.14 we obtain:

Proposition 1.8 Let $X_\bullet$ be a simplicial scheme quasi-projective over a scheme S , with projective face maps. Then there is a convergent spectral sequence:

$$E^1_{p,q} = K'_q(X_p) \Rightarrow K'_{p+q}(X_\bullet) .$$

Corollary 1.9 Let $X_\bullet$ be a simplicial scheme as above. Then $X^{red}_\bullet$ satisfies the same conditions and $K'_\bullet(X^{red}_\bullet) \simeq K'_*(X_\bullet)$.

Proof By [Q] this is true for each X_p , and so using the spectral sequence of the proposition must be true for $X_\bullet$.

Corollary 1.11 Let $f: X_\bullet \longrightarrow Y_\bullet$ be a projective morphism between simplicial schemes as above, such that $f: X_i \longrightarrow Y_i$ is an isomorphism for $i \leq p$. Then $K'_q(X_\bullet) \simeq K'_q(Y_\bullet)$ for $q < p$.

Proof Apply the spectral sequence.

Lemma 1.12 Let $f: X_\bullet \longrightarrow Y_\bullet$ be a proper morphism of simplicial schemes, as always quasi-projective over a base S , with projective face maps. Then if $Z_\bullet \subset Y_\bullet$ is a closed subsimplicial scheme, such that $U_\bullet = Y_\bullet - Z_\bullet$ is an open subsimplicial scheme, with $W_\bullet = f^{-1}(Z_\bullet)$ and $V_\bullet = f^{-1}(U_\bullet)$ such that $f: V_\bullet \longrightarrow U_\bullet$ is an isomorphism, then there is a Mayer-Vietoris exact sequence:

$$\xrightarrow{\partial} K'_q(W_\bullet) \longrightarrow K'_q(Z_\bullet) \oplus K'_q(X_\bullet) \longrightarrow K'_q(Y_\bullet) \xrightarrow{\partial}$$

Proof Look at the map of long exact sequences:

$$\begin{array}{ccccccc} \longrightarrow K'_q(W_\bullet) & \longrightarrow & K'_q(X_\bullet) & \longrightarrow & K'_q(V_\bullet) & \xrightarrow{\partial} \\ \downarrow f_* & & \downarrow f_* & & \downarrow f_* & \\ \longrightarrow K'_q(Z_\bullet) & \longrightarrow & K'_q(Y_\bullet) & \longrightarrow & K'_q(U_\bullet) & \xrightarrow{\partial} \end{array}$$

An example of the hypotheses of the lemma is $X_\bullet = Y^1_\bullet \cup Y^2_\bullet$ with $Y^1_\bullet$ and $Y^2_\bullet$ closed subsimplicial schemes of $X_\bullet$. Then there is a Mayer-Vietoris sequence:

$$\xrightarrow{\partial} K'_q(Y^1_\bullet \cap Y^2_\bullet) \longrightarrow K'_q(Y^1_\bullet)\oplus K'_q(Y^2_\bullet) \longrightarrow K'_q(X_\bullet) \longrightarrow ,$$

when $X_\bullet - Y^i_\bullet$ is an open subsimplicial scheme of $X_\bullet$ for $i = 1,2$. (This last condition is not actually necessary.)

§2. Chow homology groups of simplicial schemes If X is a noetherian scheme, we can filter the category $\underline{\underline{M}}(X)$ of coherent sheaves on X by the Serre subcategories $\underline{\underline{M}}_i(X)$ consisting of sheaves having dimension of support $\leq i$. From this we obtain a spectral sequence which is the filtration by dimension version of a spectral sequence due to Quillen ([Q]):

$$E^1_{r,s}(X) = \bigoplus_{x \in X_{(r)}} K_{r+s}(k(x)) \Rightarrow K'_{r+s}(X) \tag{2.0}$$

where $X_{(r)}$ is the set of points of dimension r (i.e. $\overline{\{x\}}$ is an r dimensional scheme) in X. We shall write the $E^2_{r,s}$ term of this spectral sequence as $CH_{r,-s}$. Then $CH_{p,p}$ is the usual group CH_p of cycles of dimension p modulo rational equivalence defined by Fulton ([F]). If we write $CH_{*,*}(X) = \bigoplus_{r,s} CH_{r,s}(X)$, then $CH_{*,*}$ is a covariant functor on the category of proper morphisms between schemes. For further details and proofs see $[G_1]$§8. Notice that for $s \in \mathbb{Z}$, the $E^1_{\bullet\bullet}$ term of this spectral sequence breaks up into complexes $R_*(X,s)$ with $R_i(X,s) = E^1_{i,-s}(X)$, so that

$$CH_{r,s}(X) = H_r(R_*(X,s)).$$

The complexes $R_*(X,s)$ are themselves covariant for proper morphisms, and depend only on the reduced structure of X.

If Y is a closed subscheme of X, with complement U (we shall speak of (X,Y) as a closed pair), then $R_*(Y,s)$ is, for each s, a subcomplex of $R_*(X,s)$ and we can define a complex $R_*(X,Y;s)$ via the short exact sequence:

$$0 \longrightarrow R_*(Y,s) \longrightarrow R_*(X,s) \longrightarrow R_*(X,Y;s) \longrightarrow 0.$$

Another way of defining this relative complex is:

$$R_i(X,Y;s) = \bigoplus_{x \in U \cap X_{(i)}} K_{i-s} k(x) .$$

Finally, define $CH_{r,s}(X,Y) = H_r(R_*(X,Y;s))$. Note that these groups fit into a long exact sequence (covariant for proper morphisms of closed pairs):

$$CH_{r+1,s}(X,Y) \xrightarrow{\partial} CH_{r,s}(Y) \longrightarrow CH_{r,s}(X) \longrightarrow CH_{r,s}(X,Y)$$

Lemma 2.1 Let $f: (Z,W) \longrightarrow (X,Y)$ be a proper morphism of closed pairs such that f induces an isomorphism:

$$f: Z-W \xrightarrow{\sim} X-Y .$$

Then the natural map $f_*: CH_{*,*}(Z,W) \longrightarrow CH_{*,*}(X,Y)$ is an isomorphism.

Proof In fact the stronger statement that for all s ,

$$f_*: R_*(Z,W;s) \longrightarrow R_*(X,Y;s)$$

is an isomorphism is true. Explicitly, this map is, in degree i:

$$\bigoplus_{z\in Z_{(i)}\cap(Z-W)} K_{s-i}(k(z)) \longrightarrow \bigoplus_{x\in X_{(i)}\cap(X-Y)} K_{s-i}(k(x)) .$$

We will be done therefore, if we can show that f induces a bijection:

$$Z_{(i)} \cap (Z-W) \xrightarrow{\sim} X_{(i)} \cap (X-Y)$$

i.e. for every point $x \in U$, the dimensions of its closures $\overline{\{x\}}_Z$ and $\overline{\{x\}}_X$, in Z and X respectively are the same. But the map $f\big|_{\overline{\{x\}}_Z}: \overline{\{x\}}_Z \longrightarrow \overline{\{x\}}_X$ is proper and birational, and so by [EGA] IV (5.4.1) and (5.6.6):

$$\dim \overline{\{x\}}_Z = \dim \overline{\{x\}}_X .$$

Corollary 2.2 Let X,Z,Y,W be as above. Then there is a Mayer-Vietoris style long exact sequence:

$$CH_{r+1,s}(X) \xrightarrow{\partial} CH_{r,s}(W) \longrightarrow CH_{r,s}(Y)\oplus CH_{r,s}(Z) \longrightarrow CH_{r,s}(X) .$$

Proof This follows in a standard fashion, using the lemma, from the map of long exact sequences corresponding to the map of pairs $(Z,W) \longrightarrow (X,Y)$.

Lemma 2.1 suggests that when dealing with the Chow homology of schemes of finite type over a fixed base scheme S , the right definition of dimension is:

Definition Let X be of finite type over S . If $Z \subset X$ is a closed subscheme, define $\dim_S(Z)$ to be the dimension of the closure of Z in $\overline{X}$, where $\overline{X}$ is a compactification of X over S i.e. $\overline{X}$ is proper over S , and X is open in $\overline{X}$ ($\overline{X}$ exists by a result of Nagata).

Now consider a simplicial scheme $X_\bullet$ in which all the face maps are proper. Then we can form the associated double complex, for each integer s ,

$$(i,j) \longrightarrow R_i(X_j,s)$$

and define ("$\int$" = "associated total complex"):

$$CH_{r,s}(X_\bullet) = H_r(\int R_*(X_\bullet,s))$$

$CH_{*,*}$ is evidently a covariant functor on the category of proper morphisms between simplicial schemes with proper face maps. One constructs from the double complex $R_*(X_\bullet,s)$ a convergent spectral sequence

(2.3) $$E^1_{p,q}(X_\bullet,s) = CH_{q,s}(X_p) \Rightarrow CH_{p+q,s}(X_\bullet) .$$

If $Y_\bullet \subset X_\bullet$ is a closed subsimplicial scheme, we can define for each s ,

$$CH_{r,s}(X_\bullet,Y_\bullet) = H_r(\int R_*(X_\bullet,s)/R_*(Y_\bullet,s))$$

and there is a long exact sequence, covariant for proper maps of pairs,

$$CH_{r+1,s}(X_\bullet,Y_\bullet) \xrightarrow{\partial} CH_{r,s}(Y_\bullet) \longrightarrow CH_{r,s}(X_\bullet) \longrightarrow CH_{r,s}(X_\bullet,Y_\bullet) .$$

There is a spectral sequence, analogous to (2.3):

(2.4) $$E^1_{p,q}(X_\bullet,Y_\bullet;s) = CH_{q,s}(X_p,Y_p;s) \Rightarrow CH_{p+q,s}(X_\bullet,Y_\bullet;s) .$$

By combining lemma (2.1) and the spectral sequence (2.4) we get the following excision result.

<u>Lemma 2.5</u> Let $f\colon (Z_\bullet,W_\bullet) \longrightarrow (X_\bullet,Y_\bullet)$ be a proper map of simplicial closed pairs, such that for all $i \geq 0$ $Z_i - W_i \simeq X_i - Y_i$. Then for all r,s , we have an isomorphism:

$$f_*\colon CH_{r,s}(Z_\bullet,W_\bullet) \longrightarrow CH_{r,s}(X_\bullet,Y_\bullet) .$$

<u>Corollary 2.6</u> Let $X_\bullet$, $Y_\bullet$, $Z_\bullet$, $W_\bullet$ be as above; then there is a long exact Mayer-Vietoris sequence for each r,s :

$$\longrightarrow CH_{r,s}(W_\bullet) \longrightarrow CH_{r,s}(Y_\bullet)\oplus CH_{r,s}(Z_\bullet) \longrightarrow CH_{r,s}(X_\bullet) \xrightarrow{\partial}$$

<u>Proof</u> as in (2.2).

If $f,g\colon (Z_\bullet,W_\bullet) \longrightarrow (X_\bullet,Y_\bullet)$ are homotopic maps of simplicial closed pairs, then the maps of complexes $(R_*(Z_\bullet,W_\bullet;s) = R_*(Z_\bullet,s)/R_*(W_\bullet,s))$:

$$f_*,g_*\colon \int R_*(Z_\bullet,W_\bullet,s) \longrightarrow \int R_*(X_\bullet,Y_\bullet,s)$$

are homotopic, and so induce the same maps

$$CH_{r,s}(Z_\bullet,W_\bullet) \longrightarrow CH_{r,s}(X_\bullet,Y_\bullet) .$$

Finally observe that by using the spectral sequence (2.4), we see that if $f\colon (Z_\bullet,W_\bullet) \longrightarrow (X_\bullet,Y_\bullet)$ is a proper morphism of simplicial closed pairs which for $i \leq p$ induces an isomorphism

$$Z_i - W_i \xrightarrow{\sim} X_i - Y_i$$

then for all $i < p$ and all s :

$$CH_{i,s}(Z_\bullet,W_\bullet) \simeq CH_{i,s}(X_\bullet,Y_\bullet) .$$

§3 Envelopes and Hyperenvelopes

Definition i) An envelope of a scheme X is a proper morphism $p\colon Z \longrightarrow X$ such that for every field F the map $Z(F) \longrightarrow X(F)$ is surjective. It is an easy exercise to see that this is equivalent to the definition of [F-G] .

ii) If $f\colon X \longrightarrow Y$ is a proper morphism, then a Chow envelope of X (relative to Y) is an envelope $p\colon Z \longrightarrow X$ such that the composite morphism $f\cdot p\colon Z \longrightarrow Y$ is projective.

iii) A hyperenvelope of a scheme X is a simplicial scheme such that $e\colon Z_0 \longrightarrow X$ is an envelope, and for all $i \geq 0$ the natural map

$$Z_i \longrightarrow \mathrm{Cosk}_i^X(\mathrm{sk}_{i-1}Z_\bullet)$$

is an envelope. Obviously, if $f\colon X \longrightarrow Y$ is a proper map then a Chow hyper-envelope of X relative to Y is a hyperenvelope $p\colon Z_\bullet \longrightarrow X$ such that for each Z_i for $i \geq 0$ is projective over Y .

iv) $p\colon Z_\bullet \longrightarrow X$ is a projective hyperenvelope if the Z_i are projective over X .

(3.1) Lemma i) If $p\colon Z_\bullet \longrightarrow X$ is a hyperenvelope and $f\colon Y \longrightarrow X$ is an arbitrary morphism, the induced map $Z_\bullet \times_X Y \longrightarrow Y$ is a hyperenvelope.

ii) If $p\colon Z \longrightarrow Y$ and $q\colon Y \longrightarrow X$ are envelopes, then $q\cdot p$ is an envelope.

iii) If $f\colon X \longrightarrow Y$ is a proper morphism there exists a Chow envelope $p\colon Z \longrightarrow X$ of X relative to Y and hence $\mathrm{cosk}_0^X(Z)$ is a Chow hyperenvelope of X relative to Y .

Proof i) and ii) follow immediately from the definitions, iii) is an elementary exercise in Chow's lemma (see [F-G]).

(3.2) Lemma If $f: X \longrightarrow Y$ is proper, and $p: Z_\bullet \longrightarrow X$ is a hyperenvelope, then there exists a Chow hyperenvelope $q: W_\bullet \longrightarrow X$ of X relative to Y, and a morphism $h: W_\bullet \longrightarrow Z_\bullet$ such that $p \cdot h = q$.

Proof First we construct W_0. Consider

$$\begin{array}{l} Z_0 \\ \downarrow p \\ X \xrightarrow{f} Y \end{array}$$

Choose a Chow envelope $h_0: W_0 \longrightarrow Z_0$ of Z_0 relative to Y; then W_0 is projective over Z_0 and X. Now define an s-split i-truncated simplicial scheme (the terminology is that of [SGA4]V^{bis}) $W[i]_\bullet$ augmented toward $Z_\bullet$ by induction on $i \geq 0$ as follows. Let $W[0] = W_0$; given $W[i-1]_\bullet$, let

$$\beta: W_i \longrightarrow \operatorname{Cosk}_i^{Z_\bullet}(W[i-1])_i$$

be a Chow envelope relative to $Y_\bullet$. Then the triple $(W[i-1]_\bullet, W_i, \beta)$ satisfies the conditions of [SGA4]V^{bis} (5.1.3), hence there exists a $W[i]_\bullet$ with $sk_{i-1} W[i]_\bullet = W[i-1]_\bullet$ and

$$W[i]_i = \coprod_{m \leq n} \coprod_{\substack{\text{surjective} \\ \text{decreasing maps} \\ [0,\dots,n] \to [0,\dots,m]}} W_m$$

so that $W[i]_i$ is a Chow envelope of $\operatorname{Cosk}_i^{Z_\bullet}(W[i-1])_i$.

Remark Over a field of characteristic zero (e.g. $\mathbb{C}$) we could, by the method above, construct smooth hyperenvelopes; i.e. for a simplicial scheme $X_\bullet$, there exists a hyperenvelope $p: Z_\bullet \longrightarrow X_\bullet$ with Z_i smooth for all $i \geq 0$.

Lemma 3.3 If $f: X \longrightarrow Y$ is a proper morphism, and $p: Z_\bullet \longrightarrow X$ and $q: W_\bullet \longrightarrow X$ are two Chow hyperenvelopes of X relative to Y, there exists a Chow hyperenvelope $r: V_\bullet \longrightarrow X$ of X relative to Y dominating both p and q:

$$\begin{array}{ccccc} & & V_\bullet & & \\ & \overset{\pi_1}{\swarrow} & & \overset{\pi_2}{\searrow} & \\ Z_\bullet & & \downarrow r & & W_\bullet \\ & \underset{p}{\searrow} & & \underset{q}{\swarrow} & \\ & & X & & \end{array}$$

Proof Since the coskeleton functor commutes with products, and the product of envelopes is an envelope, we may take $V_\bullet = Z_\bullet \times W_\bullet$ (the product simplicial scheme).

§4. The descent theorems

Theorem 4.1 Let $p\colon Z_\bullet \longrightarrow X$ be a projective hyperenvelope. Then $p_*\colon K'_q(Z_\bullet) \longrightarrow K'_q(X)$ is an isomorphism for all $q \geq 0$.

Proof We proceed in three steps. (Observe the similarities of this proof fo the proof of the homotopy property in [Q].)

(I) First, let us suppose that $Z_\bullet = \mathrm{Cosk}_0^X(Z)$ where $\pi\colon Z \longrightarrow X$ is an envelope, and that X is the spectrum of a field F. Since π is an envelope, Z must have an F-rational point; ie., $\pi\colon Z \longrightarrow X$ has a section $s\colon X \longrightarrow Z$. Such a section defines, in a standard fashion, a contracting homotopy of $Z_\bullet$ over X; i.e., s extends to a map $s\colon X \longrightarrow Z_\bullet$ (viewing X as a constant simplicial scheme) such that $p\cdot s = 1_X$ and $s\cdot p$ is homotopic to $1_{Z_\bullet}$. Hence for all $q \geq 0$ the map of simplicial groups

$$p_*\colon (i \longrightarrow K'_q(Z_i)) \longrightarrow (i \longrightarrow K'_q(X))$$

is a homotopy equivalence, and hence via the spectral sequence (1.8) the map $p_*\colon K'_q(Z_\bullet) \longrightarrow K'_q(X)$ is an isomorphism.

(II) If X is not reduced to a point, but we still have $Z_\bullet = \mathrm{Cosk}_0^X(Z)$, we proceed by noetherian induction over the closed subsets Y of X. If Y is a point, we are in case (I) above, so we may suppose that the theorem is true for all proper closed subsets of X. If X is not irreducible we may write $X = Y_1 \cup Y_2$; by the induction hypothesis, the theorem is true for Y_1, Y_2 and $Y_1 \cap Y_2$. Comparing Mayer-Vietoris sequences (1.12):

$$\begin{array}{ccccccccc} \longrightarrow & K'_q(Z^1_\bullet \cap Z^2_\bullet) & \longrightarrow & K'_q(Z^1_\bullet)\oplus K'_q(Z^2_\bullet) & \longrightarrow & K'_q(Z_\bullet) & \xrightarrow{\partial} \\ & \downarrow p_* & & \downarrow p_* & & \downarrow p_* & \\ \longrightarrow & K'_q(Y_1 \cap Y_2) & \longrightarrow & K'_q(Y_1)\oplus K_q(Y_2) & \longrightarrow & K'_q(X) & \xrightarrow{\partial} \end{array}$$

where $Z_\bullet^i = p^{-1}(Y_i)$, we deduce the theorem for $p: Z_\bullet \longrightarrow X$ by the five lemma. Since we may already suppose that X is reduced, we may now assume that X is integral. Since $\pi: Z \longrightarrow X$ is an envelope, there is an open dense subset U of X such that $\pi^U: \pi^{-1}(U) \longrightarrow U$ admits a section and hence argueing as in (I) above $(Z_\bullet^U = p^{-1}(U), p^U = p|_{Z_\bullet^U}$,

$$p_*^U: K_q'(Z_\bullet^U) \longrightarrow K_q'(U)$$

is an isomorphism. If $Y = X-U$, then writing $Z_\bullet^Y = p^{-1}(Y) = Z_\bullet - Z_\bullet^U$, we have a map of long exact sequences:

$$\begin{array}{ccccccc} \longrightarrow K_q'(Z_\bullet^Y) & \longrightarrow & K_q'(Z_\bullet) & \longrightarrow & K_q'(Z_\bullet^U) & \xrightarrow{\partial} & \\ \downarrow p_* & & \downarrow p_* & & \downarrow p_*^U & & \\ \longrightarrow K_q'(Y) & \longrightarrow & K_q'(X) & \longrightarrow & K_q'(U) & \xrightarrow{\partial} & . \end{array} \tag{4.2}$$

We already know p_*^U is an isomorphism; p_*^Y is an isomorphism by the induction hypothesis. Hence, again by the five lemma, we are done.

(III) Now consider the case of a general projective hyperenvelope $p: Z_\bullet \longrightarrow X$. Let us write $Z_\bullet[i] = \mathrm{Cosk}_i^X(\mathrm{sk}_i Z_\bullet)$; then we have an inverse system

$$Z_\bullet \longrightarrow \ldots \longrightarrow Z_\bullet[i] \longrightarrow Z_\bullet[i-1] \longrightarrow \ldots \longrightarrow Z_\bullet[0] \longrightarrow X$$

in which $Z_j \simeq Z_j[i]$ for $j \leq i$ and hence (1.12) $K_q'(Z_\bullet) \simeq K_q'(Z_\bullet[i])$ for $q < i$. It therefore suffices to prove that for all $i \geq 0$,

$$K_q'(Z_\bullet[i+1]) \simeq K_q'(Z_\bullet[i])$$

since $K_q(Z_0) \simeq K_q(X)$ for all $q \geq 0$ by case (II). The proof now follows as in lemma (3.3.3.2) of [SGA4, V^{bis}], replacing the words "morphisme de G-1-descente cohomologique universelle" by "envelope" throughout. Note that lemma (3.3.3.1) of op.cit. is replaced by (1.12) of §1 and that the relevant parts of proposition (3.3.1) of op.cit. are replaced by parallel statements in §2. Note that in order to apply (3.0.2.4) of op.cit. it is sufficient to know that a homotopy equivalence of simplicial schemes induces an isomorphism on K'-theory, as discussed in (I).

Theorem 4.3 Let $p: Z_\bullet \longrightarrow X$ be a (not necessarily projective) hyperenvelope. Then the natural map

$$p_*: CH_{*,*}(Z_\bullet) \longrightarrow CH_{*,*}(X)$$

is an isomorphism.

<u>Proof</u> The proof is exactly the same as that of theorem (4.1) (replacing K'_* by $CH_{*,*}$ throughout) except that in (II) we must replace the long exact sequences (4.2) by long exact sequences for s in $\mathbb{Z}$ and any closed subset Y of X :

$$\begin{array}{ccccccc}
\xrightarrow{\partial} & CH_{r,s}(Z^Y_{\cdot}) & \longrightarrow & CH_{r,s}(Z_{\cdot}) & \longrightarrow & CH_{r,s}(Z_{\cdot},Z^Y_{\cdot}) & \longrightarrow \\
 & \Big\downarrow p^Y_* & & \Big\downarrow p_* & & \Big\downarrow p_*^{(X,Y)} & \\
\xrightarrow{\partial} & CH_{r,s}(Y) & \longrightarrow & CH_{r,s}(X) & \longrightarrow & CH_{r,s}(X,Y) & \longrightarrow
\end{array}$$

Since $\pi: Z \longrightarrow X$ is an envelope, we know that there is a closed subscheme $\tilde{X}$ of Z , such that the induced map $f: \tilde{X} \longrightarrow X$ is an isomorphism over a Zariski open set U of X . Consider the diagram, where $\tilde{Z} = \tilde{X} \times_X Z$

$$\begin{array}{ccc}
\tilde{Z} & \longrightarrow & Z \\
s \Big\uparrow \Big\downarrow \tilde{\pi} & & \Big\downarrow \pi \\
\tilde{X} & \xrightarrow{f} & X
\end{array}$$

where the section s is deduced from the inclusion $\tilde{X} \quad Z$. Now define $\tilde{Z}_{\cdot} = \mathrm{Cosk}_0^{\tilde{X}}(\tilde{Z})$, with $\tilde{p}: \tilde{Z}_{\cdot} \longrightarrow \tilde{X}$ the natural map, $\tilde{Z}^{\tilde{Y}}_{\cdot} = \tilde{p}^{-1}(Y)$ where $\tilde{Y} = f^{-1}(Y = X-U)$. Observe $\tilde{X}-\tilde{Y} \simeq X-Y$ and that $\tilde{Z}_{\cdot}-\tilde{Z}^{Y}_{\cdot} \simeq Z_{\cdot}-Z^Y_{\cdot}$; so by (2.1) and (2.5) in order to prove that

$p_*^{(X,Y)}: CH_{*,*}(Z_{\cdot},Z^Y_{\cdot}) \longrightarrow CH_{*,*}(X,Y)$ is an isomorphism, we may instead prove that $\tilde{p}_*^{(\tilde{X},\tilde{Y})}$ is an isomorphism. So consider, for each s , the map of long exact sequences:

$$\begin{array}{ccccccc}
\longrightarrow & CH_{r,s}(Z^{\tilde{Y}}_{\cdot}) & \longrightarrow & CH_{r,s}(\tilde{Z}_{\cdot}) & \longrightarrow & CH_{r,s}(\tilde{Z}_{\cdot},Z^{\tilde{Y}}_{\cdot}) & \xrightarrow{\partial} \\
 & \Big\downarrow \tilde{p}^{\tilde{Y}}_* & & \Big\downarrow \tilde{p}_* & & \Big\downarrow \tilde{p}_*(\tilde{X},\tilde{Y}) & \\
\longrightarrow & CH_{r,s}(\tilde{Y}) & \longrightarrow & CH_{r,s}(\tilde{X}) & \longrightarrow & CH_{r,s}(\tilde{X},\tilde{Y}) & \xrightarrow{\partial} .
\end{array}$$

But since $\tilde{Z} \longrightarrow \tilde{X}$ and $\tilde{Z}^{\tilde{Y}} \longrightarrow \tilde{Y}$ both have sections, $\tilde{p}^{\tilde{Y}}_*$ and $\tilde{p}_*$ are both isomorphisms, and we are done (again by the five lemma).

<u>Remark</u> Using the proposition of §6, we could deduce theorem 4.1 from theorem 4.3. However since the proof of (4.1) is simpler than that of (4.3) and the construction of the spectral sequence in §6 is only sketched, I have presented both proofs.

The proof of the following corollary follows immediately from theorems (4.1) and (4.3) together with the spectral sequences (1.8) and (2.3).

Corollary 4.4 Let $f: X \longrightarrow Y$ be a proper morphism of schemes and $p: Z \longrightarrow X$ a Chow envelope of X relative to Y. Then there is an exact sequence:

$$K_0(Z\times_X Z) \longrightarrow K_0(Z) \longrightarrow K_0(X) \longrightarrow 0$$

and for all $p \geq 0$ there are exact sequences:

$$CH_p(Z\times_X Z) \longrightarrow CH_p(Z) \longrightarrow CH_p(X) \longrightarrow 0 .$$

§5. Applications

Theorem 5.1 Let $f: X \longrightarrow Y$ be a proper morphism of noetherian schemes. Then

(i) There is a natural map

$$f_*: K'_*(X) \longrightarrow K'_*(Y)$$

(ii) There is a natural map of spectral sequences (2.0):

$$E^r_{p,q}(X) \longrightarrow E^r_{p,q}(Y)$$

converging to the map of (i) above, and which at the E^2 level is the map

$$f_*: CH_{p,-q}(X) \longrightarrow CH_{p,-q}(Y) .$$

Proof i) Let $p: Z_\bullet \longrightarrow X$ be a Chow hyperenvelope of X relative to Y. Then by (1.6) we have maps, for $q \geq 0$:

$$K'_q(X) \xleftarrow{p_*} K'_q(Z_\bullet) \xrightarrow{(f\cdot p)_*} K'_q(Y)$$

and we define $f_* = (f\cdot p)_*(p_*)^{-1}$. If $q: W_\bullet \longrightarrow X$ is another Chow hyperenvelope, then by lemma (3.3) there is a Chow hyperenvelope $r: V_\bullet \longrightarrow X$ dominating p and q and we have a commutative diagram

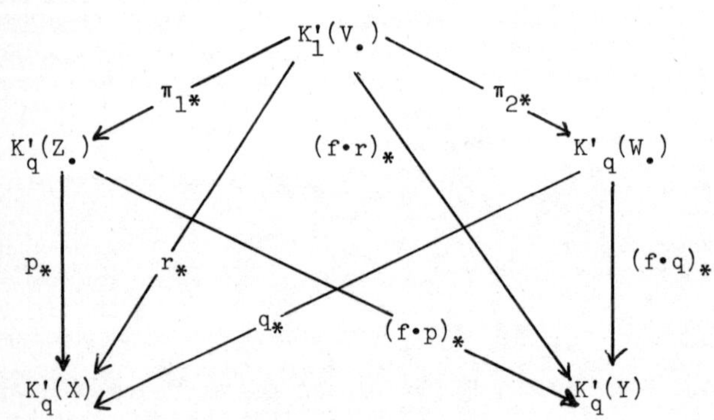

in which p_*, r_* and q_* and hence π_{1*} and π_{2*} are all isomorphisms. Therefore

$$(f\cdot p)_*(p_*)^{-1} = (f\cdot q)_*(q_*)^{-1}$$

and so f is well defined.

If $f\colon X \longrightarrow Y$, $g\colon Y \longrightarrow Z$ are two proper morphisms, then we may choose by (3.2) a Chow hyperenvelope of Y relative to Z, $p\colon W_\bullet \longrightarrow Y$ and a Chow hyperenvelope $q\colon V_\bullet \longrightarrow X$ of X relative to Z such that there is a map $\tilde{f}\colon V_\bullet \longrightarrow W_\bullet$, so that we have a commutative diagram:

$$\begin{array}{ccccc} V_\bullet & \xrightarrow{\tilde{f}} & W_\bullet & & \\ {\scriptstyle q}\downarrow & & \downarrow{\scriptstyle p} & & \\ X & \xrightarrow{f} & Y & \xrightarrow{g} & Z\ . \end{array}$$

Since $V_\bullet$ is projective over X (and therefore over Y to), the morphism $\tilde{f}$ is projective, and by construction of $W_\bullet$, $g\cdot p$ is projective; hence by (1.6) the two maps $(g\cdot p\cdot\tilde{f})_*$ and $(g\cdot p)_*\tilde{f}_*$ from $K'_q(V_\bullet)$ to $K'_q(Z)$ coincide for all $q \geq 0$. Therefore as maps $K'_q(X)$ to $K'_q(Z)$, we have

$$\begin{aligned}(g\cdot f)_* &= (g\cdot f\cdot q)_*(q_*)^{-1}\\ &= (g\cdot p\cdot\tilde{f})_*(q_*)^{-1}\\ &= (g\cdot p)_*(\tilde{f})_*(q_*)^{-1}\\ &= (g\cdot p)_*(p_*)^{-1}(p_*)(\tilde{f}_*)(q_*)^{-1}\\ &= (g\cdot p)_*(p_*)^{-1}(p\tilde{f})_*(q_*)^{-1}\\ &= (g\cdot p)_*(p_*)^{-1}(fq)_*(q_*)^{-1}\\ &= g_*f_*\end{aligned}$$

where we have only assumed $(\alpha\cdot\beta)_* = \alpha_*\beta_*$ when α and β are both projective morphisms between simplicial schemes.

(ii) Again let $p\colon Z_\bullet \longrightarrow X$ be a Chow hyperenvelope of X relative to Y. Then for $r \geq 0$, we know that if $\underline{\underline{M}}_r(X)$ is the category of coherent sheaves on X having support in dimension $\leq r$, then

$$K_q(\underline{\underline{M}}_r(X)) = \varinjlim_{\dim(T)\leq r} K'_q(T)\ .$$

The spectral sequence (2.0) is deduced from the fact that via the localization sequence, we know that

$$K_q(\underline{\underline{M}}_r(X)/\underline{\underline{M}}_{r-1}(X)) = \varinjlim_{\dim(T)\leq r} \varinjlim_{\substack{\dim(S)\leq r-1 \\ S\subset T}} K'_q(T-S)$$

$$= \bigoplus_{x\in X_{(r)}} K'_q \Bbbk(x)$$

so that we get an exact couple

$$\begin{array}{ccc} \bigoplus_{r,q\geq 0} \varinjlim_{\dim(S)\leq r-1} K'_q(S) & \longrightarrow & \bigoplus_{r,q\geq 0} \varinjlim_{\dim(T)\leq r} K'_q(T) \\ & \nwarrow \quad \swarrow & \\ & \bigoplus_{r,q\geq 0} \bigoplus_{x\in X_{(r)}} K_q(\Bbbk(x)) & \end{array}$$

However since K'_* is covariant for proper morphisms, and for subschemes T of X $\dim f(T) \leq \dim T$, this exact couple is covariant for proper morphisms. Evidently, the induced map on $E^1_{\cdot\cdot}$ and therefore on $E^2_{\cdot\cdot}$ terms of the associated spectral sequence agrees with that defined in ([G_1] §7). The naturality and uniqueness of the map of spectral sequences follows from the naturality of the map on K'_*-theory.

<u>Proposition 5.2</u> Let $f\colon X \longrightarrow Y$ be a proper morphism. Then $f_*\colon K'_*(X) \longrightarrow K'_*(Y)$ is a morphism of $K_*(Y)$ modules; <u>i.e.</u>, for all $\alpha \in K'_*(X)$ and $\beta \in K_*(Y)$:

$$f_*(\alpha\cdot f^*(\beta)) = f_*(\alpha)\cdot\beta \qquad K'_q(Y)$$

<u>Proof</u> Let $p\colon Z_\cdot \longrightarrow X$ be a Chow envelope of X relative to Y. Then if D is the diagram:

$$Z_\cdot \xrightarrow{(f\cdot p)} Y$$

for each $i \geq 0$ we have a commutative diagram ([G_1]§7.24) where $\underline{\underline{P}}(Y)$ is the category of locally free sheaves on X :

$$\begin{array}{ccc} \underline{\underline{A}}(Z_i, D) \times \underline{\underline{P}}(Y) & \xrightarrow{\otimes_{\mathcal{O}_X}(f_i\cdot p)^*(\ )} & \underline{\underline{A}}(Z_i, D \\ \Big\downarrow (f_i\cdot p)_*\times 1 & & \Big\downarrow (f_i\cdot p)_* \\ \underline{\underline{M}}(Y) \times \underline{\underline{P}}(Y) & \xrightarrow{\otimes_{\mathcal{O}_Y}} & \underline{\underline{M}}(Y) \end{array}$$

where the vertical arrows are exact, the horizontal arrows biexact. Hence we get a commutative diagram:

$$\begin{array}{ccc}
|i \longrightarrow BQ\underline{\underline{A}}(Z_i,D)|\wedge BQ\underline{\underline{P}}(Y) & \longrightarrow & |i \longrightarrow BQ\underline{\underline{A}}(Z_i,D)| \\
\Big\downarrow (f\cdot p)_* \, 1 & & \Big\downarrow (f\cdot p)_* \\
|i \longrightarrow BQ\underline{\underline{M}}(Y)|\wedge BQ\underline{\underline{P}}(Y) & \longrightarrow & |i \longrightarrow BQ\underline{\underline{M}}(Y)|
\end{array}$$

and the proof follows as in (op.cit.).

<u>Proposition 5.3</u> (Fulton and Gillet, [F-G]). Let $\tau\colon K_0' \longrightarrow CH_{*\mathbb{Q}}$ be a natural transformation defined on the category of projective morphisms between algebraic schemes over k. Then τ extends uniquely to a natural transformation on the full category of proper morphisms. ($CH_{*\mathbb{Q}}$ means Chow homology with rational coefficients.)

<u>Proof</u> Let $f\colon X \longrightarrow Y$ be a proper morphism between algebraic schemes, and let $p\colon Z_{\bullet} \longrightarrow X$ be a Chow hyperenvelope of X relative to Y. Since τ is defined on the category of projective morphisms, we have a commutative diagram of maps of complexes of abelian groups:

$$\begin{array}{ccc}
(i \longrightarrow K_0'(Z_i)) & \xrightarrow{\ \tau_{Z_{\bullet}}\ } & (i \longrightarrow CH_*(Z_i)_{\mathbb{Q}}) \\
\Big\downarrow i \longrightarrow (f\cdot p\ ) & & \Big\downarrow i \longrightarrow (f\cdot p_i)_* \\
K_0'(Y) & \xrightarrow{\ \tau_Y\ } & CH_*(Y)_{\mathbb{Q}}
\end{array}$$

and hence a commutative diagram:

$$\begin{array}{ccc}
K_0'(X) & \xrightarrow{\ \tau_X\ } & CH_*(X)_{\mathbb{Q}} \\
\Big\downarrow f_* & & \Big\downarrow f_* \\
K_0'(Y) & \xrightarrow{\ \tau_Y\ } & CH_*(Y)_{\mathbb{Q}}
\end{array}$$

where τ_X is <u>defined</u> to be $p_* \cdot \tau_{Z_{\bullet}} \cdot p_*^{-1}$. It is an easy exercise to check that τ_X is independent of the hypercover chosen.

One additional application of the ideas of §3 and §4 is the extension of the definition of the K'-theory of a simplicial scheme given in §1 to include simplicial schemes with proper face maps. Since the proofs are straightforward generalizations of those of §4, I shall omit them (see [D], §8.3 for some analagous results involving cohomological descent).

Proposition 5.4 Let $X_\bullet$ be a simplicial scheme, separated and of finite type over a scheme S, with proper face maps. Then

i) There exists a morphism $p\colon Z_\bullet \longrightarrow X_\bullet$ with $Z_\bullet$ quasiprojective over S and having projective face maps such that $p\colon Z_\bullet \longrightarrow X_\bullet$ is an envelope and for all $i \geq 1$ $p\colon Z_i \longrightarrow (\mathrm{Cosk}^{X_\bullet}_{i-1}\mathrm{sk}_{i-1}Z_\bullet)_i$ is an envelope. We say that such a $p\colon Z_\bullet \longrightarrow X_\bullet$ is a Chow hyperenvelope of $X_\bullet$ (relative to S)

ii) If $p\colon Z_\bullet \longrightarrow X_\bullet$, $q\colon W_\bullet \longrightarrow X_\bullet$ are two such Chow hyperenvelopes of $X_\bullet$, there exists a diagram

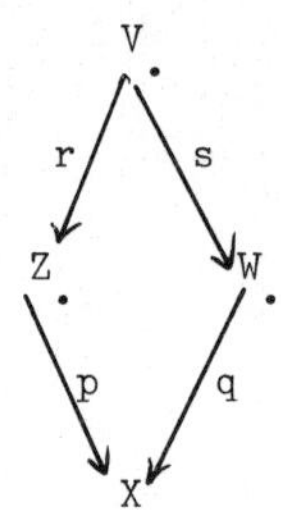

with r and s (and therefore $p \cdot r = q \cdot s$) hyperenvelopes in the same sense as in i).

iii) If we define $K'_*(X_\bullet)$ to be $K'_*(Z_\bullet)$ for $p\colon Z_\bullet \longrightarrow X_\bullet$ a Chow hyperenvelope of $X_\bullet$, then $K'_*(X_\bullet)$ is unique up to canonical isomorphism.

iv) K'_* extends to a covariant functor on the category of proper morphisms between simplicial schemes separated and of finite type over S having proper face maps.

§6. The niveau spectral sequence

Let $X_\bullet$ be a simplicial scheme, quasiprojective over a scheme S with proper face maps. Then if $\underline{\underline{A}}(X_\bullet)$ is the simplicial exact category of §1 we define an increasing filtration $\{F_p\underline{\underline{A}}(X_\bullet)\}_{p \geq -1}$ by setting $F_p\underline{\underline{A}}(X_k)$ equal to the category of coherent sheaves M in $\underline{\underline{A}}(X_k)$ such that $\dim(\overline{\mathrm{supp}(M) \cap (\mathrm{sk}_i(X_\bullet)_k - \mathrm{sk}_i(X_\bullet)_k)}) \leq p - i$ for $i = 0,\dots,p$. Applying the functor $|BQ(\ )|$ we get a filtered spectrum:

$$\dots \subset |BQF_p\underline{\underline{A}}(X_\bullet)| \subset |BQF_{p+1}\underline{\underline{A}}(X_\bullet)| \subset \dots$$

and an associated spectral sequence

$$E^1_{p,q}(X_\bullet) \Longrightarrow K'_{p+q}(X_\bullet)$$

which is easily seen to be convergent since for each $k \geq 0$, the filtration $F_p\underline{\underline{A}}(X_k)$ is finite.

Proposition 6.1 $E^2_{p,q}(X_\bullet) \cong CH_{p,-q}(X)$.

Proof (Sketch only, since we do not use this result in the rest of the paper.) First we shall discuss K-theory with supports. If S is a scheme and Φ is a family of supports on S we can consider the Serre subcategory $\underline{\underline{M}}^\Phi(S)$ of $\underline{\underline{M}}(S)$ consisting of the coherent sheaves with supports in Φ . If Φ_i $(i = 1,2)$ are two families of supports, then we have:

$$\Phi_1 \cap \Phi_2 = \{Y \subset S \mid Y \in \Phi_i \ , \ i = 1,2\}$$

$$\Phi_1 \vee \Phi_2 = \{Y = Z_1 \cup Z_2 \mid Z_i \in \Phi_i \quad i = 1,2\}$$

If we write $\underline{\underline{K}}^{\Phi_i}(S)$ for the spectrum given by Waldhausen's delooping of $BQ\underline{\underline{M}}^{\Phi_i}(S)$, then we have a Mayer-Vietoris fibration sequence coming from Quillen's localization theorem:

$$\underline{\underline{K}}^{\Phi_1 \cap \Phi_2}(S) \longrightarrow \underline{\underline{K}}^{\Phi_1}(S) \vee \underline{\underline{K}}^{\Phi_2}(S) \longrightarrow \underline{\underline{K}}^{\Phi_1 \vee \Phi_2}(S) \ .$$

If $\Psi \subset \Phi$, we shall write $\underline{\underline{K}}^{\Phi/\Psi}(S)$ for the cofibre of the natural map $K^\Psi(S) \longrightarrow K^\Phi(S)$. When we replace S by a simplicial scheme $X_\bullet$, a family of supports consists of supports Φ^n on each X_n such that if $\tau:[m] \longrightarrow [n]$ is monotone, $\Phi_n \subset X(\tau)^{-1}\Phi_m$. If the face maps of $X_\bullet$ are proper we can then form $\underline{\underline{A}}^\Phi(X_\bullet)$. If $Y_\bullet \subset X_\bullet$ is a closed subsimplicial scheme, $Y_\bullet$ will also denote the associated family of supports. We can now reinterpret the niveaux filtration. Set $\Psi_i = \{\Psi^k_i = \{Y \subset X_k \mid \dim Y \leq i\}\}_{k \geq 0}$ and $\Phi_{ij} = sk_i(X_\bullet) \cap \Psi_j$. Finally set $\Phi_p = \bigvee_{i+j=p} \Phi_{i,j}$. Then $F_p\underline{\underline{A}}(X_k) = \underline{\underline{A}}^{\Phi_p}(X_k)$. Notice that $\Phi_{i,j} \cap \Phi_{i',j'} = \Phi_{i'',j''}$ where $i'' = \inf(i,i')$, $j'' = \inf(j,j')$. Write

$$Gr_p\underline{\underline{K}}(X_k) = \underline{\underline{K}}^{\Phi_p/\Phi_{p-1}}(X_k)$$

and

$$Gr_{i,j}\underline{\underline{K}}(X_k) = \underline{\underline{K}}^{\Phi_{i,j}/(\Phi_{i-1,j}\vee\Phi_{i,j-1})}(X_k)$$

with $Gr_p\underline{\underline{K}}(X_\circ) = |k \longmapsto Gr\,\underline{\underline{K}}(X_k)|$ and $Gr_{i,j}\underline{\underline{K}}(X_\cdot) = |k \longmapsto Gr_{i,j}\underline{\underline{K}}(X_k)|$. It is easy to see, using Mayer Vietoris and induction, that

$$Gr_p\underline{\underline{K}}(X_k) \simeq \bigvee_{i+j=p} Gr_{i,j}\underline{\underline{K}}(X_k)$$

for all $k \geq 0$. If we write $X_i = \Delta_i \cup N_i$ where $\Delta_i = \bigcup_{j=0}^{i-1} s_j(X_{i-1})$ then Δ_i is closed in X_i with open complement N_i , and we may write

$$X_k = \bigcup_{\substack{i\leq n \\ [n]\longrightarrow\!\!\!>[i]}} X(\tau)(N_i)$$

where for every surjective map $\tau\colon [n] \longrightarrow [i]$ $X(\tau)(N_i)$ is a locally closed subset of X_k . A coherent sheaf M in $\underline{\underline{A}}(X_k)$ is in $F_p\underline{\underline{A}}(X_k)$ if and only if the dimension of the closure in X_k of $\sup(M) \cap X(\tau)(N_i)$ is $\leq p-i$. Let us write

$$N_{i,(j)} = \{x \in N_i \mid \dim \overline{\{x\}}_{X_i} = j\}$$

$$= N_i \cap X_{i,(j)}$$

Observe that for every surjective monotone $\tau\colon [k] \longrightarrow [i]$

$$X(\tau)(N_{i,(j)}) = X(\tau)(N_i) \cap X_{k,(j)}$$

since $X(\tau)$ is a closed immersion. By the same methods as [Q] we know that

$$\pi_q Gr_{i,j}\underline{\underline{K}}(X_k) = \bigoplus_{\tau:[k]\twoheadrightarrow[i]} \ \bigoplus_{x\in X(\tau)N_{i,j}} K_q\mathbb{k}(x) .$$

Hence the simplicial abelian group

$$k \longmapsto \pi_q Gr_{i,j}\underline{\underline{K}}(X_k)$$

is an Eilenberg-Maclane space and the spectral sequence

$$E^2_{r,s} = H_r(\pi_s Gr_{i,j}\underline{\underline{K}}(X_*)) \Longrightarrow \pi_{r+s} Gr_{i,j}\underline{\underline{K}}(X)$$

degenerates at $E^2_{\cdot\cdot}$ since

$$H_r(k \longmapsto \bigoplus_{\tau:[k]\twoheadrightarrow[i]} \ \bigoplus_{x\in X(\tau)N_{i,j}} K_s\mathbb{k}(x)) = \begin{cases} \bigoplus_{x\in N_{i,j}} K_s\mathbb{k}(x) & r = i \\ 0 & r \neq i . \end{cases}$$

Therefore

$$E^1_{p,q}(X_\cdot) = \bigoplus_{i+j=p} \bigoplus_{x\in N_{i,j}} K_{p+q-i}\Bbbk(x) .$$

To compute the differentials $E^1_{p,q} \longrightarrow E^1_{p-1,q}$, observe that for all i,j with $i+j = p$, we have the following inclusions:

$$\begin{array}{ccccc} \Phi_{p-2} & \subset & \Phi_{p-1} & \subset & \Phi_p \\ \cup & & \cup & & \cup \\ \Phi_{i-2,j}\vee\Phi_{i-1,j-1}\vee\Phi_{i,j-2} & \subset & \Phi_{i-1,j}\vee\Phi_{i,j-1} & \subset & \Phi_{i,j} . \end{array}$$

Hence the boundary map

$$\pi_s Gr_{i,j}\underline{\underline{K}}(X_\cdot) \longrightarrow \pi_s Gr_p\underline{\underline{K}}(X_\cdot) \longrightarrow \pi_{s-1}Gr_{p-1}\underline{\underline{K}}(X_\cdot)$$

factors through the boundary map

$$\pi_s Gr_{i,j}\underline{\underline{K}}(X_\cdot) \longrightarrow \pi_s \underline{\underline{K}}^{(\Phi_{i-1,j}\vee\Phi_{i,j-1})/(\Phi_{i-2,j}\vee\Phi_{i-1,j-1}\vee\Phi_{i,j-2})}(X_\cdot)$$

$$\wr\!\wr \quad \text{(by M-V)}$$

$$\pi_s Gr_{i-1,j}\underline{\underline{K}}(X_\cdot)\oplus\pi_s Gr_{i,j-1}\underline{\underline{K}}(X_\cdot) ,$$

i.e., the differentials in the spectral sequence have two components. First we have the differentials in the spectral sequence for the filtered category $\{F_p\underline{\underline{A}}(X_k)\}_{p\geq 0}$ in each degree $k \geq 0$:

$$\bigoplus_{x\in N_{i,(p-i)}} K_{p+q-i}\Bbbk(x) \xrightarrow{d''_{p-i,q}} \bigoplus_{x\in N_{i,(p-i-1)}} K_{p+q-i-1}\Bbbk(x) .$$

Second, we have differentials induced by the inclusion $sk_{p-1}X_\cdot \longrightarrow sk_pX_\cdot$:

$$\bigoplus_{x\in N_{i,(p-i)}} K_{p+q-i}\Bbbk(x) \xrightarrow{d'_{i,p+q-i}} \bigoplus_{x\in N_{i-1,(p-1)-(i-1)}} K_{(p+q-1)-(i-1)}\Bbbk(x) .$$

The total differential $E^1_{p,q} \xrightarrow{d_{p,q}} E^1_{p-1,q}$ is given by

$$d_{p,q} = \bigoplus_{i=0}^{p} d'_{i,p+q-1} + (-1)^p d''_{p-i,q} .$$

In order to compute $E^2_{\bullet,\bullet}$, observe that $E^1_{\bullet,\bullet}$, is the total complex associated to the following double complex. Consider, for each $q \geq 0$, the simplicial object $R_*(X_\bullet,q)$ in the category of complexes of abelian groups. For each $i \geq 0$, $R_i(X_\bullet,q)$ is a simplicial abelian group; if $R_i(X_*,q)$ is the associated chain complex, let $\Delta R_i(X_*,q)$ be the subcomplex of 'degenerate chains' which one sees easily is isomorphic to $R_i(\Delta_*,q)$ and hence $R_i(X_j,q)/\Delta R_i(X_j,q) \simeq R_i(N_j,q)$. Since the formation of the subcomplex of degenerate chains is functorial, we can consider the double complex $R_*(X_*,q)/\Delta R_*(X_*,q) \simeq R_*(N_*,q)$. By inspection, the total complex of this double complex is isomorphic to $E^1_{*,-q}$ and so

$$E^2_{p,q} = H_p(\int R_*(X_*,-q)/\Delta R_*(X_*,-q)).$$

But as is well known, for any simplicial abelian group $C_\bullet$, the natural map

$$C_* \longrightarrow C_*/\Delta C_*$$

is a quasi-isomorphism, and hence

$$E^2_{p,q} \cong H_p(\int R_*(X_*,-q)) \simeq CH_{p,-q}(X_\bullet) .$$

Bibliography

[B-K] A.K. Bousfield and D.M. Kan, "Homotopy Limits, Completions and Localizations" Lecture Notes in Mathematics, 304, Springer Verlag Berlin (1975).

[D] P. Deligne, Théorie de Hodge III, Publ. Math. IHES N°44 (1974), 5-78.

[EGA IV] A. Grothendieck, Etude locale des schémes et des morphismes de schémas Seconde Partie), Publ. Math. IHES N°24 (1965).

[F] W. Fulton, Rational equivalence on singular varieties, Publ. Math. IHES N°45 (1975), 147-167.

[F-G] W. Fulton and H. Gillet, Riemann-Roch for general algebraic varieties; preprint (1982).

[G_1] H. Gillet, Riemann-Roch theorems for higher algebraic K-theory, Adv. in Math., V. 40 (1981), 203-289.

[G_2] H. Gillet, Comparison of K-theory spectral sequences, with applications, in 'Algebraic K-theory, Evanston 1980', Springer Lecture Notes in Math., V. 854 (1981), 141-167.

[SGA4]V^{bis} B. Saint-Donat, Techniques de descente cohomologique, pp. 83-162 of "Théorie des Topos et cohomologie etale des Schémas", Springer Lecture Notes in Math., v270, Springer-Verlag, Berlin, (1972).

[T_1] R. Thomason, First quadrant spectral sequences in algebraic K-theory via homotopy colimits, Comm. in Alg. V10 (1982), 1589-1668.

[T_2] R. Thomason, K-theory and étale cohomology, preprint (1980).

[W] F. Waldhausen, Algebraic K-theory of generalized free products, Ann. of Math. 108 (1978), 135-256.

Department of Mathematics
Princeton University
Princeton, NJ 08544

WHAT HAPPENS TO HATCHER AND WAGONER'S FORMULA FOR $\pi_0 C(M)$ WHEN THE FIRST POSTNIKOV INVARIANT OF M IS NONTRIVIAL ?

by

Kiyoshi Igusa*

Brandeis University

Waltham, Massachusetts 02254

This paper is about algebraic K-theory, homology of rings and diffeomorphisms of smooth manifolds. Our purpose is to correct a discrepancy in a formula of Hatcher and Wagoner relating diffeomorphisms to K_2. The functor K_3 is used to resolve the discrepancy.

We now explain the original formula. Let M be a compact smooth (C^∞) manifold. Then a <u>pseudoisotopy</u> of M (rel ∂M) is defined to be a (self) diffeomorphism of $M \times I$ which fixes $M \times 0 \cup \partial M \times I$ pointwise. ($I = [0,1]$.) The space of all pseudoisotopies of M with the weak C -topology [7] is denoted $\mathcal{C}(M)$. In [3] the following statement is made. If $\dim M \geq 7$ then

$$(*) \qquad \pi_0 \mathcal{C}(M) \cong Wh_2(\pi_1 M) \oplus Wh_1^+(\pi_1 M; \mathbb{Z}_2 \oplus \pi_2 M) \quad .$$

Here Wh_2 is an algebraic K-theory functor related to K_2 and Wh_1^+ is a group homology functor. ($Wh_2(G) = K_2(\mathbb{Z}[G])/\pi_2^s(BG \cup pt)$ and $Wh_1^+(G;A) = H_0(G;A[G])/H_0(G;A)$)

There are two things wrong with Hatcher and Wagoner's formula. (Both mistakes occur in [H] so Wagoner is not responsible.) The first is that the proof assumes that the first Postnikov invariant $k_1(M)$ of M is trivial. When $k_1(M) \neq 0$ the formula is not true and we give a counterexample below. (If $\pi_1 M = \mathbb{Z}_p^3$, $\pi_2 M = \mathbb{Z}_p$ with p odd and $k_1(M)$ is the third exterior power of the generator of $H^1(\mathbb{Z}_p; \mathbb{Z}_p)$ then (*) is false.) The second mistake involves only $\pi_1 M$ and $\mathbb{Z}_2 = \pi_1^s$ = the stable 1-stem. This second mistake is corrected in [9].

The purpose of this paper is to explain how $\pi_0 \mathcal{C}(M)$ varies with $k_1(M)$. In

*Research supported by NSF Grant MCS 79-02939. The author is a Sloan Fellow.

chapter 8 we examine the proof of [5] in detail and see exactly where the Postnikov invariant assumption $k_1(M) = 0$ is used and we see what happens when $k_1(M) \neq 0$. The variation of $\pi_0 C(M)$ with $k_1(M)$ is seen to be related to $K_3\mathbb{Z}[\pi_1 M]$ and the exact relationship is given by some equation involving the Steinberg group $St(\mathbb{Z}[\pi_1 M])$. In chapter 6 we show how these equations are related to R. K. Dennis's map from algebraic K-theory to Hochschild homology. Thus the correct formula for $\pi_0 C(M)$ is the following exact sequence.

$$K_3\,\mathbb{Z}[\pi_1 M] \xrightarrow{\chi} Wh_1^+(\pi_1 M;\, \mathbb{Z}_2 \oplus \pi_2 M) \longrightarrow \pi_0 C(M) \longrightarrow Wh_2(\pi_1 M) \longrightarrow 0$$

Here χ is the sum of the Grassmann invariant $K_3\,\mathbb{Z}[\pi_1 M] \to Wh_1^+(\pi_1 M;\, \mathbb{Z}_2)$ explained in [9] and the following composition of maps.

$$K_3\,\mathbb{Z}[\pi_1 M] \xrightarrow[h_3]{} H_3(\mathbb{Z}[\pi_1 M];\, \mathbb{Z}[\pi_1 M]) \xrightarrow[\kappa[\pi_1 M]\,\cap]{} H_0(\mathbb{Z}[\pi_1 M];\, \pi_2 M[\pi_1 M]) \twoheadrightarrow Wh_1^+(\pi_1 M;\, \pi_2 M)$$

Here the first map is Dennis's map (defined in chapter 1), H_* is Hochschild homology and $\kappa[\pi_1 M]$ is an element of $H^3(\mathbb{Z}[\pi_1 M];\, \pi_2 M[\pi_1 M])$ related to $\kappa = k_1(M)$. ($\kappa[\pi_1 M] = \ell^3(\kappa)$ is defined in 6.d.) Since R. K. Dennis never published his work we give an exposition of his theory in the first half of the paper.

Here is a list of chapter headings and a brief explanation of what is contained in each chapter. Each chapter also begins with an introduction and a table of contents. (To find the true page numbers add the numbers given to the number of this page.)

1. Dennis's Hochschild homology invariant for algebraic K-theory 1

2. Homology of group extensions 11

3. Hochschild cohomology 16

4. K-theory of square zero ideals 20

5. Products in K-theory and Hochschild homology 26

6. Computations using the Steinberg group 32

7. The algebraic counterexample 43

8. Application to pseudoisotopies 44

9. A topological version of the counterexample 63

Chap. 1. We define Quillen's algebraic K-groups K_nR, the Hochschild homology groups

$H_n(R;M)$ and Dennis's map $h_n : K_nR \to H_n(R;R)$.

Chap. 2. Let $A \to E \to G$ be an exact sequence of groups where A is abelian. We define a map $b_n: H_{n+1}(G,E;\mathbb{Z}) \to H_{n-1}(G;A)$.

Chap. 3. We define the Hochschild cohomology groups $H^n(R;M)$ and the cap product $\cap: H^s(R;M) \otimes H_n(R;R) \to H_{n-s}(R;M)$.

Chap. 4. We define the algebraic K-theory of an ideal $I \subset R$ and when $I^2 = I$ we define a map $g_n: K_nI \to H_{n-1}(R/I;I)$. The definition involves the map b_n of Ch. 2.

Chap. 5. We discuss the relationship between Loday's product formula for algebraic K-theory and the exterior (shuffle) product in Hochschild homology.

Chap. 6. We do computations in K_1, K_2, K_3 and the Steinberg group in order to get a formula for Dennis's map which may be applied to pseudoisotopy theory.

Chap. 7. We give an example due to R. K. Dennis showing that his map is nontrivial. This later leads to a counterexample for the original formula (*) above.

Chap. 8. We examine Hatcher's argument in [5] to see what happens when $k_1(M) \neq 0$. The topology leads us to an algebraic formula derived in Ch. 6.

Chap. 9. We give a simplified explanation of the counterexample using only K_2 instead of K_3.

The author would like to thank R. Keith Dennis and A. Hatcher for explaining their work to him. A lot of the algebraic ideas in this paper are due to Dennis and the geometric argument of chapter 8 is essentially due to Hatcher. The author would like to take credit for the typing of this manuscript.

Chapter 1. Dennis's Hochschild homology invariant for algebraic K-theory.

In this chapter we shall define natural transformations $h^k_n: K^k_nR = \pi_n BGL_k(R)^+ \to H_n(R;R)$ and show that they are compatible with the stabilization maps $K^k_nR \to K^{k+1}_nR$ induced by the inclusions $GL_k(R) \subset GL_{k+1}(R)$. This will allow us to define a map on the limit $h_n : K_nR = \pi_n BGL_\infty(R)^+ \to H_n(R;R)$.

Here are the section titles:

1.a. Definition of K^k_nR. 1

1.b. Definition of $H_n(R;M)$. 1

1.c. The bar construction for $H_n(G;A)$. 2

1.d. Proof that $H_n(\mathbb{Z}[G];M) = H_n(G;M)$. 3

1.e. Morita invariance of $H_n(R;M)$. 3

1.f. Stabilization of Dennis's map. 6

1.g. Some splitting theorems. 7

1.a. Definition of $K_n^k R$.

Let R be an associative ring with 1. Let $M_k(R)$ be the ring of all $k \times k$ matrices with coefficients in R and let $GL_k(R)$ be the group of invertible elements of $M_k(R)$. Let $E_k(R)$ denote the subgroup of $GL_k(R)$ generated by the elementary matrices denoted by $e_{ij}(r)$, where $i \neq j$ and $r \in R$, which are the matrices with 1's on the diagonal, an r in the ij-position and 0's elsewhere.

If $k \geq 3$ then $E_k(R)$ is perfect, i.e., it is its own commutator subgroup. Thus we may perform the Quillen "plus construction" on the pair $(GL_k(R), E_k(R))$. This gives a space $BGL_k(R)^+$ and a map $p: BGL_k(R) \to BGL_k(R)^+$ which is uniquely determined up to homotopy equivalence by the condition that p induces an isomorphism in homology with integral coefficients and $\pi_1 BGL_k(R)^+ = GL_k(R)/E_k(R)$. The unstable K-groups of R for $n \geq 1$, $k \geq 3$ are defined by $K_n^k R = \pi_n BGL_k(R)^+$.

Let $GL(R) = GL_\infty(R)$ be the direct limit of the maps $GL_k(R) \to GL_{k+1}(R)$ given in the standard way by inserting a 1 in the lower right corner. We shall consider these maps to be inclusions. Let the ordinary stable K-groups of R for $n \geq 1$ be given by $K_n R = \pi_n BGL(R)^+ = \varinjlim_k K_n^k R = K_n R$. If $n = 0$ let $K_0 R$ be defined in the standard way [12].

Definition 1.a. (R.K. Dennis)

For $1 \geq n$ and $3 \leq k < \infty$ the Hochschild homology invariant for $K_n^k R$ is defined by the following composition.

$$K_n^k R = \pi_n BGL_k(R)^+ \xrightarrow[H]{} H_n(GL_k(R); \mathbb{Z}) \tag{1}$$

$$\xrightarrow[D_*]{} H_n(GL_k(R); M_k(R)) \tag{2}$$

$$\xrightarrow{\approx} H_n(\mathbb{Z}[GL_k(R)];\ M_k(R)) \tag{3}$$

$$\longrightarrow H_n(M_k(R);\ M_k(R)) \tag{4}$$

$$\xrightarrow{\approx} H_n(R;\ R) \tag{5}$$

(1) This is the Hurewicz map.

(2) This is the map in homology induced by the coefficient map $D\colon \mathbb{Z} \to M_k(R)$ determined by $D(1) = I_k$ = the identity matrix.

(3) This is the standard isomorphism between the homology of a group and the Hochschild homology of the corresponding integral group ring. (See 1.d.)

(4) The inclusion $GL_k(R) \subset M_k(R)$ induces a ring map $\mathbb{Z}[GL_k(R)] \to M_k(R)$. This is the induced map in Hochschild homology.

(5) This is the standard isomorphism described in 1.e below coming from the fact that R and $M_k(R)$ are Morita equivalent.

The map described above will be denoted by $h_n^k(R)$ or just $h_n^k\colon K_n^k R \to H_n(R;R)$. After (1) the map will be denoted with a "+" as follows. $h_n^{k+}(R)$ or $h_n^{k+}\colon H_n(GL_k(R)) \to H_n(R;R)$.

1.b. Definition of $H_n(R;\ M)$.

If M is an R-bimodule and $n \geq 0$ let $C_n(R;M) = \overbrace{R \otimes R \otimes \ldots \otimes R}^{n} \otimes M$ (all tensor products are over $\mathbb{Z}$ unless otherwise stated) and let $d_n\colon C_n(R;M) \to C_{n-1}(R;M)$ be given by the equation:

$$\begin{aligned} d_n(r_1 \otimes \ldots \otimes r_n \otimes m) = {} & r_2 \otimes \ldots \otimes r_n \otimes mr_1 \\ & + \sum_{i=1}^{n-1} (-1)^i\, r_1 \otimes \ldots \otimes r_i r_{i+1} \otimes \ldots \otimes r_n \otimes m \\ & + (-1)^n\, r_1 \otimes \ldots \otimes r_{n-1} \otimes r_n m \end{aligned}$$

Then the n-th <u>Hochschild homology</u> of R with coefficients in M is defined to be the n-th homology group of the complex $(C_*(R;M), d_*)$ and it is denoted by $H_n(R;M)$. Note that $H_0(R;M)$ is the cokernel of the map $d_1\colon R \otimes M \to M$ given by $d_1(r \otimes m) = mr - rm$. With this we can define the Hochschild homology invariant for K_0R.

<u>Definition 1.b.1.</u> (Hattori [6] and Stallings [13])

Suppose that P is a finitely generated projective left R-module. Then the <u>Hattori-Stallings trace</u> of P is defined to be

$$\operatorname{tr} P = \operatorname{tr}(1_P \oplus 0_Q) \in H_0(R;R)$$

where Q is an R-module such that $P \oplus Q \cong R^n$ for some finite n and 0_Q denotes the zero map $Q \to Q$.

<u>Proposition 1.b.2.</u> ([6], [13]) The Hattori-Stallings trace of P depends only on P.

<u>Proof</u>: Let $T = 1_P \oplus 0_Q$. Then as an element of R the trace of T depends on the choice of the isomorphism $P \oplus Q \cong R^n$. However the equivalence class of $\operatorname{tr} T$ in $H_0(R;R)$ is easily seen to be well-defined. If Q' is another R-module with $P \oplus Q' \cong R^m$ then we have: $\operatorname{tr}(1_P \oplus 0_Q) = \operatorname{tr}(1_P \oplus 0_Q \oplus 0_{R^m}) = \operatorname{tr}(1_P \oplus 0_Q \oplus 0_P \oplus 0_{Q'}) = \operatorname{tr}(1_P \oplus 0_{R^n} \oplus 0_{Q'}) = \operatorname{tr}(1_P \oplus 0_{Q'})$. □

<u>Definition 1.b.3.</u> Let $h_0(R)$ or simply h_0 denote the natural map $h_0\colon K_0R \to H_0(R;R)$ given by $h_0([P]-[Q]) = \operatorname{tr} P - \operatorname{tr} Q$. This is well-defined because of the following obvious property of the trace: $\operatorname{tr}(P \oplus P') = \operatorname{tr} P + \operatorname{tr} P'$.

<u>Example 1.b.4.</u> (Hattori) Let G be a finite group and V a finite dimensional representation of G over $\mathbb{C}$. Then V is a projective module over the group ring $\mathbb{C}[G]$ and

$$\operatorname{tr} V = \frac{1}{|G|} \sum_{g \in G} \chi_V(g)\, g$$

where $\chi_V\colon G \to \mathbb{C}^*$ is the character of the representation V. This formula shows that

$$1_{\mathbb{C}} \otimes h_0(\mathbb{C}[G])\colon \mathbb{C} \otimes K_0\mathbb{C}[G] \longrightarrow H_0(\mathbb{C}[G];\mathbb{C}[G])$$

is an isomorphism.

1.c. <u>The bar construction for $H_n(G;A)$.</u>

Let G be a group and A, B two left G-modules. Then $A \otimes_G B$ denotes the quotient of $A \otimes B$ by the diagonal left action of G (i.e., $a \otimes b \cong ga \otimes gb$.) Let

$C_n(G)$ be the free left G-module generated by the set G^n and let $d_n: C_n(G) \to C_{n-1}(G)$ be given by

$$
\begin{aligned}
d_n[g_1, \dots, g_n] &= g_1[g_2, \dots, g_n] \\
&+ \sum_{i=1}^{n-1} (-1)^i [g_1, \dots, g_i g_{i+1}, \dots, g_n] \\
&+ (-1)^n [g_1, \dots, g_{n-1}] .
\end{aligned}
$$

Then $H_n(G;A)$ is the n-th homology of the complex $(C_*(G) \otimes_G A;\ d_* \otimes 1_A)$. We shall denote $C_n(G) \otimes_G A$ by $C_n(G;A)$.

1.d. Proof that $\mathcal{H}_n(\mathbb{Z}[G]; M) \cong H_n(G; \overline{M})$.

We repeat the standard argument. (See for example [11].)

Theorem 1.d. Let G be a group and M a $\mathbb{Z}[G]$-bimodule. Let $\overline{M}$ be M considered as a left G-module by the conjugation action $g \cdot m = g m g^{-1}$. Then there is a natural isomorphism of chain complexes $f_p: C_p(G;\overline{M}) \xrightarrow{\approx} \mathcal{H}_p(\mathbb{Z}[G]; M)$ given by

$$
f_p([g_1, \dots, g_p] \otimes m) = g_1 \otimes \dots \otimes g_p \otimes (g_1 \dots g_p)^{-1} m
$$

Proof: One can easily verify that f_p is a chain map. Let $\ell_p: C_p(\mathbb{Z}[G]; M) \to C_p(G;\overline{M})$ be defined by

$$
\ell_p\Big(\sum_{i_1} g_{i_1} \otimes \dots \otimes \sum_{i_p} g_{i_p} \otimes m\Big) = \sum_{i_1, \dots, i_p} [g_{i_1}, \dots, g_{i_p}] \otimes g_{i_1} \dots g_{i_p} m
$$

Then $\ell_n f_n = 1$ and $f_n \ell_n = 1$. □

1.e. Morita invariance of $\mathcal{H}_n(R; M)$.

Let R and S be two rings with 1. We say that R and S are Morita equivalent if there is an R-S-bimodule A and an S-R-bimodule B such that $A \otimes_S B \cong R$ and $B \otimes_R A \cong S$ as bimodules. This implies that $A \otimes_S$: S-Mod → R-Mod is an isomorphism of categories, its inverse being given by $B \otimes_R$. Similarly $\otimes_R A$: Mod-R → Mod-S and $B \otimes_R - \otimes_R A$: R-Mod-R → S-Mod-S are isomorphisms of categories. Note that A is projective in R-Mod and in Mod-S since it corresponds to S and R in S-Mod and

Mod-R respectively. Similarly B is projective over both rings.

Lemma 1.e.1. Let M be a left R-module. If $M \otimes R$ is considered as an R-bimodule in the standard way then

$$H_i(R; M \otimes R) \cong \begin{cases} M & \text{if } i = 0 \\ 0 & \text{if } i \neq 0 \end{cases}$$

Proof: Let $\varepsilon: H_0(R; M \otimes R) \cong M \otimes R \to M$ be the augmentation map given by $\varepsilon(x \otimes r) = rx$. We can define a chain contraction of the augmented complex by $f_n(r_1 \otimes \ldots \otimes r_n \otimes x \otimes r) = (r \otimes r_1 \otimes \ldots \otimes r_n \otimes x \otimes 1)$ and $\eta(x) = x \otimes 1$. □

Lemma 1.e.2. Let M be a left R-module and A a projective right R-module. Then

$$H_i(R; M \otimes A) \cong \begin{cases} A \otimes_R M & \text{if } i = 0 \\ 0 & \text{if } i \neq 0 \end{cases}$$

Proof: $H_*(R; M \otimes A)$ is the homology of the complex $C_*(R; M \otimes A) \cong A \otimes_R C_*(R; M \otimes R)$ where R acts on $C_n(R; M \otimes R) = R \otimes \ldots \otimes R \otimes M \otimes R$ by left multiplication on the last (right hand) factor. The lemma follows from 1.e.1 and the exactness of $A \otimes_R$. □

By right-left symmetry we also have:

Lemma 1.e.3. Let M be a right R-module and A a projective left R-module. Then

$$H_i(R; A \otimes M) \cong \begin{cases} M \otimes_R A & \text{if } i = 0 \\ 0 & \text{if } i \neq 0 \end{cases}$$ □

Theorem 1.e.4. Let R, S be rings with 1. Let A be an R-S-bimodule which is projective over both rings and let B be any S-R-bimodule. Then there is a natural sequence of isomorphisms

$$F_n: H_n(R; A \otimes_S B) \xrightarrow{\approx} H_n(S; B \otimes_R A)$$

which vary functorially with the 4-tuple (R,A,S,B).

Corollary 1.e.5. (R.K. Dennis) Let M be an R-bimodule. Then a chain homotopy equivalence $f^k_*: C_*(M_k(R); M_k(M)) \to C_*(R;M)$ can be given by the formula

$$t^k_n(X^1 \otimes X^2 \otimes \ldots \otimes X^n \otimes Y) = \sum_{a,b,\ldots,z=1}^{k} (x^1_{ab} \otimes x^2_{bc} \otimes \ldots \otimes x^n_{yz} \otimes y_{za})$$

where x^i_{pq} and y_{pq} are the pq entries of X^i and Y.

<u>Proof:</u> Let R^k be the set of $k \times 1$ matrices with coefficients in R and let M^k be the set of $1 \times k$ matrices with coefficients in M. Then the 4-tuples (R,R,R,M) and $(M_k(R), R^k, R, M^k)$ satisfy the conditions of theorem 1.e.4. Furthermore there is a map $G: (R,R,R,M) \to (M_k(R), R^k, R, M^k)$ given by

$$G(r,s,t,m) = \left(D(r), \begin{pmatrix} s \\ 0 \\ \vdots \\ 0 \end{pmatrix}, t, (m\, 0 \dots 0)\right)$$

where

$$D(r) = \begin{pmatrix} r & & & 0 \\ & r & & \\ & & \ddots & \\ 0 & & & r \end{pmatrix}$$

Thus we get the following commuting square.

$$\begin{array}{ccc} H_n(R;\, M) & \xrightarrow[F_n]{\approx} & H_n(R;\, M) \\ \downarrow G_1 & & \downarrow G_2 \\ H_n(M_k(R);\, M_k(M)) & \xrightarrow[F_n']{\approx} & H_n(R;\, M) \end{array}$$

It is easy to see that G_2 is the identity map (and thus G_1 is an isomorphism) and that G_1 is the map in Hochschild homology induced by the map $g: (R,M) \to (M_k(R), M_k(M))$ given by $g(r,m) = (D(r), T(m))$ where $T(m)$ is the $k \times k$ matrix with an m in the upper left corner and 0's elsewhere. Thus at the chain level G_1 is given by $g_n: C_n(R;M) \to C_n(M_k(R); M_k(M))$, $g_n(r_1 \otimes \dots \otimes r_n \otimes m) = D(r_1) \otimes \dots \otimes D(r_n) \otimes T(m)$. Clearly $f_n^k g_n$ is the identity map so f_*^k must be a chain equivalence. □

<u>Proof of 1.e.4:</u> Let (C_{pq}, d_*^1, d_*^2) be the bicomplex given by

$$\begin{aligned} C_{pq} &= \underbrace{R \otimes \dots \otimes R}_{p-1} \otimes A \otimes \underbrace{S \otimes \dots \otimes S}_{q-1} \otimes B \\ &= C_{p-1}(R;\, A \otimes \underbrace{S \otimes \dots \otimes S}_{q-1} \otimes B) \\ &\cong C_{q-1}(S;\, B \otimes \underbrace{R \otimes \dots \otimes R}_{p-1} \otimes A) \end{aligned}$$

where the isomorphism is given by cyclically permuting the terms. Let $d_{pq}^1: C_{pq} \to C_{p-1,q}$ be the boundary map for the complex $C_*(R;\, A \otimes \overbrace{S \otimes \dots \otimes S}^{q-1} \otimes B)$ and let $(-1)^{p-1} d_{pq}^2: C_{pq} \to C_{p,q-1}$ be the boundary map for the complex $C_*(S;\, B \otimes \overbrace{R \otimes \dots \otimes R}^{p-1} \otimes A)$. Then clearly $d^1d^1 = 0$, $d^2d^2 = 0$ and $d^1d^2 + d^2d^1 = 0$ so we have a bicomplex. Let $X_n = \bigoplus_{p+q=n} C_{pq}$ and $d_n: X_n \to X_{n-1}$ be the map given by $d_n = d^1 + d^2$. Since

(C_{p*},d^2) has only one dimensional homology we have, by 1.e.1, a functorial chain isomorphism $(X_{p+1},d_{p+1}) \cong (H_1(C_{p*},d^2),d^1)$ given by the composition $X_{p+1} \to C_{p1} \to H_1(C_{p*},d^2)$ where the first map is projection onto a factor and the second comes from taking $H_1(C_{p*},d^2) = \mathrm{coker}(d^2_{p2}\colon C_{p2} \to C_{p1})$. Also by 1.e.1 we have a functorial isomorphism $(H_1(C_{p*},d^2),d^1) \cong (C_{p-1}(R;\ A \underset{S}{\otimes} B),d_*)$. By the symmetrical argument using 1.e.2 we get a functorial chain isomorphism $(X_{q+1},d_{q+1}) \cong (H_1(C_{*q},d^1),d^2) \cong (C_{q-1}(S;\ B \underset{R}{\otimes} A),d_*)$. When we compose one with the inverse of the other we get a functorial isomorphism $\mathcal{H}_*(R;\ A \underset{S}{\otimes} B) \cong \mathcal{H}_*(S;\ B \underset{R}{\otimes} A)$. □

1.f. Stabilization of Dennis's map.

The map h^{k+}_n is represented at the chain level by the following sequence of chain maps whose composition we shall call $\tilde{h}^{k+}_n$.

Formula 1.f.1. (R.K. Dennis)

$$
\begin{array}{llll}
 & [X^1, \ldots ,X^n] & \in C_n(GL_k(R);\ \mathbb{Z}) & \\
\longrightarrow & [X^1, \ldots ,X^n] \otimes I_k & \in C_n(GL_k(R);\ M_k(R)) & \\
\longrightarrow & X^1 \otimes \ldots \otimes X^n \otimes (X^1X^2 \ldots X^n)^{-1} & \in C_n(\mathbb{Z}[GL_k(R)];M_k(R)) & (1) \\
\longrightarrow & X^1 \otimes \ldots \otimes X^n \otimes (X^1X^2 \ldots X^n)^{-1} & \in C_n(M_k(R);\ M_k(R)) & \\
\longrightarrow & \sum_{a,\ldots,z=1}^{k} x^1_{ab} \otimes \ldots \otimes x^n_{yz} \otimes y_{za} & \in C_n(R;R) & (2)
\end{array}
$$

(1) This is the isomorphism of 1.d.

(2) This is the map given in 1.e.5.

Theorem 1.f.2. (R.K. Dennis) Let $n \geq 1$. If $j\colon H_n(GL_k(R);\ \mathbb{Z}) \to H_n(GL_{k+1}(R);\ \mathbb{Z})$ is the map induced by inclusion then $h^{k+}_n = h^{k+1+}_n j$. Consequently we have a map defined on the limit:

$$h_n^{+}(R)\colon H_n(GL(R);\ \mathbb{Z}) \to \mathcal{H}_n(R;R).$$

Corollary 1.f.3. (R.K. Dennis) If $j'\colon K^k_nR \to K^{k+1}_nR$ is the map induced by the inclusion $(GL_k(R),E_k(R)) \subset (GL_{k+1}(R),E_{k+1}(R))$ then $h^k_n = h^{k+1}_n j'$. Consequently we have a map defined on the limit:

$$h_n(R)\colon K_nR \to \mathcal{H}_n(R;R). \qquad \square$$

Proof of 1.f.2: From formula 1.f.1 one can easily see that $(\tilde{h}_n^{k+1+}\tilde{j} - \tilde{h}_n^{k+})[X^1, \ldots, X^n] = 1 \otimes \ldots \otimes 1 \otimes 1 \to \mathcal{C}_n(R;R)$ where $\tilde{j}$ is the inclusion map. So if $n \geq 1$ we have $\tilde{h}_n^{k+1+}\tilde{j} - \tilde{h}_n^{k+} = df_n + f_{n-1}d$ where $f_n[X^1, \ldots, X^n] = 1 \otimes \ldots \otimes 1 \in \mathcal{C}_{n+1}(R;R)$. $\square$

Proposition 1.f.4. Let M be an R-bimodule. Then the following composition of maps is compatible with stabilization with respect to k. $(n \geq 0.)$

$$H_n(GL_k(R);\, M_k(M)) \xrightarrow{\approx} \mathcal{H}_n(\mathbb{Z}[GL_k(R)];\, M_k(M)) \longrightarrow \mathcal{H}_n(M_k(R);\, M_k(M)) \xrightarrow{\approx} \mathcal{H}_n(R;\, M)\,.$$

Proof: At the chain level the map is given as follows.

$$\begin{array}{llll} & [X^1, \ldots, X^n] \otimes Y & \in & C_n(GL_k(R);\, M_k(M)) \\ \longrightarrow & X^1 \otimes \ldots \otimes X^n \otimes (X^1 \ldots X^n)^{-1}Y & \in & \mathcal{C}_n(M_k(R);\, M_k(M)) \\ \longrightarrow & \sum\limits_{a,\ldots,z=1}^{k} x^1_{ab} \otimes \ldots \otimes x^n_{yz} \otimes z_{za} & \in & \mathcal{C}_n(R;\, M) \end{array}$$

After stabilization the $(k+1)$-st column of $(z_{**}) = (X^1 \ldots X^n)^{-1}Y$ is trivial. Thus this chain map strictly commutes with stabilization for all $n \geq 0$. $\square$

1.g. Some splitting theorems.

Take $R = \mathbb{Z}[G]$ and let

$$\ell h_n\colon \quad K_n\mathbb{Z}[G] \longrightarrow H_n(G;\, \mathbb{Z}[G])$$

$$\ell h_n^+\colon H_n(GL(\mathbb{Z}[G]);\mathbb{Z}) \to H_n(G;\, \mathbb{Z}[G])$$

be the maps given by composing h_n and h_n^+ with the isomorphism of 1.d. These maps will be called the Dennis trace maps for K-theory and homology respectively. Let ℓh_n^k and ℓh_n^{k+} be defined analogously.

Theorem 1.g.1. (R.K. Dennis) Let G be any group and $n \geq 1$. Then the map $H_n(G;\mathbb{Z}) \to H_n(G;\mathbb{Z}[G])$ induced by the coefficient map $\mathbb{Z} \to \mathbb{Z}[G]$ which sends 1 to 1 is equal to the composition $H_n(G;\mathbb{Z}) \to H_n(GL(\mathbb{Z}[G]);\mathbb{Z}) \to H_n(G;\mathbb{Z}[G])$ where the first map is induced by the inclusion $G \subset \mathbb{Z}[G]^\times = GL_1(\mathbb{Z}[G]) \subset GL(\mathbb{Z}[G])$ and the second map is the Dennis trace map.

Proof: Examine the effect of h_n^+ and ℓ_* at the chain level:

$$\begin{array}{llll} & [g_1, \ldots, g_n] \otimes 1 & \varepsilon & C_n(G; \mathbb{Z}) \\ \longrightarrow & [g_1, \ldots, g_n] \otimes 1 & \varepsilon & C_n(GL_1(\mathbb{Z}[G]); \mathbb{Z}) \\ \xrightarrow{\tilde{h}_n^{1+}} & g_1 \otimes \ldots \otimes g_n \otimes (g_1 \ldots g_n)^{-1} & \varepsilon & C_n(\mathbb{Z}[G]; \mathbb{Z}[G]) \\ \xrightarrow{\ell_n} & [g_1, \ldots, g_n] \otimes 1 & \varepsilon & C_n(G; \mathbb{Z}[G]) \end{array}$$ □

Corollary 1.g.2. (R.K. Dennis) $H_n(G; \mathbb{Z})$ is a direct summand of $H_n(GL_k(\mathbb{Z}[G]); \mathbb{Z})$ for $1 \leq k \leq \infty$ and for all n, G. □

Corollary 1.g.3. Let $n \geq 2$ and let G be a group with $H_i(G; \mathbb{Z}) = 0$ for $0 < i < n$. Then the composition

$$K_n \mathbb{Z}[G] \xrightarrow{\ell h_n} H_n(G; \mathbb{Z}[G]) \xrightarrow{\varepsilon_*} H_n(G; \mathbb{Z})$$

is a split epimorphism if ε_* is the map induced by the augmentation map $\varepsilon: \mathbb{Z}[G] \to \mathbb{Z}$.

Proof: Since G is perfect we can perform the plus construction on BG and we get a map $BG^+ \to BGL(\mathbb{Z}[G])^+$. Now consider the following composition of maps.

$$H_n(G) \cong \pi_n BG^+ \to \pi_n BGL(\mathbb{Z}[G])^+ = K_n \mathbb{Z}[G] \to H_n(GL(\mathbb{Z}[G])) \xrightarrow{\ell h_n^+} H_n(G; \mathbb{Z}[G]) \xrightarrow{\varepsilon_*} H_n(G)$$

The first map is the inverse of the Hurewicz isomorphism. The third map is the Hurewicz map. The composition is the identity by 1.g.1 and the naturality of the Hurewicz map. □

Unfortunately this corollary does not produce any interesting topological examples since the only part of the Dennis trace map which is relevant to pseudoisotopy theory is the quotient map:

$$K_n \mathbb{Z}[G] \longrightarrow \frac{H_n(G; \mathbb{Z}[G])}{H_n(G; \mathbb{Z})} = Wh_{n+1}^+(G; \mathbb{Z})$$

Let $M_k(G)$ be the group of all $k \times k$ monomial matrices with coefficients in G. Then $M_k(G)$ is a subgroup of $GL_k(\mathbb{Z}[G])$ and thus there is an induced map in homology which satisfies the following.

Theorem 1.g.4. The following composition is trivial for all G, n and k.

$$H_n(M_k(G);\mathbb{Z}) \to H_n(GL_k(\mathbb{Z}[G]);\mathbb{Z}) \xrightarrow[\ell h_n^{k+}]{} H_n(G;\mathbb{Z}[G]) \to Wh_{n+1}^{+}(G;\mathbb{Z}).$$

Proof: Using 1.f.1 we see that the above composition does the following at the chain level.

$$\begin{array}{llll} & [X^1, \ldots, X^n] & \in & C_n(M_k(G);\mathbb{Z}) \\ \longrightarrow & \Sigma\, x^1_{ab} \otimes \ldots \otimes x^n_{yz} \otimes y_{za} & \in & C_n(\mathbb{Z}[G];\mathbb{Z}[G]) \\ \longrightarrow & \Sigma\, [x^1_{ab} \otimes \ldots \otimes x^n_{yz}] \otimes 1 & \in & C_n(G;\mathbb{Z}[G]) \\ \longrightarrow & 0 & \in & C_n(G;\mathbb{Z}[G])/C_n(G;\mathbb{Z}) \end{array}$$

where $y_{za} = ((X^1 \ldots X^n)^{-1})_{za} = (x^1_{ab} \ldots x^n_{yz})^{-1}$ since the matrices are monomial. □

Let $M(G) \subset GL(\mathbb{Z}[G])$ be the union of all the $M_k(G)$'s. Since $M_k(G)'$ is perfect for $k \geq 5$ we can construct $BM(G)^+$.

Corollary 1.g.5. The following composition is trivial.

$$\pi_n BM(G)^+ \to K_n\mathbb{Z}[G] \xrightarrow[\ell h_n]{} H_n(G;\mathbb{Z}[G]) \to Wh_{n+1}^{+}(G;\mathbb{Z}).$$ □

Corollary 1.g.6. $H_n(G;\mathbb{Q})$ is a direct summand of $\mathbb{Q} \otimes K_n\mathbb{Z}[G]$.

Proof: Use the fact that BG and $BM_k(G)$ are rationally equivalent. □

Chapter 2. Homology of group extensions.

Let G be a group and A a left G-module. Let $A \to E \overset{p}{\to} G$ be an extension of A by G. We shall construct a sequence of natural transformations of the following form.

$$b_n : H_{n+1}(G,E;\mathbb{Z}) \to H_{n-1}(G;A)$$

The maps arise from the Lyndon spectral sequence of the extension $A \to E \to G$ but knowledge of this is not required.

We shall construct b_n at the chain level using a transversal $t: G \to E$ (a set theoretic right inverse to $p: E \to G$) and show that chain maps arising from different transversals are homotopic. In 2.d and 2.e we show that b_n fits into certain commutative diagrams which will be used later. In 2.a and 2.b we review some basic notation. The section headings are:

2.a. The mapping cone construction. 0

2.b. Factor sets. 0

2.c. Definition of b_n. 0

2.d. The first diagram. 1

2.e. The second diagram. 2

2.a. The mapping cone construction.

This is taken from [11] p. 46.

Let $C_{n+1}(G,E;\mathbb{Z}) = C_n(E;\mathbb{Z}) \oplus C_{n+1}(G;\mathbb{Z})$ and define the boundary map by $d(a,b) = (-da, db + p_* a)$. Then $H_*(G,E;\mathbb{Z})$ is the homology of $C_*(G,E;\mathbb{Z})$. The map $\partial: H_{n+1}(G,E;\mathbb{Z}) \to H_n(E;\mathbb{Z})$ in the long exact homology sequence of the pair (G,E) is induced by the chain map given by $\partial(a,b) = (-1)^n a$.

2.b. Factor sets.

Let $t: G \to E$ be a transversal for p. Then every element of E can be written as $(a,g) = a\,t(g)$ where $a \in A$ and $g \in G$. Multiplication is given by $(a,g)(b,h) = (a + gb + f(g,h), gh)$ where $f: G \times G \to A$ is the factor set $f(g,h) = t(g)t(h)t(gh)^{-1}$. Of course f represents a 2-cocycle in $C^2(G;A) = \mathrm{Hom}_G(C_2(G),A)$.

2.c. Definition of b_n.

Let $c_n: C_{n+1}(G,E;\mathbb{Z}) \to C_{n-1}(G;A)$ be defined on the free generators of $C_{n+1}(G,E;\mathbb{Z})$ as follows.

$$c_n([(a_1,g_1), \ldots ,(a_n,g_n)],0) = -[g_2, \ldots ,g_n] \otimes g_1^{-1} a_1$$

$$c_n(0,[g_1, \ldots ,g_{n+1}]) = [g_3, \ldots ,g_{n+1}] \otimes g_2^{-1} g_1^{-1} f(g_1,g_2)$$

A straightforward computation shows that this is a chain map.

Exercise 2.c. Show that c_n induces an isomorphism in homology if $n = 1$ and an epimorphism if $n = 2$.

Suppose that $t'\colon G \to E$ is another transversal for p. Then $(a,g) = at(g) = at(g)t'(g)^{-1}t'(g) = (a - \alpha(g),g)'$ where $\alpha(g) = t'(g)t(g)^{-1}$. Using t' instead of t we get

$$c_n'([(a_1,g_1), \ldots ,(a_n,g_n)],0) = -[g_2, \ldots ,g_n] \otimes g_1^{-1}(a_1 - \alpha(g_1))$$
$$c_n'(0,[g_1, \ldots ,g_{n+1}]) = [g_3, \ldots ,g_n] \otimes g_2^{-1}g_1^{-1}f'(g_1,g_2)$$

where $f'(g_1,g_2) = f(g_1,g_2) + g_1\alpha(g_2) - \alpha(g_1g_2) + \alpha(g_1)$. Let $e_n\colon C_{n+1}(G,E;\mathbb{Z}) \to C_n(G;A)$ be given by $e_n(x,0) = 0$ and $e_n(0,[g_1, \ldots ,g_{n+1}]) = [g_2, \ldots ,g_{n+1}] \otimes g_1^{-1}\alpha(g_1)$. Then a straightforward computation shows that $c_{n+1}' - c_{n+1} = e_{n-1}d + de_n$. Thus c_* induces a functorial map b_* in homology.

2.d. The first diagram.

Let $0 \to A \to B \overset{q}{\to} C \to 0$ be a short exact sequence of left G-modules and let $x \in H^0(G;C)$. Since $H^0(G;C) = \mathrm{Hom}_G(\mathbb{Z}, C)$ we may regard x as an element of C which is fixed under the action of G. Let $\beta^*\colon H^0(G;C) \to H^1(G;A)$ be the connecting homomorphism in cohomology. Then $\beta^*(x)$ is represented by a crossed homomorphism $\alpha\colon G \to A$ given by $\alpha(g) = gy - y$ where y is a fixed element of $q^{-1}(x)$. This produces a homomorphism $f\colon G \to A \tilde{\times} G$ = the semidirect product by $f(g) = (\alpha(g),g)$.

Theorem 2.d. The following diagram commutes up to sign for all n. (Here β_* is the connecting homomorphism in homology and the isomorphism is given by the boundary map.)

$$\begin{array}{ccccc}
H_n(G;\mathbb{Z}) & & \xrightarrow{\quad x_* \quad} & & H_n(G;C) \\
\Big\downarrow f_* & & (-1)^n & & \Big\downarrow \beta_* \\
\dfrac{H_n(A \tilde{\times} G;\mathbb{Z})}{H_n(G;\mathbb{Z})} & \xrightarrow{\approx} & H_{n+1}(G,A \tilde{\times} G;\mathbb{Z}) & \xrightarrow{b_n} & H_{n-1}(G;A)
\end{array}$$

Proof: Let $e\colon C_{n-1}(G;\mathbb{Z}) \to C_{n-1}(G;B)$ be given by $e[g_1, \ldots ,g_{n-1}] = [g_1, \ldots ,g_{n-1}] \otimes y$. Then a chain map representing β_*x_* is given as follows. Take $[g_1, \ldots ,g_n] \in$

$C_n(G;\mathbb{Z})$. Map this to $[g_1,\dots,g_n] \otimes x \in C_n(G;C)$, pull it back to $[g_1,\dots,g_n] \otimes y \in C_n(G;B)$ then take $d([g_1,\dots,g_n] \otimes y) \in C_{n-1}(G;B)$. To get this into $C_{n-1}(G;A)$ subtract $ed[g_1,\dots,g_n]$. Computation will show that the result is $[g_2,\dots,g_n] \otimes (g_1^{-1}y - y) = -[g_2,\dots,g_n] \otimes g_1^{-1}\alpha(g_1) = c_n([(\alpha(g_1),g_1), \dots ,(\alpha(g_n),g_n)],0) = (-1)^n c_n(f_*[g_1,\dots,g_n],0)$. □

2.e. <u>The second diagram.</u>

Let $0 \to A \to B \to C \to 0$ be a short exact sequence of G-modules and let $B \to E \to G$ be an extension of B by G. Let $F = E/A$.

<u>Theorem 2.e.1</u>. The following diagram commutes up to sign for all n.

$$\begin{array}{ccccc} H_{n+1}(G,F;\mathbb{Z}) & \xrightarrow{b_n} & H_{n-1}(G;C) & & \\ \downarrow \partial & & (-1)^{n+1} & \searrow \delta & H_{n-2}(G;A) \\ H_n(F,E;\mathbb{Z}) & \xrightarrow[b_{n-1}]{} & H_{n-2}(F;A) & \nearrow p_* & \end{array}$$

∂ = the boundary map for the triple (G,F,E).

δ = the connecting homomorphism corresponding to the coefficient sequence $A \to B \to C$.

p_* = the map induced by the projection $p: F \to G$.

<u>Proof</u>: In this proof group multiplication will be written additively and the letters a, b, c, g with various indices will indicate elements of A, B, C, G respectively. Let $t: G \to F$, $u: G \to E$, $r: C \to B$ be transversals which send 0 to 0. Then each element of F can be written uniquely in the form $(c,g) = c + t(g)$. Let $s: F \to E$ be defined by $s(c,g) = r(c) + u(g)$. Then $r = s|C$ and $u = st$. Thus the formula for s can be written as

$$(*) \qquad s(c,g) = s(c) + st(g).$$

Every element of E can be written uniquely in the form $(a,(c,g)) = a + s(c,g) = a + s(c) + st(g)$. If we think of E as an extension of B by G then each element of E has the unique form $(b,g) = b + u(g) = b + st(g)$. But elements of B can be written in the unique form $(a,c) = a + r(c) = a + s(c)$ so $((a,c),g) = a + s(c) + st(g) = (a,(c,g))$. Let (a,c,g) denote either of these.

Let $f_t\colon G\times G\to C$, $f_{st}\colon G\times G\to B$, $f_s\colon F\times F\to A$ be the factor sets corresponding to the indicated transversals. That is, $f_t(g_1,g_2) = t(g_1)+t(g_2)-t(g_1+g_2)$, $f_{st}(g_1,g_2) = st(g_1)+st(g_2)-st(g_1+g_2)$ and $f_s((c_1,g_1),(c_2,g_2)) = s(c_1,g_1) + s(c_2,g_2) - s((c_1,g_1)+(c_2,g_2))$. The following formulas can easily be verified.

$$f_{st}(g_1,g_2) = (f_s(t(g_1),t(g_2)),f_t(g_1,g_2))$$

$$(\dagger)\qquad f_s((c_1,g_1),(c_2,g_2)) = s(c_1)+g_1s(c_2) - s(c_1+g_2c_2+f_t(g_1,g_2)) + f_{st}(g_1,g_2)$$

We are now ready to prove the theorem. The proof is based on the following well-known fact.

<u>Lemma 2.e.2</u>. A map between two short exact sequences of chain complexes will produce maps in homology compatible with the corresponding connecting homomorphisms. Thus a map between long exact sequences will result. □

Take the following map of short exact sequences of chain complexes where Y_* and k_* are defined below.

$$\begin{array}{ccccccccc}
0 & \longrightarrow & C_*(F,E;\mathbb{Z}) & \xrightarrow{\alpha} & Y_* & \xrightarrow{\beta} & C_*(G,F;\mathbb{Z}) & \longrightarrow & 0\\
 & & \downarrow{\scriptstyle p_*c_*} & & \downarrow{\scriptstyle k_*} & & \downarrow{\scriptstyle c_*} & & \\
0 & \longrightarrow & C_{*-2}(G;A) & \longrightarrow & C_{*-2}(G;B) & \longrightarrow & C_{*-2}(G;C) & \longrightarrow & 0
\end{array}$$

$Y_{n+1} = C_{n+1}(F,E;\mathbb{Z})\oplus C_{n+1}(G,F;\mathbb{Z}) = C_n(E;\mathbb{Z})\oplus C_{n+1}(F;\mathbb{Z})\oplus C_n(F;\mathbb{Z})\oplus C_{n+1}(G;\mathbb{Z})$

$d\colon Y_{n+1}\to Y_n$ is given by $d(w,x,y,z) = (-dw, dx+\overline{w}-y, -dy, dz+\overline{y})$ where $(\overline{\ \ })$ means image in the appropriate group.

α is the inclusion of the first factor.

β is the projection onto the second factor.

From the formula for d it should be clear that α, β are chain maps and that the connecting homorphism is $(-1)^{n+1}$ times the boundary map $\partial\colon H_{n+1}(G,F;\mathbb{Z})\to H_n(F,E;\mathbb{Z})$. Thus we need only define a chain map $k_*\colon Y_*\to C_{*-2}(G;B)$ which makes the diagram commute. Let k_{n+1} be defined on the generators of Y_{n+1} as follows.

$$k_{n+1}([(a_1,c_1,g_1),\ldots,(a_n,c_n,g_n)],0,0,0) = -[g_2,\ldots,g_n]\otimes g_1^{-1}\overline{a}_1$$

$$k_{n+1}(0,[(c_1,g_1),\ldots,(c_{n+1},g_{n+1})],0,0) = [g_3,\ldots,g_{n+1}]\otimes g_2^{-1}g_1^{-1}f_s((c_1,g_1),(c_2,g_2))$$

$$k_{n+1}(0,0,[(c_1,g_1),\ldots,(c_n,g_n)],0) = -[g_2,\ldots,g_n] \otimes g_1^{-1}s(c_1)$$

$$k_{n+1}(0,0,0,[g_1,\ldots,g_{n+1}]) = [g_3,\ldots,g_{n+1}] \otimes g_2^{-1}g_1^{-1}f_{st}(g_1,g_2)$$

Calculation will show (using equation †) that k_* is a chain map. Compatibility with p_*c_* and c_* is trivial. □

Chapter 3. Hochschild cohomology.

We define Hochschild cohomology and review its basic properties. The section headings are:

3.a. Definition of $H^n(R;M)$. 0

3.b. Interpretation of $H^2(R;M)$. 1

3.c. Interpretation of $H^1(R;M)$. 1

3.d. Interpretation of $H^0(R;M)$. 2

3.e. Long exact sequences. 2

3.f. Cap product. 2

3.g. Proof that $H^n(\mathbb{Z}[G];M) \cong H^n(G;\overline{M})$. 4

3.a. Definition of $H^n(R;M)$.

If R is a ring with 1 let $F_R\colon \mathbb{Z}\text{-Mod} \to R\text{-Mod-}R$ be the functor given by $F_R(A) = R \otimes A \otimes R$. Then we have a natural isomorphism $\mathrm{Hom}_{\mathbb{Z}}(A,M) \xrightarrow{\psi} \mathrm{Hom}_{R-R}(F_R,M)$ given by $\psi(f)(r \otimes a \otimes s) = rf(a)s$ and thus F_R is adjoint to the forgetful functor $R\text{-Mod-}R \to \mathbb{Z}\text{-Mod}$.

Let $C_n(R) = R^{\otimes(n+2)} = F_R(R^{\otimes n})$ be the tensor product of $n+2$ copies of R if $n \geq 0$ and $C_n(R) = 0$ if $n < 0$. Let $d_n\colon C_n(R) \to C_{n-1}(R)$ be given by

$$d_n(r_0 \otimes \ldots \otimes r_{n+1}) = \sum_{i=0}^{n} (-1)^i r_0 \otimes \ldots \otimes r_i r_{i+1} \otimes \ldots \otimes r_{n+1}\,.$$

Let $C^n(R;M) = \mathrm{Hom}_{R-R}(C_n(R),M) \cong \mathrm{Hom}_{\mathbb{Z}}(R^{\otimes n},M)$. Then the n-th cohomology of the cochain complex $(C^*(R;M),d^*)$ will be denote by $H^n(R;M)$ and will be called the n-th <u>Hochschild cohomology</u> group of R with coefficients in M.

The standard interpretations of H^0, H^1 and H^2 are as follows. (From [11])

3.b. Interpretation of $H^2(R;M)$.

An additive map $f\colon R \otimes R \to M$ is a 2-cocycle or linear factor set if it satisfies the condition that for all $r, s, t \in R$ we have

$$rf(s \otimes t) - f(rs \otimes t) + f(r \otimes st) - f(r \otimes s)t = 0 .$$

A linear factor set f can be used to construct a ring $M \oplus_f R$ as follows. The additive structure of $M \oplus_f R$ will be the same as that of $M \oplus R$ but the multiplication will be given as follows.

$$(m,r)(n,s) = (ms + rn + f(r \otimes s), rs)$$

The ring $M \oplus_f R$ has the following properties.

(1) Projection onto the second factor is a ring map $M \oplus_f R \to R$.

(2) The kernel of this map is a square zero ideal ($M^2 = 0$) isomorphic to M as an R-bimodule.

(3) Any ring extension $E \to R$ whose kernel is a square zero ideal isomorphic to M is equivalent to an extension $M \oplus_f R$ for some 2-cocycle f provided that $E \to R$ splits as a map of $\mathbb{Z}$-modules.

(4) The extensions $M \oplus_f R$ and $M \oplus_{f'} R$ are equivalent if and only if $[f] = [f']$ in $H^2(R;M)$.

(5) Every cohomology class contains a normalized 2-cocycle f satisfying the condition that $f(r \otimes s) = 0$ if either r or s is 1. Thus $M \oplus_f R$ contains the unit $(0,1)$.

3.c. Interpretation of $H^1(R;M)$.

If $f = 0$ then $M \oplus_f R$ is called the semidirect sum and denoted $M \oplus R$. Ring homomorphisms $R \to M \oplus R$ which are sections of the projection map $M \oplus R \to R$ correspond to elements of $H^1(R;M)$. An additive map $f\colon R \to M$ is a 1-cocycle or linear crossed homomorphism if it satisfies the condition that $rf(s) = f(r)s$ for all r, s

in R. The cocycle is normalized if $f(1) = 0$. A map f satisfying these conditions can be used to construct a ring homomorphism $g: R \to M \oplus R$ given by $g(r) = (f(r),r)$. The map g sends 1 to 1 since f is normalized.

3.d. Interpretation of $H^0(R;M)$.

We have the easy isomorphism $H^0(R;M) \cong \mathrm{Hom}_{R-R}(R;M) \cong \{x \in M \mid rx = xr$ for all $r \in R\}$.

3.e. Long exact sequences.

Let $\beta: 0 \to L \to M \to N \to 0$ be a short exact sequence of R-bimodules. If we assume that R is projective as a $\mathbb{Z}$-module or that β splits over $\mathbb{Z}$ then the induced sequences

$$0 \to C_*(R;L) \to C_*(R;M) \to C_*(R;N) \to 0$$

$$0 \to C^*(R;L) \to C^*(R;M) \to C^*(R;N) \to 0$$

will be exact sequences of chain and cochain complexes which will lead to the expected long exact sequences in Hochschild homology and cohomology. The connecting homomorphisms in homology and cohomology in the long exact sequences for β will be called β_* and β^*. Since these will appear often we make the following assumption.

Assumption 3.e.1. For the rest of this paper rings represented by the letter "R" will be assumed to be projective as $\mathbb{Z}$-modules.

Remark 3.e.2. Under the above condition we have

$$H_n(R;M) \cong \mathrm{Tor}_n^{R \otimes R^{op}}(R;M) \qquad \text{and}$$

$$H^n(R;M) \cong \mathrm{Ext}^n_{R \otimes R^{op}}(R;M)$$

since $C_*(R)$ is a projective $R \otimes R^{op}$ resolution of R.

3.f. Cap product.

Let $s \leq n$ and let M be an R-bimodule. The cap product

$$\cap: H^s(R;M) \otimes H_n(R;R) \to H_{n-s}(R;M)$$

will be defined at the chain level as follows. Suppose that $z = \sum_i r_i^1 \otimes \ldots \otimes r_i^{n+1}$ $\in C_n(R;R)$ and $f\colon C_s(R) \to M$ is an s-cochain. Then let $f \cap z \in C_{n-s}(R;M)$ be defined as follows.

<u>Formula 3.f.1.</u> $\qquad f \cap z = (-1)^e \sum_i r_i^1 \otimes \ldots \otimes r_i^{n-s} \otimes f(1 \otimes r_i^{n-s+1} \otimes \ldots \otimes r_i^{n+1})$

where $e = \dfrac{s^2 - s}{2} + ns$.

One can easily verify that

$$d_{n-s}(f \cap z) = (d_s^* f) \cap z + (-1)^s f \cap d_n z$$

and thus 3.f.1 induces a map in homology.

Let $0 \to L \to M \to N \to 0$ be a short exact sequence of R-modules. Let β_* and β^* be the connecting homomorphisms in Hochschild homology and cohomology induced by this sequence.

<u>Theorem 3.f.2.</u> Let $\kappa \in H^s(R;N)$ and $x \in H_n(R;R)$. Then

$$\beta_*(\kappa \cap x) = (\beta^*\kappa) \cap x.$$

<u>Proof</u>: Let z be an n-cycle representing x and f an s-cocycle representing κ. Then f lifts to a map $\tilde{f}$ which induces a map f' as indicated below. By definition of β^*, f' represents $\beta^*\kappa$.

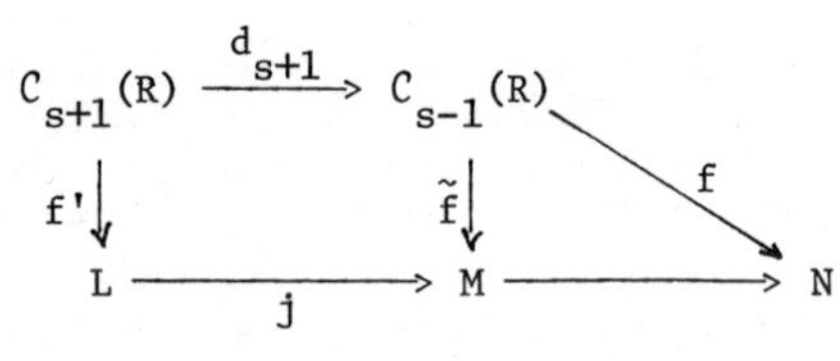

We now apply the diagram chase defining β_* to the cycle $f \cap z$. This cycle pulls back to the chain $\tilde{f} \cap z$ which maps down to $d_{n-s+1}(\tilde{f} \cap z) = (d_{s+1}{}^* \tilde{f}) \cap z = \tilde{f} d_{s+1} \cap z = jf' \cap z$ which is the inclusion map applied to $f' \cap z$. □

Together with h_n the cap product induces a map on K-theory as follows.

$$H^s(R;M) \otimes K_n R \xrightarrow{1 \cap h_n} H_{n-s}(R;M).$$

We will be particularly interested in the case $n = s$.

3.g. Proof that $H^n(\mathbb{Z}[G]; M) \cong H^n(G;\overline{M})$.

Theorem 3.g. Under the same conditions as 1.d we have a natural chain isomorphism $f^p: C^p(\mathbb{Z}[G]; M) \to C^p(G;\overline{M})$ given by

$$f^p(h)(g_0[g_1, \ldots, g_p]) = h(g_0 \otimes g_1 \otimes \ldots \otimes g_p \otimes (g_0 \ldots g_p)^{-1}).$$

Proof: One can easily verify that f^p is a cochain map. Let $\overline{\ell^p}: C^p(G;M) \to C^p(\mathbb{Z}[G]; M)$ be defined by

$$\overline{\ell^p}(h)\Big(\sum_{i_0} g_{i_0} \otimes \ldots \otimes \sum_{i_{p+1}} g_{i_{p+1}}\Big) = \sum_{i_0,\ldots,i_{p+1}} g_{i_0} h[g_{i_1}, \ldots, g_{i_p}]\, g_{i_1} \cdots g_{i_{p+1}}$$

One can easily verify that $\overline{\ell^p}$ is the inverse of f^p. □

Chapter 4. K-theory of square zero ideals.

In this chapter we discuss the relationship between Hochschild homology and the K-theory of square zero ideal. We define a map $g_n: K_n I \to H_{n-1}(E/I;I)$ analogous to the map h_n if I is a square zero ideal in E. The definition uses the map b_n of chapter 2. Here are the section headings.

4.a. Definition of K_n of an ideal. 0

4.b. The Hochschild homology invariant for the K-theory of a square zero ideal. 1

4.c. Long exact sequences. 2

4.d. $H^1(R;L) \otimes K_n R \to H_{n-1}(R;L)$. 4

4.e. $H^2(R;A) \otimes K_n R \to H_{n-2}(R;A)$. 5

4.a. Definition of K_n of an ideal.

Let E be any ring with 1 and $M \subset E$ a two sided ideal. If $n = 0$, $K_0 M$ will be as defined by Milnor in ([12], p.33.) For $1 \leq n$ and $3 \leq k \leq \infty$ let $K_n^k M$ be defined by

$$K_n^k M = \pi_{n+1}(BGL_k(E/M)^+, BGL_k(E)^+)$$

where $BGL_k(E)^+$ maps to $BGL_k(E/M)^+$ in the obvious way. We shall use the notation $K_n^k(M;E)$ when E needs to be specified. When $k = \infty$ it will be deleted from the notation.

By the long exact homotopy sequence of a pair we have a long exact sequence of the following form for $n \geq 1$.

$$\ldots \to K_{n+1}^k(E/M) \to K_n^k M \to K_n^k E \to K_n^k(E/M) \to \ldots$$

Using [12], lemma 4.1 this sequence can be extended to K_0 in the case $k = \infty$.

4.b. <u>The Hochschild homology invariant for the K-theory of a square zero ideal</u>.

Suppose now that $M^2 = 0$ and $3 \leq k < \infty$. Then the natural homomorphism $GL_k(E) \to GL_k(E/M)$ is surjective and its kernel is isomorphic to the additive group $M_k(M)$. Let $g_n^k(M)$ or simply g_n^k be defined as the following composition.

$$\begin{aligned} K_n^k M = \pi_{n+1}(BGL_k(E/M)^+, BGL_k(E)^+) &\xrightarrow{\text{Hurewicz}} H_{n+1}(GL_k(E/M), GL_k(E); \mathbb{Z}) \\ &\xrightarrow{b_n} H_{n-1}(GL_k(E/M); M_k(M)) \\ &\xrightarrow{\approx} \mathcal{H}_{n-1}(\mathbb{Z}[GL_k(E/M)]; M_k(M)) \\ &\longrightarrow \mathcal{H}_{n-1}(M_k(E/M); M_k(M)) \\ &\xrightarrow{\approx} \mathcal{H}_{n-1}(E/M;M) \end{aligned}$$

After the Hurewicz map this composition will be called $g_n^{k+}: H_{n+1}(GL_k(E/M), GL_k(E); \mathbb{Z}) \to \mathcal{H}_{n-1}(E/M;M)$. By 1.f.4 these maps are compatible with stabilization and thus they induce maps in the limit

$$g_n: \quad K_n(M;E) \longrightarrow \mathcal{H}_{n-1}(E/M;M) \quad \text{and}$$

$$g_n^+: H_{n+1}(GL(E/M), GL(E); \mathbb{Z}) \to \mathcal{H}_{n-1}(E/M;M)$$

The map g_n^{k+} is given at the chain level as follows.

<u>Formula 4.b</u>. Let $t: E/M \to E$ be a transversal. Then very element of $GL_k(E)$ can be represented uniquely as a pair $(A,X) \in M_k(M) \times GL_k(E/M)$ by $(A,X) = (I + A)t_*(X)$.

With this notation the map g_n^{k+} is given on $C_{n+1}(GL_k(E/M),GL_k(E))$ as follows.

$$
\begin{array}{lll}
(1) & ([(A^1,X^1), \dots ,(A^n,X^n)],0) & \in C_n(GL_k(E); \mathbb{Z}) \\
\longrightarrow & -[X^2, \dots ,X^n] \otimes (X^1)^{-1}A^1 & \in C_{n-1}(GL_k(E/M); M_k(M)) \\
\longrightarrow & -X^2 \otimes \dots \otimes X^n \otimes (X^1 \dots X^n)^{-1}A^1 & \in C_{n-1}(\mathbb{Z}[GL_k(E/M)]; M_k(M)) \\
\longrightarrow & -X^2 \otimes \dots \otimes X^n \otimes (X^1 \dots X^n)^{-1}A^1 & \in C_{n-1}(M_k(E/M);M_k(M)) \\
\longrightarrow & -\sum_{a,\dots,z=1}^{k} x_{ab}^2 \otimes \dots \otimes x_{yz}^n \otimes y_{za} & \in C_{n-1}(E/M;M)
\end{array}
$$

$$
\begin{array}{lll}
(2) & (0,[X^1, \dots ,X^{n+1}]) & \in C_{n+1}(GL_k(E/M); \mathbb{Z}) \\
\longrightarrow & [X^3, \dots ,X^{n+1}] \otimes (X^1X^2)^{-1}f(X^1,X^2) & \in C_{n-1}(GL_k(E/M);M_k(M)) \\
\longrightarrow & X^3 \otimes \dots \otimes X^{n+1} \otimes (X^1 \dots X^{n+1})^{-1}f(X^1,X^2) & \in C_{n-1}(M_k(E/M);M_k(M)) \\
\longrightarrow & \Sigma\, x_{ab}^3 \otimes \dots \otimes x_{yz}^{n+1} \otimes z_{za} & \in C_{n-1}(E/M);M)
\end{array}
$$

where $(y_{**}) = (X^1 \dots X^n)^{-1}A^1$, $(z_{**}) = (X^1 \dots X^{n+1})^{-1}f(X^1,X^2)$ and $f(X^1,X^2) = t_*(X^1)t_*(X^2)t_*(X^1X^2)^{-1}$

4.c. Long exact sequences.

Let $M \subset E$ be a square zero ideal. Then M is an E/M-bimodule and any submodule L of M is an ideal in E. Let $N = M/L$. Then N is a square zero ideal in E/L and the long exact sequence for the triple $(BGL_k(E/M)^+,BGL_k(E/L)^+,BGL_k(E)^+)$ produces the following.

Theorem 4.c.1. There is a long exact sequence of the following form.

$$\dots \to K_n^k L \xrightarrow{a_*} K_n^k M \xrightarrow{b_*} K_n^k N \xrightarrow{d} K_{n-1}^k L \to \dots \to K_1^k N$$

where a_*, b_* are induced by the ring maps $a\colon (E,L) \to (E,M)$ and $b\colon (E,M) \to (E/L,M/L)$ and d is the following composition

$$K_n^k N \xrightarrow{c_*} K_n^k(E/L) \xrightarrow{\partial} K_{n-1}^k L$$

where c_* is part of the long exact K-theory sequence for $(E/L,N)$ and ∂ is the boundary map for the long exact sequence for (E,L).

When $k = \infty$ this sequence extends to K_0 but this is not important here.

Theorem 4.c.2. Let $R = E/M$ be projective as a $\mathbb{Z}$-module. Then we have a map of long exact sequences of the following form which commutes up to sign. (The left two squares commute and the right hand square commutes up to the sign $(-1)^{n-1}$.)

$$\begin{array}{ccccccccc}
\cdots \longrightarrow & K_n^k L & \xrightarrow{a_*} & K_n^k M & \xrightarrow{b_*} & K_n^k N & \xrightarrow{d} & K_{n-1}^k L & \longrightarrow \cdots \\
 & \downarrow f_n & & \downarrow g_n(M) & & \downarrow g_n(N) & (-1)^{n-1} & \downarrow f_{n-1} & \\
\cdots \to & H_{n-1}(R;L) & \to & H_{n-1}(R;M) & \to & H_{n-1}(R;N) & \to & H_{n-2}(R;L) & \to \cdots
\end{array}$$

where f_n is the following composition.

$$K_n^k L \xrightarrow{g_n^k(L)} H_{n-1}(E/L;L) \to H_{n-1}(R;L)$$

Here the second map is defined since $M/L = N = \ker(E/L \to R)$ acts trivially on L.

Proof: The left two squares commute by naturality of g_n. To see the skew-commutativity of the right hand square embed it in the following diagram in which every face commutes except the pentagon which commutes up to sign by 2.e.1.

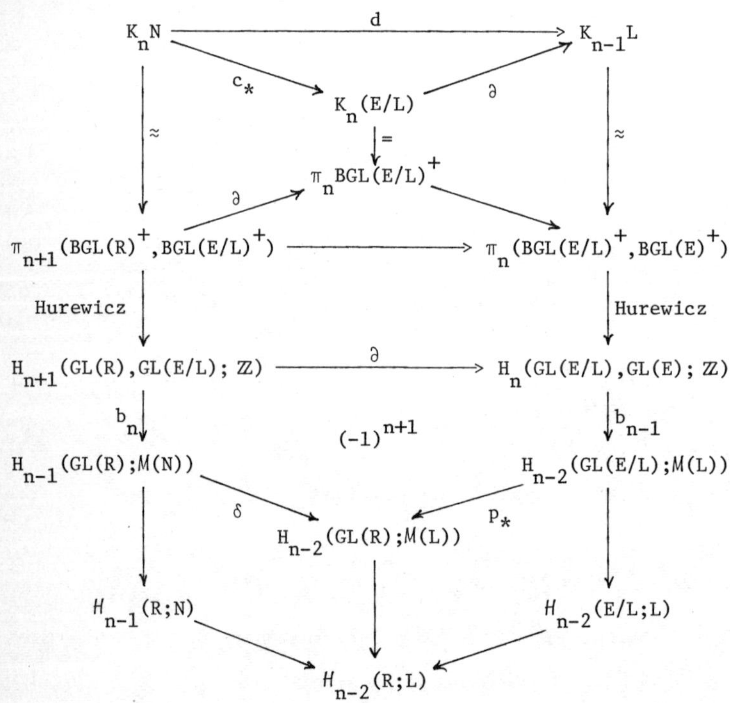

□

4.d. $\mathcal{H}^1(R;L) \otimes K_nR \to \mathcal{H}_{n-1}(R;L)$.

Let $\beta\colon 0 \to L \to M \overset{p}{\to} N \to 0$ be a short exact sequence of R-bimodules and let $x \in \mathcal{H}^0(R;N) \cong \mathrm{Hom}_{R\text{-}R}(R,N)$. We can regard x as an element of N such that $rx = xr$ for all $r \in R$. Then $\beta^*(x)$ is represented by a crossed homomorphism $\alpha\colon R \to L$ given be $\alpha(r) = ry - yr$ where y is a fixed element of $p^{-1}(x)$. This produces a ring homomorphism $f\colon R \to L \oplus R$ by $f(r) = (\alpha(r),r)$.

Theorem 4.d. The following diagram commutes up to sign for all n.

$$\begin{array}{ccc}
K_n^kR & \xrightarrow{\;h_n^k\;} & \mathcal{H}_n(R;R) \\
\downarrow f_* & & \downarrow x_* \\
K_n^k(L\oplus R) & (-1)^n & \mathcal{H}_n(R;N) \\
\downarrow \gamma & & \downarrow \beta_* \\
K_n^kL & \xrightarrow[\;g_n^k\;]{} & \mathcal{H}_{n-1}(R;L)
\end{array}$$

Here $\gamma = \partial^{-1}q$:

$$\gamma\colon K_n^k(L\oplus R) \xrightarrow{\;q\;} \frac{K_n^k(L\oplus R)}{K_n^kR} \underset{\approx}{\xleftarrow{\;\partial\;}} K_n^kL$$

Proof: By naturality of the Hurewicz map the following diagram commutes if all the horizontal arrows are Hurewicz maps.

$$\begin{array}{ccc}
K_n^kR & \longrightarrow & H_n(GL_k(R);\mathbb{Z}) \\
\downarrow f_* & & \downarrow \\
K_n^k(L\oplus R) & \longrightarrow & H_n(GL_k(L\oplus R);\mathbb{Z}) \\
\downarrow q & & \downarrow \\
\dfrac{K_n^k(L\oplus R)}{K_n^kR} & \longrightarrow & \dfrac{H_n(GL_k(L\oplus R);\mathbb{Z})}{H_n(GL_k(R);\mathbb{Z})} \\
\approx\uparrow\partial & & \approx\uparrow\partial \\
K_n^kL & \longrightarrow & H_{n+1}(GL_k(R),GL_k(L\oplus R);\mathbb{Z})
\end{array}$$

To complete the proof it suffices to show that the following diagram commutes up to sign for all $k < \infty$.

$$\begin{array}{ccccc}
H_n(GL_k(R);\mathbb{Z}) & \xrightarrow{\quad D_* \quad} & H_n(GL_k(R);M_k(R)) & \longrightarrow & \mathcal{H}_n(R;R) \\
\downarrow f_* & \searrow X_* \quad (1) & \downarrow x'_* & (2) & \downarrow x_* \\
\dfrac{H_n(GL_k(L\oplus R);\mathbb{Z})}{H_n(GL_k(R);\mathbb{Z})} & & H_n(GL_k(R);M_k(N)) & \longrightarrow & \mathcal{H}_n(R;N) \\
\approx \downarrow \partial^{-1} & (3) & \downarrow \overline{\beta}_* & (4) & \downarrow \beta_* \\
H_{n+1}(GL_k(R),GL_k(L\oplus R);\mathbb{Z}) & \xrightarrow{\;b_n\;} & H_{n-1}(GL_k(R);M_k(L)) & \longrightarrow & \mathcal{H}_{n-1}(R;L)
\end{array}$$

(1) In this triangle X_* is the map induced by the linear coefficient map $X: \mathbb{Z} \to M_k(N)$ given by $X(1) = D(x) = xI$. The triangle commutes because it is induced from a commuting triangle of coefficients.

(2) The horizontal maps are natural.

(3) This part commutes up to the sign $(-1)^n$ as a special case of 2.d since $GL_k(L \oplus R) \cong M_k(L) \tilde{\times} GL_k(R)$.

(4) The commutativity of this square follows from 2.e.2 and the proof of 1.f.4.□

Note that $\beta_* x_*: \mathcal{H}_n(R;R) \to \mathcal{H}_{n-1}(R;L)$ is the same as the cap product $[\alpha] \cap$ by 3.f.2.

4.e. $\mathcal{H}^2(R;A) \otimes K_n R \to \mathcal{H}_{n-2}(R;A)$.

Let $0 \to A \to B \to M \xrightarrow{p} N \to 0$ be an exact sequence of R-bimodules and let $L = \ker p = \operatorname{coker}(A \to B)$. Take $x \in \mathcal{H}^0(R;N)$ and $\beta_1^*(x) \in \mathcal{H}^1(R;L)$ as in 4.d. Now take $z = \beta_2^*\beta_1^*(x) \in \mathcal{H}^2(R;A)$ where β_2^* is the connecting homomorphism for the coefficient sequence $0 \to A \to B \xrightarrow{q} L \to 0$. Then z is represented by a linear factor set $\psi: R \otimes R \to A$ given as follows. Let $\overline{\alpha}: R \to B$ be an additive lifting of the crossed homomorphism $\alpha: R \to L$. Thus $q\overline{\alpha} = \alpha$. Then let $\psi(r \otimes s) = r\overline{\alpha}(s) - \overline{\alpha}(rs) + \overline{\alpha}(r)s$. Let $\overline{A}$ be A considered as an ideal in the twisted extension $A \oplus_\psi R = \overline{A} \oplus_\psi R$.

Theorem 4.e. The following diagram commutes up to sign for all n.

$$\begin{array}{ccc}
K_n^k R & \xrightarrow{\quad h_n^k \quad} & \mathcal{H}_n(R;R) \\
\partial \downarrow & (-1)^{n+1} & \downarrow [\psi] \cap \\
K_{n-1}^k \overline{A} & \xrightarrow[\quad g_{n-1}^k \quad]{} & \mathcal{H}_{n-2}(R;A)
\end{array}$$

Proof: Let $f: R \to L \oplus R$ be as in 5.d. Then $\overline{A} \oplus_\psi R$ is the pull-back along f

of $B \oplus R$. A formula for the induced ring map $\overline{f}\colon \overline{A} \oplus_{\psi} R \to B \oplus R$ is given by $\overline{f}(a,r) = (a + \overline{\alpha}(r), r)$. Now consider the following diagram.

$$\begin{array}{ccccccc}
K_nR & \xrightarrow{f_*} & K_n(L \oplus R) & \xrightarrow{\gamma} & K_nL & \xrightarrow{g_n(L)} & H_{n-1}(R;L) \\
\downarrow{\scriptstyle \partial} & & \downarrow{\scriptstyle \partial} & \swarrow{\scriptstyle d} & & (-1)^{n-1} & \downarrow{\scriptstyle \beta} \\
K_{n-1}\overline{A} & \xrightarrow[\overline{f}_*]{} & K_{n-1}A & \xrightarrow[g_{n-1}(A)]{} & H_{n-2}(L \oplus R;A) & \longrightarrow & H_{n-2}(R;A)
\end{array}$$

$$K_{n-1}\overline{A} \xrightarrow{g_{n-1}(\overline{A})} H_{n-2}(R;A)$$

The square commutes by naturality of ∂. The triangle commutes by the definition of d. The quadrilateral with $(-1)^{n-1}$ commutes up to that sign by 4.c.2. The bottom rectangle commutes by naturality of g_{n-1}. Now apply 4.d to the top row of arrows and use 3.f.2 to show that $[\psi] \cap = \beta_{2*}\beta_{1*}x_*$. □

Chapter 5. Products in K-theory and Hochschild homology.

In this chapter we review the basic notions of exterior product in algebraic K-theory and homology. The section headings are:

5.a. Definition of $\mu\colon K_nR \otimes K_mS \to K_{n+m}(R \otimes S)$ for $n, m \geq 1$. 0

5.b. The shuffle product. 1

5.c. The shuffle product in Hochschild homology. 2

5.d. Definition of $\mu\colon K_0R \otimes K_nS \to K_n(R \otimes S)$. 3

5.e. The product formula for h_n. 3

5.f. The second form of the product formula for h_n. 4

5.a. Definition of $\mu\colon K_nR \otimes K_mS \to K_{n+m}(R \otimes S)$ for $n, m \geq 1$.

We review the definition of the multiplication in K-theory as given by Loday in [10].

Let R, S be rings with 1, $p, q \geq 3$ and $\phi\colon R^p \otimes S^q \xrightarrow{\approx} (R \otimes S)^{pq}$ an isomorphism of $R \otimes S$-modules. Let f^i_{pq}, $i = 0, 1, 2$ be the group homomorphisms

$$f^i_{pq}\colon GL_p(R) \times GL_q(S) \to GL(R \otimes S)$$

given by $f^0_{pq}(A,B) = A \otimes B$, $f^1_{pq}(A,B) = A \otimes I_q$, $f^2_{pq}(A,B) = I_p \otimes B$. The homomorphisms f^i_{pq} induce maps f^i_{pq*} in homology.

$$f^i_{pq*}\colon H_n(GL_p(R) \times GL_q(S); \mathbb{Z}) \to H_n(GL(R \otimes S); \mathbb{Z})$$

They also induce maps f^{i+}_{pq} of plus constructions.

$$f^{i+}_{pq}\colon BGL_p(R)^+ \times BGL_q(S)^+ \to BGL(R \otimes S)^+$$

Using the fact that $BGL(R \otimes S)^+$ is an infinite loop space we may construct the map $\gamma_{pq} = f^{0+}_{pq} - f^{1+}_{pq} - f^{2+}_{pq}$.

$$\gamma_{pq}\colon BGL_p(R)^+ \times BGL_q(S)^+ \to BGL(R \otimes S)^+$$

Theorem 5.a.1. (Loday [10]) γ_{pq} is null homotopic on $BGL_p(R)^+ \vee BGL_q(S)^+$ and thus induces a map on the smash product.

$$\overline{\gamma}_{pq}\colon BGL_p(R)^+ \wedge BGL_q(S)^+ \to BGL(R \otimes S)^+$$

Furthermore $\overline{\gamma}_{pq}$ is compatible with stabilization of p and q. □

Definition 5.a.2. Let $\mu\colon K_nR \otimes K_mS \to K_{n+m}(R \otimes S)$ be the map in homotopy induced by $\overline{\gamma}_{\infty\infty} = \varinjlim \overline{\gamma}_{pq}$.

See [10] for the proof that μ is bilinear.

5.b. Shuffle product.

By naturality of the Hurewicz homomorphism the following diagram commutes.

$$\begin{array}{ccc}
K^p_n R \otimes K^q_m S & \xrightarrow{\text{Hurewicz}} & H_n(GL_p(R); \mathbb{Z}) \otimes H_m(GL_q(S); \mathbb{Z}) \\
\downarrow \mu & & \downarrow s \\
K_{n+m}(R \otimes S) & \xrightarrow{\text{Hurewicz}} & H_{n+m}(GL(R \otimes S); \mathbb{Z})
\end{array}$$

The map s is the composition $H_n(GL_p(R); \mathbb{Z}) \otimes H_m(GL_q(S); \mathbb{Z}) \to H_{n+m}(GL_p(R) \times GL_q(S); \mathbb{Z}) \to H_{n+m}(GL(R \otimes S); \mathbb{Z})$ where the first map is the exterior product in homology and the second map is $f^0_{pq*} - f^1_{pq*} - f^2_{pq*}$. Thus s is the sum of three maps $s = s^0 - s^1 - s^2$. However $s^1 = s^2 = 0$ when $n, m \geq 1$ so $s = s^0$.

The exterior product is given at the chain level by the <u>shuffle product</u>. (See [11] p, 243.) Thus s is given at the chain level by a map

$$s: C_n(GL_p(R); \mathbb{Z}) \otimes C_m(GL_q(S); \mathbb{Z}) \to C_{n+m}(GL(R \otimes S); \mathbb{Z})$$

given by

$$s([X^1, \ldots, X^n] \otimes [Y^1, \ldots, Y^m]) = [X^1 \otimes I_q, \ldots, X^n \otimes I_q, I_p \otimes Y^1, \ldots, I_p \otimes Y^m]$$
$$\pm \text{ other shuffles.}$$

5.c. <u>The shuffle product in Hochschild homology</u>.

Let $\bar{s}: C_n(R;A) \otimes C_m(S;B) \to C_{n+m}(R \otimes S; A \otimes B)$ be given by

$$\begin{aligned}\bar{s}((r_1 \otimes \ldots \otimes r_n \otimes a) \otimes (s_1 \otimes \ldots \otimes s_m \otimes b)) \\ = (r_1 \otimes 1) \otimes \ldots \otimes (r_n \otimes 1) \otimes (1 \otimes s_1) \otimes \ldots \otimes (1 \otimes s_m) \otimes (a \otimes b) \\ \pm \text{ other shuffles.}\end{aligned}$$

<u>Exercise 5.c.1</u>. Show that this is a chain map.

<u>Exercise 5.c.2</u>. Using formula 1.f.1 show that h_n^+ commutes with shuffle product.

<u>Corollary 5.c.3</u>. The map h_n commutes with products. This means that the following diagram commutes for $n, m \geq 1$.

$$\begin{array}{ccc} K_nR \otimes K_mS & \xrightarrow{h_n(R) \otimes h_m(S)} & H_n(R;R) \otimes H_m(S;S) \\ \downarrow \mu & & \downarrow \bar{s}_* \\ K_{n+m}(R \otimes S) & \xrightarrow{h_{n+m}(R \otimes S)} & H_{n+m}(R \otimes S; R \otimes S) \end{array}$$

<u>Exercise 5.c.4</u>. Let $M \subset R$ be a square zero ideal. Define the unlabled map below and show that the diagram commutes.

$$\begin{array}{ccc} K_nM \otimes K_mS & \xrightarrow{g_n(M) \otimes h_n(S)} & H_{n-1}(R/M;M) \otimes H_m(S;S) \\ \downarrow & & \downarrow \bar{s}_* \\ K_{n+m}(M \otimes S) & \xrightarrow{g_{n+m}} & H_{n+m-1}(\frac{R \otimes S}{M \otimes S}; M \otimes S) \end{array}$$

5.d. Definition of $\mu\colon K_0R \otimes K_nS \to K_n(R \otimes S)$.

If $n = 0$ then $\mu\colon K_0R \otimes K_0S \to K_0(R \otimes S)$ is given by $\mu([P] \otimes [Q]) = [P \otimes Q]$ where P, Q are projective modules over R, S.

Let $n \geq 1$ and let P be a finitely generated projective left R-module. We shall define multiplication by P, $\mu_P\colon K_nS \to K_n(R \otimes S)$. Let Q be an R-module such that $P \oplus Q \cong R^m$ for some m. Let $\gamma_P^k\colon GL_k(S) \to GL_{mk}(R \otimes S)$ be given by $\gamma_P^k(X) = (1_P \otimes X) \oplus (1_Q \otimes I_k)$. Then γ_P^k is a homomorphism which takes $E_k(S)$ to $E_{mk}(R \otimes S)$. Thus we have an induced map $\overline{\gamma}_P^k\colon K_n^kS \to K_n(R \otimes S)$. Since γ_P^k commutes up to conjugation with stabilization, $\overline{\gamma}_P^k$ strictly commutes with stabilization. The map $\mu([P] \otimes X) = \overline{\gamma}_P(X)$ is clearly bilinear.

5.e. The product formula for h_n.

Let $\tau\colon K_nR \to K_{n+1}R[t,t^{-1}]$ be left multiplication by $\{t\} \in K_1\mathbb{Z}[t,t^{-1}]$. Let $\overline{R}$ be R considered as a bimodule over itself. Then there is a surjective ring map $p\colon R[t,t^{-1}] \to \overline{R} \oplus R$ given by $p(\Sigma r_it^i) = \Sigma(ir_i, r_i)$,

Theorem 5.e. The following diagram commutes up to sign for all $n \geq 0$. Here γ is as in 4.d.

$$\begin{array}{ccccccc}
K_nR & \xrightarrow{\tau} & K_{n+1}R[t,t^{-1}] & \xrightarrow{p_*} & K_{n+1}(\overline{R} \oplus R) & \xrightarrow{\gamma} & K_{n+1}\overline{R} \\
\big\downarrow{\scriptstyle h_n(R)} & & & (-1)^n & & & \big\downarrow{\scriptstyle g_{n+1}(\overline{R})} \\
H_n(R;R) & & & \xrightarrow[=]{\hspace{6em}} & & & H_n(R;R)
\end{array}$$

Remark: This proves that Dennis's trace map h_n is the higher dimensional analogue of the Hattori-Stallings trace map h_0.

Proof: (We give the proof only in the case $n \geq 1$.) First apply the Hurewicz map to the top row. The result is the following: $H_nGL(R) \to H_1GL(\mathbb{Z}[t,t^{-1}]) \otimes H_nGL(R) \to H_{n+1}GL(R[t,t^{-1}]) \to H_{n+1}GL(\overline{R} \oplus R) \to H_{n+2}(GL(R), GL(\overline{R} \oplus R))$. Thus it suffices to show that the following square commutes up to sign. (There is no room on this page. Take the diagram (*) on the next page and let $k = \infty$.) By the stabilization results of 1.f this is equivalent to showing that the unstable version of (*) commutes up to sign.

$$(*)\qquad \begin{array}{ccc} H_n GL_k(R) & \xrightarrow{\ \sigma\ } & H_{n+2}(GL_k(R), GL_k(\bar{R}\oplus R)) \\ \downarrow{\scriptstyle h_n^{k+}} & & \downarrow{\scriptstyle g_{n+1}^{k+}} \\ \mathcal{H}_n(R;R) & \xrightarrow{\ =\ } & \mathcal{H}_n(R;R) \end{array}$$

Here σ is given at the chain level as follows.

$$\begin{array}{ll}
[X^1, \dots, X^n] & \varepsilon\ C_n GL_k(R) \\
\downarrow & \downarrow \\
t \otimes [X^1, \dots, X^n] & \varepsilon\ C_1 GL_k(\mathbb{Z}[t,t^{-1}]) \otimes C_n GL_k(R) \\
\downarrow & \downarrow \\
\sum_{i=0}^{n} (-1)^i [X^1 \otimes 1, \dots, X^i \otimes 1, I_k \otimes t, X^{i+1} \otimes 1, \dots, X^n \otimes 1] + & \\
 & \varepsilon\ C_{n+1} GL_k(R[t,t^{-1}]) \\
\sum_{i=0}^{n} (-1)^{i+1} [X^1 \otimes 1, \dots, X^i \otimes 1, I_k \otimes 1, X^{i+1} \otimes 1, \dots, X^n \otimes 1] & \\
\downarrow & \downarrow \\
\sum_{i=0}^{n} (-1)^i [(0,X^1), \dots, (0,X^i), (I_k,I_k), (0,X^{i+1}), \dots, (0,X^n)] + & \\
 & \varepsilon\ C_{n+1} GL_k(\bar{R}\oplus R) \\
\sum_{i=0}^{n} (-1)^{i+1} [(0,X^1), \dots, (0,X^i), (0,I_k), (0,X^{i+1}), \dots, (0,X^n)] & \\
\downarrow & \downarrow \\
(-1)^{n+1}(\text{same},0) & \varepsilon\ C_{n+2}(GL_k(R), GL_k(\bar{R}\oplus R))
\end{array}$$

Now apply c_{n+1} to the last item. We get

$$(-1)^n [X^1, \dots, X^n] \otimes I_k \ \varepsilon\ C_n(GL_k(R); M_k(R)) \qquad \square$$

5.f. The second form of the product formula for h_n.

The image of $\{t\}\ \varepsilon\ K_1 \mathbb{Z}[t,t^{-1}]$ in $\mathcal{H}_1(\mathbb{Z}[t,t^{-1}], \mathbb{Z}[t,t^{-1}])$ is the homology class of $t \otimes 1\ \varepsilon\ \mathcal{H}_1(\mathbb{Z}[t,t^{-1}], \mathbb{Z}[t,t^{-1}])$. Let $t\times\colon \mathcal{H}_n(R;R) \to \mathcal{H}_{n+1}(R[t,t^{-1}];R[t,t^{-1}])$ be left multiplication by $[t \otimes 1]$. Then as a special case of 5.c.3 we have the following commutative diagram for $n \geq 1$. (The diagram also commutes if $n = 0$.)

$$(5.f.1)\qquad \begin{array}{ccc} K_n R & \xrightarrow{\ h_n(R)\ } & \mathcal{H}_n(R;R) \\ \downarrow{\scriptstyle \tau} & & \downarrow{\scriptstyle t\times} \\ K_{n+1} R[t,t^{-1}] & \xrightarrow{\ h_n(R[t,t^{-1}])\ } & \mathcal{H}_{n+1}(R[t,t^{-1}];R[t,t^{-1}]) \end{array}$$

Let T denote the free multiplicative group generated by t. Thus $R[T] = R[t,t^{-1}]$. Then $H^1(T;\mathbb{Z}) \cong \mathbb{Z}$ is generated by the cohomology class of the 1-cocycle

$z \in C^1(T;\mathbb{Z})$ given by $z[t^n] = n$. The push-out of the sequence $C_1(T) \to C_0(T) \to \mathbb{Z} \to 0$ along the map $z\colon C_1(T) \to \mathbb{Z}$ is a sequence $\beta\colon\ \mathbb{Z} \to \mathbb{Z} \oplus \mathbb{Z} \to \mathbb{Z}$ of T-modules where T acts on $\mathbb{Z} \oplus \mathbb{Z}$ by $t(a,b) = (a + b,b)$. One can easily verify that $[z] = \beta^*(1)$ where $1 \in H^0(T;\mathbb{Z})$ is the unit of the cohomology ring $H^*(T;\mathbb{Z})$. (1 is the cohomology class of the augmentation map $\mathbb{Z}[T] \to \mathbb{Z}$.)

Consider β as a sequence of $\mathbb{Z}[T]$-bimodules by making the right action of T trivial. Apply $\otimes R[T]$ to β and we get a sequence of $R[T]$-bimodules

$$\overline{\beta}\colon\ 0 \to R[T] \to R[T] \oplus R[T] \to R[T] \to 0$$

where the right action of T is the standard one and the left action of T on $R[T] \oplus R[T]$ is given by $t(a,b) = (ta + tb,tb)$. Computation shows that the connecting homomorphism $\overline{\beta}^*\colon H^0(R[T];R[T]) \to H^1(R[T];R[T])$ sends the unit 1 to the cohomology class $[\alpha]$ of the linear crossed homomorphism $\alpha\colon R[T] \to R[T]$ given by $\alpha(\Sigma r_i t^i) = \Sigma i r_i t^i$. By 3.f.2 this means that the homology connecting homomorphism $\overline{\beta}_*\colon H_{n+1}(R[T];R[T]) \to H_n(R[T];R[T])$ given by the sequence $\overline{\beta}$ is cap product with $[\alpha]$.

We shall now compute the composition

$$H_n(R;R) \xrightarrow[t\,\times]{} H_{n+1}(R[T];R[T]) \xrightarrow[[\alpha]\,\cap]{} H_{n+1}(R[T];R[T])$$

at the chain level. Let $r_1 \otimes \ldots \otimes r_n \otimes r$ be a generator of $H_n(R;R)$. Then by the shuffle product formula we get

$$\begin{aligned} t \times (r_1 \otimes \ldots \otimes r_n \otimes r) &= (t \otimes 1) \times (r_1 \otimes \ldots \otimes r_n \otimes r) \\ &= \sum_{i=0}^{n} (-1)^i r_1 \otimes \ldots \otimes r_i \otimes t \otimes r_{i+1} \otimes \ldots \otimes r_n \otimes r\,. \end{aligned}$$

If $s_1 \otimes \ldots \otimes s_{n+1} \otimes s$ is a generator of $C_{n+1}(R[T];R[T])$ then the cap product formula 3.f.1 gives us

$$[\alpha] \cap (s_1 \otimes \ldots \otimes s_{n+1} \otimes s) = (-1)^{n+1} s_1 \otimes \ldots \otimes s_n \otimes \alpha(s_{n+1})s\,.$$

Since $R \subset R[T]$ is the kernel of α we get

(5.f.2) $$[\alpha] \cap (t \times (r_1 \otimes \ldots \otimes r_n \otimes r)) = -r_1 \otimes \ldots \otimes r_n \otimes tr$$

Combining 5.f.1 and 5.f.2 we get:

Theorem 5.f.3. The following diagram anticommutes for all $n \geq 0$ where $\bar{t}_*$ is induced by the R-bimodule morphism $\bar{t}: R \to R[T]$ given by $\bar{t}(r) = tr$.

$$\begin{array}{ccccc} K_{n+1}R[T] & \xrightarrow{\quad h_{n+1} \quad} & & \longrightarrow & H_{n+1}(R[T];R[T]) \\ \tau \uparrow & & (-1) & & \downarrow [\alpha] \cap \\ K_nR & \xrightarrow[h_n]{} & H_n(R;R) & \xrightarrow[\bar{t}_*]{} & H_{n+1}(R[T];R[T]) \end{array} \qquad \square$$

Remark 5.f.4. The augmentation map $\varepsilon: R[T] \to R$ is a left inverse for $\bar{t}$ and thus $\bar{t}_*$ is a split monomorphism.

Exercise 5.f.5. Derive 5.e from 5.f.3, 4.d and 5.f.4.

Chapter 6. Computations using the Steinberg group.

In this chapter we do computations in low dimensions (≤ 3) using the Steinberg group and Gersten's theorem that $K_3E \cong H_3St(E)$. The section titles are:

6.a. Definition of St(E). 0

6.b. Milnor's map $K_2E/M \to K_1M$. 1

6.c. $K_1M \cong H_0(E/M;M)$ 2

6.d. Integral group rings. 3

6.e. A formula for $\chi_2: H^2(G;M) \otimes K_2\,\mathbb{Z}[G] \to H_0(G;M[G])$. 4

6.f. The group P(G). 6

6.g. A formula for the map $\ell^3 \cap h_3^+: H^3(G;M) \otimes H_3GL(\mathbb{Z}[G]) \to H_0(G;M[G])$. 7

6.h. A formula for $\chi_3: H^3(G;M) \otimes K_3\,\mathbb{Z}[G] \to H_0(G;M[G])$. 8

6.i. The product formula. 9

6.a. Definition of St(E).

If E is a ring let St(E) be the group given by generators and relations as follows. St(E) is generated by symbols $x_{ij}(r)$ where $i \neq j$ are distinct positive integers and $r \in E$. The relations among the symbols $x_{ij}(r)$ are given as follows.

(0) $\quad x_{ij}(r)x_{ij}(s) = x_{ij}(r+s)$

(1) $[x_{ij}(r),x_{k\ell}(s)] = 1$ if $i \neq \ell$ and $j \neq k$

(2) $[x_{ij}(r),x_{jk}(s)] = x_{ik}(rs)$

Let $\phi\colon St(E) \to GL(E)$ be the homomorphism given by $\phi(x_{ij}(r)) = e_{ij}(r)$.

In the long exact sequence of the pair $(BGL(E)^+,BGL(E))$ we can make the following identifications functorially. (This follows easily from [2].)

$$\begin{array}{ccccccccc}
0 \to & \pi_2(BGL(E)^+) & \to & \pi_2(BGL(E)^+,BGL(E)) & \to & \pi_1(BGL(E)) & \to & \pi_1(BGL(E)^+) & \to 0 \\
 & \| \downarrow & & \approx \downarrow \psi & & \| \downarrow & & \| \downarrow & \\
0 \longrightarrow & K_2E & \longrightarrow & St(E) & \longrightarrow & GL(E) & \longrightarrow & K_1E & \longrightarrow 0
\end{array}$$

Also $K_3E \cong H_3St(E)$ and the isomorphism is given by

$$K_3E = \pi_3BGL(E)^+ \xrightarrow[\phi_*^{-1}]{\approx} \pi_3BSt(E)^+ \xrightarrow[\text{Hurewicz}]{\approx} H_3St(E)$$

In the special case $E = \mathbb{Z}[G]$ we shall take the following presentation of $St(E)$.

Generators: $x_{ij}(u)$ where $i \neq j$ are natural numbers and $u \in G$.

Relations: (I) $[x_{ij}(u),x_{k\ell}(v)] = 1$ if $i \neq \ell$ and $j \neq k$.

(II) $[x_{ij}(u),x_{jk}(v)]x_{ik}(uv)^{-1} = 1$

Notation: We usually write $x_{ij}(-u)$ for $x_{ij}(u)^{-1}$.

These generators and their inverses will be called <u>elementary operations</u>.

6.b. Milnor's map $K_2E/M \to K_1M$.

If X is a connected space let $X_{(n)}$ represent the n-coskeleton of X, that is, $X \to X_{(n)}$ is terminal among $(n+1)$-connected maps. Then we have a fibration: $BSt(E) \to BGL(E) \to BGL(E)^+_{(2)}$. If $M \subset E$ is a square zero ideal we have the following square of fibrations.

$$\begin{array}{ccccc}
F_2 & \longrightarrow & BM(M) & \longrightarrow & F_1 \\
\downarrow & & \downarrow & & \downarrow \\
BSt(E) & \longrightarrow & BGL(E) & \longrightarrow & BGL(E)^+_{(2)} \\
\downarrow & & \downarrow & & \downarrow \\
BSt(E/M) & \longrightarrow & BGL(E/M) & \longrightarrow & BGL(E/M)_{(2)}
\end{array} \tag{1}$$

Here F_1 and F_2 are the inducated homotopy fibers. Note that F_2 is connected since $p_*: St(E) \to St(E/M)$ is onto. Thus $M(M) \to \pi_1 F_1 = K_1 M$ is onto and the kernel is the image of the kernel of p_*. The long exact homotopy sequence for the horizontal fibrations look like this:

$$
(2)\qquad
\begin{array}{ccccccccccc}
 & & & & & & 1 & & & & \\
 & & & & & & \downarrow & & & & \\
 & & & & \ker p_* & \longrightarrow & M(M) & \xrightarrow{t} & K_1M & \longrightarrow & 1 \\
 & & & & \downarrow & & \downarrow & & \downarrow & & \\
1 & \longrightarrow & K_2E & \longrightarrow & St(E) & \longrightarrow & GL(E) & \longrightarrow & K_1E & \longrightarrow & 1 \\
 & & \downarrow & & \downarrow & & \downarrow & & \downarrow & & \\
1 & \longrightarrow & K_2E/M & \longrightarrow & St(E/M) & \longrightarrow & GL(E/M) & \longrightarrow & K_1E/M & \longrightarrow & 1 \\
 & & & & \downarrow & & & & & & \\
 & & & & 1 & & & & & &
\end{array}
$$

Milnor's definition of $\partial: K_2E/M \to K_1M$ comes from chasing this diagram as in the snake lemma. ([12], p.54.) It is easy for any topologist to see that this map is exactly the negative of the map given by the long exact sequence of the fibration $F_1 \to BGL(E)^+_{(2)} \to BGL(E/M)^+_{(2)}$. Thus Milnor's map is the negative of the one we are using.

6.c. $K_1M \cong H_0(E/M;M)$.

Theorem 6.c.1. Let $M \subset E$ be a square zero ideal. Then $g_1(M): K_1M \to H_0(E/M;M)$ is an isomorphism.

Corollary 6.c.2. The map $g_2(M): K_2M \to H_1(R;M)$ is an epimorphism if $E = M \oplus R$ and R is projective as a $\mathbb{Z}$-module.

Proof: Let P be a projective R-bimodule which maps onto M and let L be the kernel of the epimorphism $P \to M$. Considering L as an ideal in $P \oplus R$ we have a map of long exact sequences as in 4.c.

$$
\begin{array}{ccccccccc}
\cdots & \longrightarrow & K_2M & \longrightarrow & K_1L & \longrightarrow & K_1P & \longrightarrow & \cdots \\
 & & {\scriptstyle g_2(M)}\downarrow & (-1) & \downarrow{\scriptstyle \approx} & & \downarrow{\scriptstyle \approx} & & \\
\cdots & \longrightarrow & H_1(R;M) & \longrightarrow & H_0(R;L) & \longrightarrow & H_0(R;P) & \longrightarrow & \cdots
\end{array}
$$

Here $H_0(R;L) \cong H_0(M \oplus R;L)$ since M acts trivially on L. The commutativivy of this diagram up to sign proves the corollary. $\square$

Proof of 6.c.1: It is an easy exercise to show that K_1M is isomorphic to $H_0(E/M;M)$ and that t in 6.b.2 is given by the trace. What is not immediate is why $g_1(M)$ gives this isomorphism. So take $A \in M(M)$. Then $A \in M_k(M)$ for some finite $k \geq 3$. We shall use formula 4.b.1.

$$\begin{array}{lll} & [A] & \in C_1(M_k(M);\mathbb{Z}) \\ \longrightarrow & ([(A,I_k)],0) & \in C_1(GL_k(E);\mathbb{Z}) \\ \xrightarrow[4.b.1]{} & -\mathrm{tr}\,A & \in C_0(E/M;M) \end{array}$$

Thus according to our sign conventions $g_1(M)t(A) = -\mathrm{tr}\,A$. □

6.d. Integral group rings.

We shall now restrict to the case $R = \mathbb{Z}[G]$. If M is a left G-module then $M[G] = M \otimes \mathbb{Z}[G]$ is given the structure of a $\mathbb{Z}[G]$-bimodule by $g(a \otimes b)h = ga \otimes gbh$. A natural homomorphism $\ell^r: H^r(G;M) \to H^r(\mathbb{Z}[G]; M[G])$ can be defined as follows.

An element $\kappa \in H^r(G;M)$ is represented by an r-fold extension of left G-modules:

$$(*) \qquad 0 \to M \to E_r \to \dots \to E_1 \to \mathbb{Z} \to 0$$

Tensoring this with $\mathbb{Z}[G]$ we get an r-fold extension of $\mathbb{Z}[G]$-bimodules:

$$(**) \qquad 0 \to M[G] \to E_r[G] \to \dots \to E_1[G] \to \mathbb{Z}[G] \to 0$$

which represents an element $\ell^r(\kappa) = \kappa[G] \in \mathrm{Ext}^r_{\mathbb{Z}[G]-\mathbb{Z}[G]}(\mathbb{Z}[G];M[G]) \cong H^r(\mathbb{Z}[G];M[G])$.

Theorem 6.d.1. The map ℓ^r is a homomorphism given at the cochain level by linearly extending the following formula. (Compare this with 3.g.)

$$\ell^r(f)(g_0 \otimes \dots \otimes g_{r+1}) = g_0 f[g_1, \dots, g_r] \otimes g_0 g_1 \dots g_{r+1}.$$

Proof: If $f: C_r(G) \to M$ is an r-cochain then $\ell^r(f)$ is the following composition.

$$C_r(\mathbb{Z}[G]) \xrightarrow{\approx} C_r(G) \otimes \mathbb{Z}[G] \xrightarrow{f[G]} M[G]$$

where the chain isomorphism is given by

$$g_0 \otimes \dots \otimes g_{r+1} \longrightarrow g_1[g_1, \dots, g_r] \otimes g_0 g_1 \dots g_{r+1}.$$

If f is a cocycle representing κ then there is a map of chain complexes from the augmented complex $\dots \to C_0(G) \to \mathbb{Z}$ to (*) above which is f on $C_r(G)$ and 1 on $\mathbb{Z}$. Tensoring everything with $\mathbb{Z}[G]$ and composing with the chain isomorphism given above we get a map of chain complexes:

$$\begin{array}{ccccccccc} \dots \to & C_r(\mathbb{Z}[G]) & \to \dots \to & C_0(\mathbb{Z}[G]) & \to & \mathbb{Z}[G] & \to 0 \\ & \downarrow \ell^r(f) & & \downarrow & & \downarrow = & \\ 0 \longrightarrow & M[G] & \longrightarrow \dots \longrightarrow & E_1[G] & \longrightarrow & \mathbb{Z}[G] & \to 0 \end{array}$$

This shows that $\ell^r(f)$ represents $\kappa[G]$. □

We shall sometimes denote the cochain $\ell^r(f)$ by f_ℓ.

Definition 6.d.2. Let χ_n^r be defined by the following composition and let $\chi_n = \chi_n^n$.

$$\begin{aligned} H^r(G;M) \otimes K_n \mathbb{Z}[G] &\xrightarrow[\ell^r \otimes h_n]{} H^r(\mathbb{Z}[G];M[G]) \otimes H_n(\mathbb{Z}[G];\mathbb{Z}[G]) \\ &\xrightarrow[\cap]{} H_{n-r}(\mathbb{Z}[G];M[G]) \\ &\xrightarrow[\approx]{} H_{n-r}(G;M[G]) \end{aligned}$$

where G acts on $M[G]$ by conjugation.

6.e. A formula for $\chi_2\colon H^2(G;M) \otimes K_2 \mathbb{Z}[G] \to H_0(G;M[G])$.

An element of $H^2(G;M)$ is represented by a factor set $f\colon G \times G \to M$. We shall assume that f is normalized which is equivalent to saying that it comes from a transversal which preserves 1. This implies that $f(1,g) = f(g,1) = 0$ for all $g \in G$. By 6.d.1 the linearized version $f_\ell\colon \mathbb{Z}[G] \otimes \mathbb{Z}[G] \to M[G]$ is given by $f_\ell(g_1 \otimes g_2) = f(g_1,g_2) \otimes g_1 g_2$.

Formula 6.e.1. Let $x \in K_2 \mathbb{Z}[G]$ be represented by a product of elementary operations $x = x^1 x^2 \dots x^n \in \ker \phi \subset St(\mathbb{Z}[G])$. Then

$$\chi_2([f] \otimes x) = -\sum_{k,p} f_\ell(r^k_{pi_k} \otimes u_k) s^k_{j_k p}$$

where $x^k = x_{i_k j_k}(u_k)$, $u_k \in \pm G$ and $(r^k_{pq}) = \phi(x^1 \ldots x^{k-1}) = (s^k_{qp})^{-1}$.

Proof: The factor set f_ℓ determines an extension $E = M[G] \oplus_{f_\ell} \mathbb{Z}[G]$. By theorem 4.e $\chi_2([f] \otimes x) = -g_1^k(M[G])\partial x$ where ∂ is the boundary map $K_2 \mathbb{Z}[G] \to K_1 M[G]$ for the extension E. If we use Milnor's formula for ∂ and use trace instead of $g_1^k(M[G])$ $g_1^k(M[G])$ the sign will change twice and thus we get

$$\chi_2([f] \otimes x) = -\text{tr}\, \partial_{\text{Milnor}} x.$$

To compute $\text{tr}\, \partial_{\text{Milnor}} x$ we chase x around the diagram 6.b.2 as follows.

$$
\begin{array}{lll}
 & x = x_{i_1 j_1}(u_1) \ldots & \in K_2 \mathbb{Z}[G] \subset St(\mathbb{Z}[G]) \\
\leftarrow & x_{i_1 j_1}(0,u_1) \ldots & \in St(E) \\
\rightarrow & e_{i_1 j_1}(0,u_1) \ldots & \in GL(E) \\
\leftarrow & \text{same} & \in \mathcal{M}(M[G]) + I \\
\rightarrow & \text{tr}(e_{i_1 j_1}(0,u_1) \ldots e_{i_n j_n}(0,u_n) - I) \in \mathcal{H}_0(G;M[G]) &
\end{array}
$$

where I is the identity matrix. To compute this last term we need to following.

Formula 6.e.2. Let $x^1, \ldots, x^n$ be a sequence of elementary operations of the form $x^k = x_{i_k j_k}(0,u_k) \in St(E)$. Let ψ: $St(E) \to GL(\mathbb{Z}[G])$ be the obvious map. Then the pq-entry of the matrix $\phi(x^1 \ldots x^n) \in GL(E)$ is given by

$$(*) \qquad t_{pq} = \Big(\sum_{k=1}^{n} f_\ell(r^k_{pi_k} \otimes u_k)s^k_{j_k q}, r^{n+1}_{pq}\Big)$$

where
$$(r^k_{**}) = \psi(x^1 \ldots x^{k-1})$$
$$(s^k_{**}) = \psi(x^{k+1} \ldots x^n)$$

Proof: By induction on n. Let n = 1. Then $\phi(x^1) = (0,\psi(x^1))$ and $t_{pq} = (0,r^1_{pq})$ for all p, q since $f_\ell(1 \otimes u_1) = 0$.

Now suppose that the formula holds for n. Let $x^{n+1} = x_{ab}(0,v)$. Then the pq-entry of $\phi(x^1 \ldots x^{n+1})$ is given by

$$\bar{t}_{pq} = \begin{cases} t_{pq} & \text{if } q \neq b \\ t_{pb} + t_{pa}(0,v) & \text{if } q = b \end{cases}$$

Using (*) for t_{pa} we see that

$$(**)\qquad t_{pa}(0,v) = \Big(\sum_{k=1}^{n} f_\ell(r^k_{pi_k} \otimes u_k)s^k_{j_k a}v + f_\ell(r^{n+1}_{pa} \otimes v), r^{n+1}_{pa}v\Big)$$

Now look at the right hand side of (*). When n increases to $n+1$ we acquire one extra summand $(f_\ell(r^{n+1}_{pa} \otimes v)\overline{s}^{n+1}_{bq}, 0)$ where $\overline{s}^{n+1}_{**}$ is the identity matrix. The second coordinate changes from r^{n+1}_{pq} to

$$r^{n+2}_{pq} = \begin{cases} r^{n+1}_{pq} & \text{if } q \neq b \\ r^{n+1}_{pb} + r^{n+1}_{pa}v & \text{if } q = b \end{cases}$$

Each $s^k_{j_k q}$ changes to

$$\overline{s}^k_{j_k q} = \begin{cases} s^k_{j_k q} & \text{if } q \neq b \\ s^k_{j_k b} + s^k_{j_k a}v & \text{if } q = b \end{cases}$$

This agrees with the three things in (**) so the formula is proved. □

Formula 6.e.2 can now be used to compute $\mathrm{tr}(e_{i_1 j_1}(0,u_1) \ldots e_{i_n j_n}(0,u_n) - I)$ and we get formula 6.e.1. □

If $a = \sum_k u_k$, $u_k \in \pm G$, then note that $x_{ij}(a) = \prod_k x_{ij}(u_k)$ is a well-defined element of $St(\mathbb{Z}[G])$.

Theorem 6.e.3. Formula 6.e.1 is valid even if the condition $u_k \in \pm G$ is replaced by the condition $u_k \in \mathbb{Z}[G]$.

Proof: If $x^k = x_{ij}(a+b)$ we shall show that the formula is unchanged when x^k is replaced by $x_{ij}(a)x_{ij}(b)$. Only one term changes: $-\sum_p f_\ell(r^k_{pi} \otimes (a+b))s^k_{jp}$ will change to $-\sum_p f_\ell(r^k_{pi} \otimes a)s^k_{jp} - \sum_p f_\ell(\overline{r}^k_{pi} \otimes b)\overline{s}^k_{jp}$ where $(\overline{r}^k_{**}) = (r^k_{**})e_{ij}(a)$ and $(\overline{s}^k_{**}) = e_{ij}(-a)(s^k_{**})$. Thus $\overline{r}^k_{pi} = r^k_{pi}$ and $\overline{s}^k_{jp} = s^k_{jp}$. The term has not changed. □

6.f. The group P(G).

Let G be a group. If Z is any set let $\mathbb{Z}[G]\langle Z\rangle$ be the free left G-module generated by the elements of Z enclosed in square brackets. If G is given by generators and relations by $G = \langle X|Y\rangle$ then an exact sequence of the following kind can be constructed.

$$\mathbb{Z}[G]\langle Y\rangle \xrightarrow[d_2]{} \mathbb{Z}[G]\langle X\rangle \xrightarrow[d_1]{} \mathbb{Z}[G] \xrightarrow[\varepsilon]{} \mathbb{Z} \longrightarrow 0$$

where ε is the augmentation map, d_1 is given by $d_1[x] = x - 1$ and d_2 is given by $d_2[x_1 \ldots x_n] = [x_1] + x_1[x_2] + \ldots + x_1 \ldots x_{n-1}[x_n]$ with the convention that $[x^{-1}] = -x^{-1}[x]$. This is the beginning of a $\mathbb{Z}[G]$-free resolution of $\mathbb{Z}$. The bar constuction is an example of such a resolution where $X = G$ and the elements $[x_1,x_2]$ of G^2 correspond to the relations $(x_1)(x_2)(x_1x_2)^{-1} = 1$.

<u>Proposition 6.f.</u> Let $P(G) = \ker d_2$ and let M be any G-module. Then the following sequences are exact.

(1) $\quad 0 \to H_3G \to H_0(G;P(G)) \xrightarrow{d_{2*}} H_0(G;\mathbb{Z}[G]\langle Y\rangle)$

(2) $\quad \mathrm{Hom}_G(\mathbb{Z}[G]\langle Y\rangle;M) \xrightarrow[d_2^*]{} \mathrm{Hom}_G(P(G);M) \to H^3(G;M) \to 0$

Elements of $\mathrm{Hom}_G(P(G);M)$ will be called <u>presentation cocycles</u>. Elements of $P(G)$ can be represented pictorially [9] but this is not important here.

6.g. <u>A formula for the map $\ell^3 \cap h_3^+ \colon H^3(G;M) \otimes H_3GL(\mathbb{Z}[G]) \to H_0(G;M[G])$.</u>

An element of $H^3(G;M)$ can be represented by a normalized 3-cocycle $f\colon G \times G \times G \to M$. This means that $f(g_1,g_2,g_3) = 0$ if one of the g_i's is 1. Let $f_\ell\colon \mathbb{Z}[G] \otimes \mathbb{Z}[G] \otimes \mathbb{Z}[G] \to M[G]$ be the linearization of f given by $f(g_1 \otimes g_2 \otimes g_3) = f(g_1,g_2,g_3) \otimes g_1g_2g_3$. Combining formula 1.f.1 for h_3^+ and 3.f.1 for the cap product, the formula for $\ell^3 \cap h_3^+$ at the chain level is as follows.

$$f \otimes [g^1,g^2,g^3] \qquad \varepsilon\ C^3(G;M) \otimes C_3(GL_k(\mathbb{Z}[G]);\mathbb{Z})$$

$$\xrightarrow{\ell^3 \cap h_3^{k+}} f_\ell \otimes \sum_{a,b,c,d} r^1_{ab} \otimes r^2_{bc} \otimes r^3_{cd} \otimes s_{da} \qquad \varepsilon\ C^3(\mathbb{Z}[G];M[G]) \otimes C_3(\mathbb{Z}[G];\mathbb{Z}[G])$$

$$\longrightarrow \quad (*) \quad \sum_{a,b,c,d} f_\ell(r^1_{ab} \otimes r^2_{bc} \otimes r^3_{cd})s_{da} \qquad \varepsilon\ C_0(\mathbb{Z}[G];M[G]) \cong M[G]$$

where r^i_{**} are the entries of the matrix g^i, s_{**} are the entries of $(g^1g^2g^3)^{-1}$ and $(-1)^e = (-1)^{12} = 1$.

The formula (*) can be used to define a map $F\colon C_3(GL(\mathbb{Z}[G])) \to M[G]$ by $F(g^0[g^1,g^2,g^3]) = (*)$, i.e. ignore g^0. Since $f_\ell \cap h_3^+$ is a chain map up to sign the following composition is trivial.

$$C_4(GL(E)) \xrightarrow{d_4} C_3(GL(E)) \xrightarrow{F} M[G] \longrightarrow H_0(G;M[G])$$

Consequently we get an induced map $F_P\colon P(GL(E)) \to H_0(G;M[G])$ where $P(GL(E))$ is the kernel of $d_2\colon C_2(GL(E)) \to C_1(GL(E))$ and $E = \mathbb{Z}[G]$.

Proposition 6.g. F_P is the restriction of the additive map $F_C\colon C_2(GL(\mathbb{Z}[G])) \to H_0(G;M[G])$ given on the additive generators of $C_2(GL(\mathbb{Z}[G])$ as follows.

(**) $$F_C(g^1[g^2,g^3]) = -\sum_{a,b,c,d} f\,(r^1_{ab} \otimes r^2_{bc} \otimes r^3_{cd})s_{da}$$

where r^i_{**} are the entries of g^i and s_{**} are the entries of $(g^1g^2g^3)^{-1}$.

Proof: Just compute:

$$\begin{aligned} F_Cd_3(g^0[g^1g^2g^3]) &= -\Sigma f_\ell(r^0_{ab}r^1_{bc} \otimes r^2_{cd} \otimes r^3_{de})s_{ea} \\ &\quad +\Sigma f_\ell(r^0_{ab} \otimes r^1_{bc}r^2_{cd} \otimes r^3_{de})s_{ea} \\ &\quad -\Sigma f_\ell(r^0_{ab} \otimes r^1_{bc} \otimes r^2_{cd}r^3_{de})s_{ea} \\ &\quad +\Sigma f_\ell(r^0_{ab} \otimes r^1_{bc} \otimes r^2_{cd})r^3_{de}s_{ea} \\ &= \Sigma\, r^0_{ab}f_\ell(r^1_{bc} \otimes r^2_{cd} \otimes r^3_{de})s_{ea} \qquad \text{since } f_\ell \text{ is a cocycle} \\ &= \Sigma\, f_\ell(r^1_{bc} \otimes r^2_{cd} \otimes r^3_{de})s_{ea}r^0_{ab} \\ &= F(g^0[g^1,g^2,g^3]) \end{aligned}$$

□

6.h. A formula for $\chi_3\colon H^3(G;M) \otimes K_3\mathbb{Z}[G] \to H_0(G;M[G])$.

Let $K_3\mathbb{Z}[G]$ be identified with $H_3St(\mathbb{Z}[G])$ in such a way that the Hurewicz map $K_3\mathbb{Z}[G] \to H_3GL(\mathbb{Z}[G])$ corresponds to the map in homology induced by ϕ. Let f and f_ℓ be as above. We shall derive a formula for the cohomology class $\chi_f = \chi_3([f],-) \in H^3(St(\mathbb{Z}[G]);H_0(G;M[G]))$.

If $x \in H_3St(\mathbb{Z}[G])$ then $\chi_3([f],x) = f_\ell \cap h_3^+\phi_*(x)$ where ϕ_* is the map on H_3 induced by ϕ. Thus we will derive a formula for ϕ_* and then apply the formulas in 6.f above. We first choose a map of partial resolutions as follows. (Next page) Here ϕ_0 is the ring map induced by ϕ, $\phi_1[x] = [\phi(x)]$ if $x \in X_{St}$, $\phi_2[[a,b]] = [\phi(a),\phi(b)] - [\phi(b),\phi(a)]$ and $\phi_2[[a,b]c^{-1}] = [\phi(a),\phi(b)] - [\phi(c),\phi(b)] - [\phi(bc),\phi(a)]$.

$$\begin{array}{ccccccccc}
\mathbb{Z}[St]\langle Y_{St}\rangle & \xrightarrow{d_2} & \mathbb{Z}[St]\langle X_{St}\rangle & \xrightarrow{d_1} & \mathbb{Z}[St] & \longrightarrow & \mathbb{Z} & \longrightarrow & 0 \\
\downarrow{\scriptstyle \phi_2} & & \downarrow{\scriptstyle \phi_1} & & \downarrow{\scriptstyle \phi_0} & & \downarrow{\scriptstyle =} & & \\
\mathbb{Z}[GL]\langle GL \times GL\rangle & \xrightarrow[d_2]{} & \mathbb{Z}[GL]\langle GL\rangle & \xrightarrow[d_1]{} & \mathbb{Z}[GL] & \longrightarrow & \mathbb{Z} & \longrightarrow & 0
\end{array}$$

One can easily check that this diagram commutes. Let ϕ_3: $P(St) \to P(GL)$ be the map induced by ϕ_2 on the kernel of the d_2's. The cohomology class χ_f is represented by the presentation cocycle $\chi_f^P = F_P\phi_3$ which is the restriction to $P(St(\mathbb{Z}[G]))$ of the additive map $\chi_f^C = F_C\phi_2$: $\mathbb{Z}[St]\langle Y_{St}\rangle \to H_0(G;M[G])$ given on the additive generators of $\mathbb{Z}[St]\langle Y_{St}\rangle$ as follows.

<u>Formula 6.h.</u> (1) $\chi_f^C(t[y]) = 0$ if $y = [x_{ij}(u), x_{k\ell}(v)]$

(2) $\chi_f^C(t[[x_{ij}(u), x_{jk}(v)]x_{ik}(-uv)]) = -\sum_p f\,(r_{pi} \otimes u \otimes v)s_{kp}$

where $(r_{**}) = \phi(t) = (s_{**})^{-1}$.

<u>Proof:</u> (1) $\chi_f^C(t[[x,y]]) = F_C(\phi(t)[\phi(x),\phi(y)] - \phi(t)[\phi(y),\phi(x)] =$

$$-\sum\{f_\ell(r^1_{ab} \otimes r^2_{bc} \otimes r^3_{cd})s_{da} - f_\ell(r^1_{ab} \otimes r^3_{bc} \otimes r^2_{cd})s_{da}\} = 0$$

since each summand is zero. (Both terms between the braces are zero unless $b = c = d$ in which case they are still zero since f_ℓ is normalized.)

(2) $\chi_f^C(t[[x_{ij}(u), x_{jk}(v)]x_{ik}(-uv)])$

$= F_C(\phi(t)[e_{ij}(u), e_{jk}(v)] - \phi(t)[e_{ik}(uv), e_{jk}(v)] - \phi(t)[e_{jk}(v)e_{ik}(uv), e_{ij}(u)])$

$= -\sum_p f_\ell(r_{pi} \otimes u \otimes v)s'_{kp} + 0 + 0$

where $(r_{**}) = \phi(t)$ and $(s'_{**}) = e_{jk}(-v)e_{ij}(-u)\phi(t)^{-1}$. But $s'_{kp} = s_{kp}$ where $(s_{**}) = \phi(t)^{-1}$ since $k \neq i$. □

6.i. <u>The product formula.</u>

Let β: $0 \to \mathbb{Z} \to \mathbb{Z} \oplus \mathbb{Z} \to \mathbb{Z} \to 0$ be the short exact sequence of $\mathbb{Z}[T]$-modules given in 5.f. Let β_0 be the same sequence considered as a sequence of $G \times T$ - modules, G acting trivially. Then take $\overline{\beta}_0 = \beta_0[G \times T] = \beta_0 \otimes \mathbb{Z}[G \times T]$. By 5.f.3 we have the following anticommuting diagram.

(a)

$$\begin{array}{ccccc} K_2\,\mathbb{Z}[G] & \xrightarrow{h_2} & H_2(\mathbb{Z}[G];\,\mathbb{Z}[G]) & \overset{\bar{t}_*}{\searrow} & \\ \downarrow{\scriptstyle\tau} & & & & \\ K_3\,\mathbb{Z}[G\times T] & \xrightarrow{h_3} & H_3(\mathbb{Z}[G\times T];\,\mathbb{Z}[G\times T]) & \xrightarrow{\bar{\beta}_{0*}} & H_2(\mathbb{Z}[G\times T];\,\mathbb{Z}[G\times T]) \end{array}$$

Suppose that $\kappa \in H^2(G;M)$. Then κ is represented by a two-fold extension of G-modules as follows.

$$0 \to M \to E_2 \to E_1 \to \mathbb{Z} \to 0$$

If N is the kernel of $E_1 \to \mathbb{Z}$ then this breaks up into two short exact sequences:

$$\beta_2':\quad 0 \to M \to E_2 \to N \to 0$$

$$\beta_1':\quad 0 \to N \to E_1 \to \mathbb{Z} \to 0$$

and we have that $\kappa = \beta_2'^*\beta_1'^*(1)$. Let β_1, β_2 be the same sequences considered as $G\times T$ -modules, the action of T being trivial. This produces the following commutative diagram where $\bar{\beta}_i = \beta_i[G\times T]$ and $\bar{\beta}_i' = \beta_i'[G]$.

(b)

$$\begin{array}{ccc} H_2(\mathbb{Z}[G];\,\mathbb{Z}[G]) & \xrightarrow{\bar{\beta}'_{2*}\bar{\beta}'_{1*}} & H_0(\mathbb{Z}[G];M[G]) \\ \downarrow{\scriptstyle\bar{t}_*} & & \downarrow{\scriptstyle\bar{t}_*} \\ H_2(\mathbb{Z}[G\times T];\,\mathbb{Z}[G\times T]) & \xrightarrow{\bar{\beta}_{2*}\bar{\beta}_{1*}} & H_0(\mathbb{Z}[G\times T];M[G\times T]) \end{array}$$

Combine the two diagrams (a) and (b) and we get a larger diagram which produces the following equation for any $x \in K_2\,\mathbb{Z}[G]$.

(c) $$\bar{t}_*\bar{\beta}'_{2*}\bar{\beta}'_{1*}h_2(x) = -\bar{\beta}_{2*}\bar{\beta}_{1*}\bar{\beta}_{0*}h_3\tau(x)$$

This can be rewritten as follows.

<u>Formula 6.i.</u> $$\bar{t}_*\chi_2(\kappa,x) = -\chi_3([z]\times\kappa,\tau(x))$$

where $[z]$ is the generator of $H^1(T;\,\mathbb{Z})$ given in 5.f and "$\times$" is the exterior product in group cohomology.

<u>Proof</u>: By definition of χ_2 we have $\bar{t}_*\chi_2(\kappa,x) = \bar{t}_*(\ell^2(\kappa) \cap h_2(x)) = \bar{t}_*(\ell^2(\beta_2'^*\beta_1'^*(1)) \cap h_2(x)) = \bar{t}_*\bar{\beta}_2'^*\bar{\beta}_1'^*(1 \cap h_2(x)) = \bar{t}_*\bar{\beta}_2'^*\bar{\beta}_1'^*h_2(x) = -\bar{\beta}_{2*}\bar{\beta}_{1*}\bar{\beta}_{0*}h_3\tau(x) = -\bar{\beta}_{2*}\bar{\beta}_{1*}\bar{\beta}_{0*}(1 \cap h_3\tau(x)) = -\ell^3(\beta_2{}^*\beta_1{}^*\beta_0{}^*(1)) \cap h_3\tau(x) = -\chi_3(\beta_2{}^*\beta_1{}^*\beta_0{}^*(1),h_3\tau(x))$. Thus we need only show that $\beta_2{}^*\beta_1{}^*\beta_0{}^*(1) = \kappa\times[z]$. But $\beta_2{}^*\beta_1{}^*$ is cup product with

$\kappa \times 1$ and $\beta_0^*(1) = 1 \times [z]$. □

Chapter 7. The algebraic counterexample.

Any nonzero element of $Wh_1^+(G;A) = H_0(G;A[G])/H_0(G;A)$ which is in the image of $H^3(G;A) \otimes K_3\mathbb{Z}[G]$ under the map $\chi_3^+ = p\chi_3$ where p is the projection of $H_0(G;A[G])$ onto $Wh_1^+(G;A)$ is a counterexample to the original statement of Hatcher and Wagoner. In 7.a we give an example due to R.K. Dennis of a nontrivial element in the image of χ_2^+ and in 7.b we use the product formula for h_n to get an example for χ_3^+.

7.a. The example in $K_2\mathbb{Z}[G]$.

Let p be an odd integer ≥ 3. Let $G = \mathbb{Z}_p \times \mathbb{Z}_p = \langle x,y \mid x^p, y^p, [x,y]\rangle$. Let $A = \Lambda^2 G \cong \mathbb{Z}_p$ considered as a left G-module with the trivial action. Let $f\colon G \times G \to A$ be the factor set given by $f(a,b) = a \wedge b$. Then f represents a cohomology class $\kappa = [f] \in H^2(G;A)$. Let $a = x(y-1) \in \mathbb{Z}[G]$ and $b = 1 + y + y^2 + \ldots + y^{p-1} \in \mathbb{Z}[G]$. Then $ab = 0$ and so $[e_{12}(a), e_{21}(b)] = 1$ in $GL(\mathbb{Z}[G])$. Consequently $[x_{12}(a), x_{21}(b)] \in \ker\phi \cong K_2\mathbb{Z}[G]$.

Theorem 7.a. $\chi_2^+([f],[x_{12}(a),x_{21}(b)]) \neq 0$. In fact it is equal to $2\sum_{i=1}^{p} g \otimes xy^i$ where $g = x \wedge y$ is the generator of $\Lambda^2 G$.

Proof: We use theorem 6.e.3 to compute $\chi_2([f], x_{12}(a)x_{21}(b)x_{12}(-a)x_{21}(-b))$. The formula gives this as the sum of the following four terms.

(1) $$-\sum_q f_\ell(r^1_{q1} \otimes a)s^1_{2q} = 0$$

(2) $$-\sum_q f_\ell(r^2_{q2} \otimes b)s^2_{1q} = -f_\ell(a \otimes b) = -f_\ell((xy - x) \otimes (1 + y + \ldots + y^{p-1})) = \sum_{i=1}^{p} [f_\ell(x \otimes y^{i+1}) - f_\ell(xy \otimes y^i)] = \sum_{i=1}^{p} [(x \wedge y^{i+1}) \otimes xy^{i+1} - (x \wedge y^i) \otimes xy^{i+1}]$$
$$= \sum_{i=1}^{p} (x \wedge y) \otimes xy^{i+1}$$

(3) $$-\sum_q f_\ell(r^3_{q1} \otimes a)s^3_{2q} = -f_\ell(b \otimes a) = -f_\ell(a \otimes b) = \sum_{i=1}^{p} (x \wedge y) \otimes xy^{i+1}$$

(4) $\quad - \sum_q f_\ell(r^4_{q2} \otimes (-b))s^4_{1q} = 0$

Thus $\chi_2([f],[x_{12}(a),x_{21}(b)]) = 2 \sum_{i=1}^{p} (x \wedge y) \otimes xy^{i+1} \neq 0$ in $Wh_1^{\,+}(G;A)$. □

The element $[x_{12}(a),x_{21}(b)]$ is also denoted by $\langle a,b \rangle$.

7.b. An example in $K_3[G \times T]$.

Let $\bar{t}_*^{\,+}$ be the map induced by $\bar{t}_*$ as indicated in the following diagram.

$$\begin{array}{ccccccc} H_0(G;A) & \longrightarrow & H_0(G;A[G]) & \longrightarrow & Wh_1^{\,+}(G;A) & \to & 0 \\ \downarrow{\scriptstyle \approx} & & \downarrow{\scriptstyle \bar{t}_*} & & \downarrow{\scriptstyle \bar{t}_*^{\,+}} & & \\ H_0(G \times T;A) & \longrightarrow & H_0(G \times T;A[G \times T]) & \longrightarrow & Wh_1^{\,+}(G \times T;A) & \to & 0 \end{array}$$

Then by formula 6.i we have

(1) $$\chi_3^{\,+}([z] \times \kappa, \tau(x)) = -\bar{t}_*^{\,+}\chi_2^{\,+}(\kappa,x)$$

This formula together with 7.a gives the following.

(2) $$\chi_3^{\,+}([z] \times [f],\{t\}\cdot\langle a,b\rangle) = 2 \sum_{i=1}^{p} (x \wedge y) \otimes xy^i t$$

Another version of this example is as follows. Let $G = \mathbb{Z}_p \times \mathbb{Z}_p \times \mathbb{Z}_p$ and let $\xi^3 \in H^3(G;\mathbb{Z}_p)$ be the third exterior power of the generator $\xi \in H^1(\mathbb{Z}_p;\mathbb{Z}_p)$. If p is odd and ≥ 3 then $\chi_3^{\,+}(\xi^3,\langle a,b\rangle\cdot\{z\}) = 2 \sum_{i=1}^{p} xy^i z \in \mathbb{Z}_p[G]/\mathbb{Z}_p[1] \cong Wh_1^{\,+}(G;\mathbb{Z}_p)$. By applying symmetries the image of $\chi_3^{\,+}(\xi^3,-)$ is seen to be very large.

Chapter 8. Application to pseudoisotopies.

8.a. Introduction.

Let M be a compact smooth (C^∞) manifold of dimension ≥ 6. Let $G = \pi_1 M$ and $A = \pi_2 M$ and let $\kappa = k_1(M) \in H^3(G;A)$ be the first Postnikov invariant of M. Let $\mathcal{C}(M) = \mathrm{Diff}(M \times I \text{ rel } M \times 0 \cup \partial M \times 0)$ be the pseudoisotopy space of M. What Hatcher and Wagoner actually proved in [3] is the following.

Theorem 8.a.1. (Hatcher-Wagoner) There is a natural exact sequence as follows.

$$Wh_1^+(G;A) \oplus Wh_1^+(G;\mathbb{Z}_2) \xrightarrow{j_1 + j_2} \pi_0 C(M) \xrightarrow{\lambda} Wh_2(G) \to 0$$

Furthermore if the first Postnikov invariant of M is trivial and a section $\sigma: M_{(1)} \to M_{(2)}$ of the fibration $B^2A \to M_{(2)} \to M_{(1)} = BG$ is chosen then a retraction $\sigma_*: \pi_0 C(M) \to Wh_1^+(G;A)$ of j_1 is obtained. Here $Wh_2(G) = K_2\mathbb{Z}[G]/\pi_2 BM(G)^+$.

In this chapter we shall examine Hatcher's construction of σ_* (as explained in [5] and in personal communications to the author.) We shall see exactly where the assumption that $k_1(M) = 0$ is used. Furthermore we shall see shat happens when $\kappa = k_1(M)$ is nonzero. This is the purpose of the entire paper. We shall prove:

Theorem 8.a.2. There is a natural exact sequence as follows.

$$K_3\mathbb{Z}[G] \xrightarrow{\chi_3^+(\kappa,-) + \chi_G} Wh_1^+(G;A) \oplus Wh_1^+(G;\mathbb{Z}_2) \xrightarrow{j} \pi_0 C(M) \xrightarrow{\lambda} Wh_2(G) \to 0$$

where $\chi_G: K_3\mathbb{Z}[G] \to Wh_1^+(G;\mathbb{Z}_2)$ is the Grassmann invariant.

Strictly speaking the Grassmann invariant [9] comes first in logical order since it depends only on $G = \pi_1 M$. However the reader should have no trouble following the arguments here without knowledge of the Grassmann invariant.

We shall also investigate to what extent the map σ_* depends on the choice of $\sigma: M_{(1)} \to M_{(2)}$ and we shall prove:

Theorem 8.a.3. Suppose that $\sigma^1, \sigma^2: M_{(1)} \to M_{(2)}$ are two sections of the map $M_{(2)} \to M_{(1)}$ and suppose that σ^1 and σ^2 differ by the difference class $\upsilon \in H^2(G;A)$. Then the retractions $\sigma^i_*: \pi_0 C(M) \to Wh_1^+(G;A)$ differ by the function $\chi_2^{Wh}(\upsilon,\lambda(-))$ where $\chi_2^{Wh}: H^2(G;A) \otimes Wh_2(G) \to Wh_1^+(G;A)$ is the map induced by χ_2^+ using 1.g.5.

Here is the outline of the rest of this chapter:

8.b. Review of Cerf theory. 1

8.c. The section $\sigma: BG^2 \to X$. 2

8.d. Choices needed to define $\sigma_*: \pi_0 C(M) \to Wh_1^+(G;A)$. 3

8.e. Definition of $\sigma_*(f_t,v_t,c)$. 5

8.f. $\sigma_*(f_t,v_t,c)$ is independent of the choice of c. 8

8.g. Invariance of $\sigma_*(f_t, v_t)$ under lens-shaped homotopies of (f_t, v_t). 10

8.h. The difference obstruction. 11

8.i. Proof of the main theorem. 15

8.j. $\sigma_*: \pi_0 C(M) \to Wh_1^+(G;A)$ is well-defined. 16

8.b. Review of Cerf theory.

In this section we review briefly the function theory approach to pseudoisotopies. This is often called "Cerf theory" after its originator [1] although many good ideas were added in [3].

If M is a compact smooth manifold and $I = [0,1]$, let $F(M)$ denote the space of all smooth maps $M \times I \to I$ which agree with the projection map in a neighborhood of $\partial(M \times I)$. Let $E(M)$ denote the subspace of $F(M)$ consisting of functions without critical points. When these spaces are equipped with the weak topology [7] $F(M)$ is contractible and $E(M) \simeq C(M)$. Thus $\pi_0 C(M) \cong \pi_1(F(M), E(M))$. Consequently an element of $\pi_0 C(M)$ is represented by a one-parameter family of functions $f_t: M \times I \to I$, $t \in I$. And $[f_t] = 0$ in $\pi_0 C(M)$ if there is an admissible deformation of f_t which eliminates all the singularities.

Using jet transversality one sees the following. Transversality of the first derivative to 0 implies that the singular set is a smooth 1-manifold. Transversality of the second derivative to the stratification of the set of $n \times n$ matrices according to rank says that the singular points are all nondegenerate except for a finite number of cancellation (birth-death) points.

It is shown in [4] that f_t can be deformed do that all critical points lie in the two middle indices i and $i + 1$ where $i = [m/2]$ if $m = \dim M \geq 6$. To study f_t further we must choose a gradient-like vector field v_t for f_t. This means that $v_t = -\text{grad}\, f_t$ with respect to some metric on $M \times I$ which varies with t. We shall assume that $-v_t(x)$ is the standard unit vector $(0,1)$ in the I direction if x is near $\partial(M \times I)$.

If the trajectory of v_t goes from a Morse critical point of index a down to a Morse point of index b we shall call this trajectory an a/b intersection. Cerf showed that trajectories leading up and down from birth-death points can be made disjoint from all other critical points. Transversality shows that there are no i/i+1 intersections, there are only a finite number of i+1/i+1 intersections (these are called handle additions) and for a fixed t there are only a finite number of i+1/i intersections called incidence points. There are also i/i handle additions but these can be eliminated by sliding them up to i+1/i+1 handle additions.

If $x \in M \times I$ is a Morse point of f_t then the stable sphere of x is the collection of all trajectories of v_t leading away from x. The stable sphere of x can be equipped with a topology and a smooth structure in an obvious way so that it forms a sphere of dimension one less than the index of x. Similarly the unstable sphere of s is the collection of all trajectories of v_t leading into x.

Suppose now that (f_t, v_t) has no handle additions. Then an element of $Wh_1+ Wh_1^+(G;A \oplus \mathbb{Z}_p)$ can be associated to (f_t, v_t) in the following way. The condition of no handle additions implies that each critical point set component S of f_t has exactly two degenerate critical points, one birth and one death point. Also S is null homotopic in $X = M \times I \times I$. The incidence points between the upper and lower index Morse point arcs in S form a framed 1-manifold J_S (the framing is explained in [9] and is not important in this paper) and a choice of null homotopies of S in X produces a map $J_S \to X^{S^1}$ which sends J_S to $X_0^{S^1}$ = the component of X^{S^1} containing the trivial loop. Summing over all S we get an element of $\Omega_1^{fr}(X^{S^1}, X_0^{S^1}) \cong Wh_1^+(G;A \oplus \mathbb{Z}_2)$. Ignoring the framing of J_S we get an element of $Wh_1^+(G;A)$. It is easy to see that this is an invariant of (f_t, v_t) independent of the choice of null homotopies of S in X. It is also easy to see that any element of $Wh_1^+(G;A \oplus \mathbb{Z}_2)$ is the invariant of some family (f_t, v_t) without handle additions. Also any two such families with the same $Wh_1^+(G;A \oplus \mathbb{Z}_2)$ invariant represent the same element of $\pi_0 C(M)$. This is the definition of the map $j_1 + j_2$ in 8.a.1. Given $\sigma\colon BG^2 \to X$ we shall show how the $Wh_1^+(G;A)$ summand of this invariant can be extended to more general families thus resulting in a map $\sigma_*\colon \pi_0 C(M) \to Wh_1^+(G;A)$.

8.c. The section $\sigma\colon BG^2 \to X$.

Take $X = M \times I \times I$ as before and consider the cannonical map $p\colon X \to X_{(1)} \simeq BG$ of X into its 1-coskeleton. Let BG be the standard CW-complex corresponding to the normalized bar construction for G. Then there exists a homotopy right inverse $\sigma\colon BG^2 \to X$ of p over the 2-skeleton BG^2 of BG. If σ is extendable to BG^3 then the composition $BG^2 \to X \to X_{(2)}$ is uniquely extendable to BG (up to homotopy.) And conversely any map $BG \to X_{(2)}$ which induces the identity on π_1 is homotopic to a map derived in this way. The specification of the map $\sigma\colon BG^2 \to X$ is equivalent to making the following choices.

(1) Choose a base point $* \in X$.

(2) For each $u \in G$ choose a loop $\sigma_1(u)\colon (\Delta^1,\partial) \to (X,*)$ representing $u \in \pi_1 X$. Take $\sigma_1(1)$ to be the trivial loop.

(3) For every pair $u, v \in G$ choose a map $\sigma_2(u,v)\colon \Delta^2 \to X$ whose faces are $d_0 = \sigma_1(u)$, $d_1 = \sigma_1(uv)$ and $d_2 = \sigma_1(v)$. Let $\sigma_2(u,1) = \sigma_1(u)s_0$ and $\sigma_2(1,v) = \sigma_1(v)s_1$ where $s_j\colon \Delta^2 \to \Delta^1$ is the j-th degeneracy map.

Now consider a generic one-parameter family of functions $f_t\colon M \times I \to I$, $t \in I$, together with a gradient-like vector field v_t. We shall assume that (f_t,v_t) is a <u>lens-shaped family</u> which means that it satisfies the following three conditions.

(a) f_t has critical points only in the middle two indices i, $i+1$, $i = [\dim M/2]$.

(b) (f_t,v_t) has only $i+1/i+1$ handle additions, i.e., it has no i/i handle additions.

(c) The third condition will be explained later.

Given any triple (σ,f_t,v_t) where $\sigma\colon BG^2 \to X$ is extendable to BG^3 and (f_t,v_t) is a lens-shaped family we shall associate an element of $Wh_1^+(G;A)$. Since every element of $\pi_0 C(M)$ is represented by a lens-shaped family we will get a map $\sigma_*\colon \pi_0 C(M) \to Wh_1^+(G;A)$. The argument will be to make more choices which we will call "c" and then to define $\sigma_*(f_t,v_t,c)$ and show that it is independent of the choice of c.

8.d. Choices needed to define $\sigma_*\colon \pi_0 C(M) \to Wh_1^+(G;A)$.

To define $\sigma_*(f_t,v_t)$ we must make the following choices.

(1) Choose a path from each birth point to the base point $* \in X$.

(2) Number the birth points 1, 2, ... , k.

(3) Choose an orientation for the negative eigenspace of $D^2(f_t)_x$ at each Morse point $(x,t) \in (M \times I) \times I$. These orientations should vary continuously along a path of Morse points.

With these choices we may define an incidence matrix $r(t) = (r_{pq}(t)) \in GL_k(\mathbb{Z}[G])$ for all but a finite number of critical t's in I. For each critical t we may associate an elementary operation $x(t) = x_{pq}(u)$, $u \in \pm G$, which has the property that $r(t+\varepsilon) = r(t-\varepsilon)e_{pq}(u)$ for small $\varepsilon > 0$. We may perform the choices in (3) so that $r(0)$ is the identity matrix. We will also assume that $r(1)$ is the identity matrix. <u>This is the third condition for (f_t, v_t) to be lens-shaped.</u> Let $t_1 < t_2 < \ldots < t_n$ represent the critical t's. Then the condition $r(1) = I_k$ implies that the product $x(t_1)\ldots x(t_n)$ considered as an element of $St(\mathbb{Z}[G])$ is in the kernel of the map $\phi\colon St(\mathbb{Z}[G]) \to GL(\mathbb{Z}[G])$ and thus represents an element of $K_2\mathbb{Z}[G]$. The image of this element in $Wh_2(G)$ is $(-1)^{i+1}\lambda([f])$. The lens-shaped assumption also implies that we can make the following choice.

(4) Choose a path $\gamma(z)$ from each critical point z to the base point of X so that $\gamma(z)$ varies continuously with z and agrees with choice (1) when z is a birth point.

Let t_j be a critical parameter value with $x(t_j) = x_{pq}(\varepsilon u)$, $\varepsilon = \pm 1$, $u \in G$. This means that a trajectory ξ_j of the vector field v_{t_j} goes from z_q down to z_p where z_s is the s-th upper index Morse point. It also implies that $u = [\gamma(z_p)\xi_j\gamma(z_q)^{-1}]$.

(5) Choose a homotopy from $\gamma(z_p)\xi_j\gamma(z_q)^{-1}$ to $\sigma_1(u)$, or equivalently choose a homotopy from $\xi_j\gamma(z_q)^{-1}\sigma_1(u)^{-1}$ to $\gamma(z_p)^{-1}$.

Let c represent the choices (1) - (5) above. We shall define $\sigma_*(f_t,v_t,c) \in Wh_1^+(G;A)$ and we shall show that it does not depend on the choice of c assuming that σ is extendable to BG^3. Under the same assumption we show that $\sigma_*(f_t,v_t)$ is invariant under homotopies of f_t and v_t. Our detailed proof will also show what happens when σ is not extendable to BG^3.

8.e. Definition of $\sigma_*(f_t, v_t, c)$.

We may assume that all birth points occur at the same t-coordinate, say a, and that all death points occur at the t-coordinate b. Let y_p denote the p-th lower index Morse point of f_t, $t \in (a,b)$. The unstable sphere of y_p will be called the p-th unstable sphere. Taking the union over all $t \in I' = [a+\delta, b-\delta]$ where $\delta > 0$ is infinitesmal we get a 1-parameter family of spheres which we call the p-th unstable cylinder. This cylinder will be identified with $S_p^{m-i} \times I' = S^{m-i} \times I' \times p \subset S^{m-i} \times I' \times \{1,2,\ldots,k\}$.

Let $J_{pq} \subset S_p^{m-i} \times I'$ be the closure of the set of all trajectories going from z_q down to y_p. These are the pq-incidence points. By transversality J_{pq} is a compact 1-manifold with boundary, the boundary being a finite set of points with t-coordinates equal to one of $t_0, t_1, \ldots, t_{n+1}$ where $t_0 = a + \delta$, $t_{n+1} = b - \delta$ and the other t_j's are as before. Let J_{pq} be the set of all points in J_{pq} with t-coordinate equal to $t_0, \ldots, t_{n+1}$. By transversality we may assume that J_{pq} is a finite set.

For $x \in J_{pq} - \delta J_{pq}$ let $u(x) \in G$ be the homotopy class $u(x) = [\gamma(y_p)\xi(x)\gamma(z_q)^{-1}]$. If x is not a critical point of the projection map $J_{pq} \to I'$ then there is a sign $\varepsilon(x) = \pm 1$ such that x contributes the monomial $\varepsilon(x)u(x)$ to the pq-entry of the incidence matrix $r(x) = r(t(x))$ where $t(x)$ is the I'-coordinate of x. If we interpret $\varepsilon(x)$ as an orientation for $J_{pq} - \delta J_{pq}$ by associating the tangent vector $\varepsilon(x)dt$ to x we get an orientation which varies continuously on all of $J_{pq} - \delta J_{pq}$.

Let $\mu(x) \in \mathbb{Z}[G]$ be the qp-entry of the matrix $s(x) = r(x)^{-1}$. If $w \in G$ let $\mu(x,w)$ be the coefficient of w in $u(x)\mu(x)$, i.e.,

$$u(x)\mu(x) = \sum_{w \in G} \mu(x,w)w.$$

Also let $\Omega_w X$ be the w-component of ΩX. Define a continuous function $\Psi^w: J_{pq} - \delta J_{pq} \to \Omega_w X$ for each $w \in G$ by the following equation.

$$\Psi^w(x) = \gamma(y_p)\xi(x)\gamma(z_q)^{-1}\sigma_1(u(x))^{-1}\sigma_1(w) = \Psi^1(x)\sigma_1(w)$$

Now consider $J_{pq} - \delta J_{pq}$ as a disjoint union of open intervals and compactify it to get a disjoint union of closed intervals (attach a point at each end.) Call this compactification $\tilde{J}_{pq}$. If $x \in \delta J_{pq}$ there are at most two points in $\tilde{J}_{pq}$ corresponding to x. These will be called x_L, x_R where x_L is the limit in $\tilde{J}_{pq}$ of points in $J_{pq} - \delta J_{pq}$ converging to x from the left (i.e., smaller t-values) and x_R is the right limit. Note that the functions ξ, Ψ^w extend uniquely to continuous function on all of J_{pq} and thus induce continuous functions on $\tilde{J}_{pq}$. Also the locally constant functions r, s, μ extend uniquely to locally constant functions on $\tilde{J}_{pq}$.

Since $A[G] \cong H_1(\Omega X)$ maps onto $H_0(G;A[G]) \cong H_1(X^{S^1})$ which maps onto $Wh_1^+(G;A)$ we may define $\sigma_*(f_t,v_t,c)$ as the image in $Wh_1^+(G;A)$ of the homology class in $H_1(X^{S^1})$ given by

$$(*) \qquad (-1)^i \Sigma\, \mu(x,w)\overline{\Psi}^w$$

where the summation runs over all w in G, all components of $\tilde{J}_{pq}$ and all p, q. The maps $\overline{\Psi}^w$ are derived from Ψ^w as follows.

Let $\overline{J}_{pq}$ be the mapping cylinder of the inclusion map $\partial\tilde{J}_{pq} \to \tilde{J}_{pq}$. Thus $\overline{J}_{pq}$ is equal to $\tilde{J}_{pq}$ with an interval adjoined to each boundary point. The map $\overline{\Psi}^w : \overline{J}_{pq} \to \Omega_w X$ will be an extension of the map $\Psi^w: J_{pq} \to \Omega_w X$. The extension will be defined by choosing a homotopy $H(x,w)$ from $\Psi^w(x)$ to $\sigma_1(w)$ for each $x \in \partial\tilde{J}_{pq}$.

We divide the elements of δJ_{pq} into three classes. First there are the points $x \in \delta J_{pq}$ which lie in the interior of J_{pq}. To each of these points is associated the left and right limits x_L, $x_R \in \partial\tilde{J}_{pq}$. These <u>interior points</u> have t-coordinate equal to one of the critical values t_j for $1 \leq j \leq n$. The second class of points in δJ_{pq} are those that occur at $t = t_0$ and $t = t_{n+1}$. These will be called <u>end points</u>. The remaining points of δJ_{pq} lie in the third class. These will be called the <u>T-points</u>.

Let $x \in \delta J_{pq}$ be an interior point and take $w \in G$. Then the homotopy $H(x,w) = H(x_L,w) = H(x_R,w)$ is chosen as follows.

(1) Choose a null homotopy of the loop $\Psi^1(x) = \gamma(y_p)\xi(x)\gamma(z_q)^{-1}\sigma_1(u(x))^{-1}$.

(2) For each $w \in G$ multiply this homotopy by $\sigma_1(w)$. This gives a homotopy from $\Psi^w(x)$ to $\sigma_1(w)$. Let this be $H(x,w)$.

Suppose that x occurs at the critical t-value t_j and $x(t_j) = x_{ab}(\varepsilon v)$. If $q \neq a$ then $\mu(x_L) = \mu(x_R)$ and thus the choice in (1) above does not affect the homology class of (*). If $q = a$ then $x \in \delta J_{pa} \subset S_p^{m-i} \times I'$ is equal to a T-point $y \in \delta J_{pb}$. The points x and y are related by the equation $\xi(y) = \xi(x)\xi_{t_j}$ as indicated in the following drawings.

(3)

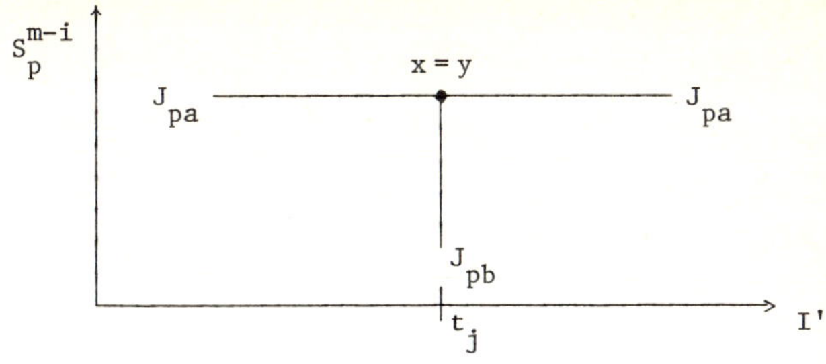

(4)

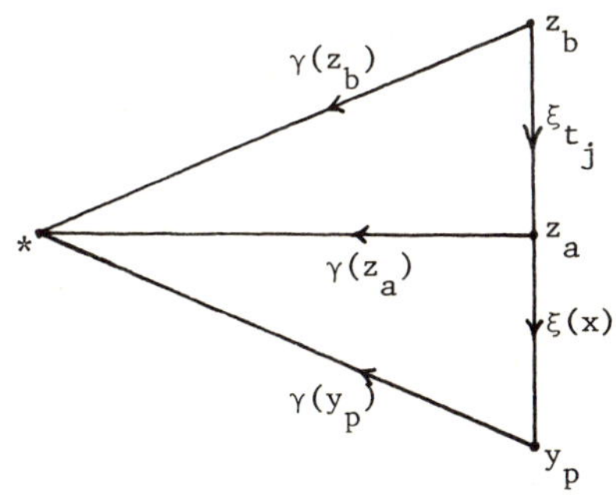

$$\xi(y) = \xi(x)\xi_{t_j}$$

Let $H(y,w)$ be the composition of the following homotopies.

(5)
$$\begin{aligned}
\Psi^w(y) &= \gamma(y_p)\xi(y)\gamma(z_b)^{-1}\sigma_1(u(y))^{-1}\sigma_1(w) \\
&= \gamma(y_p)\xi(x)\xi_j\gamma(z_b)^{-1}\sigma_1(u(x)v)^{-1}\sigma_1(w) \\
&\underset{\alpha}{\simeq} \gamma(y_p)\xi(x)\xi_j\gamma(z_b)^{-1}\sigma_1(v)^{-1}\sigma_1(u(x))\sigma_1(w) \\
&\underset{\beta}{\simeq} \gamma(y_p)\xi(x)\gamma(z_a)^{-1}\sigma_1(u(x))^{-1}\sigma_1(w) \\
&\simeq \sigma_1(w) \qquad \text{(by } H(x,w))
\end{aligned}$$

where α is the inverse of the homotopy $\sigma_2(u(x),v)$ and β is the homotopy given in 8.d.5. These two homotopies do not depend on the choice of H. Since $u(y) = u(x)v$ and $\mu(x_L) = \mu(x_R) + \varepsilon v\mu(y)$ we have $\mu(x_L,w) = \mu(x_R,w) + \varepsilon\mu(y,w)$ for all $w \in G$. This implies that the sum (*) is independent of the choice of $H(x,w)$. (The sign ε is taken care of by the orientation of $\overline{J}_{pq}$.)

All T-points are derived in the above manner and all end points are irrelevant as we shall now show. If $x \in \delta J_{pq}$ is an end point we must have $p = q$, $u(x) = 1$, $r(x) = I_k$, $s(x) = I_k$ and $\mu(x) = 1$. Thus $\mu(x,w) = 0$ for all $w \neq 1$ in G. In the homology class (*) these points and their components in J_{pq} are irrelevant since they lie in the kernel of the map $H_0(G;A[G]) \to Wh_1^+(G;A)$. This completes the definition of $\sigma_*(f_t,v_t,c)$.

8.f. $\sigma_*(f_t,v_t,c)$ is independent of the choice of c.

Lemma 8.f.1. $\sigma_*(f_t,v_t,c)$ is independent of the choice made in 8.d.5.

Proof: Let j be fixed and suppose that $x(t_j) = x_{ab}(\varepsilon v)$. Suppose that the homotopy in 8.d.5 is altered by the element $\alpha \in A = \pi_2 X$. Then the only thing that changes in the definition of $\sigma_*(f_t,v_t,c)$ is the homotopy β in 8.e.5. This changes by $u(x)\cdot\alpha \in A$. This means that the homology class 8.e(*) changes by $(-1)^i\varepsilon\varepsilon(x)\mu(y,w)$ $u(x)\cdot\alpha \otimes w$ for each $x \in J_{*a}$ at $t = t_j$ and each $w \in G$. The total change can be computed as follows where the summation is over all $x \in J_{*a}$ and all $w \in G$.

$$(-1)^i \sum_{x,w} \varepsilon\varepsilon(x)\mu(y,w)u(x)\cdot\alpha \otimes w$$

$$= (-1)^i \sum_{x,w} \varepsilon\varepsilon(x)\mu(y,w)v^{-1}\cdot\alpha \otimes u(y)^{-1}wu(x)v \qquad \text{(conjugation by } u(y) = u(x)v)$$

$$= (-1)^i \sum_{x,w} \varepsilon v^{-1}\cdot\alpha \otimes [\varepsilon(x)\mu(y,w)u(y)^{-1}wu(x)]v$$

$$= (-1)^i\varepsilon v^{-1}\cdot\alpha \otimes [\Sigma_p \Sigma_{x\in J_{pa}} \varepsilon(x)\mu(y)u(x)]v \qquad \text{(by definition of } \mu(y,w))$$

$$= (-1)^i\varepsilon v^{-1}\cdot\alpha \otimes [\Sigma_p s_{bp}r_{pa}]v \qquad \text{(since } \mu(y) = s_{bp} \text{ and } r_{pa} = \Sigma_{x\in J_{pa}} \varepsilon(x)u(x))$$

$$= 0 \qquad \text{(since } s = r^{-1} \text{ and } a \neq b)$$ □

Lemma 8.f.2. $\sigma_*(f_t,v_t,c)$ is independent of the choices made in 8.d.1 and 8.d.4 provided that σ is extendable to BG^3.

Proof: Note first that in 8.e we did not use $\gamma(z)$ for z in a neighborhood of the birth-death points. So suppose γ' is another choice of paths homotopic to γ on the set of Morse points of f_t. This homotopy composed with the homotopy 8.d.5 gives a homotopy from $\gamma'(z_p)\xi_j\gamma'(z_q)^{-1}$ to $\sigma_1(u)$. Using this choice of 8.d.5 the homology class 8.e(*) is the same as before.

Now suppose that γ' differs from γ on the p-th component by $u_p \varepsilon G$. By the comments in the previous paragraph we may assume that $\gamma'(z_q) = \sigma_1(u_q)\gamma(z_q)$ and $\gamma'(y_p) = \sigma_1(u_p)\gamma(y_p)$. Thus for every x we have

$$r_{pq}(x)' = u_p r_{pq}(x) u_q^{-1}$$

$$\mu(x) = s_{qp}(x)' = u_q s_{qp}(x) u_p^{-1} = u_q \mu(x) u_p^{-1}$$

$$u(x)' = u_p u(x) u_q^{-1}$$

$$u(x)'\mu(x)' = u_p u(x)\mu(x) u_p^{-1} = \sum_w \mu(x,w) u_p w u_p^{-1}$$

$$\mu(x,w)' = \mu(x, u_p^{-1} w u_p)$$

(3) $$\Psi^w(x)' = \sigma_1(u_p)\gamma(y_p)\xi(x)\gamma(z_q)^{-1}\sigma_1(u_q)^{-1}\sigma_1(u_p u(x) u_q^{-1})^{-1}\sigma_1(w)$$

We shall show that if $\mu(x,w)'[\overline{\Psi}^w(x)']$ (with x varying in some component of $\overline{J}_{pq}$) is one term in the new homology class then it is equal to the term in the original class given by $\mu(x,\underline{w})u_p[\overline{\Psi}^{\underline{w}}(x)]u_p^{-1}$ where $\underline{w} = u_p^{-1} w u_p$. Thus the invariant will be unchanged in $Wh_1^+(G;A)$.

The homology class $u_p[\overline{\Psi}^{\underline{w}}(x)]u_p^{-1}$ is represented by the loop $\sigma_1(u_p)\overline{\Psi}^{\underline{w}}(x)$ $\sigma_1(u_p)^{-1}$. This loop contains the path

(4) $$\sigma_1(u_p)\Psi^{\underline{w}}(x)\sigma_1(u_p)^{-1} = \sigma_1(u_p)\gamma(y_p)\xi(x)\gamma(z_q)^{-1}\sigma_1(u(x))^{-1}\sigma_1(\underline{w})\sigma_1(u_p)^{-1}$$

together with homotopies of this path (when $x \varepsilon \partial\tilde{J}_{pq}$) to the path $\sigma_1(u_p)\sigma_1(\underline{w})\sigma_1(u_p)^{-1}$. To find these homotopies first choose a null homotopy of $\Psi^1(x) = \gamma(y_p)\xi(x)\gamma(z_q)^{-1}$ $\sigma_1(u(x))^{-1}$ for each $x \varepsilon \partial\tilde{J}_{pq}$ as in 8.e.1. This gives rise to the following null homotopy of $\Psi^1(x)'$ for the same boundary point x.

$$\sigma_1(u_p)\gamma(y_p)\xi(x)\gamma(z_q)^{-1}\sigma_1(u_q)^{-1}\sigma_1(u_p u(x) u_q^{-1})^{-1}$$

(5) $$\simeq \sigma_1(u_p)\gamma(y_p)\xi(x)\gamma(z_q)^{-1}\sigma_1(u(x))^{-1}\sigma_1(u_p)^{-1}$$

$$\simeq \sigma_1(u_p)\sigma_1(u_p)^{-1} \qquad \text{(by the null homotopy of } \Psi^1(x))$$

$$\simeq 0$$

where the homotopy (5) is given by the homotopy $\sigma_1(u_q)^{-1}\sigma_1(u_p u(x) u_q^{-1})^{-1} \simeq$

$\sigma_1(u(x))^{-1}\sigma_1(u_p)^{-1}$ given uniquely up to homotopy by σ_2.

In this paragraph we shall prove that $\mu(x,w)'[\bar{\Psi}^w(x)'] = \mu(x,\underline{w})u_p[\bar{\Psi}^{\underline{w}}(x)]u_p^{-1}$ in the case in which both boundary points are interior in J_{pq}. For this we choose any homotopy from $\sigma_1(\underline{w}) = \sigma_1(u_p^{-1}wu_p)$ to $\sigma_1(u_p)^{-1}\sigma_1(w)\sigma_1(u_p)$. This gives a homotopy of $\sigma_1(u_p)\sigma_1(\underline{w})\sigma_1(u_p)^{-1}$ to $\sigma_1(w)$. Using this homotopy we may deform the loop $\sigma_1(u_p)\bar{\Psi}^{\underline{w}}(x)\sigma_1(u_p)^{-1}$ in X^{S^1} to the following new loop. The path (4) is changed to the path $\sigma_1(u_p)\gamma(y_p)\xi(x)\gamma(z_q)^{-1}\sigma_1(u(x))^{-1}\sigma_1(u_p)^{-1}\sigma_1(w)$. The boundary points have the homotopies to $\sigma_1(w)$ given by first contracting $\Psi^1(x) = \gamma(y_p)\xi(x)\gamma(z_q)^{-1}\sigma_1(u(x))^{-1}$ and then cancelling $\sigma_1(u_p)\sigma_1(u_p)^{-1}$. This new loop is homotopic to the loop $\bar{\Psi}^w(x)'$ in the following way. Along the path $\Psi^w(x)'$ perform the homotopy used in (5) above.

$$\sigma_1(u_p)\gamma(y_p)\xi(x)\gamma(z_q)^{-1}\sigma_1(u_q)^{-1}\sigma_1(u_pu(x)u_q^{-1})^{-1}\sigma_1(w)$$
$$\simeq \sigma_1(u_p)\gamma(y_p)\xi(x)\gamma(z_q)^{-1}\sigma_1(u(x))^{-1}\sigma_1(u_p)^{-1}\sigma_1(w).$$

This extends to the boundary points in an obvious way thus concluding our proof for interior points.

We now consider the T-points. Given the choices 8.d.5 for γ we can make the following choices of homotopies for γ': $\sigma_1(u_a)\gamma(z_a)\xi_j\gamma(z_b)^{-1}\sigma_1(u_b)^{-1} \simeq \sigma_1(u_a)\sigma_1(u)\sigma_1(u_b)^{-1} \simeq \sigma_1(u_auu_b^{-1})$. The last homotopy is given uniquely up to homotopy by σ_2 because of the existence of σ_3. Examining the two loops $\bar{\Psi}^w(y)'$ and $\sigma_1(u_p)\bar{\Psi}^{\underline{w}}(y)\sigma_1(u_p)^{-1}$ we see that σ_2 gives a homotopy between them:

$$\Psi^w(y)' = \sigma_1(u_p)\gamma(y_p)\xi(y)\gamma(z_b)^{-1}\sigma_1(u_b)^{-1}\sigma_1(u_pu(y)u_b^{-1})^{-1}\sigma_1(w)$$
$$\sigma_1(u_p)\Psi^{\underline{w}}(y)\sigma_1(u_p)^{-1} = \sigma_1(u_p)\gamma(y_p)\xi(y)\gamma(z_b)^{-1}\sigma_1(u(y))^{-1}\sigma_1(u_p^{-1}wu_p)\sigma_1(u_p)^{-1}$$ □

<u>Lemma 8.f.6</u>. $\sigma_*(f_t,v_t,c)$ is independent of the choices made in 8.d.2 and 8.d.3.

<u>Proof</u>: A change in the numbering (8.d.2) or the orientations (8.d.3) has no affect on the geometric choices (the paths and the homotopies.) The only things that change are the matrices which change by conjugation. This has no affect on the homology class 8.e(*). □

We have proved the following.

<u>Theorem 8.f.7</u>. If σ is extendable to BG^3 then $\sigma_*(f_t,v_t,c)$ is independent of c.□

8.g. The difference obstruction.

In this section we shall consider f_t and v_t to be fixed and consider how $\sigma_*(f_t,v_t)$ varies with the choice of σ. Thus let $\sigma, \sigma'\colon BG^3 \to X$ be two map which induce the identity map on π_1. Since σ and σ' are homotopic when restricted to BG^1 we may assume that they are actually equal on BG^1. Thus the choices in 8.c.1 and 8.c.2 are uniquely chosen and two choices σ_2 and σ_2' are given for the map σ_2 of 8.c.3. Using the standard construction of the difference cochain, the maps σ_2 and σ_2' produce a map $\sigma_2' - \sigma_2 = \tau(u,v)\colon S^2 \to X$ for every pair u, v in G such that the homotopy class $h(u,v) = [\tau(u,v)] \in A = \pi_2 X$ gives a normalized factor set $h\colon G \times G \to A$. Let $h_\ell\colon \mathbb{Z}[G] \otimes \mathbb{Z}[G] \to A[G]$ represent the corresponding linear factor set.

<u>Theorem 8.g</u>. $\sigma_*'(f_t,v_t) - \sigma_*(f_t,v_t) = \chi_2^{Wh}([h],\lambda(f_t,v_t))$

<u>Proof</u>: Let c represent the choices 8.d.1 - 5 for σ. Then c is also a suitable set of choices for σ' since $\sigma_1' = \sigma_1$. The only difference between $\sigma_*(f_t,v_t,c)$ and $\sigma_*'(f_t,v_t,c)$ lies in the homotopy 8.e.5. which changes by $h(u(x),v)$. Therefore the change in the homology class 8.e(*) is given as follows where the summation is over all $x \in J_{*a}$, all $w \in G$ and $j = 1, 2, \ldots, n$. (Note that a, b, ε, v depend on j since $x(t_j) = x_{ab}(\varepsilon v)$.)

$$
\begin{aligned}
&(-1)^i \sum_{x,w,j} \varepsilon\varepsilon(x)\mu(y,w)h(u(x),v)w \\
= \quad &(-1)^i \sum_{x,w,j} h_\ell(\varepsilon(x)u(x) \otimes \varepsilon v)\mu(y,w)v^{-1}u(x)^{-1}w \\
= \quad &(-1)^i \sum_{x,j} h_\ell(\varepsilon(x)u(x) \otimes \varepsilon v)s_{bp} \\
= \quad &(-1)^i \sum_{j,p} h_\ell(r_{pa} \otimes \varepsilon v)s_{bp} \\
= \quad &(-1)^{i+1}\chi_2^{Wh}([h],(-1)^{i+1}\lambda[f_t]) \\
= \quad &\chi_2^{Wh}([h],\lambda[f_t]).
\end{aligned}
$$

We used formula 6.e.1 and the fact that $x(t_1)\ldots x(t_n) = (-1)^{i+1}\lambda[f_t]$. □

This completes the proof of 8.a.3 modulo the invariance of $\sigma_*(f_t,v_t)$.

8.h. Invariance of $\sigma_*(f_t,v_t)$ under lens-shaped homotopies of (f_t,v_t).

A lens-shaped homotopy of (f_t,v_t) will mean a two parameter family of functions and gradient-like vector fields (f_{ts},v_{ts}) satisfying the following conditions.

(0) There is a finite set $S \subset \text{int } I$ such that if s_0 is a fixed element of $I - S$ then (f_{ts_0},v_{ts_0}) is a lens shaped family.

If s_0 is a critical value of s, i.e., an element of S then one of the following four things happens at s_0. ( <=> $\emptyset$)

(1) A component of the critical set of f_{ts} is either cancelled or created as s goes from $s_0 - \varepsilon$ to $s_0 + \varepsilon$.

(2) A pair of consecutive handle additions of the form $x_{ab}(v)$ and $x_{ab}(-v)$ is either cancelled or created as the value of s passes s_0.

(3) Two commuting handle additions pass each other.

(4) Two noncommuting handle additions of the form $x_{ab}(u)$ and $x_{bc}(v)$ pass each other and a third handle addition $x_{ac}(uv)$ is either created or destroyed depending on which of the two original handle additions started on the left.

Theorem 8.h.5. If $\sigma\colon BG^2 \to X$ is fixed and extendable to BG^3 and if (f_{ts},v_{ts}) is a lens-shaped homotopy then $\sigma_*(f_{t0},v_{t0}) = \sigma_*(f_{t1},v_{t1})$.

Proof: The proof is by case by case analysis. It is not difficult to see that $\sigma_*(f_{ts},v_{ts})$ does not change as s varies between two consecutive elements of S. We will show that $\sigma_*(f_{ts},v_{ts})$ does not change at any of the four types of critical values of s.

In case (1) the value of $\sigma_*(f_{ts},v_{ts})$ does not change because a lens-shaped critical point set component on the verge of being cancelled will have only incidence points x with $u(x) = 1$.

In case (2) it can be seen that the $Wh_1^+(G;A)$ invariant does not change if the choices 8.d.5 for the two consecutive handle additions $x_{ab}(v)$ and $x_{ab}(-v)$ are chosen compatibly. The interesting part of the verification of this case is illustrated in 8.h.6. The line segment $\overline{AD}$ is broken at the two points x_1 and x_2.

Figure 8.h.6. These are subsets of the unstable cylinder $S_p^{m-i} \times I'$ before and after the deformations of cases 2, 3, 4. The Greek letters indicate values of $\varepsilon(x)\mu(x,w)$.

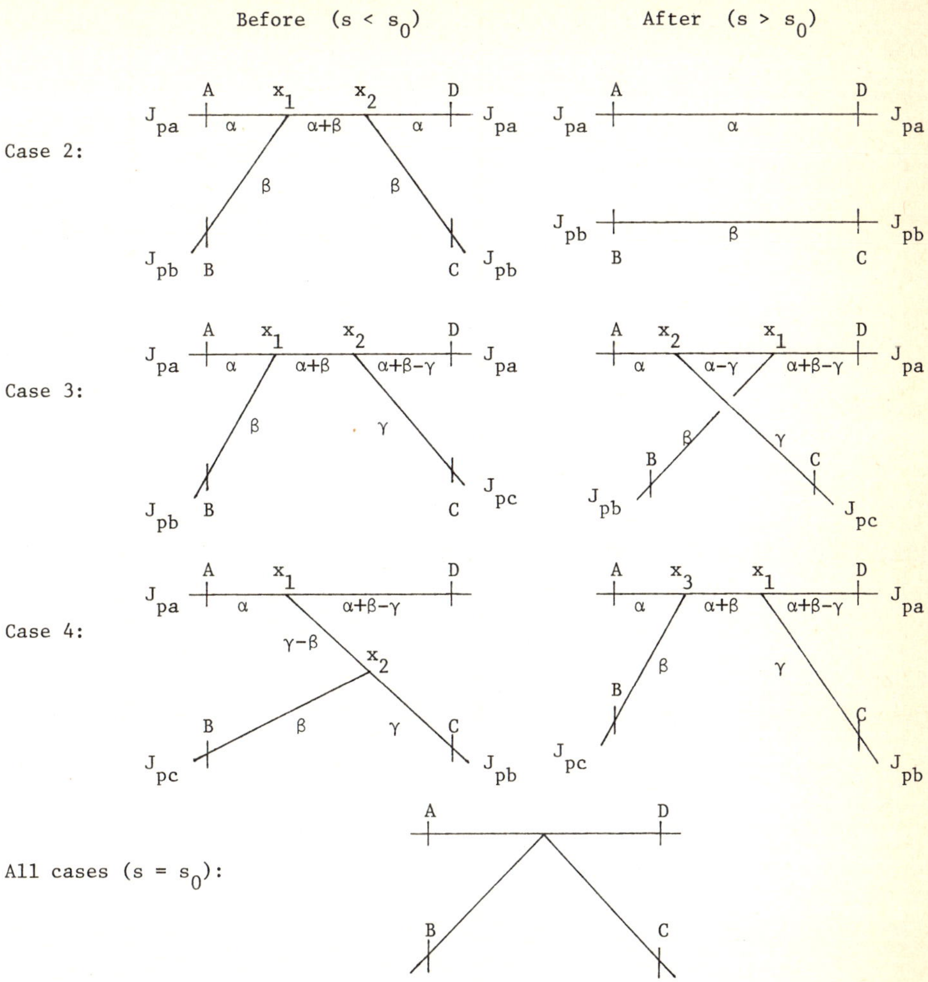

The function Ψ^w is continuous on $\overline{AD}$ but $\overline{\Psi}^w$ is given by taking paths from $\Psi^w(x_1)$ and $\Psi^w(x_2)$ to $\sigma_1(w)$. The paths are identical for the right and left limits at both points so the α contribution to the invariant 8.e(*) is not changed by the deformation. The function Ψ^w is not continuous on the path Bx_1x_2C but the compatible choice of 8.d.5 implies that the composition of the three loops $\overline{\Psi}^w$ (component of B), $\overline{\Psi}^w(\overline{x_1x_2})$, $\overline{\Psi}^w$ (component of C) is homotopic to the loop $\overline{\Psi}^w$ (component of B and C) in the "after" picture. Thus the β contribution does not change.

In case (3) the $Wh_1^+(G;A)$ invariant does not change if the choices 8.d.5 for the two commuting homotopies are chosen in such a way that they vary continuously with the deformation parameter s. The verification is trivial except when the two handle additions are of the form $x_{ab}(u)$ and $x_{ac}(v)$. This case is illustrated in 8.h.6.3 above. The α contribution to 8.e(*) is unchanged by the deformation for the same reason as in case (2). The β and γ contributions are unchanged by the continuous choice of 8.d.5.

Case (4) is the final and most important case. This is the only part of the proof which requires the assumption that σ be extendable to BG^3. Suppose that the two handle additions are $x_{ab}(u)$ on the left and $x_{bc}(v)$ on the right (i.e., the latter has a larger t-value.) When they cross the handle addition $x_{ac}(uv)$ is created. Let the homotopies in 8.d.5 for the handle additions $x_{ab}(u)$ and $x_{bc}(v)$ be chosen continuously with respect to s and let them be denoted by H_1 and H_2. Then all the contributions to the sum 8.e(*) are unchanged by the deformation except for those involving the new handle addition $x_{ac}(uv)$. This situation is illustrated in 8.h.6.4.

One choice of 8.d.5 for the new handle addition can be given as follows. Let ξ_1, ξ_2 and $\xi_3 = \xi_1\xi_2$ be the trajectories corresponding to the handle additions $x_{ab}(u)$, $x_{bc}(v)$, $x_{ac}(uv)$ respectively. Take a homotopy from $\gamma(z_a)\xi_3\gamma(z_c)^{-1}$ to $\sigma_1(uv)$ as follows. (See also figure 8.h.7.)

$$\begin{aligned}
\gamma(z_a)\xi_3\gamma(z_c)^{-1} &= \gamma(z_a)\xi_1\xi_2\gamma(z_c)^{-1} \\
&\simeq \gamma(z_a)\xi_1\gamma(z_b)^{-1}\sigma_1(v) && \text{(by } H_2\text{)} \\
&\simeq \sigma_1(u)\sigma_1(v) && \text{(by } H_1\text{)} \\
&\simeq \sigma_1(uv) && \text{(by } \sigma_2(u,v)\text{)}
\end{aligned}$$

Look now at figure 8.h.6.4. The α contribution to 8.e(*) is unchanged for the reason given in case (2). The γ contribution is unchanged by continuity of the choice of H_1. Thus the β contribution is the only thing that might change. In the "before" picture of 8.h.6.4 the β contribution is Bx_2x_1D. In the "after" picture it is Bx_3x_1D. We will use the notation y_1, y_2, y_3 to indicate x_1, x_2, x_3 considered as limit points of J_{pb}, J_{pc}, J_{pc} respectively. By continuity we then have the following equalities at $s = s_0$: $x_1 = x_3$, $x_2 = y_1$, $y_2 = y_3$, $\xi(y_1) = \xi(x_1)\xi_1$ and

Figure 8.h.7.

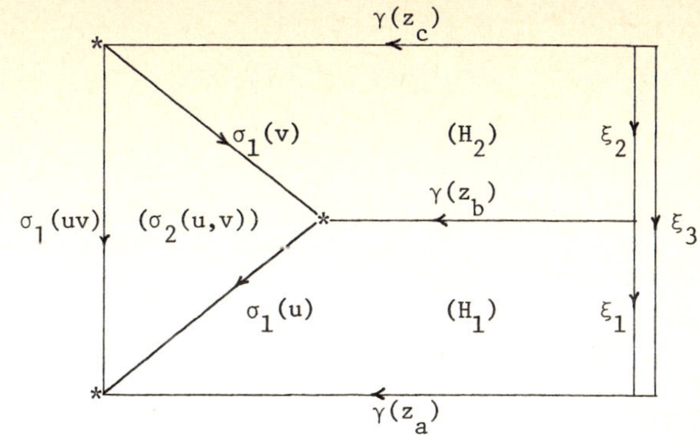

$\xi(y_2) = \xi(x_1)\xi_3 = \xi(x_1)\xi_1\xi_2 = \xi(y_1)\xi_2$.

We now examine the "before" picture. Let $x = u(x_1)$ and choose a homotopy H_0 from $\gamma(y_p)\xi(x_1)\gamma(z_a)^{-1}$ to $\sigma_1(x)$. We also use H_0 to indicate the corresponding null homotopy of $\overline{\Psi}^1(x_1) = \gamma(y_p)\xi(x_1)\gamma(z_a)^{-1}\sigma_1(x)^{-1}$. This gives a homotopy from $\overline{\Psi}^w(x_1) = \overline{\Psi}^1(x_1)\sigma_1(w)$ to $\sigma_1(w)$ for any $w \in G$, thus connecting $\overline{\Psi}^w(\overline{x_1 D})$ to $\sigma_1(w)$ in the prescribed way.

The prescribed method (8.e.5) of connecting $\Psi^w(y_1)$ to $\sigma_1(w)$ is given by taking the following null homotopy of $\Psi^1(y_1)$. (See figure 8.h.8.)

$$\begin{aligned}
\Psi^1(y_1) &= \gamma(y_p)\xi(y_1)\gamma(z_b)^{-1}\sigma_1(xu)^{-1} \\
&= \gamma(y_p)\xi(x_1)\xi_1\gamma(z_b)^{-1}\sigma_1(xu)^{-1} \\
&\simeq \gamma(y_p)\xi(x_1)\xi_1\gamma(z_b)^{-1}\sigma_1(u)^{-1}\sigma_1(x)^{-1} && \text{(by } \sigma_2(x,u)) \\
&\simeq \gamma(y_p)\xi(x_1)\gamma(z_a)^{-1}\sigma_1(x)^{-1} && \text{(by } H_1) \\
&\simeq * && \text{(by } H_0)
\end{aligned}$$

This will be taken as the null homotopy of $\Psi^1(x_2) = \Psi^1(y_1)$.

The prescribed method of connecting $\Psi^w(y_2)$ to $\sigma_1(w)$ is given by taking the null homotopy of $\Psi^1(y_2) = \gamma(y_p)\xi(x_1)\xi_1\xi_2\gamma(z_c)^{-1}\sigma_1(xuv)^{-1}$ given by H_0, H_1, H_2 and $\sigma_2(x,u)$, $\sigma_2(xu,v)$ which represent the back faces of the tetrahedron in 8.h.8. In the "after" picture the null homotopy of $\Psi^1(y_3) = \Psi^1(y_2)$ is given by H_0, H_1, H_2 and $\sigma_2(u,v)$, $\sigma_2(x,uv)$ which represent the two front faces of the tetrahedron in 8.h.8. Assuming that σ is extendable to BG^3 these two homotopies are homotopic thus completing the proof of 8.h.5. □

Figure 8.h.8.

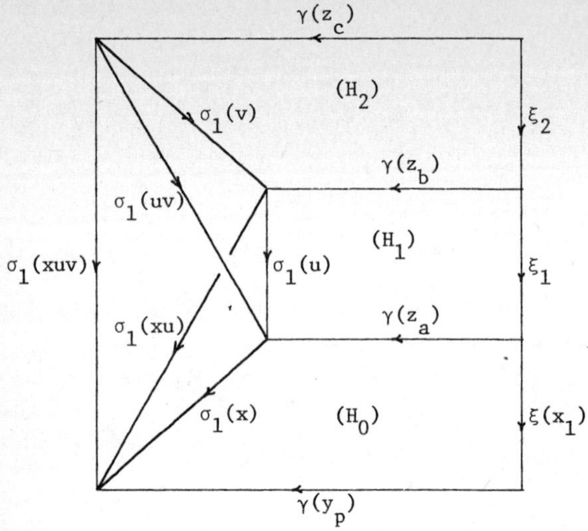

8.i. Proof of the main theorem.

Suppose now that $\sigma\colon BG^2 \to X$ is not extendable to BG^3. Then to each triple $(x,u,v) \in G^3$ we get a mapping $\partial\Delta^3 \to X$ given by σ_1 on the 1-simplices and σ_2 on the 2-simplices as indicated by the tetrahedron in 8.h.8. Taking the upper left corner as the base point this gives an element of $A = \pi_2 X$ which we denote by $h(x,u,v)$. By the standard argument h is a normalized 3-cocycle representing the first Postnikov invariant of X: $[h] = k_1(X) \in H^3(G;A)$. (With the appropriate sign and orientation conventions.)

Consider the proof of 8.h.5. The β contribution in 8.h.6.4 changes from the back null homotopy of 8.h.8 to the front null homotopy. This means that the β contribution changes by $h(x,u,v) \otimes w$. Since this counts $(-1)^i \varepsilon(y_2)\mu(y_2,w)$ times in the sum 8.e(*) the change in this sum is given by

$$\begin{aligned}
&(-1)^i \sum_{p,w,x_1 \in J_{pa}} \varepsilon(y_2)\mu(y_2,w)h(u(x_1),u,v) \otimes w \\
= \; &(-1)^i \sum_{p,x_1} \varepsilon(y_2)h(u(x_1),u,v) \otimes u(y_2)\mu(y_2) \\
= \; &(-1)^i \sum_{p,x_1} \varepsilon(x_1)h(u(x_1),u,v) \otimes u(x_1)uvs_{cp} \\
= \; &(-1)^i \sum_{p,x_1} h_\ell(\varepsilon(x_1)u(x_1) \otimes u \otimes v)s_{cp} \\
= \; &(-1)^i \sum_{p} h_\ell(r_{pa} \otimes u \otimes v)s_{cp}
\end{aligned}$$

where $h_\ell = h[G]$ is the linearization of h defined in 6.d.

Proof of 8.a.2: Take an element (x,y) in the kernel cf $j_1 + j_2\colon Wh_1^+(G;A) \oplus Wh_1^+(G;\mathbb{Z}_2) \to \pi_0 C(M)$. Then this element is represented by a lens shaped family (f_t,v_t) with no handle additions. Since $[f_t]$ is trivial in $\pi_0 C(M)$ there is a two parameter family (f_{ts},v_{ts}) which is a null homotopy of (f_t,v_t). By [8] we may assume that (f_{ts},v_{ts}) is a lens-shaped null homotopy. The handle addition pattern of (f_{ts},v_{ts}) represents an element P of $K_3\mathbb{Z}[G]$ in a way described in [7]. The computation above together with formula 6.h shows that $x = (-1)^i \chi_3^+(k_1(M),P)$. The result of [9] is that $y = (-1)^i \chi_G(P)$.

Conversely, given any element P of $K_3\mathbb{Z}[G]$ one can construct a lens-shaped homotopy (f_{ts},v_{ts}) which has a handle addition pattern corresponding to P and thus represents a lens shaped null homotopy of a family (f_t,v_t) representing $(-1)^i$ times $(\chi_3^+(k_1(M),P),\chi_G(P))$ in $Wh_1^+(G;A) \oplus Wh_1^+(G;\mathbb{Z}_2)$. □

8.j. $\sigma_*\colon \pi_0 C(M) \to Wh_1^+(G;A)$ is well-defined.

In this section we complete the proof of 8.a.1 (and thus of 8.a.3) by showing that $\sigma_*(f_t,v_t)$ depends only on the homotopy class of f_t and thus σ_* is a well-defined retraction of j_1. The argument is more of less the same as the one given in [5] with the additional assumption that σ_2 is extendable to BG^3.

Suppose that (f_{t0},v_{t0}) and (f_{t1},v_{t1}) are two lens-shaped families for $M \times I$ representing the same element of $\pi_0 C(M)$. Then we may assume that there is a homotopy (f_{ts},v_{ts}), $s \in I$, from (f_{t0},v_{t0}) to (f_{t1},v_{t1}) and a finite critical set $S \subset I$ such that for a fixed $s \notin S$ the family (f_{ts},v_{ts}) is generic and satisfies the following two conditions. First f_t has critical points only in the middle indices i, $i+1$. Second (f_{ts},v_{ts}) has only $i{+}1/i{+}1$ handle additions. The justification for this assumption will be explained later.

Let $s_0 \in S$. Then at $s = s_0$ one of the following things happens.

(1)-(4) Same as 8.h(1)-(4).

(5) A death and birth point are cancelled. (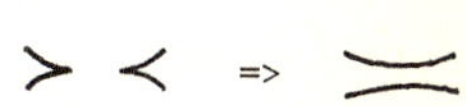)

(6) Two Morse points are cancelled at t_0 creating a death point at $t_0 - \varepsilon$ and

a birth point at $t_0 + \varepsilon$. (this is the reverse of deformation 5.)

In order for deformations 8.h(1)-(4) to make complete sense here one must make the choices 8.d.1 - 3. These choices may be taken to vary continuously through the first four deformations but they are obviously discontinuous for deformations 5,6.

To show that $\sigma_*(f_{t0},v_{t0}) = \sigma_*(f_{t1},v_{t1})$ we must define $\sigma_*(f_{ts},v_{ts})$ for every s not in the critical set S and show that it is the same for all s. To define $\sigma_*(f_{ts},v_{ts})$ one must make the choices 8.d.1 - 5 for the family (f_{ts},v_{ts}) with the modification that $\gamma(z)$ need not be continuous at birth-death points. The definition of $\sigma_*(f_{ts},v_{ts},c_s)$ works as before except for the proof of 8.f.2 which must be modified to allow the $\gamma(y)$'s and $\gamma(z)$'s to vary independently. This can be accomplished by taking the same proof and changing the notation: wherever the symbol "u_q" appears replace it by "v_q".

The proof of 8.h.5 works here to show that $\sigma_*(f_{ts},v_{ts})$ does not change at deformations of type 1 - 4. For deformation 6 take one choice of c for $s < s_0$ and extend it through s_0 in a continuous manner except for the numbering (8.d.2). The invariant $\sigma_*(f_{ts},v_{ts})$ is obviously unchanged. By symmetry deformation 5 does not change the invariant.

We now give a brief explanation for the assumptions made about the deformation (f_{ts},v_{ts}). It was shown in [8] that if two lens-shaped families (f_{t0},v_{t0}) and (f_{t1},v_{t1}) were homotopic and had the same $K_2\,\mathbb{Z}[G]$ invariant then suitable "suspensions" of both of them were homotopic by a lens-shaped homotopy. If the K_2-invariants are not the same the difference between the two K_2-invariants must be an element, say X, of $\Omega_2^{fr}(BG)$ since f_{t0} and f_{t1} have the same $Wh_2(G)$ invariant. However, elements of $\Omega_2^{fr}(BG)$ are very easy to describe and given any such element, say Y, one can easily construct a deformation of (f_{t0},v_{t0}) to a lens-shaped family $(f_{t\frac{1}{2}},v_{t\frac{1}{2}})$ whose K_2-invariant differs from that of (f_{t0},v_{t0}) by Y. Take $Y = X$ and we have (after suspension) that $(f_{t\frac{1}{2}},v_{t\frac{1}{2}})$ is lens-shaped homotopic to (f_{t1},v_{t1}).

Chapter 9. A topological version of the counterexample.

Theorem 8.a.2 together with the algebraic counterexample in 7.b produces a counterexample to the original statement of Hatcher and Wagoner. However the proofs involved the functor K_3 and its relationship to two parameter families of functions. For the sake of finding a counterexample this is not necessary. We construct in this chapter a counterexample which uses only K_2, the elementary example in 7.a and 8.g.

9.a. Construction of the manifold M.

In 7.a we gave an example of a group G, a G-module A on which G acts trivially and elements $\kappa \in H^2(G;A)$ and $s \in K_2\mathbb{Z}[G]$ such that $\chi_2^+(\kappa,s) \neq 0$ in $Wh_1^+(G;A)$. In our example G and A also satisfy the condition that BG and $B^2A = K(A,2)$ are 2-equivalent to finite 3-complexes. Given any G, A, κ, s satisfying these conditions we shall construct a compact manifold M with $\pi_1 M = G \times \mathbb{Z}$, $\pi_2 M = A$, $k_1(M) = \kappa \times \xi$ (ξ being the generator of $H^1(\mathbb{Z};\mathbb{Z})$) and show using 8.g that the image of $\chi_2^+(\kappa,s)$ in $Wh_1^+(G \times \mathbb{Z}; A)$ represents zero in $\pi_0 C(M)$.

Let $h: BG \times B^2A \to BG \times B^2A$ be the map given by $h(x,y) = (x,\mu(\bar{\kappa}(x),y))$ where $\bar{\kappa}: BG \to B^2A$ is the classifying map for κ and μ is the H-space multiplication map for $B^2A \simeq \Omega K(A,3)$. Let X be a finite 3-complex which is 2-equivalent to $BG \times B^2A$. Then there is a cellular map $H: X \to X$ corresponding to h.

Let Y be the quotient of $X \times I$ by the identification $(x,1) \sim (H(x),0)$. By embedding Y in a finite Euclidean space and taking a closed regular neighborhood we will get a compact manifold M which looks like a product near $X \times 1/2$. More precisely M contains a codimension one submanifold N such that both N and $M - N$ are homotopy equivalent to N and N has a tubular neighborhood which we identify with $N \times [-1,2]$. Let $*$ be the base point of N.

9.b. Construction of a null homotopy of $\chi_2^+(\kappa,s)$ in $\pi_0 C(M)$.

If $f: M \times I \to I$ is a smooth map the <u>support</u> of f will mean the closure of the set of all $x \in M$ such that $f(x,y) \neq y$ for some $y \in I$. Note that if W is a codimension-0 submanifold of M then every element f of $F(W)$ extends uniquely to an element $\bar{f}$ of $F(M)$ with support in W. If (f_t,v_t) is a family of functions with vector fields on $M \times I$ the <u>support</u> of (f_t, v_t) will mean the closure of the set of all points $x \in M$ such that for some $y \in I$ and $t \in I$ we have either $f_t(x,y) \neq y$ or $v_t(x,y) \neq (0,-1)$.

Let (f_t^1, v_t^1, c^1) be a lens-shaped family on $N \times [-1,0] \times I$ together with a choice c^1 for 8.d.1 - 5 such that the handle additions of (f_t^1, v_t^1, c^1) are x_1, $x_2, \dots, x_n$ with product equal to s in $St(\mathbb{Z}[G])$. Let (f_t^2, v_t^2, c^2) be a lens-shaped family on $N \times [1,2] \times I$ together with a choice c^2 for 8.d.1 - 5 such that the handle additions on (f_t^2, v_t^2, c^2) are $x_n^{-1}, \dots, x_1^{-1}$ with product s^{-1} in $St(\mathbb{Z}[G])$. We shall assume that the critical points of f_t^1 occur at $t < 1/2$ and that those of f_t^2 occur at $t > 1/2$.

Let $\sigma^1 \colon BG^2 \to N = N \times 0$ be any map which is extendable to BG^3 and which induces the identity on π_1. Then we have

<u>Lemma 9.b.1</u>. $\sigma^1{}_*(\tilde{f}_t^2, \tilde{v}_t^2) = \sigma^1{}_*(\overline{f_t^2}, \overline{v_t^2}) \pm \chi_2{}^+(\kappa, s)$ where $(\tilde{f}_t^2, \tilde{v}_t^2)$ and $(\overline{f_t^2}, \overline{v_t^2})$ are the extensions of (f_t^2, v_t^2) to $M - N \times (0,1)$ and $N \times [-1,2]$ respectively.

<u>Proof</u>: Let $(\overline{\overline{f_t^2}}, \overline{\overline{v_t^2}})$ be the extension of (f_t^2, v_t^2) to all of M. Let $\sigma^2 \colon BG^2 \times S^1 \to M$ be a map which extends σ^1 and which induces the identity on π_1. Then we have

$$\sigma^1{}_*(\overline{f_t^2}, \overline{v_t^2}) = \overline{\overline{\sigma^2}}{}_*(\overline{\overline{f_t^2}}, \overline{\overline{v_t^2}}, \bar{\gamma})$$

$$\sigma^1{}_*(\tilde{f}_t^2, \tilde{v}_t^2) = \sigma^2{}_*(\overline{\overline{f_t^2}}, \overline{\overline{v_t^2}}, \tilde{\gamma})$$

if the paths $\bar{\gamma}$ are chosen to lie in $N \times [-1,2] \times I \times I$ and the paths $\tilde{\gamma}$ are chosen to lie in the complement of $N \times (0,1) \times I \times I$. (Since σ^2 is not extendable to $BG^3 \times S^1$, $\sigma^2{}_*$ depends on the choice of γ but not on the other choices in c as we showed in 8.f.)

Now take the base point of M and move it around $\sigma_1^2(S^1)$. The affect is to conjugate all terms by the generator of $\mathbb{Z}$. Since this commutes with everything the terms are unchanged. However $\bar{\gamma}$ changes to $\tilde{\gamma}$ and σ^2 changes to its conjugate σ^3 which differs from σ^2 by the difference obstruction $\kappa \times 1 \in H^2(G \times \mathbb{Z}; A)$. Thus by 8.g we have

$$\sigma^2{}_*(f_t^2, v_t^2, \bar{\gamma}) = \sigma^3{}_*(f_t^2, v_t^2, \tilde{\gamma}) = \sigma^2{}_*(f_t^2, v_t^2, \tilde{\gamma}) \pm \chi_2{}^+(\kappa, s) \qquad \square$$

Theorem 9.b.2. There exist two lens-shaped families without handle additions on $M \times I$ which are homotopic to each other but have different $Wh_1^+(G \times \mathbb{Z};A)$ invariants.

Proof: Let (f_t^3, v_t^3) be the lens-shaped family on $M \times I$ given by

$$(f_t^3(x,y), v_t^3(x,y)) = \begin{cases} (y,(0,-1)) & \text{if } x \notin N \times ([-1,0] \cup [1,2]) \\ (f_{2t}^1(x,y), v_{2t}^1(x,y)) & \text{if } x \in N \times [-1,0], \quad t \leq 1/2 \\ (f_1^1(x,y), v_1^1(x,y)) & \text{if } x \in N \times [-1,0], \quad t \geq 1/2 \\ (y,(0,-1)) & \text{if } x \in N \times [1,2], \quad t \leq 1/2 \\ (f_{2t-1}^2(x,y), v_{2t-1}^2(x,y)) & \text{if } x \in N \times [1,2], \quad t \geq 1/2 \end{cases}$$

We make the assumption that $(f_t^1(x,y), v_t^1(x,y))$ is independent of t for t sufficiently close to 1 and that $(f_t^2(x,y), v_t^2(x,y)) = (y,(0,-1))$ for t sufficiently close to 0. The two families of the theorem will both be homotopic to (f_t^3, v_t^3).

Let (f_t^4, v_t^4) be the restriction of (f_t^3, v_t^3) to $N \times [-1,2]$. We shall deform (f_t^4, v_t^4) to a lens-shaped family without handle additions. Let $p_1, p_2, \ldots, p_k$ be the death points of (f_t^4, v_t^4) with $t < 1/2$ where the numbering is given by c^1. Let $q_1, \ldots, q_k$ be the birth points of (f_t^4, v_t^4) with $t > 1/2$ numbered according to c^2. Let all the critical points be oriented (8.d.3) according to c^1 and c^2. Now perform the deformation which for each $a = 1, 2, \ldots, k$ cancels the birth point q_a with the death point p_a along a path homotopic to $\gamma_2(q_a)^{-1}\gamma_1(p_a)$ making sure the orientations agree. The resulting family which we shall call (f_t^5, v_t^5, c^5) will have handle additions $x_1, x_2, \ldots, x_n, x_n^{-1}, \ldots, x_1^{-1}$ which can be cancelled in pairs. This gives a family (f_t^6, v_t^6) without handle additions with $Wh_1^+(G;A)$ invariant equal to $\sigma^1{}_*(f_t^6, v_t^6) = \sigma^1{}_*(f_t^5, v_t^5)$ (by 8.h.5 case 1.) $= \sigma^1{}_*(f_t^1, v_t^1) + \sigma^1{}_*(\overline{f_t^2}, \overline{v_t^2})$.

Similarly let (f_t^7, v_t^7) be the restriction of (f_t^3, v_t^3) to $M - N \times (0,1)$. For each $a = 1, 2, \ldots, k$ cancel q_a with p_a along a path homotopic to $\gamma_2(q_a)^{-1}\xi^{-1}\gamma_1(p_a)$ making sure the orientations agree, where ξ is a loop representing the positive generator of $\mathbb{Z}$. The resulting family which we call (f_t^8, v_t^8, c^8) will have handle additions $x_1, x_2, \ldots, x_n, x_n^{-1}, \ldots, x_1^{-1}$ since $\mathbb{Z}$ commutes with G in $G \times \mathbb{Z}$. Of course c^8 is the extension of c^1 but not of c^2. Cancel the handle additions as before and we get a family (f_t^9, v_t^9) without handle additions with $Wh_1^+(G;A)$ invariant equal to $\sigma^1{}_*(f_t^9,v_t^9) = \overline{\sigma^1{}_*(f_t^8,v_t^8)}$ (by 8.h.5 case 1) $=$ $\sigma^1{}_*(f_t^1,v_t^1) + \sigma^1{}_*(\tilde{f}_t^2,\tilde{v}_t^2) = \sigma^1{}_*(f_t^1,v_t^1) + \sigma^1{}_*(\overline{f_t^2},\overline{v_t^2}) \pm \chi_2{}^+(\kappa,s)$ (by the lemma.) □

References

[1] J. Cerf, La stratification naturelle des espaces de fonctions differéntiable réelles et le théorème de la pseudo-isotopie, Publ. Math. I.H.E.S. 36 (1970).

[2] S.M. Gersten, K_3 of a ring is H_3 of the Steinberg group, Proc. Amer. Math. Soc. 37 (1973), 366-368.

[3] A. Hatcher and J. Wagoner, Pseudo-isotopies of compact manifolds, Asterisque 6 (1973), Soc. Math. de France, Paris.

[4] A. Hatcher and J. Wagoner, Pseudo-isotopies of nonsimply connected manifolds and the functor K_2, Asterisque 6 (1973), 8-238, Soc. Math. de France, Paris.

[5] A. Hatcher, The second obstruction for pseudo-isotopies, Asterisque 6 (1973), 239-275, Soc. Math. de France, Paris.

[6] A. Hattori, Rank element of a projective module, Nagoya J. Math. 25 (1965), 113-120.

[7] M. Hirsch, Differential topology, Grad. Texts in Math. 33, Springer-Verlag, New York, 1976.

[8] K. Igusa, Thesis, Princeton University, 1979.

[9] K. Igusa, The generalized Grassmann invariant, (preprint).

[10] J-L. Loday, K-théorie algébrique et représentations de groupes, Ann. Sci. Ecole Norm. Sup. (4) 9 (1976), 309-377.

[11] S. MacLane, Homology, Springer-Verlag, New York, 1967.

[12] J. Milnor, Introduction to algebraic K-theory, Ann. of Math. Studies No. 72 (1971), Princeton University Press, Princeton.

[13] J. Stallings, Centerless groups - an algebraic formulation of Gottlieb's theorem, Topology 4 (1965), 129-134.

A NOTE ON EXCISION FOR K_2

Wilberd van der Kallen

Summary. We consider Ruth Charney's excision theorem ([C]) for the special case of K_2. We describe a different proof for this special case. It follows from this proof that if one requires excision only on the K_2 level, one may weaken the condition somewhat. (Recall from [S] that the same holds if one requires excision only on the K_1 level). We also give a counterexample to a stronger statement: We give an example of an ideal J with $J = J^2$ for which excision fails.

1.1. Let J be an associative ring without unit and let m be a positive integer. Following Charney we say that J is an m-excision ideal for K_2 if the following holds: For any associative ring with unit A that contains J as a (2-sided) ideal, the map

$$K_2(\mathbb{Z}\oplus J,J)\otimes\mathbb{Z}[\tfrac{1}{m}] \to K_2(A,J)\otimes\mathbb{Z}[\tfrac{1}{m}]$$

is an isomorphism.(1)

1.2. Recall that $J\otimes_J J$ is the quotient of $J\otimes_{\mathbb{Z}} J$ by the subgroup generated by the elements $xy\otimes z - x\otimes yz$ with $x,y\ z \in J$. Multiplication in J defines a map

$$\mu : J\otimes_J J\otimes\mathbb{Z}[\tfrac{1}{m}] \to J\otimes\mathbb{Z}[\tfrac{1}{m}]$$

Theorem. (cf.[C]). If μ is a bijection then J is an m-excision ideal for K_2.

1.3. Proof. Assume the hypothesis of the theorem. Let J be an ideal in A. We view $K_2(A,J)$ as the kernel of $St(A,J) \to GL(J)$ where $St(A,J)$ is the group $St(A^{(\infty)},J,A^{(\infty)})$ of the appendix to [V]. It follows from [G - W] that

$$K_2(\mathbb{Z}\oplus J,J)\otimes \mathbb{Z}\,[\tfrac{1}{m}] \to K_2(A,J)\otimes \mathbb{Z}\,[\tfrac{1}{m}]$$

is surjective. Remains to show that the kernel of $K_2(\mathbb{Z}\oplus J,J) \to K_2(A,J)$ is m-torsion. Fix an element α of this kernel. Put

$$J_s = \{j \in J \mid m^s j \in J^2\}$$

so that J is the union of the increasing sequence of ideals J_s. Choose s so that α comes from $K_2(\mathbb{Z}\oplus J_s,J_s)$ and has trivial image in $St(A,J_s)$. Define $\overline{St}(\mathbb{Z}\oplus J)$ to be the quotient of $St(\mathbb{Z}\oplus J)$ by the m-torsion subgroup of $K_2(\mathbb{Z}\oplus J)$. The theorem follows from:

1.4. <u>Lemma</u>. There is a set theoretical map $\tau : St(A,J_s) \to \overline{St}(\mathbb{Z}\oplus J)$ such that the composite with $K_2(\mathbb{Z}\oplus J_s,J_s) \to St(A,J_s)$ equals m^s times the natural map $K_2(\mathbb{Z}\oplus J_s,J_s) \to \overline{St}(\mathbb{Z}\oplus J)$.

<u>Proof of Lemma</u>. Let F be the free group on G(F), where G(F) is the generating set used in the definition of $St(A^{(\infty)},J_s,A^{(\infty)})$, and let $R = \ker(F \to St(A,J_s))$. For each $x \in G(F)$, the m^s-th power of its matrix image lies in the image $E(\mathbb{Z}\oplus J)$ of $St(\mathbb{Z}\oplus J)$. Replacing m by m^s we may and shall further assume $s = 1$. Given n distinct elements $x_1,\dots,x_n$ in G(F), choose an integer N such that the matrix images $mat(x_i)$ all lie in $GL_N(J) \subseteq GL(J)$. Choose $y_i \in \overline{St}(\mathbb{Z}\oplus J)$ with matrix image

$$\begin{pmatrix} mat(x_i) & 0 & 0 \\ 0 & 1_{iN} & 0 \\ 0 & 0 & mat(x_i)^{-1} \end{pmatrix}$$

and define a homomorphism $\langle x_1,\ldots,x_n\rangle \to \overline{St}(\mathbb{Z}\oplus J)$ sending x_i to y_i. Restrict this homomorphism to the commutator subgroup of $\langle x_1,\ldots,x_n\rangle$. This restriction φ is characterised by the property:

Let $x,x' \in \langle x_1,\ldots,x_n\rangle$, $y,y' \in \overline{St}(\mathbb{Z}\oplus J)$, $M \in \mathbb{N}$, such that the matrix image of y is

$$\begin{matrix} N \updownarrow \\ M \updownarrow \end{matrix} \begin{pmatrix} \mathrm{mat}(x) & 0 \\ 0 & P \end{pmatrix}$$

for some $P \in GL_M(J)$ and the matrix image of y' is

$$\begin{matrix} N \updownarrow \\ M \updownarrow \\ \\ \end{matrix} \begin{pmatrix} \mathrm{mat}(x') & 0 & 0 \\ 0 & 1_M & 0 \\ 0 & 0 & Q \end{pmatrix}$$

for some $Q \in GL(J)$. Then $\varphi([x,x']) = [y,y']$.

(Compare the construction of Milnor's pairing in [M] §8 and use that we have factored out m-torsion in $K_2(\mathbb{Z}\oplus J)$, including the Steinberg symbols $\{\mathrm{mat}(x_i),\mathrm{mat}(x_i)\}$.)

Using this characterisation we extend φ to all of $[F,F]$ by varying $\{x_1,\ldots,x_n\}$. Let H be the free subgroup of F generated by m-th powers of elements of $G(F)$. For each $X(v,j,w)$ in $G(F)$, choose p_i, $q_i \in J$ so that $\sum_i p_iq_i = mj$ (recall $s = 1$) and put

$$\psi(X(v,j,w)^m) = \prod_i X(vp_i,1,q_iw) \in \overline{St}(\mathbb{Z}\oplus J).$$

This defines a homomorphism $\psi : H \to \overline{St}(\mathbb{Z}\oplus J)$. It agrees with φ on $H \cap [F,F] = [H,H]$. We extend φ to $H[F,F]$ by putting $\varphi(xy) = \psi(x)\varphi(y)$ for $x \in H$, $y \in [F,F]$. Define $\tau : F \to \overline{St}(\mathbb{Z}\oplus J)$ by $\tau(x) = \varphi(x^m)$.

One shows that $\tau(x\ r\ y) = \tau(xy)\ \tau(r)$ for $x,y \in F$, $r \in R$.

Thus if τ annihilates R, τ factors through $St(A,J_s)$ and the lemma easily follows. To show that τ annihilates R indeed, one treats each of the defining relations listed in the appendix to [V]. To deal with the third, for instance, recall the hypothesis of the theorem and use that for $X(v,j,w) \in G(F)$ there is a homomorphism

$$J \otimes_J J \to St(\mathbb{Z} \oplus J)$$

sending $p \otimes q$ to $X(vp,1,qw)$.

2. The counterexample. It is commutative. Put $R_r = \mathbb{Z}[T_r,\varepsilon]/(T^{2^r},\varepsilon^2)$. Embed R_r into R_{r+1} by sending T_r to T^2_{r+1}, ε to ε. Let $R = \lim_{\substack{\to \\ r}} R_r$, $J = \lim_{\substack{\to \\ r}} J_r$, with $J_r = T_r R_r$. Clearly $J = J^2$ so that

$J \otimes_J J \to J$ is surjective. Nevertheless $K_2(\mathbb{Z} \oplus J, J) \to K_2(R,J)$ is not injective: Consider $\alpha = \langle T_1, \varepsilon T_1 \rangle \langle \varepsilon T_1, -T_1 \rangle \in K_2(\mathbb{Z} \oplus J, J)$. Its image in $K_2(R,J)$ vanishes, by an easy computation. But suppose α vanishes. Then $\langle T_1, T_1\varepsilon \rangle \langle \varepsilon T_1, -T_1 \rangle$ must vanish in $K_2(\mathbb{Z} \oplus J_r)$ for some r. However, recall that we have a Chern class $K_2(\mathbb{Z} \oplus J_r) \to \Omega^2_{\mathbb{Z} \oplus J_r}$ sending $\langle a,b \rangle$ to $\pm(1 \pm ab)^{-1} da \wedge db$. (The reader may choose conventions and then determine the correct signs). Straightforward computation shows that the image in $\Omega^2_{\mathbb{Z} \oplus J_r}$ of our element is non-zero (This image is not even torsion).

(1) Charney has now replaced $\mathbb{Z}[\frac{1}{m}]$ by an arbitrary subring of $\mathbb{Q}$. Our theorem generalizes similarly.

References

[C] R.M. Charney, A note on excision in K-theory. These proceedings.

[G-W] S.C. Geller and C.A. Weibel, $K_1(A,B,I)$. To appear in J. reine angew. Math.

[M] J. Milnor, Introduction to Algebraic K-theory, Annals of Math. Studies, Vol. 72. Princeton: University Press 1971

[S] R.G. Swan, Excision in Algebraic K-theory, J. Pure and Appl. Algebra 1 (1971), 221-252.

[V] T. Vorst, Localization of the K-theory of polynomial extensions, Math. Ann. 244, 33-53 (1979).

The K-Theory of Triangular Rings and Orders

M. E. Keating

Let I be a twosided projective ideal of a ring R and let

$$A = \begin{bmatrix} R & \cdots & R \\ I & & \vdots \\ \vdots & & \vdots \\ I & \cdots \; I & R \end{bmatrix}$$

be a 'triangular tiled' ring of $n \times n$ matrices over R; that is, A is the subring of the full matrix ring over R which consists of all those matrices having subdiagonal entries in I. When $I = 0$, A is the ring of upper triangular matrices. The main result of this paper is that

$$K_*(A) \cong K_*(R) \oplus (n-1)K_*(R/I), \quad * = 0,1,\ldots .$$

We start by considering the K-theory of a trivial extension T of a ring D by a bimodule U when U has finite homological dimension as left T-module. The existence of a transfer map from $K_*(D)$ to $K_*(T)$ allows us to show that the split surjection of $K_*(T)$ to $K_*(D)$ is an isomorphism. In the case T is the ring of $n \times n$ upper triangular matrices over a ring S, we obtain

$$K_*(T) \cong n\,K_*(S).$$

An analysis of the evident surjection of A onto a triangular ring T leads to the main result.

When R is an order over an integral domain, we also show that

$$SK_*(A) \cong SK_*(R) \oplus (n-1)K_*(R/I).$$

As an application, we show that if $\mathcal{o}$ is the ring of p-adic integers and π is a metacyclic group of order pq, then $K_3(\mathcal{o}\pi)$ has $(q-1)K_3(Z/p)$ as a quotient group.

Our arguments use Quillen's definition of K-theory in terms of exact categories of modules as given in [Q]. We will exploit the fact that if W is a X-Y-bimodule (notation ${}_XW_Y$) over the rings X and Y, then the functor $W \otimes_Y -$ converts left Y-modules into left X-modules, and, in favourable circumstances, will be an exact functor on various categories of Y-modules. The exact categories of importance to us are $\mathbb{P}(Y)$, the category of all finitely generated projective left Y-modules, and $\mathbb{P}_\infty(Y)$, the category of all those finitely generated left Y-modules which have a finite resolution whose terms belong to $\mathbb{P}(Y)$. The functor $W \otimes -$ will always be exact on $\mathbb{P}(Y)$, but some further condition is required for exactness on $\mathbb{P}_\infty(Y)$.

The basic facts about modules which we need can all be found in [AF]. We adopt the convention that homomorphisms are written opposite scalars. Thus, given a left X-module W (notation ${}_XW$), the endomorphism ring $Y = \mathrm{End}({}_XW)$ acts on the right of W, giving rise to a bimodule structure ${}_XW_Y$.

I wish to thank the referee for pointing out that the arguments in this paper are not as obvious as they first appeared to me.

1.1. Let D be a ring and U a D-D-bimodule. The trivial extension $T = D \ltimes U$ is the set $D \times U$ with the obvious addition, and multiplication given by

$$(d,u).(d',u') = (dd', ud'+du').$$

Clearly U is an ideal of T with $U^2 = 0$.

We assume that, when regarded as left T-module, U belongs to $\mathbb{P}_\infty(T)$. Although the T-structure of U is in principle determined by its D-structure, it is not straightforward to specify conditions which are equivalent to our hypothesis. This question is discussed in [FGR, Chapter IV] and [PR].

1.2. Since D is both a subring and quotient ring of T, there are exact functors

$$s = T\otimes_D - : \mathbb{P}(D) \to \mathbb{P}(T)$$

and

$$q = D\otimes_T - : \mathbb{P}(T) \to \mathbb{P}(D),$$

which induce homomorphisms

$$s_* : K_*(D) \to K_*(T)$$

and $$q_* : K_*(T) \to K_*(D).$$

Clearly, $$q_*s_* = 1 \quad \text{on} \quad K_*(D).$$

There are also inclusions

$$i : \mathbb{P}(T) \to \mathbb{P}_\infty(T)$$

and $$j : \mathbb{P}(D) \to \mathbb{P}_\infty(T).$$

By Quillen's resolution theorem [Q, Theorem 3, Corollary 1]

$$i_* : K_*(T) \to K_*(\mathbb{P}_\infty(T))$$

is an isomorphism.

We define the transfer t_* by

$$t_* = (i_*)^{-1}j_* : K_*(D) \to K_*(T).$$

1.3. There is an exact sequence of exact functors

$$0 \to U\otimes_T - \to T\otimes_T - \to D\otimes_T - \to 0$$

from $\mathbb{P}(T)$ to $\mathbb{P}_\infty(T)$ which by [Q, Theorem 2, Corollary 1] gives

$$i_* = (U\otimes_T -)_* + j_*q_*.$$

Put $$u_* = i_*^{-1}(U\otimes_T -)_*$$

Then $$1 = u_* + t_*q_* \text{ on } K_*(T);$$

thus $$t_* = (1-u_*)s_*$$

and $$t_*q_* = s_*q_* - u_*s_*q_*.$$

But $i_*u_*s_*q_*$ is induced by the functor $U\otimes_T T\otimes_D D\otimes_T -$, which is equivalent to $U\otimes_T -$; so $u_*s_*q_* = u_*$ and

$$s_*q_* = 1 \text{ on } K_*(T).$$

We have proved the following result.

<u>Theorem A</u>.

The homomorphisms s_* *and* q_* *are mutually inverse isomorphisms between* $K_*(D)$ *and* $K_*(T)$.

1.4. Consideration of the Whitehead group $K_1(T)$ shows that when Theorem A holds the D-D-bimodule U must be highly assymetric. Suppose for simplicity that D is local (i.e., D/rad D is Artinian simple). By [Sw, Theorem 2.1], $K_1(T,U)$ can be identified as the additive group U modulo the subgroup W generated by all expressions du-ud, d in D and u in U. Since

$$K_1(T) \cong K_1(D) \oplus K_1(T,U),$$

a consequence of Theorem A is that U = W.

An example where the Theorem fails is that of the ring of dual numbers over a commutative ring D; then $T = D \ltimes D$, $K_1(T,U) \cong U$ and D has infinite homological dimension over T.

2.1. Let R and S be rings and put $D = R \oplus S$. An R-S-bimodule U is naturally a D-D-bimodule and the trivial extension $T = D \ltimes U$ can be interpreted as the upper triangular matrix ring $\begin{bmatrix} R & U \\ 0 & S \end{bmatrix}$. Since $T = \begin{bmatrix} R \\ 0 \end{bmatrix} \oplus \begin{bmatrix} U \\ S \end{bmatrix}$ as left T-module, R is projective as left T-module. Thus, if U belongs to $\mathbb{P}_\infty(R)$ it is also in $\mathbb{P}_\infty(T)$, and by Theorem A, there is an isomorphism

$$q_* : K_*(T) \to K_*(R) \oplus K_*(S).$$

2.2. Suppose further that U is finitely generated projective as right S-module. Then $\begin{bmatrix} U \\ S \end{bmatrix}$, which is the right S-module $U \oplus S$ rewritten, is a projective generator in the category of finitely generated right S-modules and hence S is Morita equivalent to

$M = \text{End} \begin{bmatrix} U \\ S \end{bmatrix}_S$ [AF, Chapter 6].

Write $N = \text{End}(U_S)$ and $U^* = \text{Hom}(U_S, S_S)$. In the evident matrix notation,

$$M = \begin{bmatrix} N & U \\ U^* & S \end{bmatrix} .$$

The functors $\begin{bmatrix} U \\ S \end{bmatrix} \otimes_S - : \mathbb{P}(S) \to \mathbb{P}(M)$

and $[\, U^* \; S \,] \otimes_M - : \mathbb{P}(M) \to \mathbb{P}(S)$

are mutually inverse (up to natural transformations) and therefore induce mutually inverse isomorphisms between $K_*(S)$ and $K_*(M)$.

We wish to compute the effect m_* of $(M\otimes_T -)_*$ on $K_*(T)$ after the identification of $K_*(T)$ with $K_*(R) \oplus K_*(S)$. The composition

$$K_*(R) \oplus K_*(S) \xrightarrow{s_*} K_*(T) \xrightarrow{(M\otimes_T -)_*} K_*(M) \xrightarrow{([U^* \; S]\otimes_M -)_*} K_*(S)$$

is induced by the functor

$$[U^* \; S]\otimes_M M \otimes_T T \otimes_D - \cong [U^* \; S]\otimes_D - ;$$

thus on $K_*(R)$ the map is induced by $U^*\otimes_R -$, and on $K_*(S)$ by $S\otimes_S -$. Injecting $K_*(S)$ into $K_*(R) \oplus K_*(S)$ as the second component, we see that the endomorphism

$$m_* : K_*(R) \oplus K_*(S) \to K_*(R) \oplus K_*(S)$$

is given by

$$m_*(a,b) = (0, (U^*\otimes_R -)a + b).$$

2.3. In particular, let $T = \begin{bmatrix} S & \cdots & S \\ & \ddots & \vdots \\ 0 & & S \end{bmatrix}$ be the ring of upper triangular $n \times n$ matrices over S. Taking R to be the corresponding ring of $(n-1)\times(n-1)$ matrices and U the 'column' module of length n-1 over S, which is clearly projective both as left R-module and right S-module, we have an evident inductive description of T which leads to an isomorphism

$$K_*(T) \cong nK_*(S).$$

One each component of the diagonal subring $S \oplus \ldots \oplus S$ of R, the restriction of the functor $U^*\otimes_R-$ acts as $S\otimes_S-$, since $U^* = [S\ldots S]$. Hence the effect of $(U^*\otimes_R-)_*$ on $(n-1)K_*(S)$ is summation of coordinates :

$$(a_1,\ldots,a_{n-1}) \to a_1+\ldots+a_{n-1}$$

The ring M is now the full ring of n×n matrices over S. The above discussion shows that there is a commutative diagram

$$\begin{array}{ccc} K_*(T) & \xrightarrow{(M\otimes_T-)_*} & K_*(M) \\ \| \wr & & \| \wr \\ nK_*(S) & \longrightarrow & K_*(S) \end{array}$$

the top map being that obtained in (2.2) and the bottom map being summation of coordinates. The corresponding endomorphism m_* of $nK_*(S)$ is

$$m_*(a_1,\ldots,a_n) = (0,\ldots,0,\ a_1+\ldots+a_n).$$

For future reference, we note the following.

Proposition B.

The kernel of $(M\otimes_T-)_* : K_*(T) \to K_*(M)$ *is isomorphic to* $(n-1)K_*(S)$.

3.1. Let I be a left and right projective ideal of the ring R and let $A = \begin{bmatrix} R & \cdot & \cdot & \cdot & R \\ I & \cdot & & & \cdot \\ & \cdot & \cdot & & \cdot \\ & & \cdot & \cdot & \cdot \\ I & & I & & R \end{bmatrix}$, and $E = \begin{bmatrix} R & \cdot & \cdot & \cdot & R \\ & & & & \\ R & \cdot & \cdot & \cdot & R \end{bmatrix}$ be rings of n×n matrices.

Put H = IE, S = R/I, T = A/H and M = E/H; the rings T and M are as in (2.3).

Lemma.

The following hold.

i) H *and* E *are left and right* A*-projective.*

ii) M *is left and right* T*-projective.*

iii) $E\otimes_A(E/A) = 0$ *and* $E\otimes_A E \cong E$ *naturally.*

iv) $M\otimes_T(M/T) = 0$ *and* $M\otimes_T M \cong M$ *naturally.*

v) $E\otimes_A H \cong H$ *and* $E\otimes_A T \cong M$ *as bimodules.*

Proof.

(i) and (ii) are obvious and (iii) is similar to (iv).

As left T-module, M/T has generators $\{\bar{e}_{nj},\ j=1,\dots,n-1\}$, where $\{e_{ij}\}$ is a set of matrix units for M (that is, e_{ij} has the identity element in the i,j-th place and zero elsewhere). In $M\otimes_T M/T$, $1\otimes\bar{e}_{nj} = 1\otimes e_{nn}\bar{e}_{nj} = e_{nn}\otimes\bar{e}_{nj} = e_{n1}e_{1n}\otimes\bar{e}_{nj} = e_{n1}\otimes e_{1n}\bar{e}_{nj} = e_{n1}\otimes\bar{e}_{1j} = 0$.

The second assertion follows from the exact sequence

$$0 \to M\otimes_T T \to M\otimes_T M \to M\otimes_T(M/T) \to 0.$$

For (v), note that $E\otimes_A H$ can be identified with the ideal EH of E and then apply $E\otimes_A -$ to the exact sequence

$$0 \to H \to A \to T \to 0.$$

The reader will note that many of the results which follow hold for any triple H, A, E, which satisfies the conclusions of the Lemma; the triangular form of A and T is needed only when Proposition B is invoked. The projectivity hypothesis on the ideal I and its consequence for H are used crucially on (3.3). It is possible to relax the left-handed requirement on I to "I belongs to $\mathbb{P}_\infty(R)$".

3.2. Let $f_E : \mathbb{P}(E) \to \mathbb{P}(A)$ and $f_M : \mathbb{P}(M) \to \mathbb{P}(T)$ be the inclusions, and define

$$\varepsilon = f_E\cdot(E\otimes_A -) : \mathbb{P}(A) \to \mathbb{P}(A)$$

and

$$\mu = f_M\cdot(M\otimes_T -) : \mathbb{P}(T) \to \mathbb{P}(T).$$

We clearly have $\varepsilon_*^2 = \varepsilon_*$ on $K_*(A)$ and $\mu_*^2 = \mu_*$ on $K_*(T)$; further $\varepsilon_* K_*(A) \cong K_*(E)$ and $\mu_* K_*(T) \cong K_*(M)$.

Write $DK_*(A) = \ker \varepsilon_*$ and $DK_*(T) = \ker \eta_*$ (this is in analogy with the case E is a maximal order containing A)

We note the following consequence of the definitions.

Proposition C.

$$K_*(A) \cong DK_*(A) \oplus K_*(E)$$

and
$$K_*(T) \cong DK_*(T) \oplus K_*(M).$$

By Morita equivalence (2.2), we know that $K_*(E) \cong K_*(R)$ and $K_*(M) \cong K_*(S)$. We will prove the following result.

Proposition D.

$$DK_*(A) \cong DK_*(T).$$

Combining this with Proposition B, we obtain our main result.

Theorem E.

$$K_*(A) \cong K_*(R) \oplus (n-1)K_*(S).$$

3.3. Proposition D is proved by establishing four identities between various maps induced by the functors at our disposal. Let

$$i_A : \mathbb{P}(A) \to \mathbb{P}_\infty(A)$$

and
$$i_T : \mathbb{P}(T) \to \mathbb{P}_\infty(T)$$

be the inclusions. Since

$$0 \to A\otimes_A - \to E\otimes_A - \to E/A\otimes_A - \to 0$$

is an exact sequence of exact functors $\mathbb{P}(A) \to \mathbb{P}_\infty(A)$, we have

$$(i_A f_E(E\otimes_A -))_* = (i_A(A\otimes_A -))_* + (E/A\otimes_A -)_* .$$

Put
$$\eta_* = (-i_A)_*^{-1}(E/A\otimes_A -)_* \;;$$

then
$$\eta_* = 1 - \varepsilon_*.$$

Similarly, if
$$\nu_* = (-i_T)_*^{-1}(M/T\otimes_T -)_*,$$

$$\nu_* = 1 - \mu_*.$$

We now invoke the projection p_* and transfer t_* between $K_*(A)$ and $K_*(T)$. The projection is induced by the functor

$$T\otimes_A - : \mathbb{P}(A) \to \mathbb{P}(T).$$

For the transfer, we have a commuting diagram of inclusions of categories

$$\begin{array}{ccc} \mathbb{P}(A) & \xrightarrow{i_A} & \mathbb{P}_\infty(A) \\ & {}^{j}\nearrow & \uparrow h \\ \mathbb{P}(T) & \xrightarrow{i_T} & \mathbb{P}_\infty(T) \end{array}$$

and $$t_* = (i_A)_*^{-1} j_*, \text{ by definition.}$$

Now $$t_* . p_* . \eta_* = -(i_A)_*^{-1} j_*(T\otimes_A -)_*(E/A\otimes_A -)_*$$

$$= -(i_A)_*^{-1} \ (A\otimes_A -)_* - (H\otimes_A -)_* \ (E/A\otimes_A -)_*$$

$$= -(i_A)_*^{-1}(E/A\otimes_A -)_*,$$

since $$H\otimes_A(E/A) \cong H\otimes_E E\otimes_A(E/A) = 0.$$

Thus $$t_* p_* \eta_* = \eta_* . \qquad \ldots \quad \text{(i)}$$

Also, $$(i_T)_* p_* t_* \nu_* = -(A/H\otimes_A -)_*(i_A)_*^{-1} j_*(i_T)_*^{-1}(M/T\otimes_T -)_*$$

$$= -(A/H\otimes_A -)_*(i_A)_*^{-1} h_*(M/T\otimes_T -)_*$$

$$= -\{(A\otimes_A -)_* - (H\otimes_A -)_*\}(M/T\otimes_T -)_*$$

$$= -(M/H\otimes_T -)_*,$$

(notice that $A\otimes_A -$, $H\otimes_A -$ are exact on any category of A-modules),

hence $$p_* t_* \nu_* = \nu_* . \qquad \ldots \quad \text{(ii)}$$

Next, $$p_* \eta_* = -(T\otimes_A -)(i_T)^{-1}(E/A\otimes_A -)_*$$

$$= -(T\otimes_A -)\{(E\otimes_A -)_* - (A\otimes_A -)_*\}$$

$$= -\{(T\otimes_A E\otimes_A -)_* - (T\otimes_A -)_*\}$$

$$= -\{(M\otimes_A -)_* - (T\otimes_A -)_*\}$$

$$= -(i_T)^{-1}(M/T\otimes_T T\otimes_A -);$$

therefore $$p_* \eta_* = \nu_* p_* . \qquad \ldots \quad \text{(iii)}$$

Finally, $\eta_* t_* = (-i_A)_*^{-1}(E/A\otimes_A -)_*(i_A)_*^{-1}j_*$

$$= (-i_A)_*^{-1}\{(E\otimes_A -)_* - (A\otimes_A -)_*\}j_*$$

$$= (-i_A)_*^{-1}j_*\{(M\otimes_T -)_* - (T\otimes_T -)_*\},$$

since E and A act on T-modules via the surjections of E onto M and of A onto T. Thus,

$$\eta_* t_* = t_* \nu_* \qquad \ldots \quad \text{(iv)}$$

Using formulae (i) - (iv) it is now easy to see that p_* and t_* induce mutually inverse isomorphisms between $\eta_* K_*(A) \cong DK_*(A)$ and $\nu_* K_*(T) \cong DK_*(T)$.

4.1. Keeping our previous notation and hypothesis, we suppose further that R is an order over an integral domain o with quotient field k, and that kI = kR, so that kA = kE. Let $\mathbb{H}(A)$ and $\mathbb{H}(E)$ respectively denote the full subcategories of $\mathbb{P}_\infty(A)$ and $\mathbb{P}_\infty(E)$ whose objects are the o-torsion modules, and let the functors

$$h_A : \mathbb{H}(A) \to \mathbb{P}_\infty(A)$$

and

$$h_E : \mathbb{H}(E) \to \mathbb{P}_\infty(E)$$

be the inclusions.

Clearly, we have (by restriction) exact functors

$$E\otimes_A - : \mathbb{H}(A) \to \mathbb{H}(E)$$

and

$$\overline{f}_E : \mathbb{H}(E) \to \mathbb{H}(A),$$

and hence an idempotent operator, analoguous to that in (3.2),

$$\overline{\varepsilon}_* = (\overline{f}_E)_*(E\otimes_A -)_* : K_*(\mathbb{H}(A)) \to K_*(\mathbb{H}(A)),$$

such that $\overline{\varepsilon}_* K_*(\mathbb{H}(A)) \cong K_*(\mathbb{H}(E))$.

We write $DK_*(\mathbb{H}(A)) = \ker \overline{\varepsilon}_*$.

<u>Theorem E</u>.

$$DK_*(\mathbb{H}(A)) \cong DK_*(A).$$

Proof.

Let $c_* = (i_A)_*^{-1}(h_A)_* : K_*(\mathbb{H}(A)) \to K_*(A)$

and $d_* = (i_E)_*^{-1}(h_E)_* : K_*(\mathbb{H}(E)) \to K_*(E)$.

Invoking the naturality of the localization sequence [G], there is a commutative diagram with exact rows as follows :

$$
\begin{array}{ccccccccc}
\cdots \to & K_{*+1}(kA) & \to & K_*(\mathbb{H}(A)) & \xrightarrow{c_*} & K_*(A) & \to & K_*(kA) & \to \cdots , \\
(\mathrm{I}) & \| & & \downarrow & & \downarrow & & \| & \\
\cdots \to & K_{*+1}(kA) & \to & K_*(\mathbb{H}(E)) & \xrightarrow{d_*} & K_*(E) & \to & K_*(kA) & \to \cdots ,
\end{array}
$$

from which we see that c_* induces a homomorphism from $DK_*(\mathbb{H}(A))$ to $DK_*(A)$, also denoted c_*.

It is evident that the A-module E/A belongs to $\mathbb{H}(A)$ and $E \otimes_A (E/A) = 0$ (3.1, Lemma,(iii)). Thus the functor

$$E/A \otimes_A - : \mathbb{P}(A) \to \mathbb{H}(A)$$

induces a homomorphism

$$b_* : DK_*(A) \to DK_*(\mathbb{H}(A))$$

such that

$$\eta_* = -c_* b_*, \quad \text{where } \eta_* \text{ is as in (3.3).}$$

On the other hand,

$$-b_* c_* = (A \otimes_A -)_* - (E \otimes_A -)_*,$$

since $$0 \to A \otimes_A - \to E \otimes_A - \to E/A \otimes_A - \to 0$$

is an exact sequence of exact functors on $\mathbb{P}(A)$ in which the first two terms are already exact on $\mathbb{H}(A)$; thus

$$-b_* c_* = 1 - \bar{\varepsilon}_* .$$

Since η_* and $1-\bar{\varepsilon}_*$ act as the identity on $DK_*(A)$ and $DK_*(\mathbb{H}(A))$ respectively, the theorem follows.

4.2. Write $SK_*(A) = \ker(K_*(A) \to K_*(kA))$, etc.

Theorem F.

$$SK_*(A) \cong SK_*(E) \oplus DK_*(A)$$
$$\cong SK_*(R) \oplus (n-1)K_*(S).$$

Proof.

By Theorem E, $DK_*(A)$ is contained in $SK_*(A)$. Thus the commutative diagram (I) of (4.1) together with Proposition C give the first isomorphism.

The second isomorphism follows as in Theorem E, on noting that the Morita equivalence between E and R, which is induced by the module [R ... R] (2.2), is compatible with the Morita equivalence between kE and kR, so that there is a commutative diagram

$$\begin{array}{ccc} K_*(E) & \longrightarrow & K_*(kE) \\ \| \wr & & \| \wr \\ K_*(R) & \longrightarrow & K_*(kR) \end{array}$$

4.3. In the classical case, kR is Artinian semisimple and our definition of SK_* coincides with the usual definition. When R is a maximal order over a Dedekind domain, computations of $SK_*(R)$ have been made in [K1] and [K2]. If, in addition, I is a square-free product of prime ideals of R, then A is a hereditary order [Re, (39.14), (40.50)] and we obtain a more general version of computations made in [K3].

5. Let π be the non-abelian metacyclic group of order pq, where p and q are primes and q divides p-1. Let $\pi = \langle s,t \mid s^p = 1 = t^q, t^{-1}st = s^r, (r,p) = 1 \rangle$.

Write $\Sigma = 1 + s + \dots + s^{p-1}$, $J = \Sigma . o\pi$ and $I = (1-s)o\pi$ where o is the ring of p-adic integers. It is known [K4] that the twisted group ring $A = o\pi/J$ has the form

$$A = \begin{bmatrix} R & \cdot & \cdot & \cdot & R \\ P & \cdot & & & \cdot \\ \cdot & & \cdot & & \cdot \\ \cdot & & & \cdot & \cdot \\ P & \cdot & \cdot & P & R \end{bmatrix}, \quad q \times q \text{ matrices,}$$

with R the fixed ring of $o[P\sqrt{1}]$ under the action of <t> and P its maximal ideal. Note that $R/P = o/p = Z/p$.

There is a commutative diagram of rings and ideals

$$\begin{array}{ccccc} & & J & \longrightarrow & po[t] \\ & & \downarrow & & \downarrow \\ I & \longrightarrow & o\pi & \longrightarrow & o[t] \\ \downarrow & & \downarrow & & \downarrow \\ \bar{I} = \frac{I+J}{J} & \longrightarrow & A & \longrightarrow & (Z/p)[t] \end{array}$$

and hence [GL] a commutative diagram

$$\begin{array}{ccccc} K_3(o\pi, I) & \longrightarrow & K_3(o\pi) & \longrightarrow & K_3(o[t]) \\ \downarrow & & \downarrow & & \downarrow w \\ K_3(A, \bar{I}) & \longrightarrow & K_3(A) & \longrightarrow & K_3(Z/p[t]) \\ \downarrow & & \downarrow & & \downarrow \\ K_2(o\pi, I, J) & \xrightarrow{z} & K_2(o\pi, J) & \longrightarrow & K_2(o[t], po[t]) \end{array}$$

The map z is known to be 0 [GL, p212]. The ring $o[t]$ is the direct sum of q copies of o and $Z/p[t]$ that of q copies of Z/p; thus w is a surjection by [Sh, Theorem 4.3]. It is now easy to see that $K_3(o\pi)$ maps surjectively to $K_3(A) = K_3(R) \oplus (q-1)K_3(Z/p)$; it is well known that $K_3(Z/p)$ is cyclic of order p^2-1.

References

[AF]. F.W. Anderson and K. R. Fuller, Rings and Categories of Modules, Graduate Texts in Mathematics 13, (Springer, Berlin, 1974).

[FGR]. R.M. Fossum, P.A. Griffith and I. Reiten, Trivial Extensions of Abelian Categories, Lecture Notes in Mathematics 456 (Springer, Berlin 1975).

[G]. D. Grayson, "Higher algebraic K-theory II (after D. Quillen)" Algebraic K-theory, Evanston 1976, 217-240, Lecture Notes in Mathematics 551 (Springer, Berlin, 1976).

[GL]. D. Guin-Waléry et J.L. Loday, "Obstruction à l'excision en K-theorie algebrique", Algebraic K-theory, Evanston 1980, 179-216, Lecture Notes in Mathematics 854 (Springer, Berlin, 1981).

[K1]. M.E. Keating, "Values of tame symbols on division algebras", J. London Math. Soc. (2) 14 (1976), 25-30.

[K2]. M.E. Keating, "A transfer map in K-theory", J. London Math. Soc. (2), 16 (1977), 38-42.

[K3]. M.E. Keating, "On the K-theory of tiled orders", J. Algebra 43 (1976), 193-197.

[K4]. M.E. Keating, "Whitehead groups of some metacyclic groups and orders", J. Algebra 22(1972), 332-349.

[PR]. I. Palmer and J.E. Roos, "Explicit formulae for the global homological dimensions of trivial extensions of rings", J. Algebra 27 (1974), 380-413.

[Q]. D. Quillen, "Higher Algebraic K-theory I", Algebraic K-theory I 85-145, Lecture Notes in Mathematics 341 (Springer, Berlin, 1973).

[Re]. I. Reiner, Maximal Orders, (Academic Press, London, 1975).

[Sh]. C.C. Sherman, "Some splitting results in the K-theory of rings", Amer. J. Math. 101 (1979), 609-632.

[Sw]. R. G. Swan, "Excision in algebraic K-theory", J. Pure and Applied Algebra, 1, (1971), 221-252.

Mathematics Department
Imperial College
London SW 7
England

Present address:
Mathematics Department
National University of Singapore
Kent Ridge
Singapore 0511

The K_2 of a 1-fold stable ring

Frans Keune

1. Introduction

(1.1) A commutative ring R is called *1-fold stable* if R satisfies the stable range condition SR_2, i.e. for all unimodular $(a,b) \in R^2$ there exists a $c \in R$ such that $a+bc \in R^*$ (the group of units of R). More generally, a commutative ring R is called *k-fold stable* (where $k \in \mathbb{N}$) if for every k-tuple of unimodular pairs (a_i,b_i) $(i = 1,\ldots,k)$ there exists a $c \in R$ such that $a_i+bc_i \in R^*$ for $i = 1,\ldots,k$.

(1.2) In [2] Van der Kallen gave a presentation for the K_2 of a 3-fold stable ring. This presentation is in terms of Dennis-Stein symbols $\langle a,b\rangle$. In [3] I introduced symbols $\langle a,b,c\rangle$ and showed that the K_2 of a 1-fold stable ring is generated by these symbols. Moreover, I conjectured that the three types of relations I derived in the same article are sufficient for this type of rings. In this paper I prove this conjecture.

(1.3) Throughout the paper R is a fixed 1-fold stable ring.

(1.4) A Theorem of Dennis [1] and Vasershtein [6] implies that the canonical homomorphism $K_2(n,R) \to K_2(n+1,R)$ is surjective for $n \geq 2$, whereas from a theorem of Suslin and Tulenbaev [5] it follows that it is injective for $n \geq 3$. Equivalently, the canonical map $K_2(3,R) \to K_2(R)$ is an isomorphism and moreover $K_2(R)$ is contained in the image of $St(2,R)$ in $St(R)$.

(1.5) In order to show that the homomorphism $C(R) \to K_2(R)$ (see Section 3) is an isomorphism, I will reconstruct the group $St(3,R)$ in such a way that it contains $C(R)$ as a subgroup. This is done by reconstructing consecutively the inverse images in $St(3,R)$ of the following subgroups of $SL(3,R)$ $(= E(3,R))$, where we indicate a subgroup of $SL(3,R)$ by the shape of its elements:

$$\begin{pmatrix}1&0&0\\0&1&0\\0&0&1\end{pmatrix},\begin{pmatrix}*&0&0\\0&*&0\\0&0&1\end{pmatrix},\begin{pmatrix}*&0&0\\*&*&0\\0&0&1\end{pmatrix},\begin{pmatrix}*&*&0\\*&*&0\\0&0&1\end{pmatrix},\begin{pmatrix}*&*&*\\*&*&*\\0&0&1\end{pmatrix},\begin{pmatrix}*&*&*\\*&*&*\\*&*&*\end{pmatrix}.$$

These inverse images are respectively $K_2(3)$ (more precisely $K_2(3,R)$), H, L, St'(2), G, and St(3). The models we will construct for these groups are denoted by C (more precisely C(R)), $\tilde{H}$, $\tilde{L}$, $\widetilde{St}'(2)$, $\tilde{G}$, and $\widetilde{St}(3)$. The last group will turn out to be isomorphic to St(3) and, since it contains C, this will imply that C(R) and $K_2(R)$ are isomorphic.

(1.6) In [4] Kolster gives generators for the kernel of St(2) → St(3). If one could prove the corresponding relations in $\widetilde{St}'(2)$, this would show that $\widetilde{St}'(2) \to St(3)$ is injective. I have not been able to do so directly in the stage of the proof where $\widetilde{St}'(2)$ is constructed.

2. The Steinberg-group of R

(2.1) Since R is commutative an involution (= anti-automorphism of order 2) $x \mapsto x^*$ of the group St(n,R) is defined by

$$x_{ij}(r)^* = x_{n-j+1,n-i+1}(r).$$

In SL(n,R) it induces the reflection in the second diagonal of a matrix. Many of the groups we consider will have an involution which is induced by, or at least strongly related to, this involution for some n (in fact for n = 2 or n = 3). These involutions will also be denoted as *.

(2.2) As I showed in [3] the group $K_2(R)$ is generated by elements

$$\langle a,b,c\rangle_* = x_{12}(-a)x_{21}(b)x_{12}(-c)x_{21}(1-ab)x_{12}(-(1-bc))w_{12}(1)$$

where $a,b,c \in R$ such that $1-a-c+abc = 0$. I also showed that they satisfy the following identities:

(C1) $\langle a,b,c\rangle_* = \langle b,c,1-ab\rangle_*$,

(C2) $\langle a,b,c\rangle_* = \langle c,b,a\rangle_*^{-1}$,

(C3) $\langle a,bc,d\rangle_*\langle b,ca,e\rangle_* = \langle ab,c,d+ae\rangle_*$;

they hold whenever the left hand side is defined. Dennis-Stein symbols and Steinberg symbols are special cases:

$$\langle a,b\rangle_* = x_{21}(\tfrac{-b}{1-ab})x_{12}(-a)x_{21}(b)x_{12}(\tfrac{a}{1-ab})h_{12}(1-ab)^{-1} = \langle a,b,\tfrac{1-a}{1-ab}\rangle_*,$$

where $1-ab \in R^*$, and

$$\{c,a\}_* = h_{12}(ac)h_{12}(c)^{-1}h_{12}(a)^{-1} = \langle c,\tfrac{1-a}{c}\rangle_*$$

where $a,c \in R^*$. The following identities for these symbols are well known (they are also consequences of (C1), (C2), (C3)):

(D1) $\langle a,b\rangle_* = \langle b,a\rangle_*^{-1}$,

(D2) $\langle a,b\rangle_*\langle a,c\rangle_* = \langle a,b+c-abc\rangle_*$,

(D3) $\langle a,bc\rangle_* = \langle ab,c\rangle_*\langle ac,b\rangle_*$,

(S1) $\{c,a\}_* = \{a,c\}_*^{-1}$,

(S2) $\{c,a_1a_2\}_* = \{c,a_1\}_*\{c,a_2\}_*$,

(S3) $\{a,1-a\}_* = 1$.

The involution $*$ on $St(2,R)$ induces inversion on $K_2(2,R)$ and hence also on $K_2(R)$.

(2.3) Every $h \in H$ has a unique representation of the form

$$h_{12}(u)x$$

where $u \in R^*$ and $x \in K_2(R)$. The multiplication in H is determined by the fact that the elements of $K_2(R)$ are central together with the following rule

$$h_{12}(u)h_{12}(v) = h_{12}(uv)\{u,v\}_* \qquad (u,v \in R^*).$$

H is a subgroup of the image $St'(2)$ of $St(2)$ in $St(3)$. The involution $*$ on $St(2)$ induces on H the following

$$(h_{12}(u)x)^* = h_{12}(\tfrac{1}{u})x^{-1}.$$

For each $h \in H$ there is a unique $u \in R^*$ such that $hh_{12}(u)^{-1} \in K_2(R)$. The map $h \mapsto u$ is a surjective homomorphism from H to R^*, and is denoted by σ.

(2.4) Every $\ell \in L$ has a unique representation of the form

$$x_{21}(r)h$$

where $r \in R$ and $h \in H$. The group L is the semi-direct product of H and the subgroup of all $x_{21}(r)$, where $r \in R$. The action of H on this subgroup is determined by

$$hx_{21}(r)h^{-1} = x_{21}\left(\frac{r}{\sigma(h)^2}\right).$$

On L we have an involution * as follows:

$$(x_{21}(r)h)^* = h^*x_{21}(r) = x_{21}(\sigma(h)^2 r)h^*.$$

(2.5) Proposition. *Every element of* St'(2) *can be written in the form*

$$x_{12}(a)\ell\, x_{12}(c),$$

where $a,c \in R$ *and* $\ell \in L$.

Proof. The elements $x_{12}(r)$, $r \in R$, together with $w_{12}(1)$ generate St'(2). It suffices to show that the set of elements of type $x_{12}(a)\ell\, x_{12}(c)$ is invariant under left multiplication by $x_{12}(r)$ and $w_{12}(1)$. For $x_{12}(r)$ this is clear, so it remains to prove this for the element $w_{12}(1)$. Put $\ell = x_{21}(b)h$, where $b \in R$ and $h \in H$. Since R is 1-fold stable there is a $d \in R$ such that $-a+d(1+ab)$ is a unit of R. Put $u = -a+d(1+ab)$. Then the element

$<-\frac{a}{u},bu,\frac{d}{u}>_*$ is defined:

$$<-\frac{a}{u},bu,\frac{d}{u}>_* = x_{12}(\frac{a}{u})x_{21}(bu)x_{12}(-\frac{d}{u})x_{21}(1+ab)x_{12}(-(1-bd))w_{12}(1).$$

Conjugate by $h_{13}(u)$:

$$<-\frac{a}{u},bu,\frac{d}{u}> = x_{12}(a)x_{21}(b)x_{12}(-d)x_{21}(\frac{1+ab}{u})x_{12}(-(1-bd)u)w_{12}(u).$$

Then conjugate by $w_{12}(1)$:

$$<-\frac{a}{u},bu,\frac{d}{u}> = w_{12}(1)x_{12}(a)x_{21}(b)x_{12}(-d)x_{21}(\frac{1+ab}{u})x_{12}(-(1-bd)u)h_{12}(u).$$

So $w_{12}(1)x_{12}(a)x_{21}(b)$ can also be written as

$$x_{12}(\frac{1-bd}{u})x_{21}(-(1+ab)u)h_{12}(u)^{-1}<-\frac{a}{u},bu,\frac{d}{u}>_*\; x_{12}(d).$$

Therefore,

(A) $$w_{12}(1)x_{12}(a)x_{21}(b)h\; x_{12}(c) =$$

$$x_{12}(\frac{1-bd}{u})x_{21}(-(1+ab)u)h_{12}(u)^{-1}h<-\frac{a}{u},bu,\frac{d}{u}>_*\; x_{12}(c+\frac{d}{\sigma(h)^2}).$$

(2.6) Proposition. *Let* $a,a',b,b',c,c' \in R$ *and* $h,h' \in H$.

The elements

$$x_{12}(a)x_{21}(b)h\, x_{12}(c) \quad \textit{and} \quad x_{12}(a')x_{21}(b')h'x_{12}(c')$$

of St'(2) *are equal if and only if the following holds:*

the element $p = a'-a$ *satisfies:*

$$1-pb \in R^*,$$

$$b' = \frac{b}{1-pb},$$

$$h' = h_{12}(1-pb)h\langle p,b\rangle_*,$$

$$c' = c - \frac{p}{(1-pb)\sigma(h)^2}.$$

Proof. Suppose $x_{12}(a)x_{21}(b)h\, x_{12}(c) = x_{12}(a')x_{21}(b')h'x_{12}(c')$.
Then $x_{21}(-b')x_{12}(p)x_{21}(b)x_{12}((c-c')\sigma(h)^2)hh'^{-1} = 1$. The result follows easily by comparing underlying matrices.

(2.7) Note that the Propositions (2.5) and (2.6) imply that $K_2(R)$ is generated by symbols of type $\langle a,b,c\rangle_*$.

(2.8) Lemma. *Let* $a,a',b,b',c,c' \in R$ *and* $h,h' \in H$ *such that*

$$w_{12}(1)x_{12}(a)x_{21}(b)h\, x_{12}(c) = x_{12}(a')x_{21}(b')h'x_{12}(c')$$

then there exists a $d \in R$ *such that* $-a+d(1+ab) = u \in R^*$ *and*

$$a' = \frac{1-bd}{u},\ b' = -(1+ab)u,\ c' = c+\frac{d}{\sigma(h)^2},\ h = h_{12}(u)^{-1}h\langle -\frac{a}{u},bu,\frac{d}{u}\rangle_*.$$

Proof. By (2.5) there exist d,u such that $-a+d(1+ab) \in R^*$ and (A) in the proof of (2.5) holds. Now replace d by $d-\frac{pu^2}{1+pu(1+ab)}$, where $p = a'-\frac{1-bd}{u}$.

(2.9) Lemma. *Let* $a,a',b,b',c,c' \in R$ *and* $h,h' \in H$ *such that*

$$x_{12}(a)x_{21}(b)h\, x_{12}(c) = x_{12}(a')x_{21}(b')h'x_{12}(c').$$

If $a = a'$, *then* $b = b'$, $c = c'$ *and* $h = h'$.

If $c = c'$, *then* $a = a'$, $b = b'$ *and* $h = h'$.

Proof. This follows from (2.6).

(2.10) The group G is isomorphic to the semi-direct product of the group St'(2) and the subgroup of St(3) of all elements $x_{13}(p)x_{23}(q)$, where $p,q \in R$. The action of G on this subgroup is given by

$$z\, x_{13}(p)x_{23}(q)z^{-1} = x_{13}(p')x_{23}(q')$$

where
$$\begin{pmatrix} p' \\ q' \end{pmatrix} = \phi(z)\begin{pmatrix} p \\ q \end{pmatrix},$$

$\phi(z)$ denoting the underlying matrix in SL(2,R) of z.

(2.11) For each $r \in R$ an involution τ_r of St(3) is defined by

$$\tau_r(x) = (x_{31}(-r)x\, x_{31}(r))^* \quad (= x_{31}(r)x^* x_{31}(-r))$$

(* is the involution of St(3) described in (2.1)).

(2.12) <u>Definition</u>. For $r \in R$: $G_r = G \cap \tau_r(G)$.

(2.13) <u>Proposition</u>. *Let* $r \in R$ *and* $x \in St(3)$. *Then* $x \in G_r$ *if and only if the underlying matrix* $\phi(x)$ *of* x *in* SL(3,R) *is of the form*

$$\begin{pmatrix} 1-rp & * & p \\ -rq & * & q \\ 0 & 0 & 1 \end{pmatrix}.$$

Proof. If $x \in G_r$, then clearly the matrix $\phi(x)$ is of the desired form. Conversely, if $\phi(x)$ is of this form, the same holds for $\phi(\tau_r(x))$, so $\tau_r(x) \in G$. Then we have $x = \tau_r(\tau_r(x)) \in G \cap \tau_r(G)$.

(2.14) <u>Proposition</u>. *Let* $y,y',z,z' \in G$ *and* $r,r' \in R$. *Then the elements* $yx_{31}(r)z^*$ *and* $y'x_{31}(r')z'^*$ *of* St(3) *are equal if and only if* $r = r'$, $y^{-1}y' \in G_r$, *and* $\tau_r(y^{-1}y') = z'^{-1}z$.

<u>Proof</u>. If $y\, x_{31}(r)z^* = y'x_{31}(r')z'^*$, then comparison of the (3.1) entry in the underlying matrices gives $r = r'$, and so

$$y^{-1}y' = x_{31}(r)(z'^{-1}z)^* x_{31}(-r) = \tau_r(z'^{-1}z).$$

The converse is trivial.

(2.15) <u>Proposition</u>. *Let* $y \in G$ *and* $r \in R$. *Then there exists an* $x \in G_r$ *such that*

$\phi(yx)$ *has a zero* (2,3) *entry and a* (2,2) *entry in* R^*.

Proof. First multiply y from the right with an $x_{12}(p)$, $p \in R$, such that the (2,2) entry of $\phi(yx_{12}(p))$ is a unit. This is possible by 1-fold stability. Then multiply from the right with the unique $x_{23}(q)$ that make the (2,3) entry zero. Finally multiply from the right with $x_{21}(-rq)$. Note that $x_{12}(p)$ and $x_{21}(-rq)x_{23}(q)$ both belong to G_r.

(2.16) Proposition. *Every element of* St(3) *can be factored as*

$$yx_{31}(r)z^*$$

where $y,z \in G$ *and* $r \in R$.

Proof. Since G and $w_{23}(1)$ generate St(3), it clearly suffices to show that the set of elements of type $yx_{31}(r)z^*$ is invariant under left multiplication by $w_{23}(1)$. By (2.14) and (2.15) we may assume that the (2,3) entry of $\phi(y)$ vanishes and that the (2,2) entry is invertible. The matrix $\phi(y)e_{31}(r)$ is then of type

$$\begin{pmatrix} * & * & * \\ * & \circledast & 0 \\ * & 0 & 1 \end{pmatrix}$$

where $\circledast$ stands for an invertible element of R. Since $K_2(3) \subset G$ we can make computations on matrix level:

$$\begin{pmatrix} 1 & 0 & 0 \\ 0 & 0 & 1 \\ 0 & -1 & 0 \end{pmatrix}\begin{pmatrix} * & * & * \\ * & \circledast & 0 \\ * & 0 & 1 \end{pmatrix} = \begin{pmatrix} * & * & * \\ * & 0 & 1 \\ * & \circledast & 0 \end{pmatrix} = \begin{pmatrix} * & * & * \\ * & \circledast & 0 \\ * & 0 & 1 \end{pmatrix}\begin{pmatrix} 1 & 0 & 0 \\ 0 & 0 & \circledast \\ 0 & \circledast & 0 \end{pmatrix}$$

$$= \begin{pmatrix} * & * & * \\ * & \circledast & 1 \\ 0 & 0 & 1 \end{pmatrix}\begin{pmatrix} 1 & 0 & 0 \\ 0 & 1 & 0 \\ * & 0 & 1 \end{pmatrix}\begin{pmatrix} 1 & 0 & 0 \\ 0 & 0 & \circledast \\ 0 & \circledast & 0 \end{pmatrix}.$$

3. The group C(R)

(3.1) The group C(R) is defined in terms of generators and relations as the Abelian group on the generators

$$\langle a,b,c\rangle, \text{ where } a,b,c \in R \text{ and } 1-a-c+abc = 0,$$

which are subject to the following defining relations:

(C1) $\langle a,b,c\rangle = \langle b,c,1-ab\rangle$,

(C2) $\langle a,b,c\rangle = \langle c,b,a\rangle^{-1}$,

(C3) $\langle a,bc,d\rangle\langle b,ca,e\rangle = \langle ab,c,d+ae\rangle$.

A consequence of these relations is

(C4) $\langle a,b,(1-ac)d\rangle\langle a,c,(1-ab)d\rangle = \langle a,b+c-abc,d\rangle$.

Special types of elements are

$$\langle a,b\rangle = \langle a,b,\tfrac{1-a}{1-ab}\rangle \quad \text{for } a,b \in R,\ 1-ab \in R^*,$$

$$\{c,a\} = \langle c,\tfrac{1-a}{c}\rangle \quad \text{for } a,c \in R^*.$$

They satisfy the relations (D1), (D2), (D3), (S1), (S2), (S3) which where given in (2.2) for the symbols $\langle a,b\rangle_*$ and $\{c,a\}_*$. We also use frequently the relation

(S4) $\{-a,a\} = 1$

which is a consequence of the other relations.

(3.2) On $C(R)$ an involution * is defined by $\kappa^* = \kappa^{-1}$ for $\kappa \in C(R)$. Note that on generators $\langle a,b,c\rangle^* = \langle c,b,a\rangle$.

(3.3) There is a unique homomorphism $\pi : C(R) \to K_2(3)$ which sends $\langle a,b,c\rangle$ to $\langle a,b,c\rangle_*$. It is surjective because of (2.7). Clearly for all $\kappa \in C(R)$ we have $\pi(\kappa^*) = \pi(\kappa)^*$.

(3.4) The problem is: are the relations (C1), (C2), (C3) sufficient, i.e. is the map $\pi : C(R) \to K_2(R)$ injective? It is therefore not surprising that in the proof of this we often have to show that a certain identity holds in $C(R)$, or, what is the same, that it is a consequence of (C1), (C2) and (C3).

When an element $\langle a,b,c\rangle$ is considered it is advisable to consider the whole cycle $a,b,c,\ 1-ab,\ 1-bc$: the element is determined by three consecutive elements in the cycle (relation C1), and reversion of the order in the cycle corresponds to inversion of the element (relation C2). In the proofs that are to follow, the use of (C1) or (C2) is never explicitly mentioned. In relation (C3) the cycles of the three elements involved all contain 1-abc, so when (C3) is used, this indicates how (C1) and (C2) have been used in advance.

4. The group $\tilde{H}$

(4.1) As a set $\tilde{H}$ is defined as $R^{*} \times C(R)$. The multiplication is defined as follows:

$$(u,\kappa)(u',\kappa') = (uu',\kappa\kappa'\{u,u'\})$$

($u,u' \in R$; $\kappa,\kappa' \in C(R)$). The relations (S1) and (S2) imply that we obtain in this way a group.

(4.2) On $\tilde{H}$ an involution $*$ is defined by

$$(u,\kappa)^{*} = (\frac{1}{u},\kappa^{-1}).$$

(4.3) We will write $\tilde{h}_{12}(u)$ for $(u,1)$ and κ for $(1,\kappa)$; note that $\kappa \mapsto (1,\kappa)$ is an embedding of $C(R)$ into $\tilde{H}$. In this notation every element has a representation

$$\tilde{h}_{12}(u)\kappa$$

with unique $u \in R^{*}$ and $\kappa \in C(R)$. The homomorphism $\tilde{h}_{12}(u)\kappa \mapsto u$ from $\tilde{H}$ to R^{*} will be denoted by σ.

(4.4) From the description of H in (2.3) it follows that the map

$$\pi : \tilde{H} \to H, \ \tilde{h}_{12}(u)\kappa \mapsto h_{12}(u)\pi(\kappa)$$

is a surjective homomorphism. Moreover, for all $\tilde{h} \in H$ we have

$$\pi(\tilde{h}^{*}) = \pi(\tilde{h})^{*}.$$

5. The group $\tilde{L}$

(5.1) Let $\tilde{H}$ operate on R as follows

$$(r,\tilde{h}) \mapsto \frac{r}{\sigma(\tilde{h})^{2}} \qquad (r \in R, \ \tilde{h} \in \tilde{H}).$$

The group $\tilde{L}$ is defined to be the semi-direct product $R \rtimes \tilde{H}$.

(5.2) On $\tilde{L}$ an involution * is defined by

$$(r,\tilde{h})^* = (\sigma(\tilde{h})^2 r, \tilde{h}^*).$$

(5.3) We will write $\tilde{x}_{21}(r)$ for $(r,1)$ and $\tilde{h}$ for $(0,\tilde{h})$. Every element of $\tilde{L}$ then has a representation

$$\tilde{x}_{21}(r)\tilde{h}$$

with unique $r \in R$ and $\tilde{h} \in \tilde{H}$.

(5.4) From the description of L in (2.4) it follows that the map

$$\pi : \tilde{L} \to L, \quad \tilde{x}_{21}(r)\tilde{h} \mapsto x_{21}(r)\pi(\tilde{h})$$

is a surjective homomorphism. For all $\tilde{\ell} \in \tilde{L}$ we clearly have

$$\pi(\tilde{\ell}^*) = \pi(\tilde{\ell})^*.$$

6. The group $\widetilde{St}'(2)$

(6.1) The map $\pi : R \times \tilde{L} \times R \to St'(2)$ defined by $\pi(a,\tilde{\ell},c) = x_{12}(a)\pi(\tilde{\ell})x_{12}(c)$ is surjective. This follows from (2.5). A relation $\sim$ on $R \times \tilde{L} \times R$ is defined by

$(a,\tilde{x}_{21}(b)\tilde{h},c) \sim (a',\tilde{x}_{21}(b')\tilde{h}',c')$ if and only if for $p = a'-a$ we have

$$1-pb \in R^*,\ b' = \frac{b}{1-pb},\ c' = c - \frac{p}{(1-pb)\sigma(\tilde{h})^2},$$

and $\tilde{h}' = \tilde{h}_{12}(1-pb)\tilde{h}\langle p,b\rangle$.

Then from (2.6) it follows that if $(a,\tilde{\ell},c) \sim (a',\tilde{\ell}',c')$, then $\pi(a,\tilde{\ell},c) = \pi(a',\tilde{\ell}',c')$.

(6.2) Proposition. *The relation $\sim$ on $R \times \tilde{L} \times R$ is an equivalence relation.*

Proof. (i) Reflexivity is easy (take $p = 0$).

(ii) Symmetry. Suppose $(a,\tilde{x}_{21}(b)\tilde{h},c) \sim (a',\tilde{x}_{21}(b')\tilde{h}',c')$. Since $\pi(a,\tilde{x}_{21}(b)\tilde{h},c) = \pi(a',\tilde{x}_{21}(b')\tilde{h}',c') \in St'(2)$ we have by (2.6) only to prove that for $q = -p$ (= $a-a'$):

$$\tilde{h} = \tilde{h}_{12}(1-qb')\tilde{h}'\langle q,b'\rangle.$$

The identity in C(R) to prove is

$$\{1-pb,-1\}\langle p,b\rangle\langle -p,\frac{b}{1-pb}\rangle = 1.$$

Now

$$\langle p,b\rangle\langle -p,\frac{1}{1-pb}\rangle = \langle p,b\rangle\langle p,\frac{-b}{1-pb}\rangle\langle\frac{pb}{1-pb},-1\rangle$$

$$= \langle p,0\rangle\langle\frac{pb}{1-pb},-1\rangle = \{-1,1-pb\}.$$

(iii) Transitivity. Suppose

$$(a,\tilde{x}_{21}(b)\tilde{h},c) \sim (a',\tilde{x}_{21}(b')\tilde{h}',c') \sim (a'',\tilde{x}_{21}(b'')\tilde{h}'',c'').$$

Because of (2.6) it suffices to prove

$$\{\frac{1-(p+q)b}{1-pb},1-pb\}\langle p,b\rangle\langle q,\frac{b}{1-pb}\rangle = \langle p+q,b\rangle,$$

where $p = a'-a$ and $q = a''-a'$.

We have

$$\langle p,b\rangle\langle q,\frac{b}{1-pb}\rangle = \langle p,b\rangle\langle\frac{q}{1-pb},b\rangle\langle bq,\frac{1}{1-pb}\rangle$$

$$= \langle p+q,b\rangle\langle bq,\frac{1}{1-pb}\rangle = \langle p+q,b\rangle\{\frac{1-(p+q)b}{1-pb},\frac{1}{1-pb}\}.$$

(6.3) <u>Definition</u>. $X = R \times \tilde{L} \times R/\sim$. The class represented by $(a,\tilde{\ell},c)$ will be denoted by $[a,\tilde{\ell},c]$. The map π from $R \times \tilde{L} \times R$ to St'(2) induces a map $\pi : X \to St'(2)$.

(6.4) For every $r \in R$ a permutation $\underline{x}_{12}(r)$ of X is defined by

$$\underline{x}_{12}(r)[a,\tilde{\ell},c] = [a+r,\tilde{\ell},c]$$

and a permutation $\bar{x}_{12}(r)$ of X is defined by

$$\bar{x}_{12}(r)[a,\tilde{\ell},c] = [a,\tilde{\ell},c+r].$$

These permutations are clearly well-defined and they satisfy

$$\pi(\underline{x}_{12}(r)(x)) = x_{12}(r)\pi(x),$$

$$\pi(\bar{x}_{12}(r)(x)) = \pi(x)x_{12}(r),$$

for all $x \in X$.

(6.5) We will next define a transformation $\underline{w}_{12}(1)$ of X which satisfies

$$\pi(\underline{w}_{12}(1)(x)) = w_{12}(1)\pi(x)$$

for all $x \in X$. The proof of (2.5) makes clear how to define $\underline{w}_{12}(1)$. The problem is to show that it is well-defined. Therefore we first define $\underline{w}_{12}(1)$ on representatives, i.e. as a map

$$\underline{w}_{12}(1) \ : \ R \times \tilde{L} \times R \to X.$$

For $a,b,c \in R$, $\tilde{h} \in \tilde{H}$ put

$$\underline{w}_{12}(1)(a,\tilde{x}_{21}(b),\tilde{h},c) =$$
$$[\frac{1-bd}{u},\tilde{x}_{21}(-(1+ab)u)\tilde{h}_{12}(u)^{-1}\tilde{h}<-\frac{a}{u},bu,\frac{d}{u}>,c+\frac{d}{\sigma(h)^2}]$$

where $d,u \in R$ such that $-a+d(1+ab) = u \in R^*$.

We first show that this map is well-defined: so if also $d',u' \in R$ are such that $-a+d'(1+ab) = u' \in R^*$, then for $p = \frac{1-bd'}{u'} - \frac{1-bd}{u}$ $(= \frac{d-d'}{uu'})$ we have to show that

$$\tilde{h}_{12}(1+p(1+ab)u)\tilde{h}_{12}(u)^{-1}\tilde{h}<-\frac{a}{u},bu,\frac{d}{u}><p,-(1+ab)u>$$
$$= \tilde{h}_{12}(u')^{-1}\tilde{h}<-\frac{a}{u'},bu,\frac{d}{u'}>.$$

This easily reduces to the identity

$$\{u'-u\}<-\frac{a}{u},bu,\frac{d}{u}><\frac{d-d'}{uu'},-(1+ab)u> = <-\frac{a}{u'},bu',\frac{d'}{u'}>.$$

We have

$$<-\frac{a}{u},bu,\frac{d}{u}><\frac{u}{u'},-ab\frac{u'}{u},-pu> = <-\frac{a}{u'},bu',\frac{d'}{u'}>,$$

so the identity to be proved becomes

$$\{u',-u\}<p,-(1+ab)u> = <-pu,1+ab>,$$

but

$$<-pu,1+ab> = <p,(1+ab)(-u)><-u,p(1+ab)>$$
$$= <p,-(1+ab)u>\{-u,\frac{u}{u'}\}.$$

(6.6) Proposition. *The map* $\underline{w}_{12}(1) \ : \ R \times \tilde{L} \times R \to X$ *induces a transformation of* X.

Proof. Let $(a,\tilde{x}_{21}(b)\tilde{h},c) \in R \times \tilde{L} \times R$, and let $p \in R$ be such that $1-pb \in R^*$. We have to show that

$$\underline{w}_{12}(1)(a,\tilde{x}_{21}(b)\tilde{h},c)$$
$$= \underline{w}_{12}(1)(a+p,\tilde{x}_{12}(\frac{b}{1-pb})\tilde{h}_{12}(1-pb)\tilde{h}<p,b>,c-\frac{p}{(1-pb)\sigma(h)^2}).$$

As always this reduces to an identity to be proved in $C(R)$. This identity is

$$\langle -\frac{a}{u}, bu, \frac{d}{u}\rangle = \{1-pb,u\}\langle p,b\rangle\langle -\frac{a+p}{(1-pb)u}, bu, \frac{p+d-pbd}{u}\rangle,$$

where $d,u \in R$ are such that $-a+d(1+ab) = u \in R^*$.

The last element on the right hand side is equal to (by (C4)):

$$\langle bu,\frac{p}{u}\rangle\langle bu,\frac{d}{u},1+ab\rangle = \langle bu,\frac{p}{u}\rangle\langle -\frac{a}{u},bu,\frac{d}{u}\rangle.$$

It remains to prove

$$\{1-pb,u\}\langle p,b\rangle\langle bu,\frac{p}{u}\rangle = 1,$$

but

$$\langle bu,\frac{p}{u}\rangle = \langle b,p\rangle\langle u,\frac{bp}{u}\rangle = \langle b,p\rangle\{u,1-pb\}.$$

(6.7) Definition. The transformation $\underline{w}_{12}(1)$ of X is defined by

$$\underline{w}_{12}(1)[a,\tilde\ell,c] = \text{the class of } \underline{w}_{12}(1)(a,\tilde\ell,c).$$

In (6.10) we will show that it is actually a permutation. Also a transformation $\bar{w}_{12}(1)$ of X is defined:

$$\bar{w}_{12}(1)(x) = (\underline{w}_{12}(1)(x^*))^*$$

where * denotes the following permutation of order 2 of X:

$$[a,\tilde\ell,c]^* = [c,\ell^*,a]$$

(6.8) The transformations $\underline{w}_{12}(1)$ and $\bar{w}_{12}(1)$ satisfy

$$\pi(\underline{w}_{12}(1)(x)) = w_{12}(1)\pi(x),$$

$$\pi(\bar{w}_{12}(1)(x)) = \pi(x)w_{12}(1),$$

for all $x \in X$.

(6.9) Definition. For each $\tilde h \in \tilde H$ there is a permutation $\underline{\tilde h}$ of $R \times \tilde L \times R$ defined by

$$\underline{\tilde h}(a,\tilde\ell,c) = (a\sigma(\tilde h)^2, \tilde h\tilde\ell, c)$$

(Note that $\underline{\tilde h}^{-1} = \underline{\tilde h^{-1}}$). It is easily verified that $\underline{\tilde h}$ induces a permutation of X. This permutation is also called $\underline{\tilde h}$. Also for each $\tilde h \in \tilde H$ we have a permutation $\bar{\tilde h}$ of X defined by

$$\bar{\tilde h}(x) = (\underline{\tilde h^{\wedge}}(x^{\wedge}))^{\wedge}.$$

Then we have again

$$\pi(\underline{\tilde{h}}(x)) = \pi(\tilde{h})\pi(x),$$

$$\pi(\bar{\underline{\tilde{h}}}(x)) = \pi(x)\pi(\widetilde{h^*}) \quad (= \pi(x)\pi(\tilde{h})^*).$$

(6.10) It is easily verified that $\underline{w}_{12}(1)^2 = \underline{\tilde{h}}_{12}(-1)\ \underline{\{-1,-1\}}$ so that in particular $\underline{w}_{12}(1)$ and also $\bar{w}_{12}(1)$ are permutations of X.

(6.11) Definition. $\widetilde{St}'(2)$ is defined as the group of permutations of X generated by all

$$\begin{array}{ll} \underline{x}_{12}(r) & (r \in R), \\ \underline{\tilde{h}} & (\tilde{h} \in \tilde{H}), \\ \underline{w}_{12}(1). & \end{array}$$

The group $\widetilde{St}'(2)^*$ is defined as the group of permutations of X generated by all

$$\begin{array}{ll} \bar{x}_{12}(r) & (r \in R), \\ \bar{\underline{\tilde{h}}} & (\tilde{h} \in \tilde{H}), \\ \bar{w}_{12}(1). & \end{array}$$

(6.12) Lemma. *The groups* $\widetilde{St}'(2)$ *and* $\widetilde{St}'(2)^*$ *act transitively on* X.

Proof. $[a,\tilde{x}_{21}(b)\tilde{h},c] = \underline{x}_{12}(a)[0,\tilde{x}_{21}(b)\tilde{h},c]$,

$$[0,\tilde{x}_{21}(b)\tilde{h},c] = \underline{w}_{12}(1)^{-1}[1-b,\tilde{x}_{21}(-1)\tilde{h},c+\frac{1}{\sigma(\tilde{h})^2}]$$
$$= \underline{w}_{12}(1)^{-1}\underline{x}_{12}(-b)[1,\tilde{x}_{21}(-1)\tilde{h},c+\frac{1}{c(\tilde{h})^2}]$$
$$= \underline{w}_{12}(1)^{-1}\underline{x}_{12}(-b)\underline{w}_{12}(1)[0,\tilde{h},c],$$

$$[0,\tilde{h},c] = \underline{\tilde{h}}[0,1,c] = \underline{\tilde{h}}[c,1,0] = \underline{\tilde{h}}\ \underline{x}_{12}(c)[0,1,0].$$

Hence for every $x \in X$ there is an $\alpha \in \widetilde{St}'(2)$ such that $\alpha(x) = [0,1,0]$. This shows that $\widetilde{St}'(2)$ acts transitively on X. Similarly for $\widetilde{St}'(2)^*$.

(6.13) Lemma. *For* $r \in R$, $\kappa \in C(R)$ *and* $u \in R^*$ *we have*

(i) $\underline{\kappa}$ *is central in* $\widetilde{St}'(2)$,

(ii) $\underline{\tilde{h}}_{12}(u)\underline{x}_{12}(r) = \underline{x}_{12}(u^2r)\underline{\tilde{h}}_{12}(u)$,

(iii) $\underline{w}_{12}(1)\underline{\tilde{h}}_{12}(u) = \underline{\tilde{h}}_{12}(\frac{1}{u})\underline{w}_{12}(1)$.

Proof. (i) and (ii) are trivial. We will prove (iii). Let $x \in X$. We have to prove that

$$\underline{w}_{12}(1)\underline{\tilde{h}}_{12}(u)(x) = \underline{\tilde{h}}_{12}(\tfrac{1}{u})\underline{w}_{12}(1)(x).$$

Put $x = [a,\tilde{x}_{21}(b)\tilde{h},c]$, where $a,b,c \in R$, $\tilde{h} \in \tilde{H}$, and let $d,v \in R$ be such that $-a+d(1+ab) = v \in R^*$. Then

$$\underline{\tilde{h}}_{12}(\tfrac{1}{u})\underline{w}_{12}(1)(x) = \underline{\tilde{h}}_{12}(\tfrac{1}{u})[\frac{1-bd}{v},\tilde{x}_{21}(-(1+ab)v)\tilde{h}_{12}(v)^{-1}\tilde{h}<-\frac{a}{v},bv,\frac{d}{v}>,c+\frac{d}{\sigma(\tilde{h})^2}]$$

$$= [\frac{1-bd}{u^2v},\tilde{x}_{21}(-(1+ab)u^2v)\tilde{h}_{12}(\tfrac{1}{u})\tilde{h}_{12}(v)^{-1}\tilde{h}<-\frac{a}{v},bv,\frac{d}{v}>,c+\frac{d}{\sigma(\tilde{h})^2}],$$

and

$$\underline{w}_{12}(1)\underline{\tilde{h}}_{12}(u)(x) = \underline{w}_{12}(1)[au^2,\tilde{x}_{21}(\frac{b}{u^2})\tilde{h}_{12}(u)\tilde{h},c].$$

Since $-au^2+du^2(1+au^2.\frac{b}{u^2}) = u^2v$, we arrive at an expression which is almost the same as the one for $\underline{\tilde{h}}_{12}(\frac{1}{u})\underline{w}_{12}(1)(x)$. It remains to prove that

$$h_{12}(\tfrac{1}{u})h_{12}(v)^{-1} = h_{12}(u^2v)^{-1}h_{12}(u).$$

This leads to $\{u^2v,\frac{1}{u}\} = \{u,v\}$, which is easy to prove.

(6.14) <u>Proposition</u>. *The permutations in* $\widetilde{St}'(2)$ *commute with the permutations in* $\widetilde{St}'(2)^*$.

Proof. It suffices to check this for generators. It is trivial in all cases but one, in which case it is difficult: we have to prove that

$$\underline{w}_{12}(1)\bar{w}_{12}(1) = \bar{w}_{12}(1)\underline{w}_{12}(1).$$

Let $x \in X$, say $x = [a,\tilde{x}_{21}(b)\tilde{h},c]$. Since

$$x = \underline{\tilde{h}}[\frac{a}{\sigma(\tilde{h})^2},\tilde{x}_{21}(b\sigma(\tilde{h})^2),c]$$

and (say $\tilde{h} = \tilde{h}_{12}(u)\kappa$, $u \in R^*$, $\kappa \in C(R)$)

$$\underline{w}_{12}(1)\bar{w}_{12}(1)\underline{\tilde{h}} = \underline{w}_{12}(1)\underline{\tilde{h}}\bar{w}_{12}(1) = \underline{\tilde{h}}_{12}(\tfrac{1}{u})\underline{\kappa}\;\underline{w}_{12}(1)\bar{w}_{12}(1)$$

$$\bar{w}_{12}(1)\underline{w}_{12}(1)\underline{\tilde{h}} = \bar{w}_{12}(1)\underline{\tilde{h}}_{12}(\tfrac{1}{u})\underline{\kappa}\;\underline{w}_{12}(1) = \underline{\tilde{h}}_{12}(\tfrac{1}{u})\underline{\kappa}\;\underline{w}_{12}(1)\bar{w}_{12}(1),$$

we may assume that x is of the form

$$[a,\tilde{x}_{21}(b),c].$$

Choose $d,e \in R$ such that

$$-a+d(1+ab) = u \in R^*$$

$$-c+e(1+bc) = v \in R^*.$$

Then

$$\underline{w}_{12}(1)(\alpha) = [\frac{1-bd}{u}, \tilde{x}_{21}(-(1+ab)u)\tilde{h}_{12}(u)^{-1}\kappa_1, c+d],$$

$$\bar{w}_{12}(1)(\alpha) = [a+e, \tilde{x}_{21}(-(1+bc)\frac{1}{v})\tilde{h}_{12}(\frac{1}{v})^{-1}\kappa_2^{-1}, \frac{1-be}{v}],$$

where

$$\kappa_1 = <-\frac{a}{u}, bu, \frac{d}{u}> \quad \text{and} \quad \kappa_2 = <-\frac{c}{v}, bv, \frac{e}{v}>.$$

Next we compute $\bar{w}_{12}(1)\underline{w}_{12}(1)(x)$: choose e' such that

$$v' = -(c+d) - e'\,\frac{a+c+abc}{u} \in R^*$$

(note that $1+(c+d)(-(1+ab)\frac{1}{u}) = -\frac{1}{u}(a+c+abc))$, then

$$\bar{w}_{12}(1)\underline{w}_{12}(1)(x) =$$

$$[\frac{1-bd}{u}+\frac{e'}{u^2}, \tilde{x}_{21}(-(a+c+abc)\frac{u}{v'})\tilde{h}_{12}(u)^{-1}\tilde{h}_{12}(\frac{1}{v'})^{-1}\kappa_1\kappa_3^{-1}, \frac{u+(1+ab)e'}{uv'}],$$

where $\kappa_3 = <-\frac{c+d}{v'}, -(1+ab)\frac{v'}{u}, \frac{e'}{v'}>$.

To compute $\underline{w}_{12}(1)\bar{w}_{12}(1)(x)$ we have to choose a $d' \in R$ such that

$$u' = -(a+e)-d'\,\frac{d+c+abc}{v} \in R^*.$$

In view of (2.8) we can take the unique d' such that

$$\frac{1-be}{v} + \frac{d'}{v^2} = \frac{u+(1+ab)e'}{uv'}.$$

Then it remains to show that

$$\tilde{h}_{12}(u)^{-1}\tilde{h}_{12}(\frac{1}{v'})^{-1}\kappa_1\kappa_3^{-1} = \tilde{h}_{12}(u')^{-1}\tilde{h}_{12}(\frac{1}{v})^{-1}\kappa_2^{-1}\kappa_4,$$

where $\kappa_4 = <-\frac{a+e}{u}, -(1+bc)\frac{u'}{v}, \frac{d'}{u'}>$. Note that because of the choice of d' we have $u' = \frac{uv}{v'}$.

The following identity in C(R) has to be proved:

$$\{v',u\}\,\kappa_1\kappa_3^{-1} = \{v,u'\}\kappa_2^{-1}\kappa_4.$$

The choice of d' and e' depend on the choices of d and e, but we are still to make for d and e more special choices in order to facilitate computations. For that purpose choose $f \in R$ such that

$$t = b+f(1+ab)(1+bc) \in R^*.$$

Then it follows that we can take

$$d = \frac{1+af(a+bd)}{t}, \quad e = \frac{1+cf(1+ab)}{t},$$

and as a result we have $u = v = \frac{1}{t}$, so

$$\{v,u'\} = \{\frac{1}{t}, \frac{1}{t^2 v'}\} = \{\frac{1}{t},-\frac{1}{t}\}^2\{\frac{1}{t},\frac{1}{v'}\} = \{v',u\}.$$

The identity to be proved then becomes

$$\kappa_1\kappa_3^{-1} = \kappa_2^{-1}\kappa_4.$$

Now by (C4):

$$\kappa_1^{-1}\kappa_2^{-1} = \langle\frac{t}{t},-(a+c+abc)t,\frac{f}{t}\rangle.$$

Put $A = -(1+ab)tv'$, $B = -(1+bc)tu'$, $C = 1+f(a+c+abc)$, $D = 1+e'(1+ab)t$, $E = 1+d'(1+bc)t$, then by some elementary calculations

$$CA = -\frac{a+e}{u'}, \quad BC = -\frac{c+d}{v'}, \quad E+BD = \frac{b}{t}.$$

So we have

$$\begin{aligned}\kappa_3^{-1}\kappa_4^{-1} &= \langle A,BC,D\rangle\langle B,CA,E\rangle = \langle AB,C,E+BD\rangle \\ &= \kappa_1^{-1}\kappa_2^{-1}.\end{aligned}$$

(6.15) Proposition. *The action of* $\widetilde{St}'(2)$ *on* X *is simply transitive.*

Proof. Let $x \in X$, and $\alpha,\beta \in \widetilde{St}'(2)$ such that $\alpha(x) = \beta(x)$.

Then $\gamma\alpha(x) = \gamma\beta(x)$ for all $\gamma \in \widetilde{St}'(2)^*$, and by Proposition (6.14)

$$\alpha(\gamma(x)) = \beta(\gamma(x)) \quad \text{for all } \gamma \in \widetilde{St}'(2)^*.$$

Since $\widetilde{St}'(2)^*$ acts transitively on X, we have

$$\alpha(y) = \beta(y) \quad \text{for all } y \in X.$$

Hence $\alpha = \beta$.

(6.16) Define for $r \in R$ the element $\underline{x}_{21}(r) \in \widetilde{St}'(2)$ as follows

$$\underline{x}_{21}(r) = \underline{w}_{12}(1)\underline{x}_{12}(-r)\underline{w}_{12}(1)^{-1}.$$

The following rules are easily verified:

$$\underline{x}_{21}(r)\underline{x}_{21}(s) = \underline{x}_{21}(r+s),$$
$$\underline{w}_{21}(1)\underline{x}_{21}(r)\underline{w}_{12}(1)^{-1} = \underline{x}_{12}(-r),$$
$$\underline{\tilde{h}}\ \underline{x}_{21}(r)\underline{\tilde{h}}^{-1} = \underline{x}_{21}\left(\frac{r}{\sigma(\tilde{h})^2}\right).$$

(6.17) The group $\tilde{L}$ embeds in $\widetilde{St}'(2)$ as follows

$$\tilde{x}_{21}(b)\tilde{h} \mapsto \underline{x}_{21}(b)\underline{\tilde{h}}.$$

It clearly is a homomorphism. Injectivity follows from (6.15) and

$$\underline{x}_{21}(b)\underline{\tilde{h}}[0,1,0] = [0,\tilde{x}_{21}(b)\tilde{h},0]$$

We will henceforth consider $\tilde{H}$ ($\subset \tilde{L}$) to be a subgroup of $\widetilde{St}'(2)$, i.e. we will write $\tilde{h}$ for $\underline{\tilde{h}}$.

(6.18) $\widetilde{St}'(2)$ is our model for $St'(2)$. A surjective homomorphism

$$\pi : \widetilde{St}'(2) \to St'(2)$$

is defined as follows

$$\pi(\alpha) = \pi(\alpha[0,1,0]),$$

where the last π is the one occuring in definition (6.3).

Surjectivity follows from the surjectivity of $\pi : X \to St'(2)$ together with the transitivity of the action of $\widetilde{St}'(2)$ on X. In order to show that $\pi : \widetilde{St}'(2) \to St'(2)$ is a homomorphism it suffices to show that $\pi(\beta\alpha) = \pi(\beta)\pi(\alpha)$ for generators β of $\widetilde{St}'(2)$. The generators of $\widetilde{St}'(2)$ were all constructed in such a way that

$$\pi(\beta(x)) = \pi(\beta)\pi(\alpha)$$

for all $x \in X$. Hence

$$\pi(\beta\alpha) = \pi(\beta(\alpha[0,1,0])) = \pi(\beta)\pi(\alpha[0,1,0]) = \pi(\beta)\pi(\alpha).$$

(6.19) The transformation * on X induces an involution * on $\widetilde{St}'(2)$: for $\alpha \in \widetilde{St}'(2)$, α^* is defined by

$$\alpha^*[0,1,0] = (\alpha[0,1,0])^*.$$

Then $\pi(\alpha^*) = \pi(\alpha^*[0,1,0]) = (\pi(\alpha[0,1,0]))^* = \pi(\alpha)^*$.

(6.20) Proposition. (i) *Let* $u,v \in R^*$, *then*

$$\{u,v\} = \tilde{\underline{h}}_{12}(u)\tilde{\underline{h}}_{12}(v)\tilde{\underline{h}}_{12}(uv)^{-1} \quad (\text{in } \widetilde{St}'(2)).$$

(ii) *Let* $a,b \in R$ *such that* $1-ab \in R^*$ *then*

$$\langle a,b\rangle = \underline{x}_{21}(\tfrac{-b}{1-ab})\underline{x}_{12}(-a)\underline{x}_{21}(b)\underline{x}_{12}(\tfrac{a}{1-ab})\tilde{\underline{h}}_{12}(1-ab)^{-1}.$$

(iii) *Let* $a,b \in R$ *be such that* $1-a-c+abc = 0$, *then*

$$\langle a,b,c\rangle = \underline{x}_{12}(-a)\underline{x}_{21}(b)\underline{x}_{12}(-c)\underline{x}_{21}(1-ab)\underline{x}_{12}(-(1-bc))\underline{w}_{12}(1).$$

Proof. (i) This is clear: $\tilde{H} \to \widetilde{St}'(2)$, $\tilde{h} \mapsto \underline{\tilde{h}}$, is a homomorphism.

(ii) By (6.1)

$$[0,\tilde{x}_{21}(b),0] = [a,\tilde{x}_{21}(\tfrac{b}{1-ab})\tilde{h}_{12}(1-ab)\langle a,b\rangle,-\tfrac{a}{1-ab}]$$

so

$$\underline{x}_{21}(b) = \underline{x}_{12}(a)\underline{x}_{21}(\tfrac{b}{1-ab})\tilde{\underline{h}}_{12}(1-ab)\langle a,b\rangle\underline{x}_{12}(-\tfrac{a}{1-ab}).$$

(iii) This follows from:

$$\underline{w}_{12}(1)[-a,\tilde{x}_{21}(b),0] = [1-bc,\tilde{x}_{21}(-(1-ab))\langle a,b,c\rangle,c].$$

7. The group $\tilde{G}$

(7.1) Definition. The group $\tilde{G}$ is the semi-direct product $R^2 \rtimes \widetilde{St}'(2)$, the action of $\widetilde{St}'(2)$ on R^2 being

$$(\alpha,\binom{p}{q}) \mapsto \phi(\alpha)\binom{p}{q},$$

where $\phi(\alpha)$ denotes the underlying matrix in $SL(2,R)$ of $\pi(\alpha)$. The product in $\tilde{G}$ then is

$$(\binom{p_1}{q_1},\alpha_1)(\binom{p_2}{q_2},\alpha_2) = (\binom{p_1}{q_1})+\phi(\alpha_1)\binom{p_2}{q_2},\alpha_1\alpha_2).$$

We introduce the following notation:

$$\underline{x}_{13}(p) = (\binom{p}{0},1), \quad \underline{x}_{23}(q) = (\binom{0}{q},1)$$

and consider $\widetilde{St}'(2)$ as a subgroup of $\tilde{G}$ (i.e. we write α in stead of $(0,\alpha)$). An element of $\widetilde{G}$ then takes the form

$$\underline{x}_{13}(p)\underline{x}_{23}(q)\underline{x}_{12}(a)\underline{x}_{21}(b)\widetilde{h}\,\underline{x}_{12}(c),$$

where $p,q,a,b,c \in R$ and $\widetilde{h} \in \widetilde{H}$.

Note that by definition we have the usual relations:

$$\underline{x}_{ij}(r)\underline{x}_{ij}(s) = \underline{x}_{ij}(r+s) \qquad (r,s \in R),$$

$$[\underline{x}_{ij}(r),\underline{x}_{jk}(s)] = \underline{x}_{ik}(rs),$$

$$[\underline{x}_{ij}(r),\underline{x}_{kl}(s)] = 1 \quad \text{when} \quad i \neq \ell,\ j \neq k,$$

$$\widetilde{h}\,\underline{x}_{13}(p) = \underline{x}_{13}(p\sigma(\widetilde{h}))\widetilde{h} \qquad (p \in R),$$

$$\widetilde{h}\,\underline{x}_{23}(q) = \underline{x}_{23}\left(\frac{q}{\sigma(\widetilde{h})}\right)\widetilde{h} \qquad (q \in R).$$

(7.2) The map $\pi : \widetilde{St}'(2) \to St'(2)$ $(\subset St(3))$ extends to a surjective homomorphism

$$\pi : \widetilde{G} \to G$$

as follows

$$\pi(\underline{x}_{13}(p)\underline{x}_{23}(q)\alpha) = x_{13}(p)x_{23}(q)\pi(\alpha).$$

8. The groups $\widetilde{G}_r$

(8.1) Definition. Let $r \in R$. The subgroup $\widetilde{G}_r$ of $\widetilde{G}$ is defined to be the inverse image of G_r under $\pi : \widetilde{G} \to G$. It then also is the subgroup of $\widetilde{G}$ consisting of all elements of which the underlying matrix is of the form

$$\begin{pmatrix} 1-pr & * & p \\ 1-qr & * & q \\ 0 & * & 1 \end{pmatrix}.$$

(8.2) Subgroups of $\widetilde{G}_r$ are

$$A = \{\underline{x}_{12}(a) \mid a \in R\}.$$

$$B_r = \{\underline{x}_{21}(-qr)\underline{x}_{23}(q) \mid q \in R\},$$

$$C_r = \{\underline{x}_{13}(p)\widetilde{h} \mid p \in R \text{ and } \sigma(\widetilde{h}) = 1-pr\}.$$

(8.3) Proposition. $\widetilde{G}_r = AC_rB_rA$.

Proof. Follow the process described in the proof of (2.15).

(8.4) Definition. Let $r \in R$. Then a map

$$\tau_r : A \times C_r \times B_r \times A \to \tilde{G}_r$$

is defined as follows

$$\tau_r(\underline{x}_{12}(a), \underline{x}_{13}(p)\tilde{h}_{12}(1-pr)\kappa, \underline{x}_{21}(-qr)\underline{x}_{23}(q), \underline{x}_{12}(c))$$

$$= \underline{x}_{21}(-cr)\underline{x}_{23}(c)\underline{x}_{12}(q)\underline{x}_{13}(p)\tilde{h}_{12}(1-pr)<p,r>\kappa^{-1}\underline{x}_{21}(-ar)\underline{x}_{23}(a).$$

(8.5) Proposition. τ_r *induces a transformation of* $\tilde{G}_r$.

Proof. Let $\alpha \in \tilde{G}_r$, say

$$\alpha = \underline{x}_{12}(a)\underline{x}_{13}(p)\tilde{h}_{12}(1-pr)\kappa\ \underline{x}_{23}(q)\underline{x}_{21}(-qr)\underline{x}_{12}(c)$$

and if also

$$\alpha = \underline{x}_{12}(a')\underline{x}_{13}(p')\tilde{h}_{12}(1-p'r)\kappa'\ \underline{x}_{23}(q')\underline{x}_{21}(-q'r)\underline{x}_{12}(c')$$

then, by putting these expressions into standard form, it is easy to see that for $s = a'-a$ we have

$$a' = a+s,\ p' = p - \frac{sq}{u},\ \kappa' = \{v,u\} < s, -\frac{qr}{u^2} > \kappa,$$

$$q' = qv,\ c' = c - \frac{s}{u^2 v},$$

where $u = 1-pr$, $v = 1 + s\frac{qr}{u^2} \in R^*$.

After some rearranging, the formula to be proved in $\tilde{G}_r$ becomes

$$\underline{x}_{12}(q)\underline{x}_{13}(p)\tilde{h}_{12}(1-rp)<p,r> =$$

$$= \underline{x}_{21}(\frac{rs}{u^2 v})\underline{x}_{23}(-\frac{s}{u^2 v})\underline{x}_{12}(qv)\underline{x}_{13}(p - \frac{sq}{u})\tilde{h}_{12}(uv)<-\frac{qr}{u^2},s>$$

$$\{u,v\}<p - \frac{sq}{u},r>\underline{x}_{21}(-sr)\underline{x}_{23}(s).$$

We only have to consider the $\widetilde{St}'(2)$ component of $\tilde{G}$, so it remains to show that

$$\underline{x}_{21}(\frac{rs}{u^2 v})\underline{x}_{12}(qv)\underline{x}_{21}(-\frac{rs}{u^2 v^2})\underline{x}_{12}(-qv^2)\tilde{h}_{12}(\frac{1}{v})^{-1}\cdot$$

$$\cdot\tilde{h}_{12}(\frac{1}{v})\tilde{h}_{12}(uv)\tilde{h}_{12}(u)^{-1}\{u,v\}<-\frac{qr}{u^2},s><p-\frac{sq}{u},r><r,p> = 1$$

This leads to the following identity in $C(R)$ to be proved:

$$<-\frac{qr}{u^2},s><p-\frac{sq}{u},r><r,p><-qv,-\frac{rs}{u^2v^2}>\{-u^2,v\} = 1$$

By (D2) (and (D1)):

$$<p,r><-\frac{sq}{u^2},r> = <p-\frac{sq}{u},r>$$

and by (D3):

$$<-\frac{sq}{u^2},r><-\frac{rq}{u^2},s> = <-\frac{q}{u^2},rs>.$$

So it remains to prove

$$<-\frac{q}{u^2},rs><-qv,-\frac{rs}{u^2v^2}>\{-u^2,v\} = 1.$$

By (D3):

$$<-qv,-\frac{rs}{u^2v^2}> = <-q,-\frac{rs}{u^2v}><\frac{qvrs}{u^2},\frac{1}{v}>$$

$$= <-q,-\frac{rs}{u^2v}>\{\frac{1}{v},v\}.$$

The identity is now reduced to

$$<-\frac{q}{u^2},rs><-q,-\frac{rs}{u^2v}>\{u^2,v\} = 1.$$

Again by (D3):

$$<-q,-\frac{rs}{u^2v}> = <\frac{q}{u^2},\frac{rs}{v}>\{v,-u^2\}.$$

So we have to prove

$$<-\frac{q}{u^2},rs><\frac{q}{u^2},\frac{rs}{v}>\{v,-1\} = 1.$$

But

$$<-\frac{q}{u^2},rs> = <-\frac{q}{u^2v},rsv><-\frac{qrsv}{u^2},\frac{1}{v}>$$

$$= <-\frac{q}{u^2v},rsv>\{v,\frac{1}{v}\}$$

and

$$<\frac{q}{u^2},\frac{rs}{v}> = <-\frac{qrs}{u^2},-\frac{1}{v}><-\frac{q}{u^2v},-rs>$$

$$= <-\frac{q}{u^2v},-rs>,$$

so

$$<-\frac{q}{u^2},rs><\frac{q}{u^2},\frac{rs}{v}>\{v,-1\}$$

$$= <-\frac{q}{u^2v},rsv><-\frac{q}{u^2v},-rs> = <-\frac{q}{u^2v},0> = 1.$$

(8.6) Definition. We define $\tau_r : \tilde{G}_r \to \tilde{G}_r$ as the map induced by $\tau_r : A \times C_r \times B_r \times A \to \tilde{G}_r$.

(8.7) Proposition. τ_r *is an involution of the group* $\tilde{G}_r$.

Proof. First we prove that τ_r is an anti-homomorphism, i.e. $\tau_r(\alpha\beta) = \tau_r(\beta)\tau_r(\alpha)$ for all $\alpha,\beta \in \tilde{G}_r$. This is trivial when $\alpha \in A$ or $\beta \in A$, and it is easy to verify when $\alpha \in C_r$ or $\beta \in C_r$. Put

$$\alpha = \alpha_1\gamma_1\beta_1\alpha_1' \quad \text{and} \quad \beta = \alpha_2\beta_2\gamma_2\alpha_2'$$

where $\alpha_1,\alpha_1',\alpha_2,\alpha_2' \in A$; $\beta_1,\beta_2 \in B_r$; $\gamma_1,\gamma_2 \in C_r$. (Note that $\beta^{-1} \in AC_rB_rA$.) Then:

$$\tau_r(\alpha\beta) = \tau_r(\alpha_2')\tau_r(\gamma_2)\tau_r(\beta_1\alpha_1'\alpha_2\beta_2)\tau_r(\gamma_1)\tau_r(\alpha_1)$$

whereas

$$\tau_r(\beta)\tau_r(\alpha) = \tau_r(\alpha_2')\tau_r(\gamma_2)\tau_r(\beta_2)\tau_r(\alpha_1'\alpha_2)\tau_r(\beta_1)\tau_r(\gamma_1)\tau_r(\alpha_1).$$

It remains to show that for $\alpha \in A$ and $\beta_1,\beta_2 \in B_2$ the following holds:

$$\tau_r(\beta_1\alpha\beta_2) = \tau_r(\beta_2)\tau_r(\alpha)\tau_r(\beta_1).$$

Put $\beta_1 = \underline{x}_{21}(-ar)\underline{x}_{23}(a)$, $\alpha = \underline{x}_{12}(-b)$, $\beta_2 = \underline{x}_{21}(-cr)\underline{x}_{23}(c)$.

Then

$$\beta_1\alpha\beta_2 = \underline{x}_{13}(-bc)\underline{x}_{23}(a+c+abcr)\underline{x}_{21}(-ar)\underline{x}_{12}(-b)\underline{x}_{21}(-cr).$$

We bring $\underline{x}_{21}(-ar)\underline{x}_{12}(-b)\underline{x}_{21}(-cr)$ into standard form:

$$\begin{aligned}\underline{x}_{21}(-ar)\underline{x}_{12}(-b)\underline{x}_{21}(-cr) &= \underline{w}_{12}(1)\underline{x}_{12}(ar)\underline{x}_{21}(b)\underline{x}_{12}(cr)\underline{w}_{12}(1)\tilde{h}_{12}(-1)\\ &= \underline{x}_{12}\Big(\frac{1-bd}{u}+\frac{e}{u^2}\Big)\underline{x}_{21}\Big(-(ar+(1+abr)cr)\frac{u}{v}\Big)\tilde{h}_{12}\Big(-\frac{v}{u}\Big)\{v,u\}\kappa_1\kappa_2^{-1}\\ &\quad \underline{x}_{12}\left(\frac{1+(1+abr)\frac{e}{u}}{v}\right),\end{aligned}$$

where d, u, e and v are such that

$$u = -ar+d(1+abr) \in R^*,$$

$$v = -(cr+d)+e\Big(1-(cr+d)(1+abr)\frac{1}{u}\Big) \in R^*,$$

and κ_1, κ_2 are the following elements of C(R):

$$\kappa_1 = \langle -\frac{ra}{u}, bu, \frac{d}{u} \rangle$$

$$\kappa_2 = \langle -\frac{cr+d}{v}, -(1+abr)\frac{v}{u}, \frac{e}{v} \rangle .$$

Now $\beta_1\alpha\beta_2$ is in a form such that we can determine its image under τ_r. The $\widetilde{St}'(2)$ component of this image is

$$\underline{x}_{21}\left(-\frac{1+(1+abr)\frac{e}{u}}{v}r\right)\underline{x}_{12}(-\frac{v}{u}(a+b+abcr))\tilde{h}_{12}(-\frac{v}{u})$$

$$\underline{x}_{21}(-\frac{1-bd}{u}r - \frac{e}{u^2}r)\kappa_1^{-1}\kappa_2\{u,v\}\kappa_3$$

where $\kappa_3 = \langle P,r \rangle$ with $P = -(\frac{1-bd}{u}+\frac{e}{u^2})(a+b+abcr)-bc$. We have to prove that this element of $\widetilde{St}'(2)$ is equal to $\underline{x}_{12}(c)\underline{x}_{21}(br)\underline{x}_{12}(a)$, or equivalently (conjugate by $\underline{w}_{12}(1)$):

$$\underline{x}_{12}\left(\frac{1+(1+abr)\frac{e}{u}}{v}r\right)\underline{x}_{21}(\frac{v}{u}(a+b+abcr))\tilde{h}_{12}(\frac{v}{u})\tilde{h}_{12}(-1)$$

$$\underline{x}_{12}(\frac{1-bd}{u}r + \frac{e}{u^2}r)\kappa_1^{-1}\kappa_2\{u,v\}\kappa_3 =$$

$$= \underline{w}_{12}(1)\underline{x}_{12}(c)\underline{x}_{21}(br)\underline{x}_{12}(a)\underline{w}_{12}(1)\tilde{h}_{12}(-1).$$

Next we transform the right hand side into standard form; this standard form becomes:

$$\underline{x}_{12}\left(\frac{1+(1+abr)\frac{e}{u}}{v}r\right)\underline{x}_{21}(\frac{v}{u}(a+b+abcr))\tilde{h}_{12}(\frac{v}{u})\tilde{h}_{12}(-1)$$

$$\underline{x}_{12}(\frac{1-bd}{u}r + \frac{e}{u^2}r)\kappa_4^{-1}\kappa_5\{v',u'\},$$

where $d',u',e',v' \in R$ are such that

$$-a+d'(1+ab) = u' \in R^*$$

$$-(c+d')+e'(1-(c+d')(1+rab)\frac{1}{u'}) = v' \in R^*$$

the last identity being obtained by choosing e' such that

$$\frac{1-rbd'}{u'} + e'\frac{1}{u'^2} = \frac{1-bd}{u}r + \frac{e}{u^2}r$$

(then $v' = \frac{u'v}{u}$), furthermore

$$\kappa_4 = \langle -\frac{a}{u'}, rbu', \frac{d'}{u'} \rangle \quad \text{and} \quad \kappa_5 = \langle -\frac{c+d'}{v'}, -(1+rab)\frac{v'}{u'}, \frac{e'}{v'} \rangle .$$

The identity to be proved in C(R) then becomes

$$\kappa_1^{-1}\kappa_2\{u,v\}\kappa_3 = \kappa_4^{-1}\kappa_5\{v',u'\}.$$

By (C3) we have

$$\kappa_1^{-1}\kappa_4 = \langle -\frac{abu}{u'}, \frac{ru'}{u}, 1-\frac{b}{u'}(du'-d'u)\rangle$$

$$\kappa_2^{-1}\kappa_5 = \langle AB, C, D+AE\rangle,$$

where $A = 1+\frac{(1+abr)e}{u}$, $B = 1-(c+d')(1+abr)\frac{1}{u'}$, $C = \frac{ru'}{u}$, $D = \frac{e}{v}$, $E = -\frac{c+d'}{v'}$.

Then

$$\kappa_1^{-1}\kappa_4 = \langle 1+rab, \frac{d-rd'}{u}, -\frac{abu}{u'}\rangle$$

$$\kappa_2\kappa_5^{-1} = \langle \frac{rd'-d}{v}, -(1+rab)\frac{v}{u'}, -\frac{c+d'}{v'} - \frac{e}{u'v}(a+c-abcr)\rangle$$

$(= \langle 1-C(D+AE), 1-ABC, D+AE\rangle)$.

Now again by (C3):

$$\kappa_1^{-1}\kappa_4\kappa_2\kappa_5^{-1} = \langle \bar{A}\bar{B}, \bar{C}, \bar{D}+\bar{A}\bar{E}\rangle,$$

where $\bar{A} = 1+abr$, $\bar{B} = \frac{rd'-d}{v}$, $\bar{C} = -\frac{v}{u'}$, $\bar{D} = -\frac{abu}{u'}$, $\bar{E} = -\frac{c+d'}{v'} - \frac{e}{u'v}(a+c-abcr)$. Some elementary computations show that

$$1-\bar{C}\bar{D}-\bar{C}\bar{A}\bar{E} = \frac{u}{u'}P$$

so

$$\kappa_1^{-1}\kappa_4\kappa_2\kappa_5^{-1} = \langle \frac{ru'}{u}, \frac{u}{u'}P\rangle = \langle r,P\rangle\langle \frac{u'}{u}, \frac{u}{u'}rP\rangle$$

$$= \langle r,P\rangle\{\frac{u'}{u}, -\frac{v}{u}\} = \kappa_3^{-1}\{u,v\}^{-1}\{v',u'\}$$

This proves that τ_r is an anti-homomorphism. It remains to show that $\tau_r^2 = 1$, but this is easily checked on generators.

9. The group $\widetilde{St}(3)$

(9.1) The map $\pi : \tilde{G} \times R \times \tilde{G} \to St(3)$ defined by

$$\pi(\alpha,r,\beta) = \pi(\alpha)x_{31}(r)\pi(\beta)^*$$

is surjective. This follows from (2.16). An equivalence relation $\sim$ on $\tilde{G} \times R \times \tilde{G}$ is defined as follows

$$(\alpha,r,\beta) \sim (\alpha',r',\beta') \text{ if and only if}$$
$$r = r',\ \alpha^{-1}\alpha' \in \tilde{G}_r,\ \text{and } \tau_r(\alpha^{-1}\alpha') = \beta'^{-1}\beta.$$

In other words:

$$(\alpha\gamma,r,\beta) \sim (\alpha,r,\beta\tau_r(\gamma)) \text{ when } \gamma \in \tilde{G}_r.$$

Put $Y = \tilde{G} \times R \times \tilde{G}/\sim$. The class of (α,r,β) will be denoted as $[\alpha,r,\beta]$. From (2.14) it then follows that π factors through Y. So we have a surjective map

$$\pi : Y \to St(3),$$

$$\pi[\alpha,r,\beta] = \pi(\alpha)x_{31}(r)\pi(\beta)^*.$$

(9.2) The group $\tilde{G}$ operates on Y as follows:

$$(\gamma,[\alpha,r,\beta]) \to [\gamma\alpha,r,\beta].$$

Notation:

$$A_\gamma[\alpha,r,\beta] = [\gamma\alpha,r,\beta].$$

Note that $\pi(A_\gamma(y)) = \pi(\gamma)\pi(y)$.

(9.3) We are going to define a permutation W of Y such that

$$\pi(W(x)) = w_{23}(1)\pi(x).$$

First we will define π in a subset of $\tilde{G} \times R \times \tilde{G}$, namely the set

$$F \times R \times \tilde{G}$$

where $F = \{\underline{x}_{13}(p)\underline{x}_{12}(a)\underline{x}_{21}(b)\tilde{h} \mid p,a,b \in R,\ \tilde{h} \in \tilde{H}\}$. Note that F consists of the elements in $\tilde{G}$ for which the underlying matrix is of the form

$$\begin{pmatrix} * & * & * \\ * & \circledast & 0 \\ 0 & 0 & 1 \end{pmatrix},$$

where $\circledast$ stands for a unit. The definition of $W : F \times R \times \tilde{G} \to Y$ is as follows:

$$W(\underline{x}_{13}(p)\underline{x}_{12}(a)\underline{x}_{21}(b)\tilde{h}_{12}(u)\kappa,r,\beta)$$
$$= [\underline{x}_{13}(-a)\underline{x}_{12}(p)\underline{x}_{21}(\tfrac{r}{u})\tilde{h}_{12}(u),-bu,\beta\underline{w}_{12}(1)\tilde{h}_{12}(u^{-1})].$$

From (2.15) it follows that $F\times R\times\tilde{G}\to Y$ is surjective, so the problem is to show that $W(\alpha,r,\beta) = W(\alpha\gamma,r,\beta\tau_r(\gamma)^{-1})$ if $\gamma\in\tilde{G}_r$ and $\alpha,\alpha\gamma\in F$.

(9.4) Let $\alpha\in F$ and $\gamma\in\tilde{G}_r$ such that also $\alpha\gamma\in F$. We consider the underlying matrix $\phi(\gamma)$ of γ, say

$$\phi(\gamma) = \begin{pmatrix} 1-cr & e & c \\ -dr & f & d \\ 0 & 0 & 1 \end{pmatrix}.$$

Put $\alpha = \underline{x}_{13}(p)\underline{x}_{12}(a)\underline{x}_{21}(b)\tilde{h}_{12}(u)\kappa$, where $p,a,b\in R$, $u\in R^*$ and $\kappa\in C(R)$. Then

$$\phi(\alpha) = \begin{pmatrix} (1+ab)u & \frac{a}{u} & p \\ bu & \frac{1}{u} & 0 \\ 0 & 0 & 1 \end{pmatrix}.$$

Let the (2,2) entry of $\phi(\alpha\gamma)$ be $\frac{1}{v}$ $(\in R^*)$, then it easily follows that

$$\phi(\gamma) = \begin{pmatrix} 1-cr & e & c \\ bu^2cr & 1+\frac{u}{v}cr & -bu^2c \\ 0 & 0 & 1 \end{pmatrix}$$

and $1+beu^2 = \frac{u}{v}(1-cr)$.

Then also

$$\phi(\alpha\gamma) = \begin{pmatrix} (1+ab-cr)u & ue+\frac{a}{v} & p+uc \\ bu & \frac{1}{v} & 0 \\ 0 & 0 & 1 \end{pmatrix},$$

and therefore $\alpha\gamma = \underline{x}_{13}(p+uc)\underline{x}_{12}(a+uve)\underline{x}_{21}(b\frac{u}{v})\tilde{h}_{12}(v)\kappa'$, where $\kappa'\in C(R)$. Now we have

$$\gamma = \alpha^{-1}.\alpha\gamma = \tilde{h}_{12}(u)^{-1}\underline{x}_{21}(-b)\underline{x}_{12}(uve)\underline{x}_{13}(uc)\underline{x}_{21}(\tfrac{bu}{v})\tilde{h}_{12}(v)\kappa'\kappa^{-1}.$$

Let $\beta\in\tilde{G}$. The images of (α,r,β) and $(\alpha\gamma,r,\beta\tau_r(\gamma)^{-1})$ under W are:

$$W(\alpha,r,\beta) = [\underline{x}_{13}(-a)\underline{x}_{12}(p)\underline{x}_{21}(\tfrac{r}{u})\tilde{h}_{12}(u)\kappa,-bu,\beta\underline{w}_{12}(1)\tilde{h}_{12}(u^{-1})]$$

$$W(\alpha\gamma,r,\beta\tau_r(\gamma)^{-1}) =$$

$$[\underline{x}_{13}(-(a+uve))\underline{x}_{12}(p+uc)\underline{x}_{21}(\tfrac{r}{v})\tilde{h}_{12}(v)\kappa',-bu,\beta\tau_r(\gamma)^{-1}\underline{w}_{12}(1)\tilde{h}_{12}(v^{-1})].$$

So we have to prove

(B) $$\underline{w}_{12}(1)\tilde{h}_{12}(v^{-1})\tau_{-bu}(\gamma') = \tau_r(\gamma)\underline{w}_{12}(1)h_{12}(u^{-1}),$$

where

$$\gamma' = \tilde{h}_{12}(u)^{-1}\underline{x}_{21}(-\frac{r}{u})\underline{x}_{12}(uc)\underline{x}_{13}(-uve)\underline{x}_{21}(\frac{r}{v})\tilde{h}_{12}(v)\kappa'\kappa^{-1}.$$

The following Lemma shows that we can reduce this problem to some special cases.

(9.3) Lemma. (In the notations of (9.2)). *There exist* $\gamma_1,\gamma_2,\gamma_3 \in \tilde{G}_r$ *such that* $\gamma = \gamma_1\gamma_2\gamma_3$, $\alpha\gamma_1 \in F$, $\alpha\gamma_1\gamma_2 \in F$ *and*

(i) the (2,2) entries of $\alpha\gamma_1$ and $\alpha\gamma$ are congruent modulo r

(ii) the (2,2) entries of $\alpha\gamma_1\gamma_2$ and $\alpha\gamma$ are equal.

Proof. The pair $(\frac{1}{u}+beu, rbu) \in R^2$ is unimodular because $-ecu.rbu+(ucr+v)(\frac{1}{u}+beu) = 1$. Therefore, there exists an $m \in R$ such that

$$\frac{1}{u} + beu + mrbu \in R^*.$$

Take $\gamma_1 \in F$ such that $\phi(\gamma_1) = e_{12}(e+mr)$, then the (2,2) entry of $\phi(\alpha\gamma_1)$ is $\frac{1}{u}+beu+mrbu$, and

$$\frac{1}{u} + beu + mrbu \equiv \frac{1}{u} + beu \equiv 1 + \frac{u}{v}cr \equiv 1 \pmod r.$$

For the proof of (ii) we change notation: by (i) we may assume that α and $\alpha\gamma$ have in their underlying matrices (2,2) entries which are congruent modulo r. Then we have to prove that there exist γ_2,γ_3 such that $\gamma = \gamma_2\gamma_3$ and the (2,2) entries of $\alpha\gamma_2$ and $\alpha\gamma$ are equal. Again we use the notation of (9.2), but now we have

$$v = u(1-qr)$$

for some $q \in R$. Take $\gamma_2 = \underline{x}_{13}(q)\tilde{h}_{12}(1-qr)\underline{x}_{21}(buqr)\underline{x}_{23}(-buq)$.

(9.4) By the preceding lemma it suffices to prove (B) in (9.2) in the following three cases:

Case 1: $c = 0$.

Case 2: $e = 0$.

Case 3: $u = v$.

(9.5) Proof of (B) in Case 1: c = 0.

Now $\gamma = \tilde{h}_{12}(u)^{-1}\underline{x}_{21}(-b)\underline{x}_{12}(uve)\underline{x}_{21}(\frac{bu}{v})\kappa'\kappa^{-1}$, so

$$\underline{w}_{12}(1)\gamma\underline{w}_{12}(1)^{-1} = \underline{x}_{12}(bu^2)\underline{x}_{21}(-\frac{ve}{u})\tilde{h}_{12}(\frac{u}{v})\{-v,u\}\underline{x}_{12}(-buv)\kappa'\kappa^{-1}$$

$$= \underline{x}_{21}(-e)\tilde{h}_{12}(1)\{\frac{v}{u},\frac{u}{v}\}\{-v,u\}<-bu^2,\frac{ve}{u}> = \underline{x}_{21}(-e)\kappa_1,$$

where $\kappa_1 = \{v,-u\}<-bu^2,-\frac{ve}{u}>\kappa'\kappa^{-1}$.

So $\gamma = \underline{x}_{12}(e)\kappa_1$,

and

$$\tau_r(\gamma) = \underline{x}_{21}(-re)\underline{x}_{23}(e)\kappa_1^{-1}.$$

In this case

$$\gamma' = \underline{x}_{21}(rbu^3e)\underline{x}_{23}(ru^2e)\underline{x}_{13}(-ve)\tilde{h}_{12}(\frac{v}{u})\{-v,u\}\kappa'\kappa^{-1}$$

and so

$$\tau_{-bu}(\gamma') = \underline{x}_{13}(-ve)\tilde{h}_{12}(\frac{v}{u})\underline{x}_{12}(ru^2e)\kappa_2,$$

where $\kappa_2 = \{u,-v\}<-ve,-bu>\kappa\kappa'^{-1}$.

Then

$$\underline{w}_{12}(1)\tilde{h}_{12}(\frac{1}{v})\tau_{-bu}(\gamma')\tilde{h}_{12}(\frac{1}{u})\underline{w}_{12}(1)^{-1} = \underline{x}_{21}(-re)\underline{x}_{23}(e)\kappa_2\{v,-u\}.$$

So the relation to be proved in C(R) is

$$\kappa_1^{-1} = \kappa_2\{v,-u\},$$

or

$$<-\frac{ve}{u},-bu^2>\{u,-v\} = <-ve,-bu>.$$

By (D3)

$$<-\frac{ve}{u},-bu^2> = <-ve,-bu><bve,u> = <-ve,-bu>\{\frac{v}{u},u\}.$$

(9.6) Proof of (B) in case 2: e = 0.

In this case

$$\gamma = \underline{x}_{13}(c)\tilde{h}_{12}(1-cr)\{-v,u\}\underline{x}_{21}(bucvr)\underline{x}_{23}(bucv)\kappa'\kappa^{-1}$$

and

$$\gamma' = \underline{x}_{12}(\frac{c}{v})\{1-cr,-1\}\{-v,u\}<-ru,-\frac{c}{u}>\kappa'\kappa^{-1}.$$

So

$$\tau_r(\gamma) = \underline{x}_{12}(-bucv)\underline{x}_{13}(c)\tilde{h}_{12}(1-cr)<c,r>\{u,-v\}\kappa\kappa'^{-1}$$

and

$$\tau_{-bu}(\gamma') = \underline{x}_{21}(\frac{bc}{1-cr})\underline{x}_{23}(\frac{c}{v})\{1-cr,-1\}\{u,-v\}<-\frac{c}{u},-ru>\kappa\kappa'^{-1}.$$

The formula to be proved in C(R) is

$$\{u,-v\}\{1-cr,-1\}<-\frac{c}{u},-ru> = <c,r>.$$

By (D3)

$$<-\frac{c}{u},-ru> = <c,r><-\frac{cr}{u},-u> = <c,r>\{1-cr,-u\},$$

so the formula reduces to

$$\{u,-v\}\{1-cr,-1\}\{1-cr,-u\} = 1,$$

which is easy to prove.

(9.7) Proof of (B) in case 3: $u = v$.

In this case

$$\gamma = \underline{x}_{21}(-bu^2)\underline{x}_{12}(e)\underline{x}_{13}(c)\underline{x}_{21}(bu^2)$$

and

$$\gamma' = \underline{x}_{21}(-ru)\underline{x}_{12}(\frac{c}{u})\underline{x}_{13}(-eu)\underline{x}_{21}(ru).$$

Moreover we have $beu^2 = -cr$.

We will write γ in the form $\alpha_1\gamma\beta\alpha_2$, where $\alpha_1,\alpha_2 \in A$, $\beta \in B_r$ and $\gamma \in C_r$. So choose $g \in R$ such that

$$1+cr-g\ bu^2cr \in R^*$$

and put $q = gbu^2-1$, then $1-qcr \in R^*$.

Then we obtain

$$\gamma = \underline{x}_{12}(-\frac{g+eq}{1-qcr})\underline{x}_{13}(\frac{cq}{1-qcr})\tilde{h}_{12}(\frac{1}{1-qcr})\underline{x}_{12}(\frac{rbcu^2}{1-qcr})\underline{x}_{23}(-\frac{bcu^2}{1-qcr})\underline{x}_{12}(g)\kappa_1$$

where $\kappa_1 \in C(R)$, and so

$$\tau_r(\gamma) = \underline{x}_{21}(-gr)\underline{x}_{23}(g)\underline{x}_{12}(-\frac{bcu^2}{1-qcr})\underline{x}_{13}(-\frac{cq}{1-qcr})\tilde{h}_{12}(\frac{1}{1-qcr})$$

$$\underline{x}_{21}(\frac{gr+eqr}{1-qcr})\underline{x}_{23}(-\frac{g+eq}{1-qcr})\kappa_1^{-1}<-\frac{cq}{1-qcr},r>.$$

It easily follows that

$$\tilde{h}_{12}(\tfrac{1}{u})^{-1}\underline{w}_{12}(-1)\tau_r(\gamma)\underline{w}_{12}(-1)\tilde{h}_{12}(\tfrac{1}{u}) =$$

$$\underline{x}_{12}(gru^2)\underline{x}_{13}(equ)\tilde{h}_{12}(1-qcr)\underline{x}_{21}(bc(1-qcr))\underline{x}_{23}(\frac{c(1-qcr)}{u})$$

$$\underline{x}_{12}(-ru^2\,\frac{g+eq}{1-qcr})\{u,1-qcr\}^2 <-\frac{cq}{1-qcr},r>\kappa_1^{-1}.$$

Under τ_{-bu} this maps to:

$$\underline{x}_{21}(-bru^3\,\frac{g+eq}{1-qcr})\underline{x}_{23}(-ru^2\,\frac{g+eq}{1-qcr})\underline{x}_{12}(\frac{c}{u}(1-qcr))\underline{x}_{13}(equ)$$

$$\tilde{h}_{12}(1-qcr)\underline{x}_{21}(bgu^3r)\underline{x}_{23}(gru^2)\{1-qcr,u\}^2<r,-\frac{cq}{1-qcr}>$$

$$\kappa_1<equ,-bu>.$$

We have to prove that this element of $\tilde{G}_r$ is equal to

$$\gamma' = \underline{x}_{21}(-ru)\underline{x}_{12}(\tfrac{c}{u})\underline{x}_{13}(-eu)\underline{x}_{21}(ru).$$

This leads to the following identity in $C(R)$ to be proved

$$\kappa_1\{1-qcr,u\}^2<r,-\frac{cq}{1-qcr}><equ,-bu> = <\frac{qru}{1-qcr},-\frac{c}{u}>,$$

where we still have to compute κ_1.

By (D3) we have

$$<r,\frac{-cq}{1-qcr}> = <-\frac{cr}{u},\frac{qu}{1-qcr}><\frac{qru}{1-qcr},-\frac{c}{u}>,$$

so

$$<r,\frac{-cq}{1-qcr}><-\frac{c}{u},\frac{qru}{1-qcr}> = <-\frac{cr}{u},\frac{qu}{1-qcr}>$$

$$= <-cr,\frac{q}{1-qcr}><\frac{1}{u},\frac{-qucr}{1-qcr}> = <-cr,\frac{q}{1-qcr}>\{\frac{1}{u},\frac{1}{1-qcr}\}.$$

We also have

$$<equ,-bu> = <u,eqbu><eq,-bu^2> = \{u,1-qcr\}<eq,-bu^2>,$$

so the relation reduces to

$$\kappa_1<eq,-bu^2> = <\frac{q}{1-qcr},-cr>.$$

By definition κ_1 satisfies

$$\underline{x}_{12}(-\frac{g+eq}{1-qcr})\underline{x}_{21}(rbcu^2(1-qcr))\tilde{h}_{12}(\frac{1}{1-qcr})\kappa_1\underline{x}_{12}(g) =$$

$$\underline{x}_{21}(-bu^2)\underline{x}_{12}(e)\underline{x}_{21}(bu^2).$$

κ_1 is computed by putting the right hand side in standard form. This leads to

$$\kappa_1 = <-\frac{bu^2}{t},-et,\frac{m}{t}><-\frac{n}{t}(1-qcr),\frac{1+cr}{1-qcr},\frac{(bu^2-m)(1-qcr)}{-t}>\{t,1-qcr\}$$

where $m,n,t \in R$ are such that

$$t = -bu^2+m(1+cr) \in R^*$$

and

$$\frac{1+me}{t} + \frac{n}{t^2} = -\frac{g+eq}{1-qcr}.$$

By (C4) we have

$$<-\frac{bu^2}{t},-et,\frac{m}{t}> = <-\frac{bu^2}{t},-et\ \frac{1+q}{1-cqr},\frac{m}{t}(1-cqr)><-\frac{bu^2}{t},eqt,\frac{1+\frac{bu^2}{t}}{1-qcr}> \quad ,$$

and so

$$<-\frac{bu^2}{t},-et,\frac{m}{t}><eqt,-\frac{bu^2}{t}> = <-\frac{bu^2}{t},-et\ \frac{1+q}{1-cqr},\frac{m}{t}(1-cqr)>$$

and

$$<-\frac{bu^2}{t},-et,\frac{m}{t}><eq,-bu^2> = <-\frac{bu^2}{t},-et\ \frac{1+q}{1-cqr},\frac{m}{t}(1-cqr)>\{1-cqr,t\}.$$

So it remains to prove that

$$<-\frac{bu^2}{t},-et\ \frac{1-q}{1-cqr},\frac{m}{t}(1-cqr)><\frac{bu^2cr}{t},-\frac{n}{t}(1-cqr),\frac{1+cr}{1-cqr}><-cr,\frac{q}{1-qcr}> = 1.$$

This however follows from (D3):

$$<A,BC,D><B,AC,E> = <AB,C,D+AE>$$

where

$$A = -\frac{bu^2}{t},\ B = -cr,\ C = 1+\frac{n}{t}(1+cr),\ D = \frac{m}{t}(1-cqr),$$

and $E = 1-qcr$.

Note also that $<-cr,\frac{q}{1-qcr}> = <-q,-cr>$.

(9.8) We are going to define for each $v \in R^*$ a permutation $H(v)$ of Y such that

$$\pi(H(v)y) = h_{23}(v)\pi(y) \quad \text{for all } y \in Y.$$

First define a permutation h_v of X by

$$h_v[a,\tilde{x}_{21}(b)\tilde{h},c] = [\frac{a}{v},\tilde{x}_{21}(bv)\tilde{h}\{\sigma(\tilde{h}),v\},\frac{c}{v}].$$

It is easy to verify that h_v is well-defined, note also that $h_v h_{\frac{1}{v}} = 1$. Then

$$\pi(h_v(x)) = h_{23}(v)\pi(x)h_{23}(v)^{-1} \text{ for all } x \in X.$$

Next define a permutation η_v of $\widetilde{St}'(2)$ by

$$\eta_v(\alpha) = h_v \alpha h_v^{-1} \quad \text{for all } \alpha \in \widetilde{St}'(2)$$

and extend it to $\tilde{G}$ by

$$\eta_v(\underline{x}_{13}(p)\underline{x}_{23}(q)\alpha) = \underline{x}_{13}(pv)\underline{x}_{23}(qv^2)\eta_v(\alpha),$$

for $p,q \in R$, $\alpha \in St'(2)$. Then we have

$$\pi(\eta_v(\alpha)) = h_{23}(v)\pi(\alpha)h_{23}(v)^{-1} \quad \text{for all } \alpha \in \tilde{G}.$$

The definition of $H(v)$ now becomes

$$H(v)[\alpha,r,\beta] = [\eta_v(\alpha),\frac{r}{v},\beta\tilde{h}_{12}(\frac{1}{v})].$$

The proof of the fact that this map is well-defined comes down to

$$\tau_{\frac{r}{v}}(\eta_v(\gamma)) = \tilde{h}_{12}(\frac{1}{v})^{-1}\tau_r(\gamma)\tilde{h}_{12}(\frac{1}{v}) \text{ for all } \gamma \in \tilde{G}_r,$$

which is easily verified on generators of $\tilde{G}_r$.

(9.9) Definition. The group $\widetilde{St}(3)$ is defined as the group of permutations of Y generated by all

$$A_\alpha \text{ with } \alpha \in \tilde{G}, \text{ and } W.$$

(9.10) For every permutation P of Y there is a permutation P^* defined by

$$P^*(y) = (P(y^*))^*,$$

where $y^* = [\beta,r,\alpha]$ if $y = [\alpha,r,\beta] \in Y$.

In particular we have permutations A_γ^* for all $\gamma \in \tilde{G}$:

$$A_\gamma^*[\alpha,r,\beta] = [\alpha,r,\gamma\beta]$$

and permutations W^* and $H(v)^*$.

(9.11) Proposition. *Every permutation in* $\widetilde{St}(3)$ *commutes with all* A_γ^*, *where* $\gamma \in \tilde{G}$, *and also with* $H(v)^*$, *where* $v \in R^*$.

Proof. There is only one non-trivial case:

$$WH(v)^* = H(v)^*W,$$

but still the proof of this is straightforward.

(9.12) Definition. $V = W^{-1}A_{\underline{w}_{12}(1)}W$.

(Note that $\pi(V(y)) = w_{13}(1)\pi(y)$ for all $y \in Y$).

(9.13) Lemma. $W^{-1} = H(-1)^{-1}W$ *and* $V^{-1} = A^{-1}_{\tilde{h}_{12}(-1)}H(-1)^{-1}V$.

The proof is easy.

(9.14) Proposition. *For all* $c \in R$ *we have* $VA_{\underline{x}_{23}(c)} = A_{\underline{x}_{21}(c)}V$.

Proof. We have to prove

$$VA_{\underline{x}_{23}(c)}[\alpha,r,\beta] = A_{\underline{x}_{23}(c)}V[\alpha,r,\beta]$$

for all $\alpha,\beta \in \tilde{G}$, $r \in R$. We may assume $\alpha \in F$ (see (9.1)) and by (9.11) we may assume $\beta = 1$, and moreover that α is of the form

$$\alpha = \underline{x}_{13}(p)\underline{x}_{12}(a)\underline{x}_{21}(b).$$

We will show that

$$W^*VA_{\underline{x}_{23}(c)}[\alpha,r,1] = W^*A_{\underline{x}_{21}(c)}V[\alpha,r,1].$$

First compute $W^*VA_{\underline{x}_{23}(c)}[\alpha,r,1]$. It is equal to

$$[\underline{x}_{21}(c)\underline{x}_{13}(\tfrac{1-rd}{u})\underline{x}_{12}(a\tfrac{1-rd}{u})\underline{w}_{12}(1)\{\tfrac{1}{u},-1\}\kappa_1,a,$$
$$\underline{x}_{13}(db)\underline{x}_{12}(d)\underline{x}_{21}(-\tfrac{1+pr}{u})\underline{x}_{23}(-b)\tilde{h}_{12}(u)]$$

where $u,d \in R$ satisfy

$$-p+d(1+pr) = u \in R^*$$

and κ_1 is such that

$$\underline{x}_{12}(p)\underline{x}_{21}(r) = \underline{x}_{12}(\tfrac{1-rd}{u})\underline{x}_{21}(-(1+pr)u)\tilde{h}_{12}(u)^{-1}\underline{x}_{12}(d)\kappa_1.$$

By the equivalence relation in $\tilde{G} \times R \times \tilde{G}$ it can be transformed into

$$[\underline{x}_{21}(c)\underline{w}_{12}(1)\{\tfrac{1}{u},-1\}\kappa_1,a,\underline{x}_{23}(-b)\underline{x}_{12}(d)\underline{x}_{21}(-\tfrac{1+pr}{u})\tilde{h}_{12}(u)\underline{x}_{12}(\tfrac{1-rd}{u})].$$

Then by the definition of κ_1 it is equal to

$$[\underline{x}_{21}(c)\underline{w}_{12}(1),a,\underline{x}_{23}(-b)\underline{x}_{21}(r)\underline{x}_{12}(p)w_{12}(1)]$$
$$= [w_{12}(1),a,\underline{x}_{23}(-b)\underline{x}_{21}(r)\underline{x}_{12}(p)\underline{w}_{12}(1)\underline{x}_{21}(ac)\underline{x}_{23}(-c)].$$

By a similar computation

$$W^{*}A_{\underline{x}_{21}(c)}V[\alpha,r,1] =$$

$$[\underline{w}_{12}(1),a,\underline{x}_{13}(-c)\underline{x}_{23}(-(b+rc))\underline{x}_{21}(r)\underline{x}_{12}(p-ac)\underline{w}_{12}(1)]$$

which is the same. Since W^{*} is a permutation we also have

$$A_{\underline{x}_{21}(c)}V[\alpha,r,1] = VA_{\underline{x}_{23}(c)}[\alpha,r,1].$$

(9.15) Put $X_{ij}(r) = A_{\underline{x}_{ij}(r)}$ for $r \in R$ and

$$(i,j) \in \{(1,2),(1,3),(2,1),(2,3)\}$$

and put

$$X_{31}(r) = WX_{21}(-r)W^{-1},$$

$$X_{32}(r) = WX_{23}(-r)W^{-1}.$$

(9.16) Lemma. *Conjugation by* W *and* V *on the* $X_{ij}(r)$ *is as follows:*

(1) $WX_{12}(r)W^{-1} = X_{13}(-r)$

(2) $WX_{13}(r)W^{-1} = X_{12}(r)$

(3) $WX_{21}(r)W^{-1} = X_{31}(-r)$

(4) $WX_{31}(r)W^{-1} = X_{21}(r)$

(5) $WX_{23}(r)W^{-1} = X_{32}(-r)$

(6) $WX_{32}(r)W^{-1} = X_{23}(-r)$

(7) $VX_{12}(r)V^{-1} = X_{32}(-r)$

(8) $VX_{32}(r)V^{-1} = X_{12}(r)$

(9) $VX_{21}(r)V^{-1} = X_{23}(-r)$

(10) $VX_{23}(r)V^{-1} = X_{21}(r)$

(11) $VX_{13}(r)V^{-1} = X_{31}(-r)$

(12) $VX_{31}(r)V^{-1} = X_{13}(-r)$.

Proof.

(1) and (2) are straightforward.

(3) and (5) hold by definition.

(4) and (6) follow from (3) and (5) and (9.13).

(7) and (8) follow from (1) and (5).

(9) and (10) follows from (9.13) and (9.14).

(11) and (12) follow from (2) and (4).

(9.17) Proposition. *The* $X_{ij}(r)$ *satisfy the Steinberg relations.*

Proof. The relations are trivial when they only involve $X_{ij}(r)$ with $i \neq 3$. The remaining Steinberg relations can be put in this form by conjugation with W or V.

(9.18) <u>Corollary</u>. *There is a homomorphism* $\lambda : St(3) \to \widetilde{St}(3)$ *which sends* $x_{ij}(r)$ *to* $X_{ij}(r)$.

(9.19) The composition

$$C(R) \overset{\pi}{\to} St(3) \overset{\lambda}{\to} \widetilde{St}(3) \to Y$$

sends $\kappa \in C(R)$ to $[\kappa,0,1] \in Y$, because $\lambda\pi(\kappa) = A_\kappa$, which is true on generators. This map clearly is injective, so the same holds for $\pi : C(R) \to St(3)$. Hence we proved:

(9.20) <u>Theorem</u>. *The homomorphism* $\pi : C(R) \to K_2(R)$ *which sends* $<a,b,c>$ *to* $<a,b,c>_*$ *is an isomorphism.*

(9.21) It is not hard to show that in fact $\lambda : St(3) \to \widetilde{St}(3)$ is the inverse of $\pi : \widetilde{St}(3) \to St(3)$ defined by

$$\pi(\sigma) = \pi(\sigma[1,0,1]) \quad \text{for } \sigma \in \widetilde{St}(3).$$

A few more relations in $\widetilde{St}(3)$ have to be checked.

References

[1] R.K. Dennis, "Surjective stability for the functor K_2" in: Lecture Notes Math., Vol. 353, Springer-Verlag, Berlin (1973) 85-94.

[2] W. van der Kallen, "The K_2 of rings with many units", Ann. scient. Éc. Norm. Sup. 4e serie, t.10 (1977) 473-515.

[3] F. Keune, "Another presentation for the K_2 of a local domain", J. of Pure and Applied Algebra, <u>22</u> (1981) 131-141.

[4] M. Kolster, "On injective stability for K_2", in: Lecture Notes Math., Vol. 966, Springer-Verlag, Berlin (1982) 128-168.

[5] A.A. Suslin and M.S. Tulenbaev, "Stabilization theorem for the Milnor K_2-functor", J. of Soviet Math., Vol. 17 (1981), No. 2, 1804-1819.

[6] L.N. Vasershtein, "Stabilization for the Milnor K_2-functor", Usp. Mat. Nauk, <u>38</u>, No. 1, p.224 (1975).

Mathematisch Instituut
Katholieke Universiteit
Toernooiveld
6525 ED Nijmegen
The Netherlands

On the Birch-Tate conjecture for maximal real subfields of cyclotomic fields

Manfred Kolster

The Birch-Tate conjecture (cf. [1], [9]) predicts, that for any totally real algebraic number field F with ring of integers o_F the order of the finite group $K_2(o_F)$ is given by $w_2(F)\cdot|\zeta_F(-1)|$, where ζ_F denotes the zeta-function of F and $w_2(F)$ is the number of roots of unity in the field extension $F(\sqrt{F})$. The proof of the main conjecture in Iwasawa-theory by Mazur and Wiles (cf. [4], [8]) implies, that the Birch-Tate conjecture is true up to 2-torsion for abelian number fields. In this note we show, that the 2-primary part of the conjecture holds as well in case F is the maximal real subfield of the cyclotomic field $\mathbb{Q}(\mu)$, where μ is a primitive 2^m-th root of unity, $m \geq 2$.

The proof is elementary. We compute the 2-Sylow subgroup of $K_2(o_F)$, which turns out to be isomorphic to $(\mathbb{Z}/2\mathbb{Z})^r$, $r = [F:\mathbb{Q}]$. Then we show, that 2^r is the exact 2-power dividing $w_2(F)\cdot|\zeta_F(-1)|$.

Let F be an arbitrary number field with ring of integers o_F. The following result - implicitely contained in [10], §6 - has been noticed by Browkin [2]: Let S be the set of all infinite and all dyadic primes of F, C^S the class-group of the ring of S-integers in F and c the number of complex places of F:

<u>Proposition 1</u>: The 2-rank of $K_2(o_F)$ is equal to $|S|-c-1 + 2\text{-rk}(C^S)$.

<u>Proof</u>: By Tate ([10], Thm. 6.1) any element x of order ≤ 2 in $K_2(F)$ is of the form $x = \{-1,b\}$. Now $x \in K_2(o_F)$ if and only if all tame symbols vanish on x, hence if and only if b has even valuation $v_p(b)$ at all

primes p not in S. Moreover, the group $\{b \in F^* \mid \{-1,b\} = 1\}/F^{*2}$ is elementary abelian of order 2^{1+c} (cf. [10], Thm. 6.3), hence it suffices to show, that the 2-rank of the group

$$\Delta_S := \{b \in F^* \mid v_p(b) \equiv 0 \quad (2) \quad \text{for all } p \notin S\}/F^{*2}$$

is equal to $|S| + 2\text{-rk}(C^S)$. Let C_2^S denote the subgroup of C^S of elements of order ≤ 2 and let U_S denote the group of S-units. The claim follows from the S-unit theorem via the exact sequence

$$0 \to U_S/U_S^2 \to \Delta_S \to C_2^S \to 0,$$

where the map $\Delta_S \to C_2^S$ is given by $b \to$ class of $\prod_{p \notin S} p^{\frac{1}{2}v_p(b)}$.

We fix the following notations:

μ_m primitive 2^m-th root of unity, $m \geq 2$

$E_m := \mathbb{Q}(\mu_m + \bar{\mu}_m)$

o_m ring of integers in E_m

$r_m := [E_m : \mathbb{Q}] = 2^{m-2}$

Now, in o_m there is only one dyadic prime, which is principal, hence $|S| = r_m+1$ and C^S is equal to the class-group of o_m, which has odd order by a theorem of Weber (cf. [12], [5] Satz 7). Thus Prop. 1 yields, that the 2-rank of $K_2(o_m)$ is equal to r_m. Let $\varepsilon_1, \ldots, \varepsilon_{r_m-1}$ be a system of fundamental units in E_m and let $\varepsilon_0 = -1$. Again by a theorem of Weber (cf. [5], Satz 7) the extended system $\varepsilon_0, \varepsilon_1, \ldots, \varepsilon_{r_m-1}$ of fundamental units has independant signatures at the r_m real places $v_1, \ldots, v_{r_m}$, i.e. given any product $\varepsilon = \varepsilon_{i_1} \cdot \ldots \cdot \varepsilon_{i_k}$ of distinct fundamental units there exists a real place v, such that ε is negative at v. Let c_v denote the real symbol belonging to v. Then $c_v(\{-1,\varepsilon\}) = -1$. Hence the Steinberg-symbols $\{-1,\varepsilon_i\}$, $i = 0,\dots,r_m-1$, generate a subgroup of $K_2(o_m)$, which is isomorphic to $(\mathbb{Z}/2\mathbb{Z})^{r_m}$. Since the 2-rank of $K_2(o_m)$ is equal to r_m all elements of order 2 lie in this subgroup. If $K_2(o_m)$ contained an element x of order 4, x^2 would be equal to $\{-1,\varepsilon\}$ with ε as above, hence $c_v(x^2) = -1$ for some v, which is impossible, since c_v vanishes on squares. Thus we have proved:

<u>Proposition 2</u>: The 2-Sylow-subgroup of $K_2(o_m)$ is isomorphic to $(\mathbb{Z}/2\mathbb{Z})^{r_m}$, generated by the Steinberg-symbols $\{-1,\varepsilon_i\}$, $i = 0,\dots,r_m-1$.

In view of this result we have to show that 2^{r_m} is the exact 2-power dividing $w_2(E_m)\cdot|\zeta_{E_m}(-1)|$ in order to verify the Birch-Tate-conjecture. Now we have (cf. [11], Thm. 4.3)

$$\zeta_{E_m}(-1) = \prod_{\chi} L(-1,\chi),$$

where the product is taken over the L-series belonging to even Dirichlet-characters mod 2^m. For each such character let f_χ denote the conductor. Then (cf. [11], Prop. 4.1, Thm. 4.2):

$$(*) \qquad L(-1,\chi) = -\frac{f_\chi}{2}\sum_{t=1}^{f_\chi} \chi(t)B_2\left(\frac{t}{f_\chi}\right),$$

where $B_2(x) = x^2-x+\frac{1}{6}$ is the second Bernoulli-polynomial. We get

$$\zeta_{E_{m+1}}(-1) = \zeta_{E_m}(-1)\circ \prod_{\substack{\chi \\ f_\chi=2^{m+1}}} L(-1,\chi).$$

Since we want to proceed by induction, we have to evaluate $\prod_{\substack{\chi \\ f_\chi=2^{m+1}}} L(-1,\chi)$. Thus let χ be a Dirichlet-character mod 2^{m+1} with conductor $f_\chi = 2^{m+1}$. χ is uniquely determined by its value at 5, which has to be a primitive 2^{m-1}-th root of unity, say $\chi(5) = \mu_{m-1}$.

Lemma 3: With the above notations

$$L(-1,\chi) = \sum_{i=o}^{2^{m-2}-1} b_i \mu_{m-1}^{i} \in \mathbb{Z}[\mu_{m-1}],$$

where the b_i are odd integers, independant of χ.

Corollary 4: Let $x = \sum_{i=o}^{2^{m-2}-1} b_i \mu_{m-1}^{i}$. Then

$$\zeta_{E_{m+1}}(-1) = \zeta_{E_m}(-1) \circ N_{\mathbb{Q}(\mu_{m-1})/\mathbb{Q}}(x).$$

Proof: As χ runs through the 2^{m-2} Dirichlet-characters mod 2^{m+1} of conductor $f_\chi = 2^{m+1}$, the value $\chi(5)$ runs through the Galois-conjugates of μ_{m-1}. Hence the result follows from Lemma 3.

Proof of Lemma 3: Let $t_1,\ldots,t_{2^{m-1}} \in \mathbb{Z}$ be representatives of the powers of 5 mod 2^{m+1} with $o < t_j < 2^{m+1}$. If $o < b < 2^{m+1}$ is an odd integer different from t_j for all j, then for some k $-b \equiv t_k \bmod 2^{m+1}$, hence $b = 2^{m+1}-t_k$. Since $\chi(b) = \chi(t_k)$ and $B_2(\frac{b}{f_\chi}) = B_2(\frac{t_k}{f_\chi})$, we can rewrite (*) as

$$L(-1,\chi) = -f_\chi \cdot \sum_{k=1}^{2^{m-1}} \chi(t_k) \cdot B_2(\frac{t_k}{f_\chi}).$$

Let $o < t_j < 2^m$. There is precisely one t_k, such that $\chi(t_k) = -\chi(t_j)$, given by $t_k \equiv 5^{2^{m-2}} \cdot t_j \bmod 2^{m+1}$. Now $5^{2^{m-2}} \equiv 1+2^m \bmod 2^{m+1}$, hence $t_k \equiv t_j+2^m \bmod 2^{m+1}$, hence $t_k = t_j+2^m$. Now an easy calculation shows that

$$\chi(t_j) \cdot B_2(\frac{t_j}{f_\chi}) + \chi(t_k) \cdot B_2(\frac{t_k}{f_\chi}) = \frac{\chi(t_j)}{f_\chi} \cdot (2^{m-1}-t_j).$$

But this proves the claim, since $2^{m-1}-t_j$ is odd and $\chi(t_j)$ runs - up to sign - through the different powers $\mu_{m-1}^{o}, \mu_{m-1},\ldots,\mu_{m-1}^{2^{m-2}-1}$ of μ_{m-1}, if t_j runs through the 2^{m-2} representatives with $0 < t_j < 2^m$.

Lemma 5: Let $m \geq 2$ and $x = \sum_{i=o}^{2^{m-2}-1} b_i \mu_{m-1}^i \in \mathbb{Z}[\mu_{m-1}]$ with odd integral coefficients b_i. Then 2^{r_m-1} is the exact power of 2 dividing $N_{\mathbb{Q}(\mu_{m-1})/\mathbb{Q}}(x)$.

Proof: We proceed by induction on m. The case m = 2 is trivial, hence assume $m \geq 3$. Write $x = y+\mu_{m-1}z$, where $y,z \in \mathbb{Z}[\mu_{m-1}^2] = \mathbb{Z}[\mu_{m-2}]$. Then $N_{\mathbb{Q}(\mu_{m-1})/\mathbb{Q}(\mu_{m-2})}(x) = y^2-\mu_{m-1}^2 z^2$. We show, that the coefficients of $y^2-\mu_{m-1}^2 z^2$ with respect to the integral basis given by $1,\mu_{m-1}^2,\mu_{m-1}^4,\ldots,\mu_{m-1}^{2^{m-2}-2}$ are congruent 2 mod 4. To do this we may assume, that $y = z = 1+\mu_{m-1}^2+\ldots+\mu_{m-1}^{2^{m-2}-2}$. In this case $y^2-\mu_{m-1}^2 z^2 = 2\cdot y$. Hence in general $y^2-\mu_{m-1}^2 z^2 = 2\cdot x'$, where x' has odd integer coefficients. Hence the result follows by induction.

Proposition 6: Let $m \geq 2$. Then 2^{r_m} is the exact power of 2 dividing $w_2(E_m)\cdot|\zeta_{E_m}(-1)|$.

Proof: We proceed by induction on m: If m = 2, we get the usual ζ-function, which is $-\frac{1}{12}$ at -1. Since $w_2(E_m) = 3\cdot 2^{m+1}$ for any m, the induction starts. Now

$$w_2(E_{m+1})\cdot|\zeta_{E_{m+1}}(-1)| = 2\cdot w_2(E_m)\cdot|\zeta_{E_m}(-1)|\cdot N_{\mathbb{Q}(\mu_{m-1})/\mathbb{Q}}(x)$$

by Corollary 4. By induction 2^{r_m} is the exact power of 2 dividing $w_2(E_m)\cdot|\zeta_{E_m}(-1)$ and by Lemma 5 2^{r_m-1} is the exact power of 2 dividing $N_{\mathbb{Q}(\mu_{m-1})/\mathbb{Q}}(x)$, hence $2\cdot 2^{r_m}\cdot 2^{r_m-1} = 2^{2r_m} = 2^{r_{m+1}}$ is the exact power of 2 dividing $w_2(E_{m+1})\cdot|\zeta_{E_{m+1}}(-1)|$.

References

1. Birch, B.: K_2 of global fields, Proc. Symp. Pure Math. 20 (1969), 87-95

2. Browkin, J.: The functor K_2 of the ring of integers of a number field, Banach Center Publications, vol. 9, Warszawa 1983

3. Browkin, J.; Schinzel, A.: On Sylow 2-subgroups of $K_2\mathcal{O}_F$ for quadratic number fields F; J. reine angew. Math. 331 (1982), 104-113

4. Coates, J.: The work of Mazur and Wiles on cyclotomic fields, Sém. Bourbaki 575 (Juin 1981)

5. Hasse, H.: Über die Klassenzahl abelscher Zahlkörper, Akademie-Verlag, Berlin 1952

6. Hurrelbrink, J.: On the size of certain K-groups, Comm. in Algebra 10 (1982), 1873-1889

7. Hurrelbrink, J.: $K_2(\mathcal{O})$ for two totally real fields of degree three and four, Algebraic K-theory, Proc., Oberwolfach 1980, Part I, Lect. Notes in Math. 966, Springer 1982, 112-114.

8. Mazur, B.; Wiles, A.: Class fields of abelian extensions of $\mathbb{Q}$, preprint

9. Tate, J.: Symbols in arithmetic, Actes du Congrès Int. des Mathématiciens 1970, Tome 1, Gauthier Villars, Paris 1971, 201-211

10. Tate, J.: Relations between K_2 and Galois cohomology, Inv. math. 36 (1976), 257-274

11. Washington, L.C.: Introduction to Cyclotomic Fields, Springer, New York-Heidelberg-Berlin 1982

12. Weber, H.: Theorie der Abelschen Zahlkörper, Acta math. 8 (1886), 193-263

Westf. Wilhelms-Universität
Mathematisches Institut
Einsteinstr. 62
D-4400 Münster
West-Germany

EQUIVARIANT K-THEORY AND THE COHOMOLOGY OF PROFINITE GROUPS

Aderimi O. Kuku
Department of Mathematics,
University of Ibadan,
Ibadan, NIGERIA.

Introduction

The aim of this paper is to extend the constructions and results in [4] and [5] from finite to profinite groups.

So, if π is a profinite group, T a π-set, C an exact category in the sence of Quillen, $\hat{\pi}$ the category of π-sets, then we construct, for all $n \geqslant 0$, equivariant K-functors $K_n^{\pi}(-,C)$, $K_n^{\pi}(-,C,T)$, $P_n^{\pi}(-,C,T): \hat{\pi} \to \underline{Z\text{-mod}}$.

In section two, we prove that if π is a profinite group, k a field of characteristic p, then for each $n \geqslant 0$, the Cartan homomorphism $K_n(k\pi) \to G_n(k\pi)$ induces an isomorphism $Z(\frac{1}{p}) \otimes K_n(k\pi) \tilde{\to} Z(\frac{1}{p}) \otimes G_n(k\pi)$. In the process of doing this, we prove in 2.3 that if R is a commutative ring with identity, $\pi = \varprojlim \pi_\gamma$ where $\pi_\gamma = \pi/\gamma$ and γ runs through the open normal subgroups of π, then $K_n(R\pi) = \varinjlim K_n(R\pi_\gamma)$ and $G_n(R\pi) \simeq \varinjlim G_n(R\pi_\gamma)$.

In section three, we draw attention to the fact that since $\hat{\pi}$ is a category with final object, finite sums and finite pull-backs, there is a cohomology theory for Mackey functors: $\hat{\pi} \to \underline{Z\text{-mod}}$, as a result of which some vanishing theorems are obtained for the cohomology of π and the K-functors constructed in §1.

This work was concluded and written up while I was visiting the Mathematics Department at Queen's University, Kingston, Canada, and I like to thank the department for its hospitality and financial support. Part of the work was done while I was visiting the Mathematics Department at the University of Illinois at Urbana, U.S.A. and I also would like to thank the Department for its hospitality and financial support.

§1. Let π be a profinite group. The aim of this section is to describe briefly the constructions of the algebraic K-functors: $\hat{\pi} \to \underline{Z\text{-mod}}$ we shall be involved with. These constructions are similar to the case of K-functors: $\hat{\pi} \to \underline{Z\text{-mod}}$ π finite, and the reader is referred to [4] and [5] for details.

1.1. Let I be a filtered index set (i.e. I is a partially ordered set such that for $i,j \in I$, there exists $k \in I$ such that $k \leqslant i$, $k \leqslant j$). Note that I is a category whose objects are elements of I, with exactly one morphism from i to j if $i \leqslant j$ and no morphism otherwise. Suppose that C is a category. A projectively (resp. injectively) filtered system of C-objects is a covariant (resp. contravariant) functor $F: I \to C$. A projective limit (X,μ_i) of a projectively filtered system of C-objects is a C-object X together with maps $\mu_i: X \to F_i$ such that for all $i,j \in I$ with $i < j$ and C-morphisms $\phi_{ij}:F_i \to F_j$, we have $\mu_j = \phi_{ij}\mu_i$ and moreover, given another C-object Y and morphisms $\nu_i: Y \to F_i$ with the same properties, then there exists exactly one map $\rho: Y \to X$ such that $\mu_i\rho = \nu_i$ for any $i \in I$. Note that (X,μ_i) is well-defined up to isomorphism by F. We shall denote the projective limit by $\varprojlim F$ or $\varprojlim F_i$. The injective limit $\varinjlim F$ is analogously defined.

Although we shall apply the above definition to some other situations (e.g. 2.2(c)), we mention a particular example just to fix notations. Let I be the set of open normal subgroups γ of a profinite group π, ordered by set inclusions, $F_\gamma = \pi/\gamma$, and for $\gamma \leqslant \gamma'$, $\phi_{\gamma\gamma'}: \pi/\gamma \to \pi/\gamma'$ the canonical map. Then $\pi = \varprojlim F$. Note that each π/γ is a finite group. We shall henceforth denote π/γ by π_γ.

1.2(a). Let π be a profinite group. A π-set is a finite set with discrete topology on which π acts continuously by permutations on the left i.e. $\pi \times S \to S$ is continuous. The π-sets form a category $\hat{\pi}$ with initial objects which we shall denote by ϕ, final object which we shall denote by $*\pi = \pi/\pi$, sums (disjoint union) and products (Cartesian products) and more generally finite projective and injective limits. If S,T are π-sets, we write $\hat{\pi}(S,T)$ for the set of $\hat{\pi}$-morphisms from S to T.

(b) A simple π-set has to form π/γ where γ is some open subgroup of π and so any π-set S has the form $\dot{\cup}\, \pi/\gamma$ where γ runs through some finite collection U of open subgroups of π.

(c) If S is a π-set, we can associate with S a category $\underline{S}$ as follows - the objects of $\underline{S}$ are elements of S while for $s,t \in S$, $\underline{S}(s,t) = \{(g,s) \mid g \in \pi, gs = t\}$ with composition defined for $t = gs$ by $(h,t)o(g,s) = (hg,s)$ and the identity morphism s → s given by (e,s) where e is the identity element of π.

(d) Now, let C be an exact category in the sense of Quillen [8], $[\underline{S},C]$ the category of covariant functors from $\underline{S}$ to C which factor through some finite quotient of π, that is, such that we have a functor $\varinjlim_{\gamma}[\underline{S}^{\gamma},C] \to [\underline{S},C]$ where γ runs through all open normal subgroups of π which act trivially on S and $\underline{S}^{\gamma}$ corresponds to S as a π/γ-set. If $\zeta \in [\underline{S},C]$, we shall write ζ_s for $\zeta(s)$. It can be easily checked that $[\underline{S},C]$ is an exact category (just as in the case for π finite) where a sequence $\zeta' \to \zeta \to \zeta''$ is said to be exact if it is exact fibre-wise i.e. $\zeta'_s \to \zeta_s \to \zeta''_s$ is exact for all $s \in S$. We now define $K_n^{\pi}(S,C)$ as the nth algebraic K-group associated with $[\underline{S},C]$ with fibre-wise exact sequences.

(e) Let S,T be π-sets. Just as for π finite, (see [4] or [5]) we have the notion of a sequence of functors in $[\underline{S},C]$ being T-exact and hence the definition of the nth algebraic K-group associated with the category $[\underline{S},C]$ with respect to T-exact sequences which we denote by $K_n^{\pi}(-,C,T)$.

(f) Finally, the definition of T-projective functors in $[\underline{S},C]$ is as in [4] or [5] and so we define $P_n^{\pi}(S,C,T)$ as the nth algebraic K-group associated with the full sub-category $[\underline{S},C]_T$ of T-projective functors in $[\underline{S},C]$.

We now record the following theorem whose proof is similar to that for finite groups.

<u>Theorem 1.3</u>: Let π be a profinite group, T a π-set, C an exact category. Then $K_n^{\pi}(-,C)$, $K_n^{\pi}(-,C,T)$, $P_n^{\pi}(-,C,T): \hat{\pi} \to \underline{Z\text{-mod}}$ are Mackey functors. If the pairing $C \times C \to C$: $(M,N) \to M \circ N$ is naturally associative and commutative and C has an object E such that $E \circ M = M \circ E = M$ for all $M \in C$, then $K_0^{\pi}(-,C)$ - $K_0^{\pi}(-,C,T)$ become Green functors and $K_n^{\pi}(-,C)$ becomes a $K_0^{\pi}(,C)$ - modules, while $K_n^{\pi}(-,C,T)$, $P_n^{\pi}(-,C,T)$ become $K_0^{\pi}(-,C,T)$ - modules.

<u>Remarks 1.4</u>: Since we saw in 1.2(b) that a π-set S has the form $\dot{\cup}\, \pi/\gamma$ where γ runs through a finite collection of open subgroups of π and the K-functors in 1.3, being Mackey functors, take disjoint union into direct sums, then to understand how the functors behave on a π-set S, it suffices to understand their behaviour on transitive π-sets π/γ. The interpretation of the behaviour of the functors in th latter case in terms of group-rings is similar to that for π finite and we

omit details which can be found in [4] or [5]. However, we observe that if e is the identity of π, and γ an open subgroup of π, $\underline{P}(R)$ the category of finitely generated projective modules over a commutative ring, R, we can identify $K_n^{\pi}(\pi/\gamma, \underline{P}(R), \pi/e)$ with $G_n(R\gamma)$ while we identify $P_n^{\pi}(\pi/\gamma, \underline{P}(R), \pi/e)$ with $K_n(R\gamma)$.

§2. The aim of this section is to prove the following result (2.1) which is an extension of a similar result in [4] for finite groups.

Theorem 2.1: Let π be a profinite group, k a field of characteristic p. Then for each $n \geqslant 0$, the Cartan homomorphism $K_n(k\pi) \to G_n(k\pi)$ induces an isomorphism $Z(\frac{1}{p}) \otimes K_n(k\pi) \to Z(\frac{1}{p}) \otimes G_n(k\pi)$.

Remarks 2.2(a): In view of the identifications in 1.4, we only have to prove that for each $n \geqslant 0$, the Cartan map $P_n^{\pi}(\pi/\pi, \underline{P}(k), \pi/e) \to K_n^{\pi}(\pi/\pi, \underline{P}(k), \pi/e)$ induces an isomorphism

$Z(\frac{1}{p}) \otimes P_n^{\pi}(-, \underline{P}(k), \pi/e) \simeq Z(\frac{1}{p}) \otimes K_n^{\pi}(-, \underline{P}(k), \pi/e)$

(b) Let $\pi \xrightarrow{\phi} \pi_1$, be a continuous homomorphism of profinite groups. Then ϕ induces a functor $\hat{\phi}: \hat{\pi}_1 \to \hat{\pi}$ given by $S \to S|_{\pi}$ where $S|_{\pi}$ is obtained from S by restriction of the action of π_1 on S to π via ϕ. i.e. for $s \in S$, $g \in \pi$, $gs = \phi(g)s$. Also, if $\alpha: S \to T$ is a π_1-map then α is also a π-map $\alpha|_{\pi} = \hat{\phi}(\alpha): S|_{\pi} \to T|_{\pi}$. So, we have a canonical functor $\underline{S}|_{\pi} \xrightarrow{\delta} \underline{S}$ given by $s \to s$; $(g,s) \to (\phi(g), s)$. If $\mathcal{C}$ is an exact category, then δ induces an exact functor $[S, \mathcal{C}] \to [\underline{S}|_{\pi}, \mathcal{C}]$ which associates to any $\zeta \in [\underline{S}, \mathcal{C}]$ the π-equivariant $\mathcal{C}$-bundle $\zeta|_{\pi}$ over $S|_{\pi}$. (see [6] Chapter I for definitions). Note that $\zeta|_{\pi}$ has the same fibres as ζ with π-action defined by restricting the π_1-action to π via ϕ. Hence we have a homomorphism $\delta: K_n^{\pi_1}(S, \mathcal{C}) \to K_n^{\pi}(S|_{\pi}, \mathcal{C})$ of K-groups. In particular, if $S = \pi_1/\pi_1$, then $S|_{\pi} = \pi/\pi$ and we have a homomorphism $K_n^{\pi_1}(\pi_1/\pi_1, \mathcal{C}) \to K_n^{\pi}(\pi/\pi, \mathcal{C})$. We now apply this situation to the canonical map $\pi \to \pi_\gamma$ from a profinite group π to $\pi_\gamma = \pi/\gamma$ where γ is an open normal subgroup of π and $\mathcal{C} = \underline{P}(R)$, and get a homomorphism $K_n^{\pi_\gamma}(\pi_\gamma/\pi_\gamma, \underline{P}(R)) \xrightarrow{\chi_\gamma} K_n^{\pi}(\pi/\pi, \underline{P}(R))$ where $\underline{P}(R)$ is the category of finitely generated projective R-modules.

(c) Note that the functor $G: I \to \underline{Z\text{-mod}}$ given by $G(\gamma) = K_n^{\pi_\gamma}(\pi_\gamma/\pi_\gamma, \underline{P}(R))$ is an injectively filtered system of Abelian groups since if $\gamma, \gamma' \in I$ and $\gamma \leqslant \gamma'$, then we have a homomorphism $\pi_\gamma \to \pi_{\gamma}'$, and hence a functor $\underline{\pi_\gamma/\pi_\gamma} \to \underline{\pi_{\gamma'}/\pi_{\gamma'}}$ which induces an exact functor $[\underline{\pi_{\gamma'}/\pi_{\gamma'}}, \underline{P}(R))] \to [\underline{\pi_\gamma/\pi_\gamma}, \underline{P}(R)]$ which then induces a homomorphism $K_n^{\pi_{\gamma'}}(\pi_{\gamma'}/\pi_{\gamma'}, \underline{P}(R)) \xrightarrow{\chi_{\gamma,\gamma'}} K_n^{\pi_\gamma}\pi_\gamma/\pi_\gamma, \underline{P}(R))$ of K-groups. Moreover, the homomorphisms $\chi_\gamma : K_n^{\pi_\gamma}(\pi_\gamma/\pi_\gamma, \underline{P}(R)) \to K_n^{\pi}(\pi/\pi, \underline{P}(R))$, are compatible with $\chi_{\gamma,\gamma'}$ and so we have a well-defined map

$$\lim_{\substack{\to \\ \gamma}} K_n^{\pi_\gamma}(\pi_\gamma/\pi_\gamma, \underline{P}(R)) \to K_n^{\pi}(\pi/\pi, \underline{P}(R)).$$

By a similar argument, if T is a π_γ-set, we have a homomorphism $K_n^{\pi_\gamma}(\pi_\gamma/\pi_\gamma, \underline{P}(R), T) \to K_n^{\pi}(\pi/\pi, \underline{P}(R), T|_\pi)$ and in particular, if e_γ is the identity element of π_γ, and e the identity element of π, we have homomorphisms $K_n^{\pi_\gamma}(\pi_\gamma/\pi_\gamma, \underline{P}(R), \pi_\gamma/e_\gamma) \to K_n^{\pi}(\pi/\pi, \underline{P}(R), \pi/e)$ and hence

$$\lim_{\substack{\to \\ \gamma}} K_n^{\pi_\gamma}(\pi_\gamma/\pi_\gamma, \underline{P}(R), \pi_\gamma/e_\gamma) \to K_n^{\pi}(\pi/\pi, \underline{P}(R), \pi/e).$$

Similarly we obtain

$$\lim_{\substack{\to \\ \gamma}} P_n^{\pi}(\pi_\gamma/\pi_\gamma, \underline{P}(R), \pi_\gamma/e_\gamma) \to P_n^{\pi}(\pi/\pi, \underline{P}(R), \pi/e).$$

We now have the following:

Theorem 2.3. Let π be a profinite group, $I = \{\gamma\}$ the filtered index set of open normal subgroups of π, $\underline{P}(R)$ the exact category of finitely generated projective R-modules. Then, in the notation of 2.2(c), the induced maps

(i) $\lim\limits_{\substack{\to \\ \gamma}} K_n^{\pi_\gamma}(\pi_\gamma/\pi_\gamma, \underline{P}(R)) \to K_n^{\pi}(\pi/\pi, \underline{P}(R)$

(ii) $\lim\limits_{\substack{\to \\ \gamma}} K_n^{\pi_\gamma}\pi_\gamma/\pi_\gamma, \underline{P}(R), \pi_\gamma/e_\gamma) \to K_n^{\pi}(\pi/\pi, \underline{P}(R), \pi/e)$

(iii) $\lim\limits_{\substack{\to \\ \gamma}} P_n^{\pi}(\pi_\gamma/\pi_\gamma, \underline{P}(R), \pi_\gamma/e_\gamma) \to P_n^{\pi}(\pi/\pi, \underline{P}(R), \pi/e)$

are isomorphisms.

Proof. Note that $K_n^{\pi_\gamma}(\pi_\gamma/\pi_\gamma, \underline{P}(R)) = K_n$ of the exact category $[\pi_\gamma/\pi_\gamma, \underline{P}(R)]$ where $[\pi_\gamma/\pi_\gamma, \underline{P}(R)]$ can be identified with the category $\underline{P}(R)_{\pi_\gamma}$ of π_γ-objects in $\underline{P}(R)$, i.e. with the category $[\underline{\pi}_\gamma, \underline{P}(R)]$ of functors from $\underline{\pi}_\gamma$ (considered as a category with one object with morphisms elements of π) to $\underline{P}(R)$. So,

$\varinjlim_{\gamma} K_n^{\pi_\gamma}(\pi_\gamma/\pi_\gamma, \underline{P}(R)) = \varinjlim_{\gamma} K_n(\pi_\gamma, \underline{P}(R))$. Now the functor $\gamma \to [\underline{\pi}_\gamma, \underline{P}(R)]$ from I to the category of small categories is a filtered injective system of small categories (with the homomorphisms $\pi_\gamma \xrightarrow{\phi_{\gamma\gamma'}} \pi_{\gamma'}$ inducing exact functors $[\underline{\pi}_{\gamma'}, \underline{P}(R)] \xrightarrow{\phi_{\gamma\gamma'}} [\underline{\pi}_\gamma, \underline{P}(R)]$) and moreover, the canonical homomorphisms $\pi \xrightarrow{\phi_\gamma} \pi_\gamma$ induces exact functors $[\underline{\pi}_\gamma, \underline{P}(R)] \xrightarrow{\phi_\gamma} [\underline{\pi}, \underline{P}(R)] : \zeta \to \zeta \circ \phi_\gamma$ which commute with the $\phi_{\gamma,\gamma'}$ and hence induce a map $\varinjlim_{\gamma} [\underline{\pi}_\gamma, \underline{P}(R)] \xrightarrow{\phi} [\pi, \underline{P}(R)]$ which is easily checked to be an isomorphism of categories since $\pi = \varprojlim \pi_\gamma$.

Now, $K_n^\pi(\pi, \underline{P}(R)) = K_n^\pi(\varinjlim [\pi_\gamma, \underline{P}(R)]) = \varinjlim_{\gamma} K_n[\pi_\gamma, \underline{P}(R)]$ by Quillen's result (see [8] page 96. i.e.

$K_n^\pi(\pi/\pi, \underline{P}(R) = \varinjlim_{\gamma} K_n^{\pi_\gamma}(\pi_\gamma/\pi_\gamma, \underline{P}(R)$ as required.

The proof of (b) and (c) are similar and are omitted. In fact (b) is a restatatement of (a) via the identification in 1.4.

We now prove 2.1.

Proof of 2.1. It was proved in [4] that for any finite group π', k a field of characteristic p, the canonical map $K_n(k\pi') \to G_n(k\pi')$ induces an isomorphism $Z(\frac{1}{p}) \otimes K_n(k\pi') \tilde{\to} Z(\frac{1}{p}) \otimes G_n(k\pi')$ for all $n \geq 0$, i.e.

$Z(\frac{1}{p}) \otimes P_n^\pi(\pi'/\pi', \underline{P}(k), \pi'/e') \simeq Z(\frac{1}{p}) \otimes K_n^{\pi'}(\pi'/\pi', \underline{P}(k), \pi'/e')$ where e' is the identity element of π. Now let π be a profinite group. $I = \{\gamma\}$ the filtered index set of open normal subgroups of π. Then by 2.3 we have

$Z(\frac{1}{p}) \otimes \varinjlim_{\gamma} K_n^{\pi_\gamma}(\pi_\gamma/\pi_\gamma, P(k), \pi_\gamma/e_\gamma) \simeq Z(\frac{1}{p}) \otimes K_n^\pi(\pi/\pi, \underline{P}(k), \pi/e)$

and $Z(\frac{1}{p}) \otimes \varinjlim_{\gamma} P_n^{\pi_\gamma}(\pi_\gamma/\pi_\gamma, \underline{P}(k), \pi_\gamma/e_\gamma) \simeq Z(\frac{1}{p}) \otimes P_n^\pi(\pi/\pi, \underline{P}(k), \pi/e)$

Now, consider the following commutative diagram

$$\begin{array}{ccc} Z(\frac{1}{p}) \otimes P_n^{\pi_\gamma}(\pi/\pi, \underline{P}(k), \pi/e) & \longrightarrow & Z(\frac{1}{p}) \otimes K_n^\pi(\pi/\pi, \underline{P}(k), \pi/e) \\ \uparrow & & \uparrow \\ \varinjlim_{\gamma} Z(\frac{1}{p}) \otimes P_n^{\pi_\gamma}(\pi_\gamma/\pi_\gamma, \underline{P}(k), \pi_\gamma/e_\gamma) & \to & \varinjlim_{\gamma} Z(\frac{1}{p}) \otimes K_n^{\pi_\gamma}(\pi_\gamma/\pi_\gamma, \underline{P}(k), \pi_\gamma/e_\gamma). \end{array}$$

Note that each π_γ is a finite group and so it follows from [4], that the Cartan map

$P_n^{\pi_\gamma}(\pi_\gamma/\pi_\gamma, \underline{P}(k), \pi_\gamma/e_\gamma) \longrightarrow K_n^{\pi_\gamma}(\pi_\gamma/\pi_\gamma, \underline{P}(k), \pi_\gamma/e_\gamma)$ induces an isomorphism:

$$Z(\tfrac{1}{p}) \otimes P_\gamma^{\pi_\gamma}(\pi_\gamma/\pi_\gamma, \underline{P}(k), \pi_\gamma/e_\gamma) \simeq Z(\tfrac{1}{p}) \otimes K_n^{\pi_\gamma}(\pi_\gamma/\pi_\gamma, \underline{P}(k), \pi_\gamma/e_\gamma).$$

So we have an induced isomorphism

$$\varinjlim_\gamma Z(\tfrac{1}{p}) \otimes P_n^{\pi_\gamma}(\pi_\gamma/\pi_\gamma, \underline{P}(k), \pi_\gamma/e_\gamma) \simeq \varinjlim_\gamma Z(\tfrac{1}{p}) \otimes K_n^{\pi_\gamma}(\pi_\gamma/\pi_\gamma, \underline{P}(k), \pi_\gamma/e_\gamma).$$

So the bottom arrow of the square is an isomorphism. Since the vertical arrows are also isomorphisms, it means that the top arrow is also an isomorphism.

§3. In this section, we call attention to the fact that if π is a profinite group, the category $\hat{\pi}$ of π-sets is a category with finite sums, final object and finite pull-backs and so cohomology theory can be defined for Mackey functors $\hat{\pi} \to Z\text{-mod}$. Consequently, vanishing theorems are obtained for the cohomology of the K-functors of §1 and §2 and also for the cohomology of profinite groups.

<u>3.1(a)</u>. First note that if S is any π-set, $M:\hat{\pi} \to \underline{Z\text{-mod}}$ a Mackey functor, then $M_S:\hat{\pi} \to \underline{Z\text{-mod}}$ defined by $M_S(T) = M(S\times T)$ is a Mackey functor and moreover, the projection map $pr:S\times T \to T$ define natural transformations $\theta_S:M_S \to M$ (resp $\theta^S:M \to M_S$) and M is then said to be S-projective (resp. S-injective) if θ_S(resp. θ^S) is split surjective (resp. injective). Note that M is S-projective iff M is S-injective (see [6] 4.7(d) for various other equivalent properties of S-projectivity).

(b) If $S \in \hat{\pi}$, we write $S^k = S \times \ldots \times S$ (k-times). We then obtain a simplical Amitsur complex

$$0 \longleftarrow S \overset{\longleftarrow}{\longleftarrow} S^2 \;\substack{\longleftarrow\\ \longleftarrow\\ \longleftarrow}\; \cdots \; \underset{f_i^i}{\overset{f_0^i}{\substack{\longleftarrow\\ \vdots\\ \longleftarrow}}} \; S^{i+1} \; \underset{f_{i+1}^{i+1}}{\overset{f_0^{i+1}}{\substack{\longleftarrow\\ \vdots\\ \longleftarrow}}} \; S^{i+2}$$

where $f_j^i : S^{i+1} \longrightarrow S^i$ eliminates the (j+1)th factor in S^{i+1}. We could consider the natural transformation $\theta^S:M \to M_S$ and $\theta_S: M_S \to M$ as the beginning (or the end) of an infinite "augmented" Amitsur complexes:-

$$A^*(M) = A^*_S(M) : 0 \to M \xrightarrow{d^0} M_S \xrightarrow{d^1} M_{S^2} \to \dots \to M_{S^i} \xrightarrow{d^i} M_{S^{i+1}} \to \quad \text{or}$$

$$A_*(M) = A_*^S(M) : 0 \longleftarrow M \xleftarrow{d^0} M_{S^2} \xleftarrow{d^1} M_{S^2} \longleftarrow \dots \longleftarrow M_{S^i} \leftarrow d^i \leftarrow M_{S^{i+1}} \leftarrow$$

where $d^i = \overset{i}{\Sigma}(-1)^i f_j^{i*}$ $d_i = \sum_{j=0}^{i}(-1)^j f^i_{j*}$ where by abuse of notation f_j^{i*} denotes the national transformation $f_j^{i*}:M_{S^{i+1}} \to M_{S^i}$ given by $f^i_j(T) = pr_1 \times \dots \times pr_j \times pr_{j+1} \times \dots \times pr_{i+1} \times pr_{i+2}$ and $f^i_{j*} : M_{S^i} \to M_{S^{i+1}}$ is correspondingly defined.

By cutting off the "augmentation", we have non-augmented complexes:

$$H^*(M) = H_*^S(M) : 0 \xrightarrow{d_0} M_S \xrightarrow{d^1} M_{S^2} \xrightarrow{>} \dots .$$

$$H_*(M) = H_*^S(M) : 0 \xleftarrow{d_0} M_S \xleftarrow{d_1} M_{S^2} \longleftarrow \dots .$$

w0ose (co)-homology Mackey functors are now defined by $H^i_S(M) = \ker d^{i+1}/\mathrm{Im} d^i$ and $H_i^S(M) = \ker d_i/\mathrm{Im} d_{i+1}$.

One can also splice together the complexes $A^*(M)$ and $A_*(M)$ via the augmentation to get a doubly infinite complex

$$H(M) = H_S(M) : \dots \to M_{S^2} \longrightarrow M_S \xrightarrow{\theta_S \nearrow M \searrow \theta^S} M_S \longrightarrow M_{S^2} \to \dots .$$

-(the Tate-Amitsur-complex) - whose cohomology Mackey functors will be denoted by $\hat{H}^i M = \hat{H}^i_S(M) = \ker d^{i+1}/\mathrm{Im} d^i$ $i \in Z$ and $d^0 = \theta^S\theta_S$.

(c) Let S,T be π-sets, $f:S \to *$, $g:T \to *$ the unique maps from S to $*$, T to $*$ respectively. If $M: \hat{\pi} \to$ Z-mod is a Mackey functor, we write $K_S(M) = \mathrm{Ker}(M(*) \xrightarrow{f^*} M(S))$ and $I_S(M) = \mathrm{Image}\ (M(T) \xrightarrow{f_*} M(*))$. Now, if $G : \hat{\pi} \to$ Z-mod is a Green functor such that M is a G-module, it follows from [6] 5.2, 5.3 or [3] that the pairing $G \times M \to M$ defines a pairing $H^iG \times H^iM \to H^{i+j}M$ $(i,j \geqslant 0)$ and also $\hat{H}^iG \times \hat{H}^jM \to \hat{H}^{i+j}M (i,j \in Z)$. In particular, putting $i = j = 0$, $M = G$, we see that H^0G, $\hat{H}^0G$ are Green functors and that H^iM are H^0G-modules and $\hat{H}^iM$ are $\hat{H}^0G$-modules. Hence if one can find $n \in N$ such that $n\, 1_{G(*)} \in K_SG + I_TG$, the $n\, 1_{\hat{H}^0G(*)} \in I_T\hat{H}^0G$ and so, n annihalates $K_T\hat{H}^i_SG$ for all $i \in Z$. In particular when $T = S$ (i.e. $f = g$) and hence $K_T\hat{H}^iG = \hat{H}^iG$, we have $n.\hat{H}^i_SG = 0$ for all $i \in Z$.

Our next step is to indicate how to get a suitable n which annihalates the cohomology groups in terms of the order of π. This would come from theorem 3.3 below due to A. Dress see [1]. First we need some preliminary definition and remarks.

3.2(a) A super-natural number is a formal product Πp^{n_p} where the product is taken over all prime numbers and n_p is an integer $0 \leqslant n_p \leqslant \infty$. If π is a profinite group, γ' a closed subgroup of π, then the index $(\pi:\gamma')$ of γ' in π is defined by $(\pi:\gamma') = \text{l.c.m.}\,(\pi/\gamma, \gamma'/\gamma' \cap \gamma)$ where γ ranges over all open normal subgroups of π. The order of π written $|\pi|$ is defined by $|\pi| = (\pi:1) = \text{l.c.m.}\,|\pi/\gamma|$.

(b) Let U be a family open subgroups of π^{γ}, $S_U = \dot{\bigcup}_{\gamma\varepsilon U} \pi/\gamma$. Suppose that P is a set of rational primes. Define $h_P U = \{\gamma' \leqslant \pi \mid \text{there exist } p\varepsilon P, \gamma' \leqslant \gamma, g\varepsilon\pi, \delta\varepsilon U$ with γ' open in π, γ'/γ a p-group and $g\gamma' g^{-1} \leqslant \delta\}$. If for any π-set S, we write $\pi_S = \{g\varepsilon\pi \mid gs = s \text{ for all } s \varepsilon S\}$, and $N = \bigcap_{\substack{g\varepsilon\pi\\ \delta\varepsilon U}} (g\delta g^{-1})$, then $N = \pi_{S_U}$. Note that any π-set has the form S_U, for some U.

Theorem 3.3. Let U be a finite set of open subgroups of π, P a set of rational primes, $S = S_U$, $(\pi:\pi_S)'_P$ the maximal divisor of $(\pi:\pi_S)$ which contains no prime divisor in P. Then for any Mackey functor $M:\hat{\pi} \to \underline{\text{Z-mod}}$, we have

$$(\pi:\pi_S)'_P \; M(*) \subseteq I_{S_{h_P U}}(M) + K_S(M) .$$

We state our first conclusion from the above in the following:

Theorem 3.4. Let C be an exact category such that the pairing $C \times C \to C$; $(M,N) \to M\circ N$ is naturally associative and commutative and C has an object E such that $E\circ M = M\circ E = M$ for all $M \varepsilon C$. Let S, T be π-sets, P a set of rational primes. Suppose that M is any of the functors $K_n^{\pi}(-,C)$, $K_n^{\pi}(-,C,T)$, $P_n^{\pi}(-,C,T)$, then $(\pi:\pi_S)'_P \; \hat{H}^i(M) = 0$.

Proof. By 3.3 and 3.1(d), $(\pi:\pi_S)\hat{H}^i_S G = 0$ for $G = K_0^{\pi}(-,C)$ and $G = K_0^{\pi}(-,C,T)$. Now, $K_n^{\pi}(-,C)$ is a $K_0^{\pi}(-,C)$-module and $K_n^{\pi}(-,C,T)$, $P_n^{\pi}(-,C,T)$ are $K_0^{\pi}(-,C,T)$-modules. So, for $M = K_n^{\pi}(-,C)$, $K_n^{\pi}(-,C,T)$ $P_n^{\pi}(-,C,T)$, we have $(\pi:\pi_S)'_P \hat{H}^i M = 0 \; \forall \, i \varepsilon Z$.

Our second conclusion concerns the vanishing of cohomology groups of π.

3.5. Let π be a profinite group, A any $Z\pi$-module. For $S \varepsilon \hat{\pi}$, let $\bar{A}(S) = \hat{\pi}(S,A)$. Then $\hat{\pi}(S,A)$ is an Abelian group where for $f_1, f_2 \varepsilon \hat{\pi}(S,A)$ we define $(f_1 + f_2)(s) = f_1(s) + f_2(s)$. One can prove that $\bar{A}:\hat{\pi} \to \underline{\text{Z-mod}}$ is a Mackey functor (see [6] 4.2 and 4.5 for a similar proof for π finite). If γ is an open normal subgroup of π, $\bar{A}(\pi/\gamma) = \hat{\pi}(\pi/\gamma, A) \simeq A^{\gamma} = H^0(\gamma, A)$ the zeroth cohomology group of

γ with coefficients in A. Moreover $H^i\bar{A}(*) \simeq \hat{H}^i(\pi,A)$ and the $\hat{H}^i\bar{A}$ are $\hat{H}^0\bar{A}$-modules by 3.1(d). More generally if we write $\hat{H}^i\bar{A}(S) = \hat{H}^i(S,A)$ for any π-set S, then we have

Theorem 3.6. Let π be a profinite group, S a π-set, A a $Z\pi$-module, P a set of rational primes, then $(\pi:\pi_S)'_P\ \hat{H}^i(S,A) = 0$ for all $i \in Z$.

Proof. Similar to that of 3.5 by applying 3.3 and 3.1(d) since $\hat{H}^i\bar{A}$ are $H^0\bar{A}$-modules.

Remarks 3.7. The above (3.6) applies to $\pi = \mathrm{Gal}\,(k_s/k)$ the Galois group of the separable closure k_s of a field k, $A = k_s^*$, the units of k_s.

References

1. Dress, A.W.M. - The Witt ring as a Mackey functor - Appendix A to Bielefeld notes on Representation theory of finite groups 1970.

2. Dress, A.W.M. - A relation between Burnside and Witt-rings - Appendix B to Bielefeld notes on Representation theory of finite groups 1970.

3. Dress, A.W.M. - Contributions to theory of induced representations - Springer Verlag Lecture notes. No. 342, 183-240, 1973.

4. Dress, A.W.M. and Kuku, A.O. - The Cartan map for equivariant higher algebraic K-groups - Communications in algebra 9(7) 727-746 (1981).

5. Dress, A.W.M. and Kuku, A.O. - A convenient setting for equivariant algebraic K-theory - Springer-Verlag Lecture notes - Proceedings of the Obervolfach conference 1980, (to appear).

6. Kuku, A.O. - Axiomatic theory of induced representation of finite groups - (Preprint).

7. Kuku, A.O. - Higher algebraic K-theory of group-rings and orders in algebras over number fields - communications in algebra - 10(8) 805-816 (1973).

8. Quillen, D.G. - Higher algebraic K-theory I - Springer-Verlag Lecture notes, No. 341, 77-139 (1973).

Transfer maps on K_i , $i \leq 1$, associated with certain periodic resolutions

by

Hans J. Munkholm

1. Introduction

Let $\mathbb{Z} \xrightarrow{i} \pi \xrightarrow{\varphi} \rho \to 1$ be an exact sequence of groups with a compatible action $\omega : \rho \to \{\pm 1\}$ (i.e. if $t = i(1) \in \pi$ then $gtg^{-1} = t^{\omega(\varphi(g))}$ for all $g \in \pi$). If i is monic then the integral group ring $\mathbb{Z}\rho$ is of finite homological dimension when viewed as a $\mathbb{Z}\pi$ module via φ . Consequently there is defined a "transfer map" $\varphi_i^! : K_i(\mathbb{Z}\rho) \to K_i(\mathbb{Z}\pi)$, $i \in \mathbb{Z}$, see e.g. Quillen [9, p. 111]. For $i = 0$ or 1 $\varphi_i^!$ is obtained by mapping the class $[\bar{P}]$ of a f.g. projective $\mathbb{Z}\rho$ module (resp $[\bar{\alpha}]$ of an automorphism $\bar{\alpha} : \bar{F} \to \bar{F}$ of a f.g. free $\mathbb{Z}\rho$ module) to the Euler characteristic $\Sigma_i(-1)^i[P_i]$ of a finite resolution of $\bar{P}$ by f.g. projectives over $\mathbb{Z}\rho$ (resp. $\Sigma_i(-1)^i[\alpha_i]$ of a lifting of $\bar{\alpha}$ to an automorphism α of a suitable finite resolution of $\bar{F}$).

When the map $i : \mathbb{Z} \to \pi$ is <u>not</u> monic $\mathbb{Z}\rho$ admits a periodic, but not a finite, resolution over $\mathbb{Z}\pi$, and one may consider the *"one-period-Euler-characteristics"* $[P_0]-[P_1]$, resp. $[\alpha_0]-[\alpha_1]$, for resolutions or liftings as above. It is the purpose of this note to point out that in this way one gets homomorphisms $\varphi_i^! : K_i(\mathbb{Z}\rho) \to K_i(\mathbb{Z}\pi)$ $i \leq 1$. (For $i = 0$ one has to replace projectives by projections, see section 3.)

For $i = 0,1$ these homomorphisms are the ones described by Munkholm and Ranicki [6], respectively Munkholm and Pedersen [3,4] (to which we also refer for a geometric interpretation of the maps). What is new here is the viewpoint that the maps are defined as "one-periodic-Euler-characteristics". This viewpoint leads to several questions listed in section 4.

The author wishes to thank Erik K. Pedersen and Andrew A. Ranicki for useful conversations.

2. The definition of $\varphi_1^! : K_1(\mathbb{Z}\rho) \to K_1(\mathbb{Z}\pi)$

We shall work in a more general set up. Thus let R be a ring with two given elements N, t and a given automorphism α such that

$$\text{(2.1)} \qquad \alpha(t) = t\ ,\ \alpha(N) = N\ ,\ N \text{ is central},\ N(t-1) = 0$$
$$(t-1)x = \alpha(x)(t-1) \quad \text{for all} \quad x \in R\ .$$

Then $R(t-1)$ is a two sided ideal in R. Let $\varphi : R \to \bar{R} = R/R(t-1)$ be the corresponding projection. We have maps $\nu : \bar{R} \to R$, $t-1 : R \to R$ with $\nu(\varphi(x)) = Nx$, $x \in R$, and $(t-1)(x) = (t-1)x$. We assume that

$$\text{(2.2)} \qquad 0 \to \bar{R} \xrightarrow{\nu} R \xrightarrow{t-1} R \xrightarrow{\varphi} \bar{R} \to 0 \quad \text{is}$$

an exact sequence of right R-modules

$$\text{(2.3)} \qquad \mathrm{Im}(\nu) \cap \mathrm{Im}(t-1) = \{0\}\ .$$

Example 2.4 Let π be a discrete group with a given element t of order $n < \infty$. Assume that the subgroup $\langle t\rangle$ generated by t is normal and that the action of $\rho = \pi/\langle t\rangle$ on $\langle t\rangle$ is given by $gtg^{-1} = t^{\omega(\varphi(g))}$ for some homomorphism $\omega : \rho \to \{\pm 1\}$. Further let A be an integral domain of characteristic not dividing n. We can take $R = A\pi$, the group ring, and let

$$\alpha(g) = \begin{cases} -gt^{-1}, & \omega(\varphi(g)) = -1 \\ g, & \omega(\varphi(g)) = 1 \end{cases}$$

$$N = 1+t+t^2+\cdots+t^{n-1} \quad .$$

Then it is easy to prove that (R,N,t,α) satisfies our requirements. The projection $\varphi : R = A\pi \to \bar{R} = A\rho$ becomes the homomorphism of rings induced by $\varphi : \pi \to \rho$.

The automorphism $\alpha : R \to R$ extends to an automorphism $\alpha : M_r(R) \to M_r(R)$ of the ring of $r\times r$ matrices over R and (2.1) still holds when $x \in M_r(R)$. Also we continue to write φ for the induced projection $M_r(R) \to M_r(\bar{R}) = M_r(R)/(t-1)M_r(R)$, and we denote by $\bar{\alpha} : M_r(\bar{R}) \to M_r(\bar{R})$ the automorphism induced by α . Finally $G\ell_r(R)$ is the group of units in $M_r(R)$ and $\tau : G\ell_r(R) \to K_1(R)$ takes a matrix $\beta \in G\ell_r(R)$ to its Whitehead torsion $\tau(\beta)$ in the sense of Milnor, [2].

<u>Theorem 2.5</u> *If* $\bar{\beta} \in G\ell_r(\bar{R})$ *then for any* $s \geq r$ *there exists* $\beta_1, \beta_o \in G\ell_{r+s}(R)$ *such that*

$$(2.6) \qquad \begin{array}{ccccccc} \bar{R}^r & \xrightarrow{\begin{pmatrix}\nu\\0\end{pmatrix}} & R^r\oplus R^s & \xrightarrow{\begin{pmatrix}t-1&0\\0&1\end{pmatrix}} & R^r\oplus R^s & \xrightarrow{(\varphi\ 0)} & \bar{R}^r \\ \downarrow{\scriptstyle \bar{\alpha}^{-1}(\bar{\beta})} & & \downarrow{\scriptstyle \beta_1} & & \downarrow{\scriptstyle \beta_o} & & \downarrow{\scriptstyle \bar{\beta}} \\ \bar{R}^r & \xrightarrow[\begin{pmatrix}\nu\\0\end{pmatrix}]{} & R^r\oplus R^s & \xrightarrow[\begin{pmatrix}t-1&0\\0&1\end{pmatrix}]{} & R^r\oplus R^s & \xrightarrow[(\varphi\ 0)]{} & \bar{R}^r \end{array}$$

commutes. Moreover, $\tau(\beta_o) - \tau(\beta_1)$ *depends only on* $\tau(\bar{\beta})$ *and*

$$\varphi_1^!(\tau(\bar{\beta})) = \tau(\beta_o)-\tau(\beta_1)$$

defines a homomorphism $\varphi_1^! : K_1(\bar{R}) \to K_1(R)$.

<u>Proof</u> First a remark concerning notation: We think of maps $\bar{R}^r \to \bar{R}^r$ as elements of $M_r(\bar{R})$ and understand $\bar{\alpha}^{-1}(\bar{\beta})$ in that sense.

The existence proof is very simple: Given $\bar{\beta} \in G\ell_r(\bar{R})$ we can choose $\beta,\gamma,\delta \in M_r(R)$ s.t. $\beta\gamma = 1 + (t-1)\delta$ and $\varphi(\beta) = \bar{\beta}$. Now

$$\beta_o = \begin{pmatrix} \beta & (t-1)\delta \\ 1 & \gamma \end{pmatrix} = \begin{pmatrix} 1 & \beta \\ 0 & 1 \end{pmatrix} \begin{pmatrix} 0 & -1 \\ 1 & \gamma \end{pmatrix}$$

is a (simple) isomorphism, and with $\beta_1 = \begin{pmatrix} \alpha^{-1}(\beta) & \delta \\ t-1 & \gamma \end{pmatrix}$ it is easily seen that (2.6) commutes. And it then follows from the five-lemma that β_1 is an isomorphism. In proving that, for fixed $\bar{\beta}$, $\tau(\beta_o) - \tau(\beta_1)$ is independent of the choice of β_o,β_1 we can clearly assume that $\bar{\beta} = 1$, and we have to show that $\tau(\beta_o) = \tau(\beta_1)$. This is an easy consequence of

<u>Lemma 2.7</u> (Vasershtein [10]) *If* A *and* B *are square matrices with* 1+AB *invertible then* 1+BA *is also invertible and* $\tau(1+AB) = \tau(1+BA)$.

In fact, using (2.2), (2.3) and the commutativity of (2.6), one easily shows that β_1 and β_o must have the form

$$\beta_1 = \begin{pmatrix} 1+z_{11}(t-1) & z_{12} \\ z_{21}(t-1) & z_{22} \end{pmatrix} , \qquad \beta_o = \begin{pmatrix} 1+(t-1)z_{11} & (t-1)z_{12} \\ z_{21} & z_{22} \end{pmatrix}$$

and the lemma applies with

$$A = \begin{pmatrix} z_{11} & z_{12} \\ z_{21} & z_{22}-1 \end{pmatrix} , \qquad B = \begin{pmatrix} t-1 & 0 \\ 0 & 1 \end{pmatrix}$$

Finally, it is easy to see that this does indeed define a homomorphism as claimed.

Remark The above formula for β_1 differs slightly from the corresponding formula in [4]. The reason is that here we use right modules, there we used left modules.

Acknowledgement In an earlier manuscript (for a different paper) we had a more complicated proof of theorem 2.5. We want to thank the referee of that manuscript for pointing out the applicability of Vasershtein's lemma.

3. The transfer $\varphi_o^! : K_o(\bar{R}) \to K_o(R)$

Let $P_n(R) \subseteq M_n(R)$ be the subset of *projections*, i.e. matrices X with $X^2 = X$. Let $P_\infty(R) = \bigcup_{n\geq 1} P_n(R)$ where $X \to \begin{pmatrix} X & 0 \\ 0 & 0 \end{pmatrix}$ is the embedding of $P_n(R)$ into $P_{n+1}(\bar{R})$. Call two elements $X \in P_n(R) \subseteq P_\infty(R)$ and $Y \in P_m(R) \subseteq P_\infty(R)$ *stably equivalent* if for some $\ell \geq m,n$ there is an $S \in G\ell(\ell,R)$ with $SXS^{-1} = Y$ in $M_\ell(R)$. The set of stable equivalence classes of projections form a commutative semigroup under the obvious direct sum operation. Let $K_o'(R)$ be the corresponding abelian group. If $X \in P_\infty(R)$ then $\operatorname{Im} X$ is a projective, finitely generated module so $\psi([X]) = [\operatorname{Im} X]$ defines a homomorphism $K_o'(R) \to K_o(R)$ which is easily seen to be an isomorphism. We use this isomorphism in order to think of elements of $K_o(R)$ as formal differences of stable equivalence classes of projections.

We then have the following analogue of theorem 2.5

Theorem 3.1 *Let* $\varphi : R \to \bar{R}$ *be as in section 2. For any* $\bar{X} \in P_r(\bar{R})$ *and any* $s \geq r$ *there exist* $X_o, X_1 \in P_{r+s}(R)$ *such that*

$$(3.2)\qquad \begin{array}{ccccccc}
\bar{R}^r & \xrightarrow{\binom{\nu}{0}} & R^r \oplus R^s & \xrightarrow{\begin{pmatrix} t-1 & 0 \\ 0 & 1 \end{pmatrix}} & R^r \oplus R^s & \xrightarrow{(\varphi \ \ 0)} & \bar{R}^r \\
\downarrow{\scriptstyle \bar{\alpha}^{-1}(\bar{X})} & & \downarrow{\scriptstyle X_1} & & \downarrow{\scriptstyle X_o} & & \downarrow{\scriptstyle \bar{X}} \\
\bar{R}^r & \xrightarrow[\binom{\nu}{0}]{} & R^r \oplus R^s & \xrightarrow[\begin{pmatrix} t-1 & 0 \\ 0 & 1 \end{pmatrix}]{} & R^r \oplus R^s & \xrightarrow[(\varphi \ \ 0)]{} & \bar{R}^r
\end{array}$$

commutes. Moreover, $[X_o]-[X_1] \in K_o(R)$ *depends only on* $[\bar{X}] \in K_o(\bar{R})$ *and* $\varphi_o^!([\bar{X}]) = [X_o]-[X_1]$ *defines a homomorphism* $\varphi_o^! : K_o(\bar{R}) \to K_o(R)$.

Proof We may take $s = r$ for the existence. If $\bar{X} \in P_r(\bar{R})$ then there exist $X,Y \in M_r(R)$ so that $\varphi(X) = \bar{X}$ and $X(1-X) = Y(t-1)$. Furthermore, in [6] it is shown that one can always choose Y so that $XY = Y\alpha(X)$. For such a choice

$$X^! = \begin{pmatrix} X & Y \\ t-1 & 1-\alpha(X) \end{pmatrix} \in P_{2r}(R)$$

and with $X_1 = \alpha^{-1}(X^!)$, $X_o = \begin{pmatrix} X & Y(1-t) \\ 1 & 1-X \end{pmatrix}$ we have $X_o, X_1 \in P_{2r}(R)$ and (3.2) commutative.

Uniqueness of $[X_o]-[X_1]$ for fixed $\bar{X}$ is a simple consequence of the Bass-Heller-Swan decomposition of $K_1(R[z,z^{-1}])$. This supplies, inter alia, a split injection $h : K_o(R) \to K_1(R[z,z^{-1}])$ with $h([X]) = \tau(zX+1-X)$. But the Laurent series ring $A = R[z,z^{-1}]$ together with $t,N \in R$ and the automorphism α extended to have $\alpha(z) = z$ is a new structure as in section 2 and the associated projection $A \to \bar{A} = \bar{R}[z,z^{-1}]$ is the map of Laurent series induced by $\varphi : R \to \bar{R}$. Now, if (3.2) commutes then so does

$$\begin{array}{ccccccc}
\bar{A}^r & \longrightarrow & A^r \oplus A^s & \longrightarrow & A^r \oplus A^s & \longrightarrow & \bar{A}^r \\
\downarrow{\scriptstyle \bar{\alpha}^{-1}(z\bar{X}+(1-\bar{X}))} & & \downarrow{\scriptstyle zX_1+(1-X_1)} & & \downarrow{\scriptstyle zX_o+(1-X_o)} & & \downarrow{\scriptstyle z\bar{X}+(1-\bar{X})} \\
\bar{A}^r & \longrightarrow & A^r \oplus A^s & \longrightarrow & A^r \oplus A^s & \longrightarrow & \bar{A}^r
\end{array}$$

Thus theorem 2.5 guarantees that $h([X_o]-[X_1])$, and hence also $[X_o]-[X_1]$, is independent of choices (for fixed $\bar{X}$).

And once this is known the rest is easy.

Corollary 3.3 *The transfer maps* $\varphi_o^!$ *and* $\varphi_1^!$ *commute with the Bass-Heller-Swan injections, i.e.*

$$\begin{array}{ccc} K_o(R) & \xrightarrow{h} & K_1(R[z,z^{-1}]) \\ \uparrow \varphi_o^! & & \uparrow \varphi_1^! \\ K_o(\bar{R}) & \xrightarrow[h]{} & K_1(\bar{R}[z,z^{-1}]) \end{array}$$

commutes. Consequently $\varphi_1^!$ *restricts to give a homomorphism*

$$\varphi_{-i}^! : K_{-i}(\bar{R}) \to K_{-i}(R) , \quad \textit{any} \quad i \geq 0 .$$

Proof Obvious, at least if one recalls that $K_{-i}(\bar{R})$ is defined as the intersection of the images of the various Bass-Heller-Swan injections

$$h_j : K_o(\bar{R}[z_o,z_o^{-1},\ldots,\hat{z}_j,\hat{z}_j^{-1},\ldots,z_i,z_i^{-1}]) \to$$

$$K_1(\bar{R}[z_o,z_o^{-1},\ldots,z_j,z_j^{-1},\ldots,z_i,z_i^{-1}]) , \; j = 0,1,\ldots,i .$$

(as usual $\hat{}$ denotes deletion).

Remark 1 Recently Pedersen, [11], obtained a direct description of $K_{-i}(R)$ as a K_1 construction on bounded isomorphisms of $\mathbb{Z}^{i+1}$-graded free R-modules of finite type. Using such a description one may obtain $\varphi_{-i}^!$ by mimicking theorem 2.5 directly.

Remark 2 It follows from the results of [4] that theorem 3.1 does not hold with finitely generated projective modules instead of projections, i.e. an exact sequence

$$0 \to \bar{\alpha}^{-1}(\bar{P}) \to P_1 \to P_o \to \bar{P} \to 0$$

of finitely generated R modules with $\bar{P}$ projective over $\bar{R}$ and P_1, P_o projective over R does not necessarily give a well defined $[P_o]-[P_1] \in K_o(R)$ when one fixes $\bar{P}$.

4. Properties and questions

For the case of $\varphi : \mathbb{Z}\pi \to \mathbb{Z}\rho$ induced by a projection of finite groups $\varphi : \pi \to \rho$ with cyclic kernel $\langle t\rangle$ we distinguish between the orientable $(\omega=1)$ and the non orientable $(\omega\neq 1)$ case. In the *orientable* case $\varphi_o^! = 0$, see [4]. In the *non-orientable* case the formulas

$$\varphi_i^!\varphi_*(x) = x-\alpha^{-1}(x) , \quad x \in K_i(R) , \quad i \le -1$$
$$\varphi_*\varphi_i^!(y) = y-\bar{\alpha}^{-1}(y) , \quad y \in K_i(\bar{R}) , \quad i \le -1$$

(which are easily proved) can be used to give examples where $\varphi_i^! \neq 0$, see [8].

Concerning generalizations the most obvious questions are

1°. Generalize to higher K-groups. In this respect the present, more conceptual, approach is perhaps better than the earlier direct formula approach.

2°. Can one accept periodic resolutions of a less specialized nature? Note e.g., that the condition (2.3) is violated if one considers $\varphi : A\pi \to A\rho$ where the characteristic of A divides the order of t . And, in fact, the construction in theorem (2.5) gets an indeterminacy which contains $\mathrm{Im}(K_1(N(A\pi)) \to K_1(A\pi))$.

3°. Can one generalize to cases where $\mathrm{Ker}(\varphi : R \to \bar{R})$ is more complicated? For example one might consider a periodic group H , normal in π . Then $\mathbb{Z}\rho = \mathbb{Z}[\pi/H]$ has a periodic resolution over $\mathbb{Z}\pi$. In this connection it should be noted that whenever an exact sequence of groups $1 \to \mathrm{Ker}\varphi \to \pi \to \rho \to 1$ is realized as $\pi_1(F)/\mathrm{Im}\, \pi_2(B) \to \pi_1(E) \to \pi_1(B) \to 1$ by some fibration $F \to E \to B$ then there is a geometrically defined homomorphism $\mathrm{Wh}(\rho) \to \mathrm{Wh}(\pi)$, see [1], which depends only on the exact sequence, including orientation data, and the fibre F , see [7].

4°. How does the present viewpoint fit in with the transfer maps on L-groups $\varphi_L^! : L_n(\rho) \to L_{n+1}(\pi)$ described in [5].

References:

1. D.R. Anderson, *The Whitehead torsion of a fiber homotopy equivalence*, Mich. Math. J. 21 (1974), 171-180.

2. J. Milnor, *Introduction to algebraic K-theory*, Ann. of Math. Studies, Princeton, N.J. 1971.

3. H.J. Munkholm and E.K. Pedersen, *On the Wall finiteness obstruction for the total space of certain fibrations*, Trans. Amer. Math. Soc. 76 (1980), 529-545.

4. H.J. Munkholm and E.K. Pedersen, *Whitehead transfers for S^1-bundles. An algebraic description.* Comm. Math. Helv., 56 (1981), 404-430.

5. H.J. Munkholm and E.K. Pedersen, *The S^1-transfer in Surgery theory*, to appear, Trans. Amer. Math. Soc.

6. H.J. Munkholm and A. A. Ranicki, *The projective class group transfer induced by an S^1-bundle*, to appear, Proc. Conf. Alg. Top., London, Ontario 1981.

7. E.K. Pedersen, *Universal geometric examples for transfer maps in algebraic K- and L-theory*, J. Pure Appl. Alg. 22 (1981), 179-192.

8. E.K. Pedersen and L. Taylor, *The Wall finiteness obstruction for a fibration*, Amer. J. Math. 100 (1978), 887-896.

9. D. Quillen, *Higher Algebraic K-theory; I*, Alg. K-theory I, Battelle Conf. 1972, Springer Lecture Notes vol. 341.

10. L.N. Vasershtein, *Foundations of algebraic K-theory*, Russian Math. Surveys 31 (1976), 89-156 (from Uspekhi Mat. Nauk 31 (1976), 87-149).

11. E.K. Pedersen, *On the* $K_{-i}(-)$ *functors*, preprint, Odense Universitet.

AN EXACT SEQUENCE INVOLVING $K_1(\hat{\mathbb{Z}}_p\pi)$ AND $K_2(\hat{\mathbb{Z}}_p\pi)$

Robert Oliver

In [4], an exact sequence was constructed which, for any p-group π, helps to describe both $Wh(\hat{\mathbb{Z}}_p\pi)$ and a certain quotient group $Wh_2^*(\hat{\mathbb{Z}}_p\pi)$ of $Wh_2(\hat{\mathbb{Z}}_p\pi)$. In this note we describe, and sketch the construction of, an analogous exact sequence involving $K_1(\hat{\mathbb{Z}}_p\pi)$ and $K_2^*(\hat{\mathbb{Z}}_p\pi)$ for an arbitrary finite group π. Among other things, this gives a way of detecting elements in $K_2^{top}(\hat{\mathbb{Z}}_p\pi)$; as well as giving lower bounds on the orders of $K_2^{top}(\hat{\mathbb{Z}}_p\pi)$ and $K_2(\mathbb{Z}\pi)$. The use of K_i instead of Wh_i makes the new sequence somewhat more complicated than the sequence in [4]; but it still has an intriguing regularity which seems to indicate that something more is going on.

More specifically, fix a prime p, and let A be an unramified p-ring: the ring of integers in a finite unramified extension of $\hat{\mathbb{Q}}_p$. For any finite group π, define

$$K_2^*(A\pi) = K_2(A\pi)/\cap\{\alpha_*(K_2(A\tilde{\pi})): \tilde{\pi} \text{ finite, } \alpha: \tilde{\pi} \twoheadrightarrow \pi\}$$

$$\cong K_2^{top}(A\pi)/\cap\{\alpha_*(K_2^{top}(A\pi)): \tilde{\pi} \text{ finite, } \alpha: \tilde{\pi} \twoheadrightarrow \pi\}.$$

That these two quotients are isomorphic follows as in the p-group case [4, Proposition 0.1]. Here

$$K_2^{top}(A\pi) = \varprojlim_n K_2(A/p^n[\pi]).$$

For any finite π, π_r will denote the set of p-regular elements in π: the elements of order prime to p. We will be considering the groups $H_n(\pi;A\pi)$ and $H_n(\pi;A(\pi_r))$; where π acts on $A\pi$ and $A(\pi_r)$ via conjugation. The main result is:

Theorem 1. For any unramified p-ring A and any finite group π, there is an exact sequence

$$H_3(\pi;A(\pi_r)) \to K_2^*(A\pi) \to \frac{H_1(\pi;A\pi)}{\langle g\otimes\lambda g^n\rangle} \oplus H_2(\pi;A(\pi_r)) \to H_2(\pi;A(\pi_r))$$

$$\to K_1(A\pi)_{(p)} \to H_0(\pi;A\pi) \oplus H_1(\pi;A(\pi_r)) \oplus H_0(\pi;A/2(\pi_r))$$

$$\to H_1(\pi;A(\pi_r)) \oplus H_0(\pi;A/2(\pi_r)) \to 0.$$

All of these maps are natural with respect to group homomorphisms and maps between p-rings.

Recall [5] that

$$K_2(A\pi) \cong K_1(A\pi)_{(p)} \oplus K_1(A\pi/J),$$

where $J \subseteq A$ is the Jacobson radical.

When discussing Theorem 1 in more detail, it is convenient to split the sequence into two parts. Set

$$K_1'(A\pi) = K_1(A\pi)/SK_1(A\pi).$$

For any $g \in \pi$, g_r and g_u will denote the unique elements such that g_r is p-regular, g_u has p-power order, and $g = g_r g_u = g_u g_r$. Finally, let $\Phi\colon A\pi \to A\pi$ denote the map $\Phi(\Sigma\lambda_i g_i) = \Sigma\varphi(\lambda_i)g_i^p$, where φ is the Frobenius automorphism of A ($\varphi(\lambda) \equiv \lambda^p \pmod p$). We also use Φ to denote the induced endomorphisms of $H_n(\pi;A\pi)$ and $H_n(\pi;A(\pi_r))$.

<u>Theorem 2</u>. For any unramified p-ring A and any finite group π, there is a short exact sequence

$$0 \to K_1'(A\pi)_{(p)} \xrightarrow{(\Gamma,\omega_1,\hat{\theta})} H_0(\pi;A\pi) \oplus H_1(\pi;A(\pi_r)) \oplus H_0(\pi;A/2(\pi_r))$$

$$\xrightarrow{\begin{pmatrix}\nu_0 & \Phi-1 & 0\\ \theta\Phi & 0 & \Phi-1\end{pmatrix}} H_1(\pi;A(\pi_r)) \oplus H_0(\pi;A/2(\pi_r)) \to 0.$$

Here:

(i) $\Gamma(u) = \log(u) - \frac{1}{p}\Phi(\log u)$ for any $u \in 1+J$ ($J \subseteq A\pi$ is the Jacobson radical, again, and $\log(u) \in \hat{\mathbb{Q}}_p\pi$ is the p-adic logarithm).

(ii) For any $g_i \in \pi$ and $\lambda_i \in A$ $(i = 1,\ldots,k)$:

$$\nu_0\Big(\sum_{i=1}^k \lambda_i g_i\Big) = \sum_{i=1}^k g_i \otimes \lambda_i (g_i)_r \in H_1(\pi;A(\pi_r)); \quad \text{and (if } p=2)$$

$$\theta\Big(\sum_{i=1}^k \lambda_i g_i\Big) = \sum_{i=1}^k \bar{\lambda}_i (g_i)_r \in H_0(\pi;A/2(\pi_r))$$

($\bar{\lambda}_i$ denotes the mod 2 reduction of λ_i).

(iii) If π is p-elementary – $\pi \cong \mathbb{Z}/n\times\tau$ where $p \nmid n$ and τ is a p-group (so $\pi_r = \mathbb{Z}/n$) – then for any $h_i \in \tau$, $g_i \in \mathbb{Z}/n$, and $\lambda_i \in A$ $(i = 1,\ldots,k)$:

$$\omega_1\Big(1 + \sum_{i=1}^k \lambda_i(h_i-1)g_i\Big) = \sum_{i=1}^k h_i \otimes \lambda_i g_i \in H_1(\pi;A(\pi_r)),$$

$$\omega_1\Big(1 + \sum_{i=1}^k p\lambda_i g_i\Big) = 0,$$

$$\hat{\theta}\Big(1 + \sum_{i=1}^k \lambda_i(h_i - 1)g_i\Big) = 0, \quad \text{and (if } p=2)$$

$$\hat{\theta}\Big(1 + \sum_{i=1}^k p\lambda_i g_i\Big) = \sum_{i=1}^k \bar{\lambda}_i g_i \in H_0(\pi;A/2(\pi_r)) \cong A/2[\mathbb{Z}/n].$$

We have not been able to find any simple general formula for ω_1 and $\hat{\theta}$; but they are uniquely determined by (iii) and naturality.

To make Theorem 2 a little less mysterious, note that $\mathrm{Ker}(\Gamma)$ is the torsion subgroup of $K_1'(A\pi)_{(p)}$ (since $H_0(\pi;A\pi)$ is torsion free and $H_1(\pi;A(\pi_r))$ and $H_0(\pi;A/2(\pi_r))$ are both finite). In other words, $H_0(\pi;A\pi)$ detects the free part of $K_1'(A\pi)_{(p)}$ and the other two groups detect its torsion. More specifically:

Theorem 3. $\mathrm{tors}(K_1'(A\pi))_{(p)} \cong H_1(\pi;A(\pi_r))^{\Phi} \oplus H_0(\pi;A/2(\pi_r))^{\Phi}$; where $(-)^{\Phi}$ denotes the subgroup of elements fixed by Φ.

When π is a p-group (so that $A(\pi_r) \cong A$), this reduces to the well-known formula of Wall's: that the torsion subgroup is $H_1(\pi)$ (p odd), or $H_1(\pi) \oplus \mathbb{Z}/2$ $(p=2)$. In general, the term $H_1(\pi;A(\pi_r))^{\Phi}$ detects (roughly) torsion coming from group elements in components of A; while (if $p=2$) $H_0(\pi;A/2(\pi_r))^{\Phi}$ detects torsion coming from $\{\pm 1\}$ in the various components.

When π is a p-group, Theorem 2 follows easily from [4, Proposition 2.3] or [2, Theorem 2]. If π is p-elementary - $\pi \cong \mathbb{Z}/n \times \tau$ where $p \nmid n$ and τ is a p-group - then $A\pi \cong \Sigma A_i[\tau]$ for some unramified p-rings $A_1,\dots,A_k$, and the sequence of Theorem 2 for $A\pi$ is the sum of the corresponding sequences for the $A_i[\tau]$. This proves Theorem 2 for p-elementary π; the result for general π then follows from the induction theory of Dress [1, Section 1] and results in [3, Section 2]. Induction theory is also used to show that ω_1 and $\hat{\theta}$ are well defined and uniquely determined by (iii) and naturality.

The other half of the sequence in Theorem 1 is described by:

<u>Theorem 4</u>. For any unramified p-ring Λ and any finite group π, there is an exact sequence

$$H_3(\pi;A(\pi_r)) \xrightarrow{\eta_2} K_2^*(A\pi) \xrightarrow{(\Gamma_2^*,\omega_2)} \frac{H_1(\pi;A\pi)}{\langle g\otimes\lambda g^n\rangle} \oplus H_2(\pi;A(\pi_r))$$
$$\xrightarrow{(\nu_1,\Phi-1)} H_2(\pi;A(\pi_r)) \xrightarrow{\eta_1} SK_1(A\pi) \to 0.$$

Here, Γ_2^* and ω_2 have the properties: for any $g\in\pi$ and $x\in K_1(A[Z])$, where $Z = Z_\pi(g)$:

$$\Gamma_2^*(\{g,x\}) = g\otimes\Gamma_Z(x) \in H_1(\pi;A\pi),$$

and

$$\omega_2(\{g,x\}) = g\wedge\omega_1(x) \in H_2(\pi;A(\pi_r)) \qquad (\omega_1(x)\in H_1(Z;A(Z_r))).$$

For $g_i,h_i\in\pi$ and $\lambda_i\in A$ such that $[g_i,h_i] = 1$ $(i = 1,\ldots,k)$:

$$\nu_1\Big(\sum_{i=1}^k g_i\otimes\lambda_i h_i\Big) = \sum_{i=1}^k (g_i\wedge h_i)\otimes\lambda_i(h_i)_r \in H_2(\pi;A(\pi_r)).$$

Theorem 4 is first proven for p-hyperelementary π, by directly constructing the sequence in a way similar to the construction in [4, Section 0] for the p-group case. In particular, instead of considering surjections of p-groups, one considers surjections $\alpha\colon\tilde\pi \twoheadrightarrow \pi$ where $\tilde\pi$ is p-hyperelementary and $\mathrm{Ker}(\alpha)$ is a p-group (note that $\tilde\pi_r \cong \pi_r$ in this case). Once Theorem 4 is shown for p-hyperelementary groups, the general result follows by induction theory.

Induction theory can also be applied to the groups

$$\mathrm{Ker}[K_2^{top}(A\pi) \twoheadrightarrow K_2^*(A\pi)] \quad\text{and}\quad \mathrm{Ker}[\eta_2\colon H_3(\pi;A(\pi_r)) \to K_2^*(A\pi)];$$

to show that both are generated by p-elementary induction. This reduces the problem of describing $Ker(\eta_2)$ to the p-group case, which is handled in [4, Section 4], Similarly, this reduces the problem of finding generators for $Ker[K_2^{top}(A\pi) \to K_2^*(A\pi)]$ to the case where π is a p-group. It should be noted that neither $K_2^{top}(A\pi)$ nor $K_2^*(A\pi)$ is generated by p-elementary induction.

REFERENCES

1. A. Dress, Introduction and structure theorems for orthogonal representations of finite groups, Annals of Math. 102 (1975) 291 - 325

2. R. Oliver, SK_1 for finite group rings: II, Math. Scand. 47 (1980), 195 - 231

3. R. Oliver, SK_1 for finite group rings III, Lecture Notes in Math., vol. 854, Springer Verlag (1981), 299 - 337

4. R. Oliver, Lower bounds for $K_2^{top}(\hat{\mathbb{Z}}_p\pi)$ and $K_2(\mathbb{Z}\pi)$ (to appear)

5. C. T. C. Wall, On the classification of hermetian forms, III: Complete semilocal rings, Invent. Math. 19 (1973), 59 - 71

A RELATION BETWEEN WITT GROUPS AND ZERO-CYCLES IN A REGULAR RING

William Pardon
Department of Mathematics
Duke University
Durham, NC 27706

Let R be a commutative ring with unit, suppose $1/2 \in R$, and let W(R) denote the Witt group of R. This is the Grothendieck group of isometry classes of nonsingular symmetric bilinear froms $\phi: P \times P \to R$, where P is finitely-generated projective, modulo the subgroup generated by hyperbolic forms. (It is denoted $W_0^1(R)$ or $W_0^1(CM_0(R))$ in the body of the paper.) The main results of this paper are the following theorems, A and B.

Theorem A. Let A be a regular Noetherian domain containing 1/2 and let K be its fraction field.

(i) If $\dim A \leq 3$, then the natural map $W(A) \to W(K)$ is injective.

(ii) If $\dim A \leq 4$ and A is local, then $W(A) \to W(K)$ is injective.

It was shown in [P4] that, if in part (ii) of Theorem A it is also assumed that A is essentially of finite type over a field, then the dimension restriction can be removed. This was first proved (for localization at a closed point) by Ojanguren ([O2]). He has also proved Theorem A(i) independently and by different methods in [O1].

The question of injectivity of $W(A) \to W(K)$ when A is local, seems first to have been raised by Grothendieck in a more general form, which also includes, for example, the theorem of Auslander and Goldman on the injectivity of Brauer groups, $Br(A) \to Br(K)$. An excellent survery of this and related problems has been written by Colliot-Thélène ([C1]; cf. also [CRW] and [K2].)

There is a heuristic, if somewhat technical reason for the upper bound on the dimension of the rings appearing in Theorem A. It is that if something becomes trivial at the generic point (of Spec(A)), then it ought to lift back to codimension one. It turns out that the resulting thing is trivial at all generic points of its support, so lifts back to codimension two. The process can be continued until codimension four is reached, at whose generic points symmetric bilinear forms appear again. These may be non-hyperbolic (i.e., non-zero in a Witt group), impeding further progress.

This informal explanation is made precise in (2.4). Although I don't know what happens in Theorem A(ii) for arbitrary 5-dimensional regular local rings, the co-ordinate ring A of the real variety $S^2 \times S^2$ is an example for which $W(A) \to W(K)$ is not injective ([KOS]), showing Theorem A(i) is false in dimension four. The following theorem implies there aren't any such four-dimensional examples over $\mathbb{C}$ and that the group of zero-cycles mod rational equivalence, $A_0(-)$, accounts for the examples over $\mathbb{R}$ (at least modulo odd torsion). For its full statement we need the Witt group $W^{-1}(A)$ of skew-symmetric forms (denoted $W_0^{-1}(A)$ or $W_0^{-1}(CM_0(A))$ in the body of the paper). Its definition is the obvious modification of that of $W(A)$.

Theorem B. (i) Let A be a regular Noetherian domain, of dimension four and of finite type over a field k, char k $\neq$ 2. If k is algebraically or real closed, then there is a surjection of abelian groups,

$$A_0(\text{Spec } A) \otimes \mathbb{Z}/2 \to \ker(W(A) \to W(K)) \otimes \mathbb{Z}/2$$

If the group $C_3(A) \otimes \mathbb{Z}/2 = 0$ (defined in [CF]), then this is an isomorphism.

(ii) If A is as in (i) but is 2-dimensional, then there is a surjection

$$A_0(\text{Spec } A) \otimes \mathbb{Z}/2 \to W^{-1}(A) \otimes \mathbb{Z}/2$$

If $\text{Pic}(A) \otimes \mathbb{Z}/2 = 0$, then this is an isomorphism.

For instance, if $k = \mathbb{C}$ then $A_0(\text{Spec } A)$ is divisible, so $A_0(\text{Spec } A) \otimes \mathbb{Z}/2 = 0$. On the other hand, according to [CI], $A_0(\text{Spec } A) \otimes \mathbb{Z}/2 = (\mathbb{Z}/2)^r$ when $k = \mathbb{R}$; here r is (in some cases) the number of compact topological components of the real algebraic 4-manifold defined by A. (This is the explanation for the above example where A is the co-ordinate ring of $S^2 \times S^2$.) A precise statement can be found in (3.2).

A result like Theorem B, connecting Witt groups to geometric invariants, is the motivation for a program to study quadratic forms on rings of dimension higher than one (the one-dimensional case being that of classical arithmetic interest). Indeed, the techniques of this paper can be globalized to apply to forms on schemes, for which the foundations have been laid in [K3]. For example, using the methods here, it can be shown that W(X) is a birational invariant of smooth projective surfaces X over C. This was suggested by Colliot-Thélène and Sansuc in [CS]. The papers [C2], [C3] and [CS] present another approach to the study of quadratic forms over geometric rings.

In a more algebraic vein, it is shown in §4 below that a construction of Serre and Horrocks (used by Horrocks to find indecomposable bundles on $\mathbb{P}^3$) is closely related to the surjection in part (ii) of Theorem B. The consequence ((4.5)) is a fairly explicit set of generators of $W^{-1}(A) \otimes \mathbb{Z}/2$.

As has already been suggested, the method of proof of Theorem A is to use localization sequences, comparing what happens in adjacent codimensions. A template both for this and Theorem B exists already in [O], following ideas of Gersten [G] and Claborn-Fossum [CF]. However, the technical details are necessarily different. For instance, one must first find a suitable value group V_p for forms defined on modules supported in codimension p. And not every such M can support a non-singular form $M \times M \to V_p$: M is at least reflexive in the sense that the natural map $M \to \mathrm{Hom}(M,\mathrm{Hom}(M,V_p))$ is an isomorphism. (I was led to some of these ideas by surgery theory and, to find a suitable V_p, by [Ba].)

In §1 we make choices of the value groups V_p and of the category of codimension p modules M which support non-singular forms. (In [P4] a more refined choice is used in a proof of the "Gersten conjecture" for Witt groups over rings of geometric type.). We then define the Witt groups ((1.7) and (1.12f)) and give some of their properties.

In §2 we define the maps in the localization sequences ((2.1)) and prove Theorem A(i) in (2.3)(a), Theorem A(ii) in (2.5). The definitions are needed for the proof of Theorem B and the discussion of §4, but the more technical parts of the proof of exactness are deferred to §§5-8.

In §3 Theorem B is proved and the necessary computations of $A_0(X) \otimes \mathbb{Z}/2$ cited. In §4 the connection between the Serre-Horrocks construction and Theorem B is made. It is also pointed out how Theorem A (ii) answers a question of Kustin and Miller about algebra structures on resolutions of codimension four Gorenstein ideals in a regular local ring.

§5 is devoted to a proof of the "Dévissage" theorem for Witt groups. In §§6-8 the notion of a Poincaré complex is introduced and its relation to the Witt group is carried far enough for the applications in §8 (to exactness in the localization sequences of (2.1)). This theory is due essentially to Ranicki and the exposition attempts to familiarize the reader with the considerable simplification in [R1] and [R2] brought about by the assumption that 1/2 is in the ground ring. (I am informed that a program using Poincaré complexes and localization sequences to prove Theorem A(i) is being carried out jointly by Barge, Sansuc and Vogel.)

Here are the chapter headings.

§1: Witt groups in the category of Cohen-Macaulay modules

§2: Localization sequences

§3: The Witt group and A_0

§4: Examples: skew-symmetric forms on surfaces and algebra structures on resolutions
§5: Dévissage
§6: Poincaré complexes
§7: Poincaré complexes and Witt groups
§8: Two applications of the theory of Poincaré complexes to Witt groups

Conventions

In this paper, dim refers to Krull dimension. If A is a ring and $p \subseteq A$ is a prime ideal, then ht $p := \dim A_p$; for any ideal I, ht I = $\min\{\mathrm{ht}\, p \mid p \supseteq I\}$. If M is an A-module, dim M:= $\max\{\dim A/p \mid p \in \mathrm{Supp}(M)\}$ and ht M = $\min\{\mathrm{ht}\, p \mid p \in \mathrm{Supp}(M)\}$. (There is the usual inconsistency between the two possible definitions of ht I, I an ideal; this should cause no confusion.)

If M is an A-module, an M-regular sequence $\{x_1,\ldots,x_n\} \subseteq A$ satisfies (i) $(x_1,\ldots,x_n)M \neq M$ and (ii) x_i is not a zero-divisor on $M/(x_1,\ldots,x_{i-1})M$, $i = 1,\ldots,n$. ([K, p. 84]). If M is an A-module and $I \subseteq A$ an ideal, then $\mathrm{depth}_I(M) = \max\{n \mid$ there is an M-regular sequence $\{x_1,\ldots,x_n\} \subseteq I\}$; depth M means $\mathrm{depth}_A(M)$. A <u>Cohen-Macaulay ring</u>, or CM ring, is finite-dimensional Macaulay in the sense of [N , p. 82]: it is Noetherian, finite-dimensional, and $\dim A_m = \dim A = \mathrm{depth}\, A_m$ for all maximal $m \subseteq A$. A <u>CM-module</u> M satisfies $\dim M_p = \mathrm{depth}\, M_p$, for all primes p, or is the zero module.

Finally we make the following convention: All rings will be commutative with unit, contain 1/2, and will be CM.

§1. Witt groups in the category of Cohen-Macaulay modules.

Let A be a commutative CM ring of Krull dimension n.

(1.1) Definition. For each $p \geq 0$, $CM_p(A)$ denotes the category of finitely-generated CM-modules of height p, together with the zero-module: If $M \neq 0$, then $ht(M) = p$ and for every $\mathfrak{p} \in Supp(M)$, $depth(M_\mathfrak{p}) = ht(\mathfrak{p}) - ht(M_\mathfrak{p}) = \dim M_\mathfrak{p}$.

When A is contextually specified we write CM_p for $CM_p(A)$. Here are some simple examples.

(1.2) Proposition. a) If $M \in CM_p$, it is height-unmixed. b) $M \in CM_n$ if and only if M has finite length. c) If $\{x_1,\ldots,x_p\} \subseteq A$ is an A-regular sequence, then $A/(x_1,\ldots,x_p) \in CM_p$. d) Let $M_1 \rightarrowtail M_2 \twoheadrightarrow M_3$ be a short exact sequence of finitely-generated A-modules.

(i) If $M_1, M_3 \in CM_p$, then $M_2 \in CM_p$

(ii) If $M_2, M_3 \in CM_p$, then $M_1 \in CM_p$.

(iii) If $M_1, M_2 \in CM_p$ and $ht\, M_3 > p$, then $M_3 \in CM_{p+1}$.

Proof: a) [K , Thm. 141]. b) is clear from our assumption that $\dim A_\mathcal{M} = n$ for all maximal $\mathcal{M}$. c) [M , Thm. 30(i)]. d) [K , p. 103, Ex. 14].

(1.3) From now on let A be a Gorenstein ring of dimension n such that $\dim A_\mathcal{M} = n$ for all maximal $\mathcal{M}$. Let

$$0 \to A \to E_0 \overset{d_0}{\to} E_1 \overset{d_1}{\to} \cdots \overset{d_{n-1}}{\to} E_n \overset{d_n}{\to} 0$$

be the minimal injective resolution of A over itself, so that $E_k = \coprod_{ht\, \mathfrak{q} = k} E(A/\mathfrak{q}) = \coprod_{ht\, \mathfrak{q} = k} E(A_\mathfrak{q}/\mathfrak{q} A_\mathfrak{q})$ where E denotes injective hull (See [B , §1]). Let

$$V_k(A) = \ker d_k, \quad 0 \leq k \leq n .$$

Two useful properties of the injective resolution are recorded here for later use. They follow from [B, (2.2)] and [B, (2.5)], respectively.

(1.4) For each prime $\mathfrak{p} \subseteq A$ and k, $0 \leq k \leq n$, there are isomorphisms $E_k(A)_\mathfrak{p} \overset{\cong}{\to} E_k(A_\mathfrak{p})$ commuting with the d_i's hence

$$V_k(A)_\mathfrak{p} \overset{\cong}{\to} V_k(A_\mathfrak{p}).$$

(1.5) If $x \in A$ is a non-zero-divisor on A then there are isomorphisms $E_k(A/(x)) \stackrel{\cong}{\to} (0:x) \subseteq E_{k+1}(A)$ commuting with the d_i's, $0 \leq k \leq n-1$, hence

$$V_k(A/(x)) \stackrel{\cong}{\to} (0:x) \subseteq V_{k+1}(A).$$

The next result states some simple, mostly well-known properties of CM_p. Note in particular that the non-singularity asserted in part a) is essentially the same as the isomorphism

$$M \simeq \mathrm{Ext}^p(\mathrm{Ext}^p(M,A)A)$$

proved in [AB,(4.35)].

(1.6) <u>Proposition</u>. a) If $M \in CM_p$, then $\mathrm{Hom}(M,V_p) \in CM_p$, $\mathrm{Ass}(M) = \mathrm{Ass}(\mathrm{Hom}(M,V_p))$ and the natural pairing

$$\nu: \mathrm{Hom}(M,V_p) \times M \to V_p, \quad \nu(f,m) = f(m)$$

is non-singular (both adjoints are isomorphisms).

b) If $N \in CM_{p+1}$, then there are $M_0, M_1 \in CM_p$ and a short exact sequence

$$M_1 \stackrel{\alpha}{\rightarrowtail} M_0 \twoheadrightarrow N.$$

Given any such exact sequence, there is a short exact sequence

$$M_0^{\wedge} \stackrel{\alpha^{\wedge}}{\rightarrowtail} M_1^{\wedge} \twoheadrightarrow N^{\sim},$$

where $(-)^{\wedge} := \mathrm{Hom}(-,V_p)$ and $(-)^{\sim} = \mathrm{Hom}(-,V_{p+1})$. M_0 may be assumed to be of the form $(A/(x_1,\ldots,x_p))^m$ where $\{x_1,\ldots,x_p\}$ is an A-regular sequence.

c) If $M_1 \stackrel{\alpha}{\rightarrowtail} M_2 \stackrel{\beta}{\twoheadrightarrow} M_3$ is short exact in CM_p,

$$\hat{M}_3 \stackrel{\beta^{\wedge}}{\rightarrowtail} \hat{M}_2 \stackrel{\alpha^{\wedge}}{\twoheadrightarrow} \hat{M}_1$$

is short exact in CM_p, where $\hat{M}_i = \mathrm{Hom}(M_i,V_p)$.

<u>Proof</u>: We begin with b). Let $p \geq 0$ and $N \in CM_{p+1}$. Then since $\mathrm{ht}(\mathrm{Ann}(N)) = p + 1$, there is by [K , Thm. 136] an A-regular sequence $\{x_1,\ldots,x_{p+1}\} \subseteq \mathrm{Ann}(N)$. Thus, there is a surjection j for some m, with kernel M_1,

$$M_1 \rightarrowtail (R/(x_1,\ldots,x_p))^m \overset{j}{\twoheadrightarrow} N \ .$$

Then $M_0 := (R/(x_1,\ldots,x_p))^m \in \mathcal{CM}_p$ by (1.2) and $M_1 \in \mathcal{CM}_p$ by [K , Ex. 14, p. 103].

From $V_p \rightarrowtail E_p \twoheadrightarrow V_{p+1}$ we obtain the exact sequence

$$\operatorname{Hom}(N,E_p) \to \operatorname{Hom}(N,V_{p+1}) \to \operatorname{Ext}^1(N,V_p) \to \operatorname{Ext}^1(N,E_p) .$$

Since E_p is injective, $\operatorname{Ext}^1(N,E_p) = 0$; on the other hand, $\operatorname{Ass}\operatorname{Hom}(N,E_p) = \operatorname{Ass}(E_p) \cap \operatorname{Supp}(N) = \emptyset$. Hence

$$\operatorname{Hom}(N,V_{p+1}) \overset{\simeq}{\to} \operatorname{Ext}^1(N,V_p) \ .$$

Next, from $M_1 \rightarrowtail M_0 \twoheadrightarrow N$ we obtain the exact sequence

$$\operatorname{Hom}(N,V_p) \to \operatorname{Hom}(M_0,V_p) \to \operatorname{Hom}(M_1,V_p) \to \operatorname{Ext}^1(N,V_p) \to \operatorname{Ext}^1(M_0,V_p) \ .$$

As above, $\operatorname{Hom}(N,V_p) = 0$. From $V_{p-1} \rightarrowtail E_{p-1} \twoheadrightarrow V_p$ we get $\operatorname{Ext}^1(M_0,V_p) \simeq \operatorname{Ext}^2(M_0,V_{p-1})$; continuing in this way, $\operatorname{Ext}^1(M_0,V_p) \simeq \operatorname{Ext}^{p+1}(M_0,A)$, which vanishes by [K , Thm. 217].

The proof of a) is by induction on p. By [AB, (4.12),(3.8)], if $M \in \mathcal{CM}_0$, then $M \overset{\simeq}{\to} M^{**}$ ($M^* := \operatorname{Hom}(M,A)$) and $\operatorname{Ext}^i(M^*,A) = 0$, $i > 0$. By [K , p. 163, Ex. 7] $M^* \in \mathcal{CM}_0$. This proves a) in case $p = 0$.

Let $p \geq 0$, $\operatorname{Hom}(-,V_p) = (\ )^\wedge$, $\operatorname{Hom}(-,V_{p+1}) = (-)^\sim$, and $N \in \mathcal{CM}_{p+1}$. We have $\operatorname{Ass}(N^\sim) = \operatorname{Supp}(N) \cap \operatorname{Ass}(V_{p+1}) = \operatorname{Ass}(N)$, since N is unmixed. Let $M_1 \rightarrowtail M_0 \twoheadrightarrow N$ be a resolution as in b). Then $\hat{M_0} \rightarrowtail \hat{M_1} \twoheadrightarrow N^\sim$ is also exact and $\hat{M_0}, \hat{M_1} \in \mathcal{CM}_p$ by induction. So $N^\sim \in \mathcal{CM}_{p+1}$ by (1.2)(d).

Finally, consider the diagram of exact sequences

$$\begin{array}{ccccc} M_1 & \rightarrowtail & M_0 & \twoheadrightarrow & N \\ \downarrow\simeq & & \downarrow\simeq & & \\ \hat{\hat{M_1}} & \rightarrowtail & \hat{\hat{M_0}} & \twoheadrightarrow & N^{\sim\sim} \end{array}$$

It follows that there is an isomorphism $N \overset{\simeq}{\to} N^{\sim\sim}$ which can be shown to be the right adjoint of ν.

c) is immediate from the chain of isomorphisms $\operatorname{Ext}^1(M_3,V_p) \simeq \operatorname{Ext}^2(M_3,V_{p-1}) \simeq \ldots \simeq \operatorname{Ext}^{p+1}(M_3,A) = 0$ by [K , Thm. 215].

(1.7) Now let

$$\Omega^\lambda(\mathcal{CM}_p), \quad \lambda = \pm 1$$

denote the category of pairs (M,ϕ) where $M \in CM_p$ and $\phi: M \times M \to V_p$ is a λ-symmetric, A-bilinear form which is non-singular, $Ad\phi: M \overset{\cong}{\to} Hom(M,V_p)$. For instance if $N \in CM_p$ and $N^{\wedge} := Hom(N,V_p)$ then the <u>hyperbolic form</u> $(N \oplus N^{\wedge},\phi_h) \in Q^{\lambda}(CM_p)$ is defined by $\phi_h|N \times N \equiv 0 \equiv \phi_h|N^{\wedge} \times N^{\wedge}$, and $\phi_h|N^{\wedge} \times N$ is the natural pairing ν; it is non-singular by (1.6)(a). More generally, (M,ϕ) is called a <u>lagrangian</u> if there is $N \subseteq M$ such that $\phi|N \times N \equiv 0$, the induced pairing $N \times (M/N) \to V_p$ is non-singular (both adjoints are isomorphisms) and $N, M/N \in CM_p$. N will be called a <u>sublagrangian</u>.

One may add isometry classes of objects in $Q^{\lambda}(CM_p)$ by orthogonal sum and obtain an abelian semi-group. The corresponding Grothendieck group, modulo the subgroup generated by lagrangians is denoted

$$W_0(CM_p).$$

For instance if A is regular, then $CM_0(A)$ is the category of finitely-generated projectives, and we set

$$W_0^{\lambda}(CM_0(A)) = W_0^{\lambda}(A), \tag{1.8}$$

the Witt group referred to in the introduction.

Here is a result describing the typical equivalence of elements in $W_0^{\lambda}(CM_p)$.

(1.9) <u>Proposition</u>. Let $(M,\phi) \in Q^{\lambda}(CM_p)$, $L \subseteq M$, $L \subseteq L^{\perp}$ be given such that M/L and $M/L^{\perp} \in CM_p$. Then $L^{\perp}/L \in CM_p$, and there is an induced form $(L^{\perp}/L,\psi) \in Q^{\lambda}(CM_p)$ such that in $W_0^{\lambda}(CM_p)$.

$$[L^{\perp}/L,\psi] = [M,\phi]$$

<u>Proof</u>: The same as [Pl, (3.4),(3.5)].

Next comes the corresponding K_1-functor, which compares two ways of making something in $Q^{\lambda}(CM_p)$ equal zero in $W_0^{\lambda}(CM_p)$.

(1.10) A <u>λ-formation</u> in CM_p is a triple (K,H,Δ), where $K,H \in CM_p$ and $\Delta: K \to H \oplus H^{\wedge}$ is an injection whose image is a sublagrangian of the hyperbolic form $(H \oplus H^{\wedge},\rho_h)$. For instance, if $\rho: K \to K^{\wedge}$ satisfies $\rho + \lambda\,\rho^{\wedge} = 0$, then $(K,K^{\wedge},(\rho,1))$ is a λ-formation called a <u>graph formation</u>. The collection of λ-formations is denoted

$$F^{\lambda}(CM_p), \quad \lambda = \pm 1$$

(K,H,Δ) and (K',H',Δ') are <u>isomorphic</u> if there are isomorphisms

$A: K \to K'$ and $B: H \to H'$ so that

$$\begin{array}{ccc} K & \overset{\Delta}{\to} & H \oplus H^{\wedge} \\ \downarrow A & & \downarrow B \oplus B^{\wedge -1} \\ K' & \overset{\Delta'}{\to} & H' \oplus H'^{\wedge} \end{array}$$

commutes. The (orthogonal) <u>sum</u> of these λ-formations is $(K \oplus K', H \oplus H', \Delta \oplus \Delta')$; and the zero-formation has $K = (0) = H$.

(1.11) Here are two useful properties of λ-formations (K,H,Δ): If $\Delta = (\alpha,\gamma): K \to H \oplus H^{\wedge}$ then $\gamma^{\wedge}\alpha + \lambda\alpha^{\wedge}\gamma = 0$; and the hyperbolic form on $H \oplus H^{\wedge}$ induces an isomorphism $H \oplus H^{\wedge}/\mathrm{im}(\Delta) \overset{\cong}{\to} K^{\wedge}$, by definition of sublagrangian.

(1.12) Next define two operations on $(K,H,\Delta) \in F^{\lambda}(CM_p)$ as follows.

a) Let $(\varepsilon) = (L \rightarrowtail H_1 \twoheadrightarrow H)$ be an extension in CM_p and K_1 the pullback in

$$\begin{array}{ccccc} L & \rightarrowtail & K_1 & \overset{j_1}{\twoheadrightarrow} & K \\ \downarrow = & & \downarrow \alpha_1 & & \downarrow \alpha \\ L & \rightarrowtail & H_1 & \overset{j}{\twoheadrightarrow} & H \end{array}$$

where $\Delta = (\alpha,\gamma)$. Let $\gamma_1 := j^{\wedge}\gamma j_1$ and $\Delta_1 = (\alpha_1,\gamma_1): K_1 \to H_1 \oplus H_1^{\wedge}$. Then $(K_1,H_1,\Delta_1) \in F^{\lambda}(CM_p)$, and its isomorphism class depends only on that of (K,H,Δ) and the extension (ε). It is denoted

$$\sigma_{\varepsilon}(K,H,\Delta)$$

and is called the <u>stabilization</u> of (K,H,Δ) by (ε).

b) Let $\psi: H \times H \to V_p$ be a $(-\lambda)$-symmetric A-bilinear form and set $\Delta_{\psi} = (\alpha, \gamma + (\mathrm{Ad}\,\psi)\alpha)$. Then $(K,H,\Delta_{\psi}) \in F^{\lambda}(CM_p)$ and is denoted

$$\chi(H,\psi)(K,H,\Delta).$$

It is said to be <u>isometric</u> to (K,H,Δ).

We define

$$W_1^{\lambda}(CM_p)$$

to be the Grothendieck group on isomorphism classes of elements of $F^{\lambda}(CM_p)$, modulo the subgroup generated by elements of the form

$(K,H,\Delta) - \chi(H,\psi)\cdot(K,H,\Delta)$, $(K,H,\Delta) - \sigma_\varepsilon(K,H,\Delta)$, and $(K,K^\wedge,(\rho,1))$.

(1.13) Remarks. a) This is Ranicki's definition [ ], except that the roles of H and $H^\wedge$ are reversed. This makes no difference because of (1.19) below. b) When $p = 0$ and A is regular, then CM_0 is the category of finitely generated projectives and we write in this case

$$W_1^\lambda(CM_0(A)) = W_1^\lambda(A)$$

If $\tilde{K}_0(A) = 0$, then $W_1^\lambda(A)$ is the commutator quotient of the (stabilized) group of isometries of $(A^\infty + A^{\wedge\infty}, \phi_h)$ modulo hyperbolic rotations (cf. [P1, 1.21 (b)] and matrices w_n^λ ([P1, 1.21 (a)]).

We need to discuss the above relations in $W_1^\lambda(CM_p)$. Clearly, $\chi(H,-\psi)\cdot\{\chi(H,\psi)\cdot(K,H,\Delta)\} = (K,H,\Delta)$. We next show how to invert operation (1.12)(b), as well.

(1.14) Given $(K_1,H_1,\Delta_1) \in F^\lambda(CM_p)$, suppose there is $L \in CM_p$ and an inclusion $i_1\colon L \to K_1$ such that $\gamma_1 i_1 = 0$, where $\Delta_1 = (\alpha_1,\gamma_1)$, and $\operatorname{cok}(i_1)$, $\operatorname{cok}(\alpha_1 i_1) \in CM_p$. Then $\alpha_1 i_1$ is injective (since Δ_1 is), and we set $K = K_1/i_1(L)$, $H = H_1/\alpha_1 i_1(L)$, and $\alpha\colon K \to H$ equal to the induced map. As $\Delta_1(L) \subseteq H_1$ we have $\Delta_1(L)^\perp \supseteq H_1^\perp = H_1$. From the hyperbolic form on $H_1 \oplus H_1^\wedge$ we get a form

$$H \times (\Delta_1(L)^\perp \cap H_1^\wedge) \to V_p$$

which induces an isomorphism

$$\Delta_1(L)^\perp \cap H_1^\wedge \xrightarrow{\simeq} H^\wedge.$$

Since $\operatorname{im}(\Delta_1) = \operatorname{im}(\Delta_1)^\perp$ and $\gamma_1(L) = 0$, γ_1 induces a map $K \to \Delta_1(L)^\perp \cap H_1^\wedge$, which, composed with the above isomorphism, gives $\gamma\colon K \to H^\wedge$. Setting $\Delta = (\alpha,\gamma)$ it is easily checked that $(K,H,\Delta) \in F^\lambda(CM_p)$, the destabilization of (K_1,H_1,Δ_1) by L. The reason for this terminology is the following.

(1.15) Proposition. a) Let (ε) denote the extension $L \overset{\alpha_1 i_1}{\rightarrowtail} H_1 \twoheadrightarrow H$ in the construction (1.14) above. Then $\sigma_\varepsilon(K,H,\Delta) \simeq (K_1,H_1,\Delta_1)$.

b) Given $(K,H,\Delta) \in F^\lambda(CM_p)$ and an extension $(\varepsilon) = (L \rightarrowtail H_1 \twoheadrightarrow H)$, the destabilization of $\sigma_\varepsilon(K,H,\Delta)$ by L is (K,H,Δ).

Thus the operations (1.12)(a) and (b) are reflexive and symmetric. If we call the equivalence relation generated by isomorphism plus these operations stable isometry, then $W_1^\lambda(CM_p)$ is the Grothendieck group on stable isometry classes of λ-formations, modulo the subgroup generated by graph formations.

The next proposition allows us to represent elements of $W_1^\lambda(CM_p)$ by formations instead of by their formal differences. It is analogous to the fact that if $(M,\phi) \in Q^\lambda(CM_p)$, then $(M,\phi) \oplus (M,-\phi)$ is hyperbolic.

(1.16) <u>Proposition</u>. Let $(K,H,(\alpha,\gamma)) \in F^\lambda(CM_p)$ be given. Then the formation

$$(K,H,(\alpha,\gamma)) \oplus (K,H\hat{\ },(\gamma,-\lambda\alpha))$$

is stably isometric to a graph formation

$$(K,K\hat{\ },(\rho,1)).$$

<u>Proof</u>: Let $\theta_1 = (K_1,H_1,(\alpha_1,\gamma_1))$, $\theta_2 = (K_2,\hat{H}_2,(\gamma_2,-\lambda\alpha_2))$ where $K_i = K$, $H_i = H$, $\alpha_i = \alpha$, $\gamma_i = \gamma$. Then $\theta_1 \oplus \theta_2 = (K_1 \oplus K_2, H_1 \oplus \hat{H}_2, (\alpha_1 \oplus \gamma_2, \gamma_1 \oplus -\lambda\alpha_2))$. Define the $(-\lambda)$-symmetric form

$$\psi\colon (H_1 \oplus \hat{H}_2) \times (H_1 \oplus \hat{H}_2) \to V_p$$

to be the standard hyperbolic form $(H_1 = H_2)$. Then

$$(1.17)\ \chi(H_1\oplus\hat{H}_2,-\psi)\cdot(\theta_1\oplus\theta_2) = (K_1\oplus K_2, H_1\oplus\hat{H}_2, (\alpha_1\oplus\gamma_2, (\gamma_1-\gamma_2)\oplus\lambda(\alpha_1-\alpha_2)))\ .$$

Let $i\colon K \to K_1 \oplus K_2$ be the diagonal inclusion.

Then if we denote the right side of (1.17) by $(M,N,(\beta,\delta))$, $\delta i = 0$, βi is injective, and $\operatorname{cok}(i)$, $\operatorname{cok}(\beta i) \in CM_p$. By the construction of (1.14) we may destabilize. It is now easily shown that if $(G,L,(\rho,\tau))$ is the resultant λ-formation, τ is an isomorphism and $G \simeq K$. Such a formation is isomorphic to a graph formation of the required form.

Besides allowing us to represent elements of $W_1^\lambda(CM_p)$ by actual formations, the last result shows that the following two operations on a formation (K,H,Δ) do not change its class in $W_1^\lambda(CM_p)$.

(1.18) a) Given $(\varepsilon) = (L \to \hat{H}_1 \to H\hat{\ })$, let K_1 and γ_1 be defined by pullback in

$$\begin{array}{ccccc} L & \rightarrowtail & K_1 & \xrightarrow{\ell_1} & K \\ \downarrow= & & \downarrow\gamma_1 & & \downarrow\gamma \\ L & \rightarrowtail & \hat{H}_1 & \xrightarrow{\ell} & H\hat{\ } \end{array}$$

and set $\alpha_1 := \ell^\wedge \alpha \ell_1$, $\Delta_1 := (\alpha_1,\gamma_1): K_1 \to H_1 \oplus H_1^\wedge$. Then $(K_1,H_1,\Delta_1) \in F^\lambda(CM_p)$ and is denoted ${}_\varepsilon\sigma(K,H,\Delta)$.

(1.18) b) Let $\psi: H^\wedge \times H^\wedge \to V_p$ be a $(-\lambda)$-symmetric form and set ${}_\psi\Delta = (\alpha + (\mathrm{Ad}\,\psi)\gamma,\gamma)$. Then $(K,H,{}_\psi\Delta) \in F^\lambda(CM_p)$ and is denoted $\chi(H^\wedge,\psi) \cdot (K,H,\Delta)$.

(1.19) Proposition. Let (K,H,Δ) be a λ-formation. Then elements of the form $(K,H,\Delta) - {}_\varepsilon\sigma(K,H,\Delta)$ and $(K,H,\Delta) - \chi(H^\wedge,\psi) \cdot (K,H,\Delta)$ in $W_1^\lambda(CM_p)$ are trivial.

It is now easy to deduce that the definition of $W_1^\lambda(CM_p)$ given above agrees with that of [Pl, (1.34)]. (Note that $(K,K^\wedge,(\rho, 1)) = \chi(K^\wedge,\rho) \cdot \{{}_\varepsilon\sigma(0,0,(0,0))$.) We will freely use this fact in what follows.

The idea of a formation is to compare two lagrangian structures on a non-singular form. However, somewhat awkwardly, one structure is actually hyperbolic, while the other is not in general so. This causes some serious technical difficulties (cf. (2.23)), which we will not address here. In one case it is useful to know that a lagrangian structure is always hyperbolic.

(1.20) Proposition. Let $(M,\phi) \in Q^{-1}(CM_n(A))$ where $n = \dim A$ and $1/2 \in A$. If $L \subseteq M$ is a sublagrangian, then there is an isometry $(L \oplus L^\wedge,\phi_h) \xrightarrow{\simeq} (M,\phi)$ extending the inclusion of L in M.

Proof: Suppose we have $K \subseteq M$ with $K \subseteq K^\perp$, $K \cap L = (0)$ and $\ell(K) + \ell(L) < \ell(M)$. Then we first show it is possible to choose $y \in M - (K + L)$ so that $\phi(y,K) \equiv 0$.

Choose any $y \in M - (K + L)$. Define $\phi_y: K \to E_n = V_n$ by $\phi_y(k) = \phi(y,k)$. Since E_n is injective and $i: K \to M/L$ is injective, there is a homomorphism $\phi_y': M/L \to E_n$ so that $\phi_y = \phi_y' \circ i$. Since L is a sublagrangian, $L \simeq \mathrm{Hom}(M/L,E_n)$ so there is $\ell \in L$ such that

$$\phi(\ell,x) = \phi_y'(x), \quad x \in M/L .$$

Then $y - \ell \in M - (K + L)$ and $\phi(y - \ell,K) \equiv 0$.

If $K' = K + (y)$, we may repeat the above procedure (on K' and L) until $\ell(K) + \ell(L) = \ell(M)$, in which case $K \oplus L = M$. Since $K \subseteq K^\perp$ and $L \subseteq L^\perp$ it is clear that M is hyperbolic.

Finally we give a simpler description of zero in $W_1^\lambda(CM_p)$.

(1.21) Proposition. Let $(K,H,\Delta) \in F^\lambda(CM_p)$. Then (K,H,Δ) represents zero in $W_1^\lambda(CM_p)$ if and only if it is stably isometric to a graph formation.

Proof: See (8. 1).

§2. Localization sequences.

This section provides the localization sequences and states the dévissage theorem needed for the applications.

(2.1) Theorem. Let A be a regular Noetherian domain of dimension ≤ 4 with fraction field K. Then there are exact sequences where $CM_i = CM_i(A)$, $\lambda = \pm 1$, and K_i^λ is induced by $\coprod(\otimes A_p)$,

$$\text{a}):\quad W_1^\lambda(K) \xrightarrow{L_1^\lambda} W_1^\lambda(CM_1) \xrightarrow{D_1^\lambda} W_0^\lambda(A) \xrightarrow{K_0^\lambda} W_0^\lambda(K)$$

$$\text{b}):\quad \coprod_{\mathrm{ht}\,p=1} W_0^{-\lambda}(CM_1(A_p)) \xrightarrow{L_0^{-\lambda}} W_0^{-\lambda}(CM_2) \xrightarrow{D_0^{-\lambda}} W_1^\lambda(CM_1) \xrightarrow{K_1^\lambda} \coprod_{\mathrm{ht}\,p=1} W_1^\lambda(CM_1(A_p))$$

$$\text{c}):\quad \coprod_{\mathrm{ht}\,p=2} W_1^{-1}(CM_2(A_p)) \xrightarrow{L_1^{-1}} W_1^{-1}(CM_3) \xrightarrow{D_1^{-1}} W_0^{-1}(CM_2) \xrightarrow{K_0^{-1}} \coprod_{\mathrm{ht}\,p=2} W_0^{-1}(CM_2(A_p))$$

$$\text{d}):\quad \coprod_{\mathrm{ht}\,p=3} W_0^{1}(CM_3(A_p)) \xrightarrow{L_0^{1}} W_o^{1}(CM_4) \xrightarrow{D_0^{1}} W_1^{-1}(CM_3) \xrightarrow{K_1^{-1}} \coprod_{\mathrm{ht}\,p=3} W_1^{-1}(CM_3(A_p)) .$$

The proof of the theorem is given later in this section.

Let (R,M) be a local Gorenstein ring of dimension n. Since $k(M)$ is isomorphic as an R-module to $\{e \in E_n(R) = E(R/M) \mid Me = 0\}$ ([SV,4.24]), we may choose an imbedding $k(M) \to E_n$ and use the inclusion of the category of $k(M)$-vector spaces into $CM_n(R)$ to produce a map

$$W_i^\lambda(k(M)) \to W_i^\lambda(CM_n(R)).$$

(2.2) Theorem. (Dévissage) Let (R,M) be a local Gorenstein ring of dimension n with residue class field $k(M)$. Then there is an isomorphism $(i = 0,1$ and $\lambda = \pm 1)$

$$W_i^\lambda(k(M)) \xrightarrow{\simeq} W_i^\lambda(CM_n(R)).$$

Proof: §5.

It is well-known that $W_1^\lambda(k(M)) = 0 = W_0^{-1}(k(M))$, so we have the following consequences of the two theorems:

(2.3) Corollary. a) If A is a regular Noetherian domain of dimension ≤ 4 then there are isomorphisms

$$\ker(W_0^1(A) \to W_0^1(K)) \simeq W_1^1(CM_1)$$
$$\simeq W_0^{-1}(CM_2)$$

$$\simeq W_1^{-1}(CM_3)$$

$$\simeq \operatorname{cok}\Big(\coprod_{\operatorname{ht}p=3} W_0^1(CM_3 A_p)) \overset{L_0^1}{\to} \coprod_{\operatorname{ht}p=4} W_0^1(CM_4 A_p))\Big) .$$

In particular $\ker(W_0^1(A) \to W_0^1(K)) = 0$ if $\dim A \leq 3$.

b) With A as above but $\dim A \leq 2$, there are isomorphisms

$$W_0^{-1}(A) \simeq W_1^{-1}(CM_1) \simeq \operatorname{cok}\Big(\coprod_{\operatorname{ht}p=1} W_0^1(CM_1 A_p)) \to \coprod_{\operatorname{ht}p=2} W_0^1(CM_2 A_p))\Big)$$

(2.4) Remark. The corollary furnishes one reason for the restrictions, $\dim A \leq 4$ or $\dim A \leq 2$: the obstructions to increasing the codimension p of the quadratic objects $W_i^\lambda(CM_p)$ all vanish until an element of W_0^1 is reached ($p = 4$ in a), $p = 2$ in b)). This seems to be a kind of internal periodicity. The other reason has to do with the problem of finding CM-modules (cf. (2.11)).

But it is easy now to show that the codimension four obstruction vanishes if we localize. (This presages the connection with cycles in §3.)

(2.5) Theorem. Let (R,M) be a regular local ring of dimension ≤ 4. Then $W_0^1(R) \to W_0^1(K)$ is injective.

Remark: Since R is local the conclusion is equivalent to the assertion that two symmetric forms $\phi,\psi\colon R^n \times R^n \to R$ are isometric if and only if their extensions $\phi \otimes K$ and $\psi \otimes K$ are.

Proof: From (2.3) it is sufficient to show that the composition

$$W_0^1(CM_4) \overset{\mathcal{D}_0^1}{\to} W_1^{-1}(CM_3) \overset{\mathcal{D}_1^{-1}}{\to} W_0^{-1}(CM_2)$$

is trivial.

Choose two regular parameters $x,y \in M$, so that $\bar{R}:=R/(x,y)$ is regular local, 2-dimensional, and has maximal ideal $\bar{M} := M/(x,y)$. By (1. 5) there are isomorphisms $E_i(\bar{R}) \simeq (0:(x,y)) \subseteq E_{i+2}(R)$, $i = 0,1,2$, commuting with the differentials of $E_*(\bar{R})$ and $E_*(R)$. Consequently there are commutative diagrams, $i = 0,1,2$,

$$\begin{array}{ccccc} V_i(\bar{R}) & \rightarrowtail & E_i(\bar{R}) & \twoheadrightarrow & V_{i+1}(\bar{R}) \\ \downarrow & & \downarrow & & \downarrow \\ V_{i+2}(R) & \rightarrowtail & E_{i+2}(R) & \twoheadrightarrow & V_{i+3}(R) \end{array}$$

Also there are obvious inclusions

$$CM_i(\bar{R}) \to CM_{i+2}(R), \quad i = 0,1,2$$

and

$$CM_i(\bar{R}_{\bar{p}}) \to CM_{i+2}(R_p), \quad i = 0,1,2$$

where ht $p = \imath + 2$, and $(x,y) \subseteq p$.

With these facts and the definitions of the maps $\mathcal{D}_0^1$ and $\mathcal{D}_1^{-1}$ given below, it is immediate that there is a commutative diagram

$$\begin{array}{ccccc} W_0^1(CM_2(\bar{R})) & \xrightarrow{\mathcal{D}_0^1} & W_1^{-1}(CM_1(\bar{R})) & \xrightarrow{\mathcal{D}_1^{-1}} & W_0^{-1}(CM_0(\bar{R})) \\ \downarrow & & \downarrow & & \downarrow \\ W_0^1(CM_4(R)) & \xrightarrow{\mathcal{D}_0^1} & W_1^{-1}(CM_3(R)) & \xrightarrow{\mathcal{D}_1^{-1}} & W_0^{-1}(CM_2(R)) \end{array}$$

But because $\dim \bar{R} = 2$ and $\dim R = 4$, dévissage says each of the extreme left terms is isomorphic to $W_0^1(k(M))$ and the left vertical is an isomorphism. And $W_0^{-1}(CM_0(\bar{R})) = W_0^{-1}(\bar{R})$ (cf. (1. 8)) vanishes. Hence the composition along the bottom is trivial, as claimed.

<u>Proof</u> (of (2.1)). The first exact sequence is part of the theorem in [P1]. Most of the remaining work just generalizes this, so the more routine details will be omitted. We begin with the definitions of the maps.

(2.6) <u>Definition of</u> $\mathcal{D}_0^\lambda$: $W_0^\lambda(CM_{p+1}) \to W_1^{-\lambda}(CM_p)$.

Given $(T,\mu) \in \Omega^\lambda(CM_{p+1})$, [P3, 3.12(b)] shows how to produce $(K,H,\Delta) \in F^{-\lambda}(CM_p)$:

a) Choose a resolution of T, $K \overset{\alpha}{\rightarrowtail} H \overset{j}{\twoheadrightarrow} T$ where $K,H \in CM_p$.

b) Lift μ to τ,

$$\begin{array}{ccc} H \times H & \xrightarrow{\tau} & E_p \\ \downarrow j \times j & & \downarrow d_p \\ T \times T & \xrightarrow{\mu} & V_{p+1} \end{array}$$

c) Define γ by the commutativity of the diagram

$$\begin{array}{ccc} K & \xrightarrow{\gamma} & H\hat{} := \mathrm{Hom}(H,V_p) \\ \downarrow \alpha & & \downarrow \\ H & \xrightarrow{\mathrm{Ad}\,\tau} & \mathrm{Hom}(H,E_p) \end{array}$$

and set $\Delta = (\alpha,\gamma)$.

The proof of [P3, (3.19)] shows that the class of $(K,H,\Delta) \in W_1^{-\lambda}(CM_p)$ depends only on the isometry class of (T,μ). Moreover, since the construction is additive (sending orthogonal sums to orthogonal sums), we need only check it sends lagrangians to zero in $W_1^{-\lambda}(CM_p)$. Indeed, [P3, (3.16)(b)] shows that if (T,μ) is a lagrangian, then there is a choice of $\alpha: K \to H$ and τ so that (K,H,Δ) has $\Delta = (\alpha,\gamma)$ where γ is invertible. This is isomorphic to a graph formation, so (K,H,Δ) represents zero in $W_1^{-\lambda}(CM_p)$.

(2.7) <u>Definition of</u> $L_0^\lambda: \coprod_{ht\,p=p} W_0^\lambda(CM_p(A_p)) \to W_0^\lambda(CM_{p+1}(A))$ in case

a) $p = 0$ and A is a Gorenstein domain.

b) $p = 1$ and A is Gorenstein and locally factorial.

c) $p = \dim A - 1$ or $\dim A - 2$ and A is Gorenstein.

For the duration of this section we write, when R is local CM of dimension p,

$$CM_p(R) = F(R),$$

the category of modules of finite length.

(2.8) <u>Definition</u>. Let $p \geq 0$ and $F = \coprod F_i$, where $F_i \in F(A_{q_i})$ and $q_i \in \operatorname{Spec} A$ has height p, $i = 1,\dots,n$. A <u>CM_p-lattice</u> $L \subseteq F$ is an A-module $L \in CM_p(A)$ such that $L_{q_i} = F_i$ for each i. If $(F_i,\tau_i) \in Q^\lambda(F(A_{q_i}))$, then $L \subseteq F$ is called an <u>integral lattice</u> if it is a CM_p-lattice and $\tau(L \times L) \subseteq V_p (\subseteq E_p = \coprod_{ht\,q=p} E_p(A_q))$, where $\tau = \coprod \tau_i$.

(2.9) <u>Proposition</u>. Let $(F_i,\tau_i) \in Q^\lambda(F(A_{q_i}))$ where $ht_{q_i} = p$, $i = 1,\dots,n$. Then under the conditions a), b), c) above there is an integral lattice $L \subseteq F := \coprod F_i$.

<u>Proof</u>: We may assume $n = 1$ and set $q = q_1$. To begin with we need a CM_p-lattice in F. For $p = 0$ this is obvious. If $p = 1$, we use the fact that A_q is a DVR to assume $F = A_q/(q A_q)^t$, $t \geq 1$. In this case we take $L = A/q^{(t)}$, where $q^{(t)}$ denotes the t-th symbolic power. L belongs to CM_1 because A is locally factorial.

Finally, if $p = \dim A - 1$ or $\dim A - 2$, then we need the following result, essentially due to Hochster. A proof is given in [P4].

(2.10) <u>Lemma</u>. Let M be a finitely-generated R-module, $ht\, M = p$, where R is local Gorenstein and $p = \dim R - 1$ or $\dim R - 2$. Then $\operatorname{Hom}(M,V_p) \in CM_p(R)$.

From this lemma, it follows that, choosing any finitely-generated $L \subseteq F$ where $L_q = F$, $\operatorname{Hom}(L,V_p)$ will be everywhere locally CM. This is also true of $\operatorname{Hom}(\operatorname{Hom}(L,V_p),V_p) \subseteq \operatorname{Hom}(\operatorname{Hom}(F,E_p),E_p) \simeq F$, so we have our lattice.

Now suppose L is a CM_p-lattice in F for any $p \geq 0$. Let $\{\ell_1,\ldots,\ell_s\}$ generate L and let $\{p_1,\ldots,p_m\}$ be a finite set of height $p+1$ primes such that $d_p\tau(\ell_i,\ell_j) \subseteq \coprod_{k=1}^{m} E(A/p_k)$, for some fixed i and j. Let $d_p\tau(\ell_i,\ell_j) = \Sigma\, r_k$, $r_k \in E(A/p_k)$. By [SV, (4.23)] there is an integer n_k so that $p_k^{n_k} r_k = 0$, $k = 1,\ldots,m$. Each $p_k \supseteq q$ by [SV, 4.21] so let $x_k \in p_k - q$. Then if $n = \max(n_i)$ and $z = \Pi\, x_i^n$, we get $zr_k = 0$ for all k. Since $z \notin q$, z acts isomorphically on F. Consequently, zL is still a CM_p-lattice and $d_p\tau(z\ell_i,z\ell_j) = 0$, so $\tau(z\ell_i,z\ell_j) \in V_p$. Continuing in this way (for all i and j) completes the proof.

(2.11) Remark. The existence of CM_p-lattices is essentially equivalent to the existence of small CM-modules ([Ho, Conj E'''']) about which very little is known if $p \geq 3$. It is perhaps only accidential, but if A is regular, then (2.9) applies for all p if and only if $\dim A \leq 4$, which is precisely the range of dimensions for which our quadratic form techniques are successful (see (2.4)).

(2.12) Definition. Let $(F_i,\tau_i) \in \mathcal{Q}^\lambda(F(A_{q_i}))$ be given, $\mathrm{ht}\, q_i = p$, $i = 1,\ldots,n$. Set $F = \coprod F_i$, $\tau = \coprod \tau_i\colon F \times F \to E_p$. Suppose $L \subseteq F$ is an integral lattice. The dual lattice (to L) is

$$L' = \{f \in F \mid \tau(f,L) \subseteq V_p\} .$$

Since L is integral, $L \subseteq L'$. Since τ is non-singular, there is an isomorphism

$$\begin{aligned} L' &\overset{\simeq}{\to} \operatorname{Hom}(L,V_p) \\ \ell' &\to \{\ell \to \tau(\ell',\ell)\} \end{aligned} \tag{2.13}$$

so $L' \in CM_p$ by (1.6). Using the fact that $L_{q_i} = L'_{q_i} = F$ for each i, it follows from (1.2)(d) that $L'/L \in CM_{p+1}$. Set $L'/L = M$ and define $\phi\colon M \times M \to V_{p+1}$ by

$$\phi(j\ell_1,j\ell_2) = d_p\, \tau(\ell_1,\ell_2) \tag{2.14}$$

where $j\colon L' \to M$ is the quotient map.

(2.15) Remark. The composition $L \to L' \overset{\simeq}{\to} \operatorname{Hom}(L,V_p)$ is $\operatorname{Ad}(\tau|L \times L)$. Thus the map γ constructed in (2.6)(c) is the identity (modulo the identification $L \equiv L^{\wedge\wedge}$). Conversely, if for a given $(M,\phi) \in Q^{\lambda}(CM_{p+1})$ one gets γ equal to the identity in the construction of $\mathcal{D}_0^{\lambda}$ above, then it is straightforward to verify there is a nonsingular form (F,τ) and an integral lattice $L \subseteq F$ so that $M \simeq L'/L$ and τ defines ϕ by the formula (2.14).

Returning to the construction of L_0^{λ}, one proves that $(M,\phi) \in Q^{\lambda}(CM_{p+1})$ as in [P3, p. 354]; and that the class of (M,ϕ) in $W_0^{\lambda}(CM_{p+1})$ depends only on the isometry class of (F,τ) (not on the choice of lattice L) as in [P3, 3.3].

Since the construction preserves (orthogonal) sums, it remains to show that lagrangians are sent to lagrangians. The case $p = 0$ is done in [P3, §3]. For the remaining cases we need

(2.16) Lemma. Let $i: N \to G$ be a CM_p-lattice where G is a finite length A_q-module and ht $q = p$. Then there is an injection of A-modules whose image is a CM_p-lattice, $i_*: \mathrm{Hom}(N,V_p) \to \mathrm{Hom}(G,E_p)$, such that $(i_*f)(i(n)) = f(n)$, where $f \in \mathrm{Hom}(N,V_p)$ and $n \in N$.

Proof: Use injectivity of E_p to fill in:

$$\begin{array}{ccc} N & \to & G \\ \downarrow & & \downarrow i_*f \\ V_p & \to & E_p \end{array}$$

This defines i_*f and imbeds $\mathrm{Hom}(N,V_p)$ in $\mathrm{Hom}(G,E_p)$ because inclusion induces isomorphisms $N_q \xrightarrow{\simeq} G$ and $(V_p)_q \xrightarrow{\simeq} E(A/q)$.

Now for any p, if $\lambda = -1$, then by (1.20), a lagrangian (F,τ) is hyperbolic: $(F,\tau) \simeq (G \oplus \overline{G},\phi_h)$, where $\overline{G} = \mathrm{Hom}(G,E_p)$. Setting $(-)^{\wedge} = \mathrm{Hom}(-,V_p)$, we use the lemma to find a CM_p-lattice $L = N \oplus i_*(N^{\wedge})$ in F such that $\tau|L \times L$ is the skew-symmetric hyperbolic form on $(N \oplus N^{\wedge}) \times (N \oplus N^{\wedge})$. Hence $L = L'$, so the construction produces the zero form in $Q^{\lambda}(CM_{p+1})$.

In case $p = 1$ and $\lambda = 1$, $R := A_q$ is a DVR. In this case a lagrangian over $CM_1(R)$ is an sum of hyperbolics and "unary" forms, $(R/(t^{2m}),r)$ where t is a uniformizer and $r \in (R/t^{2m})^{\times}$. (This is an exercise.) The form $(R/(t^{2m}),r)$ has a sublagrangian $t^m(R/(t^{2m}) \subseteq R/(t^{2m})$. Since the hyperbolic case was treated above, we may assume $(F,\tau) = (R/(t^{2m}),r)$.

Choose an integral lattice $A/q^{(2m)} \to R/(t^{2m})$ (compare the proof of (2.9)). Define K to be the kernel of the natural surjection $A/q^{(2m)} \to A/q^{(m)}$. Then $K \in CM_1(A)$ by (1.2) and $K_q = t^mF$, the lagrangian of (F,τ).

Consider the diagram, where $(-) = \mathrm{Hom}(-,V_p)$,

$$\begin{array}{ccccc} K & \rightarrowtail & A/q^{(2m)} & \twoheadrightarrow & A/q^{(m)} \\ \downarrow & & \downarrow & & \downarrow \\ (A/q^{(m)})^{\wedge} & \rightarrowtail & (A/q^{(2m)})^{\wedge} & \twoheadrightarrow & K \\ \downarrow & & \downarrow & & \downarrow \\ N & \rightarrowtail & M & \twoheadrightarrow & M/N \end{array}$$

in which the second row is the dual of the first (short exact by (1.6)); the upper row of verticals is induced from $\tau|A/q^{(2m)} \times A/q^{(2m)}$ and the fact that $\tau|K \times K = 0$; and the last row is the resulting sequence of cokernels. Since (F,τ) is non-singular, the upper verticals are injective, making the bottom row a short exact sequence in CM_{p+1} (cf. (1.2)(d)).

Now $(A/q^{(2m)})^{\wedge}$ is identified with the dual lattice by (2.13), so $M \in CM_{p+1}$ supports the form ϕ constructed above. By (1.6), $M/N \simeq N^{\wedge}$. In fact $\phi|N \times N \equiv 0$ and it is easily checked that ϕ induces the latter isomorphism. Thus, (M,ϕ) is a lagrangian, as required.

Finally, assume $p = \dim A - 2$ or $\dim A - 1$, and that $(F,\tau) \in Q^{\lambda}(F(A_q))$ (ht $q = p$) is a lagrangian with sublagrangian G. Choose an integral lattice $L \subseteq F$ and set $I = \mathrm{im}\{L \to F \to F/G\}$. Then $\mathrm{Ass}(I) = \{q\}$ and so for every maximal ideal M of A, depth $I_M \geq 1$. Since $L \in CM_p(A)$, it follows from (1.2)(d) that $N := \ker(L \to I) \in CM_p(A)$ if $p = \dim A - 1$. If $p = \dim A - 2$, we again get $N \in CM_p(A)$, by [K, Ex. 14, p. 103].

Now $N \subseteq N^{\perp}$ since $N \subseteq G$, so there is a commutative diagram of exact sequences

$$\begin{array}{ccccc} N & \rightarrowtail & L & \twoheadrightarrow & I \\ \downarrow & & \downarrow & & \downarrow \\ I^{\wedge} & \rightarrowtail & L' & \to & N^{\wedge} \end{array}$$

where the verticals are induced by $\mathrm{Ad}(\tau|L \times L)$, the bottom line is the dual of the top one, L' is identified with $L^{\wedge}$ by (2.13), but $L' \to N^{\wedge}$ need not be surjective. ($\mathrm{Ext}^1(I,V_p) \neq 0$ if $p = \dim A - 2$ and depth $I_M = 1$ for some maximal $M \in \mathrm{Supp}(I)$.) However, if $p = \dim A - 1$ the bottom line is exact by (1.6) and if $p = \dim A - 2$, it is exact at all primes of height $\dim A - 1$. Thus, taking cokernels of the vertical maps we are finished (as in the case $p = 1$ above), if $p = \dim A - 1$. If $p = \dim A - 2$, the sequence of cokernels

$$S \rightarrowtail L'/L \to T$$

is exact at all primes of height $\dim A - 1$. Hence to complete the proof, it suffices to apply the following lemma to the class of $(L'/L,\phi)$ in $W_0^\lambda(CM_{n-1}(A))$.

(2.17) Lemma. Let A be Gorenstein of dimension n. Then

$$W_0^\lambda(CM_{n-1}(A)) \to \coprod_{\mathrm{ht}q=n-1} W_0^\lambda(CM_{n-1}(A_q))$$

is injective.

Proof: Let $(M,\phi) \in Q^\lambda(CM_{n-1}(A))$ be a lagrangian at each $q \in \mathrm{Ass}(M)$. Consider the collection L of submodules $N \subseteq M$ such that $N \subseteq N^\perp$ and N_q is a sublagrangian of (M_q,ϕ_q) for each $q \in \mathrm{Ass}(M)$. L is not empty: if K_q is a sublagrangian of M_q and $i\colon M \to \coprod M_q$ $(q \in \mathrm{Ass}(M))$ is the canonical imbedding, then $i^{-1}(\coprod K_q)$ is such an N. Since A is noetherian and M is finitely-generated, L has a maximal element N_0. Then $N_0 = N_0^\perp$: if $x \in N_0 - N_0$, $n_1,n_2 \in N_0$ and $a_1,a_2 \in A$, then

$$\phi(n_1 + a_1x, n_2 + a_2x) = a_1\, a_2\, \phi(x,x).$$

If this is zero, then $(N_0 + Ax) \subseteq (N_0 + Ax)^\perp$, contradicting the maximality of N_0. But $(N_0)_q = ((N_0)_q)^\perp$ for each $q \in \mathrm{Ass}(M)$ so $\phi_q(x,x) = 0$, hence $\phi(x,x) = 0$. Write N for N_0.

Now consider the commutative diagram

$$\begin{array}{ccc} & N & \\ {}^{\kappa}\swarrow & & \searrow^{\alpha} \\ N^{\wedge\wedge} & \xrightarrow{\beta^\wedge} & (M/N)^\wedge \end{array}$$

where κ is the canonical map and α,β are induced by ϕ. Ad ϕ is injective, so is α, which means κ is. Since $N = N^\perp$, β is injective; this implies M/N is height-unmixed (because $N^\wedge$ is). But this means N and $M/N \in CM_{n-1}(A)$ because for M maximal, N_M (or $(M/N)_M$) can have depth zero if and only if $M \in \mathrm{Ass}(N_M)$ (or $\mathrm{Ass}(M/N)_M$). Thus κ is an isomorphism by (1.6)(a) and $\beta^\wedge$ is surjective because β was injective. This means α and β are isomorphisms, so $N \subseteq M$ is a sublagrangian.

(2.18) Definition of $\mathcal{D}_1^\lambda\colon W_1^\lambda(CM_{p+1}) \to W_0^\lambda(CM_p)$.

Let $(K,H,\Delta) \in F^\lambda(CM_{p+1})$ and let

(2.19) $$L \overset{\alpha}{\rightarrowtail} N \overset{j}{\twoheadrightarrow} H$$

be a short exact sequence with $L, N \in CM_p$ (cf. (1.6)). Dualize (2.19) to $N^\wedge \overset{\alpha^\wedge}{\rightarrowtail} L^\wedge \twoheadrightarrow H^\sim$ using (1.6) and its notation. Add the sequences, getting a resolution of $H \oplus H^\sim$,

$$(2.20) \qquad L \oplus N^\wedge \overset{\alpha \oplus \lambda\alpha^\wedge}{\rightarrowtail} N \oplus L^\wedge \overset{k}{\twoheadrightarrow} H \oplus H^\sim$$

and set $M = k^{-1}(\Delta(K))$; clearly $M \in CM_p$.

Let $Q = \mathrm{Ass}(L) = \mathrm{Ass}(N)$ and let $(-)_Q$ denote localization at the set of primes in Q. Using the isomorphisms and inclusions

$$L_Q \overset{\alpha_Q}{\underset{\simeq}{\to}} N_Q \supseteq N$$

(2.21) and

$$\mathrm{Hom}(L, V_p) \subseteq \mathrm{Hom}(L, V_p)_Q \simeq \mathrm{Hom}(L_Q, E_p)$$

define $\tau\colon (N \oplus L^\wedge) \times (N \oplus L^\wedge) \to E_p$ by

$$\tau(n_1 \oplus f_1, n_2 \oplus f_2) = f_2(\alpha_Q^{-1}(n_1)) + \lambda f_1(\alpha_Q^{-1}(n_2))$$

where $n_i \in N$ and $f_i \in L^\wedge$. Then $((N \oplus L^\wedge)_q, \tau_q) \in Q^\lambda(CM_p(A_q))$ for all $q \in Q$ and the diagram

$$(2.22) \qquad \begin{array}{ccc} (N \oplus L^\wedge) \times (N \oplus L^\wedge) & \overset{\tau}{\to} & E_p \\ \downarrow k \times k & & \downarrow d_p \\ (H \oplus H^\sim) \times (H \oplus H^\sim) & \overset{\phi_h}{\to} & V_{p+1} \end{array}$$

commutes. It follows that $\tau(M \times M) \subseteq V_p$ so $M \subseteq N \oplus L^\wedge \subseteq (N \oplus L)_Q$ is an integral lattice in $((N \oplus L^\wedge)_Q, \tau_Q)$.

Set $\psi = \tau | M \times M$. Since τ_Q is non-singular, in order to show $\psi\colon M \times M \to V_p$ is non-singular, it suffices to show $M = M'$, the dual lattice. Let $m \in M'$, so that $\tau(M, m) \in V_p$. Then $m \in N \oplus L^\wedge \subseteq (N \oplus L^\wedge)_Q$ because $(\alpha \oplus \lambda\alpha^\wedge)(L \oplus N^\wedge) \subseteq M$ and $\{(\alpha \oplus \lambda\alpha^\wedge)(L \oplus N^\wedge)\}' = N \oplus L^\wedge$. By (2.22) we get $\phi_h(\Delta(K), k(m)) = 0$. But since $\Delta(K) = \Delta(K)^\perp$, $k(m) \in \Delta(K)$; hence $m \in M$. Since $M \subseteq M'$, this proves equality.

Thus we conclude $(M, \psi) \in Q^\lambda(CM_p(A))$ and denote it by $I((K,H,\Delta);R)$, where R is the resolution (2.19). We next show the class of $I(K,H,\Delta);R)$ in $W_0^\lambda(CM_p)$ is independent of R.

Let $R_1 - (L_1 \overset{\alpha_1}{\to} N_1 \overset{j_1}{\to} H)$ be another resolution and form the pullback P,

$$\begin{array}{ccccc}
 & & L_1 & = & L_1 \\
 & & \downarrow i_1 & & \downarrow \alpha_1 \\
L & \overset{i}{\rightarrowtail} & P & \overset{\ell_1}{\twoheadrightarrow} & N_1 \\
\| & & \downarrow \ell & & \downarrow j_1 \\
L & \rightarrowtail & N & \overset{j}{\twoheadrightarrow} & H
\end{array}$$

From this we get a commutative diagram with exact rows

$$\begin{array}{ccccc}
L_1 & \overset{\alpha_1}{\rightarrowtail} & N_1 & \overset{j_1}{\twoheadrightarrow} & H \\
\uparrow pr_2 & & \uparrow \ell_1 & & \uparrow = \\
L \oplus L_1 & \overset{i_1 + i_2}{\rightarrowtail} & P & \overset{\ell j = \ell_1 j_1}{\twoheadrightarrow} & H \\
\downarrow pr_1 & & \downarrow \ell & & \downarrow = \\
L & \overset{\alpha}{\rightarrowtail} & N & \overset{j}{\twoheadrightarrow} & H
\end{array}$$

Hence it is sufficient to compare resolutions connected by

$$\begin{array}{ccccc}
L_1 & \overset{\alpha_1}{\rightarrowtail} & N_1 & \overset{j_1}{\twoheadrightarrow} & H \\
\downarrow & & \downarrow \ell & & \downarrow = \\
L & \overset{\alpha}{\rightarrowtail} & N & \twoheadrightarrow & H
\end{array}$$

Let $(M_1,\psi_1) = I((K,H,\Delta);R_1)$. Then $\ker \ell \subseteq M_1$, $\ker \ell \subseteq (\ker \ell)^{\perp}$ and the induced form on $(\ker \ell)^{\perp}/\ker \ell$ is isometric to (M,ψ). Hence by (1. 9) $[M,\psi] = [M_1,\psi_1]$ in $W_0^{\lambda}(CM_p)$, as claimed.

Next we claim that $I((K,H,\Delta);R)$ is isometric to $I(\chi(H,\psi)\cdot(K,H,\Delta);R)$ where $\psi\colon H \times H \to V_p$ is any $(-\lambda)$-symmetric form. This is shown following the argument in [Pl, p. 376].

Finally, there are choices R and R' for any extension (ε) of H so that $I((K,H,\Delta);R)$ is isometric to $I(\sigma_{\varepsilon}(K,H,\Delta);R')$. Indeed if $R = (L \to N \to H)$ and $(\varepsilon) = (I \to H_1 \to H)$, then from the pull-back diagram

$$\begin{array}{ccccc}
 & & L & \overset{=}{\to} & L \\
 & & \downarrow & & \downarrow \\
I & \rightarrowtail & P & \twoheadrightarrow & N \\
\downarrow = & & \downarrow & & \downarrow \\
I & \rightarrowtail & H_1 & \twoheadrightarrow & H
\end{array}$$

we take $R' = (L \to P \to H_1)$. Details are left to the reader (cf. [Pl, p. 376]).

(2.23) <u>Definition of</u> $L_1^{-1}: \coprod_{ht\,p=p} W_1^{-1}(CM_p(A_p)) \to W_1^{-1}(CM_{p+1})$

It is sufficient to define $L_1^{-1}|W_1^{-1}(CM_p(A_p))$ for each p of height p. Given an element $(F,G,\Gamma) \in F^{-1}(CM_p(A_p))$, we have seen in (1.20) that if $\Gamma = (\mu,\nu): F \to G \oplus \overline{G}$, $\overline{G} := \mathrm{Hom}(G,E_p)$, then Γ can be extended to an isometry

$$A = \begin{pmatrix} \mu & \sigma \\ \nu & \rho \end{pmatrix}: F \oplus \overline{F} \to G \oplus \overline{G} ,$$

of the hyperbolic forms on $F \oplus \overline{F}$ and $G \oplus \overline{G}$. It is easy to verify that

$$(2.24) \qquad \begin{pmatrix} \overline{\rho} & \overline{\sigma} \\ \nu & \mu \end{pmatrix} \begin{pmatrix} \mu & \sigma \\ \nu & \rho \end{pmatrix} = \begin{pmatrix} I & 0 \\ 0 & I \end{pmatrix} ,$$

the identity map of $F \oplus \overline{F}$.

Now choose a CM_p-lattice

$$N \oplus N^\wedge \subseteq G \oplus \overline{G}$$

where $N \subseteq G$, $N^\wedge = \mathrm{Hom}(N,V_p)$ and $N^\wedge \subseteq \overline{G}$ according to (2.16). Similarly, choose a CM_p-lattice

$$L \oplus L^\wedge \subseteq F \oplus \overline{F} ,$$

this time so that

$$A(L \oplus L^\wedge) \subseteq N \oplus N^\wedge .$$

By restricting the entries of the above matrix for A to L and $L^\wedge$ we set

$$B = A|L \oplus L^\wedge = \begin{pmatrix} m & s \\ n & r \end{pmatrix}: L \oplus L^\wedge \to N \oplus N^\wedge.$$

Let $B^\wedge: N \oplus N^\wedge \to L \oplus L^\wedge$ be given by

$$\begin{pmatrix} r^\wedge & s^\wedge \\ n^\wedge & m^\wedge \end{pmatrix}$$

Then by (2.24),

$$B^\wedge B = \begin{pmatrix} \zeta & 0 \\ 0 & \zeta^\wedge \end{pmatrix}: L \oplus L^\wedge \to L \oplus L^\wedge$$

Let $H = \mathrm{cok}\,\zeta$ ($\in CM_{p+1}$ by (1.2)), so that $H^\sim (:= \mathrm{Hom}(H,V_{p+1})) = \mathrm{cok}\,\zeta^\wedge$ by (1.6). Let $K = \mathrm{cok}\,B$ and set

$$\Delta: K \to H \oplus H^\sim$$

equal to the map cok $B \to$ cok $B^{\wedge}B$ induced by $B^{\wedge}$. Then $\mathrm{im}(\Delta)$ is a sublagrangian of the skew-symmetric hyperbolic form on $H \oplus H^{\sim}$ so $(K,H,\Delta) \in F^{-1}(CM_{p+1})$. Further details, including well-definedness, are an easy generalization of [P1, §6].

(2.25) Exactness of the sequences (2.1)(a)-(d): The exactness of the first sequence is part of the theorem of [P1]. The exactness of c) follows easily from [P1, §8]. This will use the fact that lagrangians are hyperbolic ((1.20)), so that given the above construction of the maps involved, exactness is another straightforward generalization.

Given the Remark (2.15), exactness at $W_0^{\lambda}(CM_p)$ in b) and d) is immediate from (1.21). It follows from the constructions in [P3, §3] that a λ-formation (K,H,Δ), with $\Delta = (\alpha,\gamma)$ and α injective, represents an element in $\mathrm{im}\, \mathcal{D}_0^{-\lambda}$. Thus exactness at $W_1^{\lambda}(CM_1)$ in b) follows from (8. 4).

Given what we now have, it remains to prove $W_0^1(CM_4) \to W_1^{-1}(CM_3)$ is surjective in d). (The analogous fact in b) was proved using the fact that A is a DVR at all height one primes.). Since we now also know that c) is exact and the extreme terms are zero, it is sufficient to prove that the composite

(2.26) $$W_0^1(CM_4) \to W_1^{-1}(CM_3) \to W_0^{-1}(CM_2)$$

is surjective. This is what was proved in [P3, §3] except that in place of the sequence $\{4,3,2\}$ of heights, the descent was from height 2 to height 0. The surjectivity of (2.26) is a straightforward variation.

§3. The Witt group and A_0.

To state the result we first recall the definition of the Chow group of cycles mod rational equivalence ([Ch], [Fu]). Let X be an n-dimensional scheme of finite type over a field. Let X_k denote the set of generic points of irreducible closed subsets of dimension k and Z_k the free abelian group on X_k. Let $x \in X_k$, $y \in X_{k+1}$, $x \in \overline{\{y\}}$, $f \in k(y)^x$. Then $f = g/h$, $g,h \in \mathcal{O}_{Y,x}$, where Y is the integral subscheme corresponding to y. Define a homomorphism

$$b_y\colon k(y)^x \to X_k$$

by

$$b_y(f) = \sum_{x\in\overline{\{y\}}} \text{length}(\mathcal{O}_{Y,x}/g\mathcal{O}_{Y,x})x - \sum_{x\in\overline{\{y\}}} \text{length}(\mathcal{O}_{Y,x}/h\mathcal{O}_{Y,x})x.$$

Finally, set

$$b_{k+1} := \coprod b_y\colon \coprod_{y\in X_{k+1}} k(y)^x \to Z_k$$

and

$$A_k(X) := \text{cok}\; b_{k+1},$$

the k-th Chow group.

Next is the definition of "higher class groups" from [CF]. Given an A-regular sequence $\{x_1,\ldots,x_{n-k}\}$, let

$$d(x_1,\ldots,x_{n-k}) = \sum_{\text{ht}\,q=n-k} \text{length}(A_q/(x_1,\ldots,x_{n-k}))\cdot q \in Z_k$$

where $Z_k = Z_k(\text{Spec } A)$, the free abelian group on height $n-k$ primes. Let $R_k \subseteq Z_k$ be the subgroup generated by expressions $d(x_1,\ldots,x_{n-k})$, for all A-sequences $\{x_1,\ldots,x_{n-k}\}$, and let

$$C_k = Z_k/R_k\,.$$

(This is denoted C_{n-k} in [CF]). For instance C_{n-1} is the class group of A, when A is a normal domain. If A is regular local and essentially of finite type over a field, then it is a consequence of truth of the Gersten conjecture in this case that the groups $W_*(A)$ of [CF] are trivial. Consequently, $C_*(A) \equiv 0$ also, by [CF,3.4].

(3.1) Theorem. Let A be a regular domain of finite type over a field k which is either real or algebraically closed. Let K be the fraction field of A and let $X = \text{Spec } A$. Then

a) if $\dim A = 4$, there is a surjection

$$A_0(X) \otimes \mathbb{Z}/2 \to \ker(W_0^1(A) \to W_0^1(K)) \otimes \mathbb{Z}/2$$

b) if $\dim A = 2$, there is a surjection

$$A_0(X) \otimes \mathbb{Z}/2 \to W_0^{-1}(A) \otimes \mathbb{Z}/2$$

If $C_1(A) \otimes \mathbb{Z}/2 = 0$ in either case, then the corresponding map is an isomorphism.

In some cases the left side of the surjection of a) and b) is known.

(3.2) Proposition. Let A be as in (3.1). Then

a) if $k = \bar{k}$, $A_0(X) \otimes \mathbb{Z}/2 = 0$;

b) if $k = \mathbb{R}$, and X is an open subscheme of $\bar{X}$, an integral scheme proper over $\mathbb{R}$ where $X(\mathbb{R}) = \bar{X}(\mathbb{R})$, then $A_0(X) \otimes \mathbb{Z}/2 = (\mathbb{Z}/2)^r$ where r is the number of connected components of $X(\mathbb{R})$, in the Euclidean topology of $X(\mathbb{R})$.

(3.3) Remark. a) Suppose $A = \mathbb{R}[x_1,\ldots,x_n]/I$, A is a regular domain and there is $f \in I$ such that $V(f_d) = \{0\}$ or $\emptyset$, where f_d is the homogeneous part of f of highest degree. (Here $V(f_d)$ is the set of real points x such that $f_d(x) = 0$.) Then the conditions in (3.2)(b) are satisfied on setting $X = \operatorname{Spec} A$ and $\bar{X}$ = the closure of X in $\mathbb{P}^n_{\mathbb{R}}$.

b) It is possible to make a statement in (3.2)(b) for k any real closed field (by defining "algebraic components"--cf. [DK]); and for the case where $\bar{X}(\mathbb{R}) \supsetneq X(\mathbb{R})$. These are left to the reader.

Proof of (3.1). What will be shown is the following more general fact, which is sufficient by (2.1).

(3.4) Proposition. Let A be an n-dimensional regular Noetherian domain of finite type over k, where k is either real or algebraically closed. Let $X = \operatorname{Spec} A$ and identify $CM_n(A)$ with the category of modules of finite length. Then there is a surjection

$$A_0(X) \otimes \mathbb{Z}/2 \to \operatorname{cok}(L: \coprod_{\operatorname{ht}p=n-1} W_0^1(CM_{n-1}(A_p)) \to \coprod_{\operatorname{ht}p=n} W_0^1(CM_n(A_p))) \otimes \mathbb{Z}/2.$$

where $L = L_0^1$ is defined in (2.7). If $C_1(A) \otimes \mathbb{Z}/2 = 0$, then this is an isomorphism.

Notice that (if $C_1(A) \otimes \mathbb{Z}/2 = 0$) the proposition is consistent with Bloch's formula ([Q]) and a Gersten resolution for $W(A)$ (cf. [P4]).

To simplify notation, whenever $\operatorname{ht} p = p$, we replace $CM_p(A_p)$ by $F(A_p)$, the category of A_p-modules of finite length. (This uses (1.2).) Also $W(-)$ will always mean $W_0^1(-)$. If $(M,\phi) \in Q^\lambda(F(A_p))$ is a lagrangian, then the length of M is clearly even. Thus there is a homomorphism

$$\ell_p\colon W(F(A_p)) \to \mathbb{Z}/2$$

given by $\ell_p(M,\phi)$ = length of M, mod 2; its composition with the dévissage isomorphism is the usual rank homomorphism

$$\mathrm{rk}\colon W(k(p)) \to \mathbb{Z}/2 .$$

The proof of the proposition comes down to the following three lemmas.

(3.5) Lemma. Assume $C_1(A) = 0$. Then for each prime $p \subseteq A$ of height n-1, there is an element $\coprod (M_q,\phi_q) \in \coprod_{\operatorname{ht}q=n-1} W(F(A_q))$ such that

$$\ell_q(M_q,\phi_q) = \begin{cases} 1, & q = p \\ 0, & q \neq p \end{cases}$$

and

$$L(\coprod (M_q,\phi_q)) = 0.$$

(3.6) Lemma. Let $IW(F(A_p)) = \ker \ell_p$. Then there is a commutative diagram

$$\begin{array}{ccc}
\coprod_{\operatorname{ht}q=n-1} IW(F(A_q)) & \xrightarrow{L} & \coprod_{\operatorname{ht}M=n} W(F(A_M)) \\
\simeq \downarrow \text{dévissage} & & \simeq \downarrow \text{dévissage} \\
\coprod_{\operatorname{ht}q=n-1} IW(k(q)) & & \coprod_{\operatorname{ht}M=n} W(k(M)) \\
\downarrow \text{dis} & & \downarrow \text{rk} \\
\coprod_{\operatorname{ht}q=n-1} k(q)^{x}/k(q)^{x2} & \xrightarrow{b_1 \otimes \mathbb{Z}/2} & \coprod_{\operatorname{ht}M=n} \mathbb{Z}/2
\end{array}$$

where dis and rk are the discriminant and rank mod 2, respectively.

(3.7) Lemma. For k either real or algebraically closed, there are isomorphisms

a) $\mathrm{rk}\colon W(k(M)) \otimes \mathbb{Z}/2 \xrightarrow{\simeq} \mathbb{Z}/2$,

where M is maximal in A, and

b) $I \otimes \mathbb{Z}/2 \overset{\cong}{\to} I/I^2 \otimes \mathbb{Z}/2 \overset{\cong}{\to} k(q)^x/k(q)^{x^2}$ where $I = IW(k(q))$, ht $q = n-1$ and the second isomorphism is induced by dis.

Assuming the three lemmas, the proof of (3.4) is completed as follows. Lemma (3.5) says that, if $C_1(A) = 0$, then im L = (im $L|\coprod_{htq=n-1} IW(F(A_q)))$ and Lemmas (3.6) and (3.7) then imply that, mod 2, this is $\operatorname{im}(b_1 \otimes \mathbb{Z}/2)$; taken by themselves, (3.6) and (3.7) give the surjection in (3.4).

Proof of (3.5). Let p be a fixed height $n-1$ prime. Then the hypothesis $C_1(A) \otimes \mathbb{Z}/2 = 0$ means there are A-regular sequences

$$X_i = \{x_{i,1},\dots,x_{i,n-1}\}, \quad i = 1,\dots,k$$

such that for any height $n-1$ prime q,

$$\sum_i \operatorname{length}(A_q/(X_i)_q) \equiv \begin{cases} 1 & q = p \\ 0, & q \neq p \end{cases}, \quad \text{mod } 2$$

By the reasoning for (2. 5), there is a homomorphism

$$W(CM_0(A/(Z)) \to W(CM_{n-1}(A))$$

for any A-sequence Z with n-1 elements. For each $i = 1,\dots,k$, let $\alpha_i \in W(CM_{n-1}(A))$ be the image of $\langle 1\rangle \in W(CM_0(A/(X_i)))$. Set

$$\phi = K(\Sigma\, \alpha_i)$$

where $K\colon W(CM_{n-1}(A)) \to \coprod_{htq=n-1} W(F(A_q))$ is from (2.1). It is an easy consequence of the definition of L in (2. 7), that the composition $L \circ K = 0$. This means the desired element is a representative of ϕ.

Proof of (3.6). For any maximal ideal M of A, localization at M induces the vertical homomorphisms in the commutative diagram (cf. (1. 4)) where $R = A_M$,

$$\begin{array}{ccc} \coprod_{htq=n-1} W(F(A_q)) & \overset{L}{\to} & \coprod_{htq=n} W(F(A_q)) \\ \downarrow & & \downarrow \\ \coprod_{\substack{htq=n-1 \\ q\subseteq M}} W(F(R_q)) & \overset{L}{\to} & W(F(R)). \end{array}$$

Note that $A_q = R_q$ if $q \subseteq M$ (a slight abuse of notation) and that the verticals are the consequent projections. Thus, to prove (3.6), we may replace A with its localization R. To compute $L|W(F(R_p))$ for p of height n - 1, the idea is to reduce to the 1-dimensional case by passing to R/p. However, since R/p is not, in general, Gorenstein, this does not seem to be possible. (But if dim R = 2, then R/p is Gorenstein and the following argument can be greatly simplified.) Instead we work with a complete intersection ideal in p, whose quotient is Gorenstein.

So fix p, ht p = n - 1. Let $\{x_1,\ldots,x_{n-1}\} \subseteq p$ be a regular sequence with

$$(3.8)\qquad (x_1,\ldots,x_{n-1})_p = p\,R_p,$$

and set

$$S = R/(x_1,\ldots,x_{n-1}),$$

a one-dimensional Gorenstein ring. From (1. 5) it is easy to construct a commutative diagram

$$(3.9)\qquad \begin{array}{ccccc} S & \rightarrowtail & E_0(S) & \twoheadrightarrow & E_1(S) \\ \downarrow & & \downarrow & & \downarrow \\ V_{n-1}(R) & \rightarrowtail & E_{n-1}(R) & \twoheadrightarrow & V_n(R) = E_n(R) \end{array}$$

There are inclusions

$$CM_i(S) \to CM_{i+n-1}(R),\quad i = 0,1$$

and

$$(3.10)\qquad F(S_{\bar{\pi}}) \overset{=}{\to} F(R_\pi),\quad \bar{\pi} = \pi/(x_1,\ldots,x_{n-1}) .$$

From these facts we get an exact commutative diagram of localization sequences

$$(3.11)\qquad \begin{array}{ccccc} W(CM_0(S)) & \to & \coprod_{\mathrm{ht}\bar{q}=0} W(F(S_{\bar{q}})) & \xrightarrow{L(S)} & W(F(S)) \\ \downarrow & & \downarrow & & \downarrow \\ W(CM_{n-1}(R)) & \to & \coprod_{\mathrm{ht}q=n-1} W(F(R_q)) & \xrightarrow{L = L(R)} & W(F(R)) \end{array}$$

(The exactness of the top sequence can be extracted from [Pl]). Using dévissage the right vertical is an isomorphism; it is similarly seen that the middle vertical is an inclusion to the summands corresponding to q containing $(x_1,\ldots,x_{n-1})$.

Now if Σ is the set of non-zero-divisors of S, there is a canonical map $S_\Sigma \to E_0(S)$ extending $S \to E_0(S)$, hence also an induced map $S_\Sigma/S \to E_1(S)$ of S-modules. Since

$$S \rightarrowtail S_\Sigma \twoheadrightarrow S_\Sigma/S$$

is also a minimal injective resolution of S over itself (cf. [B , 6.2]) these maps are isomorphisms making

$$\begin{array}{ccccc} S & \rightarrowtail & S_\Sigma & \twoheadrightarrow & S_\Sigma/S \\ \downarrow = & & \downarrow \simeq & & \downarrow \simeq \\ S & \rightarrowtail & E_0(S) & \twoheadrightarrow & E_1(S) \end{array}$$

commutative. We also have $E_0(S) = \coprod E(S/\bar{q})$ and (clearly) $S_\Sigma \overset{\simeq}{\to} \coprod S_{\bar{q}}$ where the sums are over the minimal primes $\bar{q}$ of S. Hence there are canonical isomorphisms

$$(3.12) \qquad S_{\bar{q}} \overset{\simeq}{\to} E(S/\bar{q}) ,$$

the localizations of $S \to E_0(S)$ at the minimal primes $\bar{q} \subseteq S$.

Next let

$$(0) = Q_1 \cap \ldots \cap Q_r$$

be the minimal primary decomposition of the zero ideal in S, with associated primes $\bar{q}_1,\ldots,\bar{q}_r$. Let

$$\bar{q}_1 = \bar{p} := p/(x_1,\ldots,x_{n-1}) .$$

Then by (3.8),

$$(3.12) \qquad Q_1 = \bar{q}_1, \quad \text{so} \quad S_{\bar{q}_1} = k(p) .$$

Since the decomposition is unmixed,

$$Q_i = \ker(S \to S_{\bar{q}_i}), \quad i = 1,\ldots,r.$$

From this there are canonical factorizations of $S \to S_{\bar{q}_i}$,

$$(3.13) \qquad \begin{array}{ccc} S & \to & S_{\bar{q}_i} \\ & \searrow \quad \swarrow & \\ & S/Q_i & \end{array}$$

We are now prepared to prove (3.6) commutes (when A is local).

It is well-known that $IW(k(p))$ $(= IW(F(R_p)))$ is generated by binary forms $\langle f,-1\rangle$, $f \in k(p) = k(\bar{p})$. Clearing its denominator we may take f to be in the image of $\bigcap_{i\neq 1} Q_i - \bar{q}_1$ in $k(p)$; similarly replace -1 by minus the square of an element of $\bigcap_{i\neq 1} Q_i - \bar{q}_1$. The result is a symmetric bilinear form

$$\phi_1\colon (S/\bar{q}_1)^2 \times (S/\bar{q}_1)^2 \to S_{\bar{q}_1}$$

such that

$$\operatorname{im} \phi_1 \subseteq \operatorname{im}((\bigcap_{i\neq 1} Q_i - \bar{q}_1) \to S_{\bar{q}_1})$$

and

$$\phi_1 \otimes S_{\bar{q}_1} \simeq \langle f,-1\rangle .$$

Similarly, for each $j = 2,\dots,r$, choose

$$\phi_j\colon (S/Q_j)^2 \times (S/Q_j)^2 \to S_{\bar{q}_j}$$

with

$$\operatorname{im} \phi_j \subseteq \operatorname{im}((\bigcap_{i\neq j} Q_i - \bar{q}_j) \to S_{\bar{q}_j}$$

and

$$\phi_j \otimes S_{\bar{q}_j} \simeq \langle f_j,-f_j\rangle$$

for some $f_j \in S_{\bar{q}_j}$. (This will use the fact that $\bigcap_{i\neq j} Q_i - \bar{q}_j \neq \emptyset$.)

Choose liftings $a \in \bigcap_{i\neq 1} Q_i - \bar{q}_1$ for f and $a_j \in \bigcap_{i\neq j} Q_i - \bar{q}_j$ for f_j. Then

$$b := a + a_2 + \dots + a_r,\quad b' := a' - a_2 \dots - a_r$$

are not zero-divisors in S. Indeed, if so, then $b \in \cup \bar{q}_j$, so $b \in \bar{q}_1$ (say). But for each $j \neq 1$, $a_j \in Q_1 \subseteq \bar{q}_1$, which implies $a \in \bar{q}_1$, contradicting the choice of a.

Now let $\psi\colon S^2 \times S^2 \to S$ have matrix $\begin{pmatrix} b & 0 \\ 0 & b' \end{pmatrix}$. Then

$$\psi \otimes S_{\bar{q}_i} \simeq \begin{cases} \begin{pmatrix} f & 0 \\ 0 & -1 \end{pmatrix}, & i = 1 \\ \begin{pmatrix} f_i & 0 \\ 0 & -f_i \end{pmatrix}, & i \neq 1 , \end{cases}$$

because $a_j \in \bigcap_{k\neq j} Q_k$ localizes to zero in $S_{\bar{q}_i}$ if $i \neq j$. Referring to (2. 8), this means we have found a lattice (S^2,ψ) for the element of $\coprod_{\operatorname{ht}\bar{q}=0} W(F(S_{\bar{q}})$ represented by $\langle f,-1\rangle \in W(k(\bar{q})) = W(F(S_{\bar{q}}))$.

Referring now to the diagram (3.11), the image $L(S)\langle f,-1\rangle \in W(F(S))$ is supported on the S-module

$$\operatorname{cok}(\operatorname{Ad} \psi\colon S^2 \to S^2) ;$$

this is immediate from the definition (2. 7) of $L(S)$. Since we have seen that the right vertical in the diagram is an isomorphism and the center one induces $W(F(S_{\bar{q}_1})) \xrightarrow{\simeq} W(F(R_p))$, we will be done if we can show that

$$\text{length}(\text{cok}(\text{Ad }\psi)) \equiv b_1(\bar{a}) \mod 2$$

where $\bar{a}$ is the image of a in $S/\bar{q}_1 = R/p$.

By a formula of Grothendieck ([Gr, IV.21.10.17.8]),

$$(3.14) \qquad \ell_S(\text{cok Ad }\psi) = \Sigma\, \ell_{S_{\bar{q}_i}}(S_{\bar{q}_i})\,\ell_{S/\bar{q}_i}(\text{cok}((\text{Ad }\psi) \otimes S/\bar{q}_i))$$

where the subscript on ℓ indicates the local ring with respect to which length ℓ is being computed. Now for each $i > 1$,

$$\psi \otimes S/\bar{q}_i = \begin{pmatrix} \bar{a}_i & 0 \\ 0 & -\bar{a}_i \end{pmatrix}$$

where $\bar{a}_i \neq 0$ and is the image of a_i in $S/\bar{q}_i$ (because $a_j \in Q_j \subseteq \bar{q}_j$ for $j \neq i$); also, by construction, $\bar{a}'$ is a square in the domain $S/\bar{q}_1 = S/\bar{p}$. Finally, $S_{\bar{q}_1} = k(\bar{q}_1)$, so $\ell_{S_{\bar{q}_1}}(S_{\bar{q}_1}) = 1$. Thus, from (3.14) we get, mod 2,

$$\ell_S(\text{cok Ad }\psi) \equiv \ell_{S/\bar{p}}(\text{cok}(\bar{a}\colon S/\bar{p} \to S/\bar{p}))$$

Since the latter is $b_1(\bar{a})$, the proof of (3.6) is complete.

<u>Proof of</u> (3.7). If k is algebraically closed, $\text{rk}\colon W(k) \xrightarrow{\cong} \mathbb{Z}/2$ and if k is real closed, $W(k) \xrightarrow{\cong} \mathbb{Z}$ by the signature. Since rank and signature are congruent mod two and since $k(M)$ is either real or algebraically closed, we get a). The second follows from [L] and (Kl, §11] and uses the fact that if $\text{ht } p = n - 1$, then $k(p)$ has transcendence degree one over k.

<u>Proof of</u> (3.2). If $k = \bar{k}$, then we claim $A_0(X)$ is divisible; this is sufficient for a). Each closed point of X is in the image of a proper map from a smooth affine curve C. Covariance of A_0 for proper maps ([Fu, 1.9]) reduces the claim to the case $X = C$. Complete C to a smooth projective curve $\bar{C}$; the existence of a Jacobian for $\bar{C}$ implies that the group of zero-cycles of degree zero, $\tilde{A}_0(\bar{C})$, is divisible. Now from the exact sequence of [Fu, 1.9],

$$A_0(\text{pts.}) \to A_0(\bar{C}) \twoheadrightarrow A_0(C),$$

so $A_0(C)$ also divisible since $A_0(\bar{C}) \cong \tilde{A}_0(\bar{C}) \oplus \mathbb{Z}$.

For (3.2)(b), tensor [Fu, 1.9] with $\mathbb{Z}/2$ to get an exact sequence

$$A_0(\bar{X} - X) \otimes \mathbb{Z}/2 \xrightarrow{i} A_0(\bar{X}) \otimes \mathbb{Z}/2 \twoheadrightarrow A_0(X) \otimes \mathbb{Z}/2 .$$

From [CI, 3.2], it follows that there is an isomorphism

$$\theta: A_0(\overline{X}) \otimes \mathbb{Z}/2 \stackrel{\approx}{\to} (\mathbb{Z}/2)^r .$$

(The argument in [CI, (3.2)(i)] requires that $\overline{X}/\mathbb{R}$ be smooth in order to prove $\tilde{A}_0(X_{\mathbb{C}})$ is divisible using Bertini's theorem. This hypothesis is avoided with the argument in the $k = \overline{k}$ case above.) In fact, looking at (3.2)(i) and the proof of (3.1) in [CI], θ is induced by the natural map $Z_0(\overline{X}) \to (\mathbb{Z}/2)^r$ which sends complex points to zero and a real point in the i-th component of $\overline{X}(\mathbb{R})$ to the i-th standard basis vector of $(\mathbb{Z}/2)^r$. But since we are assuming $(\overline{X} - X)(\mathbb{R}) = \emptyset$, $\theta \circ i = 0$. Hence $i = 0$, which means $A_0(X) \otimes \mathbb{Z}/2 \stackrel{\approx}{\to} (\mathbb{Z}/2)^r$.

(I learned the above argument for the $k = \overline{k}$ case from V. Srinivas. The reference for (3.2)(b) and a correction of its attempted proof I owe to J.-L. Colliot-Thélène.)

§4. Examples: skew-symmetric forms on surfaces and algebra structures on resolutions.

We begin by summarizing (without proofs) Ferrand's [F , §1] description of a construction due to Serre and Horrocks. Throughout, I is a height two ideal of an n-dimensional regular domain A such that $A/I \in CM_{n-2}(A)$.

Suppose given an isomorphism $\eta\colon A/I \to \operatorname{Ext}^2(A/I,A)$; then there is an extension (ε) in $\operatorname{Ext}^1(I,A)$ corresponding to $\eta(1)$ under the isomorphism

$$\operatorname{Ext}^2(A/I,A) \simeq \operatorname{Ext}^1(I,A).$$

If

$$(\varepsilon) = (A \overset{s'}{\rightarrowtail} E \overset{s}{\twoheadrightarrow} I) \tag{4.1}$$

then E is projective and there is an isomorphism $\sigma\colon \Lambda^2 E \overset{\simeq}{\to} A$ defined by $s'\sigma(x \wedge y) = s(x)y - s(y)x$. Thus, setting $h\colon E \to A$ equal to s followed by the inclusion of I into A, the top line in

$$\begin{array}{ccccccccc} A & \overset{s'}{\rightarrowtail} & E & \overset{h}{\to} & A & \twoheadrightarrow & A/I & & \\ \downarrow & & \downarrow\mathrm{Ad}\phi\downarrow & & \eta'\downarrow & & & \searrow^{\eta} & \\ \bar{A} & \overset{\bar{h}}{\rightarrowtail} & \bar{E} & \to & \bar{A} & \twoheadrightarrow & \operatorname{Ext}^1(I,A) & \simeq & \operatorname{Ext}^2(A/I,A) \end{array} \tag{4.2}$$

is (using σ') a Koszul resolution of A/I; the maps $A \to \bar{A}(=\operatorname{Hom}(A,A))$ are the obvious ones; ϕ is the non-singular skew-symmetric form $E \times E \to \Lambda^2 E \overset{\sigma}{\to} A$; the bottom line is the Ext-sequence of (ε); η' is induced by the verticals on its left; and the whole diagram is commutative.

Conversely, given rank 2 A-projective E and $h\colon E \to A$ a regular cosection (at any prime containing $\operatorname{im}(h)$ the image under h of a basis of E is a regular sequence of A) such that $\Lambda^2 E \simeq A$, we have the Koszul resolution of A/I, which can be dualized to reproduce (4.2) and hence also the isomorphism $\eta\colon A/I \overset{\simeq}{\to} \operatorname{Ext}^2(A/I,A)$.

The two constructions are inverse in an appropriate sense.

Using the canonical isomorphism $\operatorname{Ext}^2(A/I,A) \simeq \operatorname{Hom}(A/I,V_2)$, we may regard the initial data in the first construction as coming from an element $(A/I,\mu) \in Q^1(CM_2(A))$, and the resulting pair (E,ϕ) as an element of $Q^{-1}(CM_0(A))$.

(4.3) Proposition. The diagram

$$\begin{array}{ccccc} Q^1(CM_2(A)) & & \xrightarrow{SH} & & Q^{-1}(CM_0(A)) \\ \downarrow & & & & \mathcal{D}_1^{-1} \downarrow \\ W_0^1(CM_2(A)) & \xrightarrow{\mathcal{D}_0^1} & W_1^{-1}(CM_1(A)) & \xrightarrow{\mathcal{D}_1^{-1}} & W_0^{-1}(CM_0(A)) \end{array}$$

commutes, where SH is the Serre-Horrocks construction.

Proof: We recall from (2. 6) and (2.18) the definition of the maps $\mathcal{D}_0^1$ and $\mathcal{D}_1^{-1}$. Given $(A/I,\mu) \in Q^1(CM_2(A))$ choose $a \in I$, $a \neq 0$, hence a surjection $j: A/aA \to A/I$ and lift μ to τ in

$$\begin{array}{ccc} A/aA \times A/aA & \xrightarrow{\tau} & E_1 \\ \downarrow j \times j & & \downarrow d_1 \\ A/I \times A/I & \xrightarrow{\mu} & V_2 \end{array}$$

Setting $A/aA = H$, $I/aA = K$, $(\text{inc}: I/aA \to A/aA) = \alpha$ and defining γ by the commutativity of

$$\begin{array}{ccc} I/aA & \xrightarrow{\alpha} & A/aA \\ \downarrow \gamma & & \downarrow \text{Ad}\,\tau \\ H^{\wedge} =: \text{Hom}(A/aA,V_1) & \to & \text{Hom}(A/aA,E_1), \end{array}$$

the triple $(K,H,(\alpha,\gamma)) \in F^{-1}(CM_1(A))$ represents $\mathcal{D}_0^1(A/I,\mu)$.

Now let $J: A \oplus A \to H \oplus H^{\wedge} \simeq A/aA \oplus A/aA$ be the natural surjection and set

$$E' = J^{-1}(\text{im}(\alpha,\gamma)) \subseteq A \oplus A \tag{4.4}$$

Then E' is A-projective, and there is a non-singular skew-symmetric form $\phi': E' \times E' \to A$ such that $\mathcal{D}_1^{-1}[K,H,(\alpha,\gamma)] = [E',\phi'] \in W_0^1(A)$.

Now consider the commutative diagram, with exact rows and columns,

$$\begin{array}{ccccc} \text{Hom}(A/I,V_2) & & & & \\ \downarrow & & & & \\ \text{Hom}(A,V_2) & \leftarrow & \text{Hom}(A,E_1) & \leftarrow & \text{Hom}(A,V_1) \\ \downarrow & & \downarrow & & \downarrow \\ \text{Hom}(I,V_2) & \leftarrow & \text{Hom}(I,E_1) & \leftarrow & \text{Hom}(I,V_1) \\ & & & & \downarrow \\ & & & & \text{Ext}^1(A/I,V_1) \end{array}$$

Let $\mu_0 = (\text{Ad}\,\mu)(1) \in \text{Hom}(A/I,V_2)$ and let $\bar{\mu}_0$ be its image in $\text{Hom}(A,V_2)$. View $(\text{Ad}\,\tau)(1)$ as a map $A \to E_1$; it then maps to $\bar{\mu}_0$ and so its image in $\text{Hom}(I,E_1)$ lifts back to a unique element $f \in \text{Hom}(I,V_1)$, where

$$\begin{array}{ccc} I & \xrightarrow{f} & V_1 \\ \downarrow p & & \uparrow \\ I/aA & \xrightarrow{\gamma} & \mathrm{Hom}(A/aA,V_1) \end{array}$$

commutes. Further, the image of f under $\mathrm{Hom}(I,V_1) \to \mathrm{Ext}^1(A/I,V_1)$ equals the image of μ_0 under $\mathrm{Hom}(A/I,V_2) \overset{\cong}{\to} \mathrm{Ext}^1(A/I,V_1)$. From this, the commutativity of

$$\begin{array}{ccccc} & & \mathrm{Hom}(I,V_1) & \to & \mathrm{Ext}^1(I,A) \\ & & \downarrow & & \downarrow \cong \\ \mathrm{Hom}(A/I,V_2) & \overset{\cong}{\to} & \mathrm{Ext}^1(A/I,V_1) & \overset{\cong}{\to} & \mathrm{Ext}^2(A/I,A) \ , \\ \mu_0 & \longmapsto & & & \eta(1) \end{array}$$

and the definition of $\mathrm{Hom}(I,V_1) \to \mathrm{Ext}^1(I,A)$ we see that the extension (ε) corresponding to $\eta(1)$ (or to μ_0) in the Serre-Horrocks construction is the top line in the pull-back

$$\begin{array}{ccccc} A & \rightarrowtail & E & \overset{s}{\twoheadrightarrow} & I \\ \downarrow = & & \downarrow & & \downarrow f \\ A = V_0 & \rightarrowtail & E_0 & \twoheadrightarrow & V_1 \end{array}$$

From the commutative diagram

$$\begin{array}{cccccccc} A & \rightarrowtail & E_0 & \twoheadrightarrow & V_1 & & \overset{f}{\longleftarrow} & \\ \updownarrow = & & \uparrow & & \uparrow & & & \\ A & \rightarrowtail & A[a^{-1}] & \twoheadrightarrow & (0:a) & = \mathrm{Hom}(A/aA,V_1) & \overset{\gamma p}{\longleftarrow} & I \\ \uparrow = & & \uparrow xa^{-1} & & \uparrow = & & & \\ A & \overset{xa}{\rightarrowtail} & A & \twoheadrightarrow & A/aA & & & \end{array}$$

we see that (ε) is likewise the top line of the pull-back in

$$\begin{array}{ccccc} A & \rightarrowtail & E & \overset{s}{\twoheadrightarrow} & I \\ \downarrow = & & \downarrow & & \downarrow \gamma p \\ A & \overset{a}{\rightarrowtail} & A & \twoheadrightarrow & \mathrm{Hom}(A/aA,V_1) \end{array}$$

Setting h equal to s followed by the inclusion of I into A, we get a map $E \to A \oplus A$ and it is immediate that E' (from (4.4)) and E define the same submodule of $A \oplus A$. One shows routinely that $(E',\phi') = (E,\phi) := SH(A/I,\eta)$.

As a corollary of (4.3) and (3.1) we get the following description of $W_0^{-1}(A) \otimes \mathbb{Z}/2$, where A is a smooth 2-dimensional real affine k-algebra satisfying the conditions of (3.2)(b). Let $I_1,\ldots,I_r \subseteq A$ be ideals representing a basis of $A_0(\mathrm{Spec}\ A) \otimes \mathbb{Z}/2$. These can be taken to be the maximal ideals of real points in the r topological components of $(\operatorname{Spec} A)(\mathbb{R})$. Then the unary forms $\langle 1\rangle : A/I_i \times A/I_i \to \mathbb{R} \to E(A/I_i)$ give elements (ε_i) in $\operatorname{Ext}^1(I_i, A)$ as above, hence rank 2 projectives $P_1, \ldots, P_r$.

(4.5) Proposition. The skew-symmetric forms

$$P_i \times P_i \to \Lambda^2 P_i \simeq A$$

generate $W_0^{-1}(A) \otimes \mathbb{Z}/2$ as a $\mathbb{Z}/2$-vector space; they form a basis if $\operatorname{Pic}(A) \otimes \mathbb{Z}/2 = 0$.

(4.6) Remark. I don't know whether class group assumptions are necessary here or in (3.1). Observe also that no class group assumption is needed for $\Lambda^2 P_i$ to be free: by construction, the P_i come from codimension 2 in the Grothendieck filtration of $\tilde{K}_0$.

(4.6) We next give an example of the kind of forms predicted by the Serre-Horrocks construction in dimensions 2 and 4. However, the connection between the 4-dimensional example and the Koszul resolution will not be justified.

Let $A_n = \mathbb{R}[x_0, \ldots, x_n]/(\Sigma\, x_i^2 - 1)$ be the real co-ordinate ring of the n-sphere. According to [F] the upper left $2n \times 2n$ block of the following matrix defines an endomorphism $\alpha_n : (A_n)^{2n} \to (A_n)^{2n}$ whose kernel (or cokernel) generates $\tilde{K}_0(A_n)$, $n = 1, 2$, or 4.

$$\begin{pmatrix}
1-x_0 & -x_1 & -x_2 & 0 & -x_3 & 0 & 0 & -x_4 \\
-x_1 & 1+x_0 & 0 & -x_2 & 0 & -x_3 & x_4 & 0 \\
-x_2 & 0 & 1+x_0 & x_1 & 0 & -x_4 & -x_3 & 0 \\
0 & -x_2 & x_1 & 1-x_0 & x_4 & 0 & 0 & -x_3 \\
-x_3 & 0 & 0 & x_4 & 1+x_0 & x_1 & x_2 & 0 \\
0 & -x_3 & -x_4 & 0 & x_1 & 1-x_0 & 0 & x_2 \\
0 & x_4 & -x_3 & 0 & x_2 & 0 & 1-x_0 & -x_1 \\
-x_4 & 0 & 0 & -x_3 & 0 & x_2 & -x_1 & 1+x_0
\end{pmatrix}$$

Let $P_n = \ker \alpha_n$. Define $h: P_n \to A_n$ by $(b_1, \ldots, b_{2n}) \to b_1$. I claim h is a regular cosection $(n = 1, 2$, or $4)$ and will verify it when $n = 4$. To simplify notation, drop the subscript n's. First $(1 + x_0, 1 - x_0)$ is the unit ideal so $\operatorname{Spec}(A) = \operatorname{Spec}(A_{1+x_0}) \cup \operatorname{Spec}(A_{1-x_0})$. Then $(b_1, \ldots, b_8) \in P_{1-x_0}$ can be written

$$\Big(\frac{x_1}{1-x_0}b_2 + \frac{x_2}{1-x_0}b_3 + \frac{x_3}{1-x_0}b_5 + \frac{x_4}{1-x_0}b_8, b_2, b_3,$$
$$\frac{x_2}{1-x_0}b_2 - \frac{x_1}{1-x_0}b_3 - \frac{x_4}{1-x_0}b_5 + \frac{x_3}{1-x_0}b_8, b_5, \ldots\Big)$$

Hence a basis for P_{1-x_0} can be gotten by setting successively each of b_2, b_3, b_5 and b_8 equal to 1 and the others equal to zero. The image under h of this basis is $\left\{\frac{x_1}{1-x_0}, \frac{x_2}{1-x_0}, \frac{x_3}{1-x_0}, \frac{x_4}{1-x_0}\right\}$, a regular sequence in A_{1-x_0}. The ideal it generates has zero set equal to $(-1,0,0,0,0) \in S^4 - \{(1,0,0,0,0)\} = (\operatorname{Spec} A_{1-x_0})(\mathbb{R})$. Inverting $1 + x_0$ and carrying out the same procedure, the image of a basis turns out to be the unit ideal. Thus h is a regular cosection and $V(\operatorname{im}(h)) = \{(-1,0,0,0,0)\}$, a real point which generates $A_0(S^4) \otimes \mathbb{Z}/2 = \mathbb{Z}/2$.

Now consider the form

$$\phi\colon \Lambda^2 P \times \Lambda^2 P \to \Lambda^4 P \underset{\cong}{\overset{e}{\to}} A$$

where $\phi(x \wedge y, z \wedge w) = e(x \wedge y \wedge z \wedge w)$ and e is some isomorphism. The claim is that the class of $(\Lambda^2 P, \phi) \in Q^1(A)$ in $W_0^1(A)$ is the non-zero element of $\ker(W_0^1(A) \to W_0^1(K)) \otimes \mathbb{Z}/2$ (corresponding to the above generator of $A_0(S^4) \otimes \mathbb{Z}/2$). Going through the same construction, but using instead the co-ordinate ring A_2 of S^2, the class of $(P,\phi) \in Q^{-1}(A_2)$ (where P is a rank 2 projective) is the non-trivial element of $W_0^{-1}(A_2)$ by (4.5), since A_2 is a UFD. But in the 4-dimensional case, it is not known (to me) whether $C_1(A_4) \otimes \mathbb{Z}/2 = 0$, so it may be of some interest to show that the above element of $\ker(W_0^1(A) \to W_0^1(K)) \otimes \mathbb{Z}/2$ is non-trivial. (That the form $(\Lambda^2 P, \phi)$ over A_4 becomes hyperbolic over the fraction field is left to the reader.)

Kere is a sketch of the non-triviality of $[\Lambda^2 P, \phi]$ in $W_0^1(A)$, found with the help of M. Ojanguren.

First, Fossum [Ibid.] showed that the usual Serre-Swan construction gives a surjection

$$\tilde{K}_0(A) \overset{\cong}{\to} \widetilde{KO}(S^4)\ ,$$

so we can use the computation of Adams operations on the generator of $\widetilde{KO}(S^4)$ in [Hu, p. 175] together with a simple computation to show

$$[\Lambda^2 \xi] - [\varepsilon^6] = 2g \tag{4.7}$$

where ξ is the vector bundle corresponding to P, ε^6 is the 6-dimensional trivial bundle and g is the generator of $\widetilde{KO}(S^4)$. Now view $(\Lambda^2 P, \phi)$ as an element of the topological Witt group [MH, App. 1].

$$W(S^4) \overset{\cong}{\to} KO(S^4)$$

where the isomorphism sends a "bundle of forms" η to $[\eta_+] - [\eta_-]$, and $\eta_+(\eta_-)$ is the positive- (negative-) definite subbundle. Hence if $(\Lambda^2 P, \phi)$ is hyperbolic, then the corresponding topological bundle of forms is also: hence it is zero in $W(S^4)$. The isomorphism above thus says $[(\Lambda^2\xi)_+] = [(\Lambda^2\xi)_-]$ in $KO(S^4)$. Combining this with (4.7) we get

$$(4.8) \qquad g = [(\Lambda^2\xi)_+] - [\varepsilon^3] \quad \text{in} \quad \widetilde{KO}(S^4).$$

But $\pi_4(BO_3) \to \pi_4(BO)$ is not surjective (it is $\mathbb{Z} \xrightarrow{\times 2} \mathbb{Z}$ by [Hu, 5.12.11]) so $[(\Lambda^2\xi)_+] - [\varepsilon^3]$ cannot generate $\widetilde{KO}(S^4)$, contradicting (4.8).

(4.9) We now make a connection between the results of §2 (specifically, (2.5)) and the constructions in §6. This involves the theory of algebra structures on resolutions from [KM] and [BE].

Let A be a regular local ring of dimension n and I an ideal such that A/I is Gorenstein.

Let

$$F = (F_p \xrightarrow{d_p} F_{p-1} \to \quad \xrightarrow{d_1} F_0 = A \to A/I)$$

be a free resolution of A/I. A differential graded commutative algebra structure on F is a multiplication $F \otimes F \to F$, $F_iF_j \subseteq F_{i+j}$, such that if $\deg x_\ell = \ell$,

(i) $d(x_ix_j) = d(x_i)x_j + (-1)^i x_i\, dx_j$,

(ii) $x_ix_j = (-1)^{ij} x_j x_i$.

By [BE,(1.1),(1.5)] each such F supports an algebra structure (not in general associative), and if F is minimal, the multiplication $\sum_{i+j=p} F_i \otimes F_j \to F_p \simeq A$ induces isomorphisms $\sigma_k\colon F_k \to F^*_{p-k}$, $0 \le k \le p$ such that

a) $\sigma_{k-1}d_{k-1} = (-1)^{k+1} d^*_{p-k+1}\sigma_k$

and by (ii) above one also has

b) $\sigma^*_k = (-1)^{k(p-k)}\sigma_{p-k}$.

Now F has only one non-zero homology group, $H_0(F) = A/I$; hence, since A/I is Gorenstein, the only non-zero cohomology group is $H^p(F) \simeq \mathrm{Ext}^p(A/I,A) \simeq \mathrm{Hom}(A/I,V_p)$. This means that if we define $\phi_r = \sigma^{-1}_{p-r}$, $\psi(F,d,\phi)$ satisfies (6.1), and so defines a Poincare complex $(F,d,\{\psi\}) \in P^1_p(M_p)$, where $\psi + T_1\psi = \phi$ (cf. (6.2ff)).

According to [KM,(1.5)], when $p = 2q$ is even, the map $\sigma_q\colon F_q \to F^*_q$ defines a non-singular symmetric bilinear form $s\colon F_q \times F_q \to A$ which becomes hyperbolic at the generic point. By (2.5) if $n = p = 4$ (F_q,s) is also hyperbolic, making the algebra structure in [KM, Thm.

1.1] canonical in a sense.

Thus, if one believes the claim of (7.1), then the algebra structures above (when $n = p = 4$) are null-cobordant and this fact is equivalent to the algebra structures being canonical.

§5. Dévissage.

The purpose of this section is to prove Theorem (2.2). To simplify notation, we write $\mathcal{F}(R)$ for the category of R-modules of finite length, where R is local Gorenstein. If $\dim R = n$, then this is $CM_n(R)$ by (1.2).

Begin with surjectivity. Let $(M,\phi) \in \Omega^{\lambda}(\mathcal{F}(R))$ and let ℓ be the smallest integer for which $m^{\ell}M = 0$. If $\ell = 1$, there is nothing to prove, so we suppose $\ell > 1$, and set $M_i = m^i M$.

Claim. $M_{\ell-1} \subseteq (M_{\ell-1})^{\perp}$ and the naturally induced form $\phi'\colon M_{\ell-1}^{\perp}/M_{\ell-1} \times M_{\ell-1}^{\perp}/M_{\ell-1} \to E := E_n(R)$ is non-singular.

Assuming this, it follows from (1.9) that $[M_{\ell-1}^{\perp}/M_{\ell-1}, \phi'] = [M,\phi]$ in $W_0^{\lambda}(\mathcal{F}(R))$. But since $M_{\ell-1}^{\perp}/M_{\ell-1} \subseteq M/M_{\ell-1}$, $m^{\ell-1}(M_{\ell-1}^{\perp}/M_{\ell-1}) = 0$ so induction on ℓ finishes the proof of surjectivity.

To prove the first statement of the claim, let $m_1,\dots,m_k$ generate M and let $a,b \in m^{\ell-1}M$. Then $a = \Sigma\, p_i m_i$, $b = \Sigma\, q_j m_j$ where $p_i, q_j \in m^{\ell-1}$. Then $\phi(a,b) = \sum_{i,j} \phi(p_i m_i, q_j m_j) = \sum_{i,j} \phi(p_i q_j m_i, m_j) = 0$ since $q_i p_j \in m^{2(\ell-1)}$ and $2(\ell-1) \geq \ell$. To prove the second statement, we need to know that

$$K = K^{\perp\perp}$$

for any $K \subseteq M$. Clearly $K \subseteq K^{\perp\perp}$. There is an obvious bilinear form $K^{\perp} \times M/K \to E$ inducing an isomorphism $K^{\perp} \xrightarrow{\cong} \mathrm{Hom}(M/K,E)$. Since $\ell(M/K) = \ell(\mathrm{Hom}(M/K,E))$ ([B,(2.1)iv] where ℓ denotes length, it follows that $\ell(K^{\perp}) + \ell(K) = \ell(M)$. Similiarly, $\ell(K^{\perp\perp}) + \ell(K^{\perp}) = \ell(M)$, so $\ell(K) = \ell(K^{\perp\perp})$. Since $K \subseteq K^{\perp\perp}$, $K = K^{\perp\perp}$. Applying this to $K = M_{\ell-1}$ gives the injectivity of $\mathrm{Ad}\,\phi'$; surjectivity follows from this and [ibid.]. Hence ϕ' is non-singular and the claim is proved.

Next suppose given $(k(m)^n,\phi)$ whose class vanishes in $W_0^{\lambda}(\mathcal{F}(R))$. This means there are lagrangians (M_1,γ_1), $(M_2,\gamma_2) \in \Omega^{\lambda}(\mathcal{F}(R))$ and an isometry

$$(k(m)^n,\phi) \oplus (M_1,\gamma_1) \cong (M_2,\gamma_2)\ .$$

Using the technique of the claim above, we may assume $mM_1 = 0$, but it remains to show (M_1,γ_1) and (M_2,γ_2) are still lagrangians. It is evidently sufficient to show that if (M,γ) is a lagrangian and $L \subseteq M$ satisfies $L \subseteq L^{\perp}$, then the induced form on $L^{\perp}/L$ is a lagrangian. The following stronger statement suffices for this.

(5.1) Lemma. Let (M,γ) be a lagrangian and let $L \subseteq M$ satisfy $L \subseteq L^{\perp}$. Then if $K' \supseteq L$ is maximal such that $K' \subseteq (K')^{\perp}$, then

$K' = (K')^{\perp}$ and K' is a sublagrangian.

Proof: Let K be a sublagrangian of (M,γ), and set $J = K \cap K'$. Then in $J^{\perp}/J$, $K'/J \subseteq (K'/J)^{\perp}$; and from [P1, (3.5)] it follows that K' is a sublagrangian of M if and only if K'/J is a sublagrangian in $J^{\perp}/J$. Hence, working inductively on $\ell(M)$ we may assume $K' \cap K = (0)$. We have $K \cong (M/K)^{\wedge}$, so $\ell(K) = \ell(M/K)^{\wedge}) = \ell(M/K) = \ell(M) - \ell(K)$ where $(M/K)^{\wedge} := \mathrm{Hom}(M/K,E)$. Hence $\ell(M) = 2\ell(K)$. Since $K'^{\perp\perp} = K'$, the naturally induced form $K' \times (M/K'^{\perp}) \to E$ is nonsingular, so $\ell(K') = \ell((M/K'^{\perp})\ ) = \ell(M/K'^{\perp}) = \ell(M) - \ell(K'^{\perp})$.

If $\ell(K') < \ell(K)$, then $\ell(K'^{\perp}) > \ell(M) - \ell(K)$, which means $K'^{\perp} \to M \to M/K$ is not injective, so that $K'^{\perp} \cap K \neq 0$. Then since $K' \cap K = (0)$, $K'^{\perp} - K'$ contains an isotropic element, which contradicts maximality of K'. Thus $\ell(K') = \ell(K) = 1/2\ell(M)$, so $\ell(K') = \ell(K'^{\perp})$. Since $K' \subseteq K'^{\perp}$, $K' = K'^{\perp}$ and K' is a sublagrangian.

$\underline{i = 1}$. It is well-known that $W_1^{\lambda}(k(m)) = 0$ (see, e.g., [P2, (4.1)] where our W_1^{λ} is denoted $L_{\lambda}\cdot$). Hence surjectivity is enough here.

Let $[M,N,(\alpha,\gamma)] \in F^{\lambda}(F(R))$ be given. We denote by $\phi_h: (N \oplus N^{\wedge}) \times (N \oplus N^{\wedge}) \to E$ the standard λ-symmetric hyperbolic form. By stablizing ((1.12(a))) using the extension

$$\ker p \rightarrowtail (R/m^n)^{\ell} \overset{p}{\twoheadrightarrow} N ,$$

(for some n,ℓ and surjection p), we assume $N = (R/m^n)^{\ell}$. The point of the proof is to reduce n to 1, inside the class of $[M,N,(\alpha,\gamma)] \in W_1^{\lambda}(F(R))$.

We begin with some observations. First, for any k, $0 \leq k \leq n$, and with respect to the natural pairing $\nu: N^{\wedge} \times N \to E$

$$(m^{n-k}N^{\wedge})^{\perp} = m^k N$$

(5.2) and

$$m^{n-k}N^{\wedge} = (m^k N)^{\perp}$$

Second, if $\{m_{ij} | 1 \leq i \leq r,\ 1 \leq j \leq \ell\}$ is any $k(m)$-basis of $m^{n-1}N$ $(r = \dim_{k(m)} m^{n-1}/m^n)$ then there is a subset $\{f_{ij} | 1 \leq i \leq r,\ 1 \leq j \leq \ell\} \subseteq N^{\wedge}$ such that

$$f_{i'j'}(m_{ij}) = \delta_{ii'}\delta_{jj'} \in k(m) \subseteq E$$

and $\{f_{ij}\}$ is a basis of $N^{\wedge}$ mod m. To prove this use (5.2) to get a nonsingular pairing

$$N^{\wedge}/mN^{\wedge} \times m^{n-1}N \to k(m) \subseteq E.$$

The set $\{f_{ij}\}$ is thus a lifting to $N^{\wedge}$ of a dual basis in $N^{\wedge}/mN^{\wedge}$.

Now let $\{p_i \mid 1 \leq i \leq r\}$ be a $k(m)$-basis of $m^{n-1}/m^n \subseteq R/m^n$ and $\{b_1,\ldots,b_\ell\}$, an R/m^n-basis of N. Then we may take $m_{ij} = p_i b_j$ in the discussion above. Note that

$$(5.3) \qquad \delta_{ii'}\delta_{jj'} = f_{i'j'}(m_{ij}) = p_i\ f_{i'j'}(b_j)$$

In particular, for each j, each of

$$(5.4) \qquad p_i\ f_{ij} = p_k\ f_{kj} \neq 0, \quad 1 \leq i, \quad k \leq r$$

is a basis for the 1-dimensional $k(m)$-vector space it generates (because $m^{n-1}(R/m^n)^\wedge \cong \mathrm{Hom}(R/mR,E) = k(m)$ and $N/mN \cong (R/mR)^\ell$ has for a basis the classes of $b_1,\ldots,b_\ell$ mod m).

<u>Step 1</u>. is to show we may modify $(M,N,(\alpha,\gamma))$ within its class in $W_1^\lambda(F(R))$ so that

$$(5.5) \qquad m^{n-1}\ \alpha(M) = 0.$$

If this is not already so, choose $y \in M$ with $m^{n-1}\ \alpha(y) \neq 0$. Then $\alpha(y) \notin mN$, so by Nakayama's lemma, we may take $\alpha(y)$ to be part of an R/m^n-basis of N, say $\{\alpha(y) = b_1,b_2,\ldots,b_\ell\}$. Clearly, y generates a summand of M isomorphic to R/m^n.

Suppose $\gamma(y) = \sum_{i=1}^{\ell} h_i \in N^\wedge$, where h_j is characterized by: $h_j(b_k) \neq 0$ only if $j = k$. Define $r: N \to N^\wedge$ by $r(b_1) = 1/2\ h_1 + \sum_{i>1} h_i$ and $r(b_i) = 0$, $i > 2$. Set

$$\rho = r - \lambda r^\wedge : N \to N^\wedge \quad (N \equiv N^{\wedge\wedge}) .$$

Then ρ is $(-\lambda)$-symmetric and a computation shows

$$\rho(b_1) = \begin{cases} \Sigma\ h_i, & \lambda = -1 \\ \sum_{i>1} h_i, & \lambda = 1. \end{cases}$$

When $\lambda = 1$, $0 = \phi_h(\alpha(y) + \gamma(y),\alpha(y) + \gamma(y)) = 2h_1(b_1)$ so $h_1 = 0$. Thus, in general (i.e., $\lambda = \pm 1$), $\rho(b_1) = \gamma(b_1)$, which means $(\gamma - \rho\alpha)(b_1) = 0$. After an operation of type (2.11)(b) we may thus destablize using the submodule $R_y \subseteq M$ (even splitting off a formation $(R/m^n, R/m^n,(1,0)))$. Repeating the process if necessary, we eventually get (5.5).

<u>Step 2</u>. is to show $(M,N,(\alpha,\gamma))$ can be modified within its class in $W_1^\lambda(F(R))$ so that

(5.6) $$m^{n-1}\gamma(M) = 0,$$

keeping (5.5) also. Again if this is not so, choose $y \in M$ with $\gamma(y) = \sum_{i=1}^{\ell} h_i$, where h_i is characterized as above and where, say, $ph_1 \neq 0$ for some $p \in m^{n-1}$.

Suppose first that

(5.7) $$p\, h_i = 0, \quad i > 1.$$

Then since $\alpha(py) = p\alpha(y) = 0$ by (5.5), we may destabilize using the submodule $R(py) \subseteq M$. (Notice that only the first terms of the direct sum decompositions of N and $N\hat{}$ are changed in this process.) In general, we can get (5.7) by replacing b_j with

$$b_j' = b_j - \frac{p\, h_j(b_j)}{p\, h_1(b_1)} b_1, \quad j > 1$$

and setting $b_1' = b_1$. Then in terms of this new basis, if $j > 1$,

$$p\gamma(y)(b_j') = \sum_i ph_i(b_j') = ph_j(b_j) - \frac{ph_j(b_j)}{ph_1(b_1)} ph_1(b_1) = 0,$$

so (5.7) is satisfied.

Assuming the result still has $m^{n-1}\gamma(M) \neq 0$, we continue the process; this time the first term ph_1 of $p\gamma(y)$ must be zero for all $p \in m^{n-1}$ because of (5.4). This completes Step 2, and shows we may assume

(5.8) $$m^{n-1}\alpha(M) = 0 = m^{n-1}\gamma(M).$$

From this it follows that

$$m^{n-1}(N + N\hat{}) \subseteq (\alpha,\gamma)(M).$$

For $\nu(\gamma(M), m^{n-1}N) = \nu(m^{n-1}\gamma(M), N) \equiv 0$; similarly, $\nu(m^{n-1}N\hat{}, \alpha(M)) \equiv 0$. Consequently

$$\phi_h((\alpha,\gamma)M, m^{n-1}(N + N\hat{})) \equiv 0$$

so that

$$m^{n-1}(N + N\hat{}) \subseteq (\alpha,\gamma)(M)^{\perp} = (\alpha,\gamma)(M) .$$

Now let $L' = \{m \in M \mid (\alpha,\gamma)(m) \in m^{n-1}(N + N\hat{})\}$. Then $(\alpha,\gamma)|L'\colon L' \xrightarrow{\cong} m^{n-1}(N + N\hat{})$, an isomorphism of $k(m)$-vector spaces. Consider $\gamma|L'\colon L' \to m^{n-1}N\hat{}$. This is surjective with kernel L, say, and $\alpha|L: L \overset{\cong}{\to} m^{n-1}N$. Evidently, destabilizing with $L \subseteq M$ leaves a formation $(M',N',(\alpha',\gamma'))$ with $m^{n-1}N' = 0$. Now we may begin the process over again, replacing n with n - 1 at the beginning of the proof then performing Steps 1 and 2.

This completes the proof.

In one case we need to "de-localize" these results.

(5.9) Corollary. Let A be a locally factorial Gorenstein ring. Then each element of $W_1^{\lambda}(CM_1(A))$ has a representative $(K,H,(\alpha,\gamma))$ such that if $\operatorname{ht} q = 1$, then $(K_q,H_q,(\alpha_q,\gamma_q))$ is isomorphic to the orthogonal sum of formations of the form

$$(R/(x^t) \oplus R/(x^{m-t}),R/(x^m),((0,x^t),(x^{m-t},0))$$

where $R = A_q$, $(x) = qA_q$ and $0 \le t \le m$. If $\lambda = -1$, then we can take $t = m$.

Proof: There is a surjection $p: (A/q_1^{(m)} \cap \dots \cap q_r^{(m)})^n \to H$, for some $m,n > 0$ and height one primes $q_1,\dots,q_r$. Since A is locally factorial, $A/q_1^{(m)} \cap \dots \cap q_r^{(m)} \in CM_1(A)$, so $\ker p \in CM_1(A)$. After stabilization, we may therefore assume

$$H_q = (R/(t^m))^n, \quad R = A_q$$

for each $q \in \operatorname{Ass}(H)$ and $(t) = qA_q$.

Let $\lambda = 1$ and set $q = q_1$. We will carry out "Step 1" of the proof of Dévissage for $i = 1$ above, but over A. Namely, recall that γ_q was first changed to $\gamma_q + \rho\alpha_q$, where $\rho: H_q \to H_q^{\wedge}$ is the adjoint of a skew-symmetric form. To be able to lift this operation back to A, we change (α_q,γ_q) to $(s\alpha_q,s^{-1}\gamma_q)$, $s \in A - q$, an isomorphism of the formation. Now ρ becomes $s^2\rho$. For suitable s, $s^2\rho$ is the localization of the adjoint $H \to H^{\wedge}$ of a skew-symmetric form. If s is further chosen in $q_2^{(m)} \cap \dots \cap q_r^{(m)}$, then at every other height one prime, there is no change in $(K,H,(\alpha,\gamma))$.

The effect of this modification is to give a subformation $(R/(t^m),R/(t^m),(1,0))$ of $(K_q,H_q,(\alpha_q,\gamma_q))$. Further operations will be carried out below to actually split off such a formation. We will assume these operations have also been delocalized (as above, up to isomorphism of the local formation $(K_q,H_q,(\alpha_q,\gamma_q))$). The fact that $(R/(t^m))^{\wedge} \cong R/(t^m)$ (because $\dim R = 1$) means we can carry out modifications in place of Step 2 exactly like those above from Step 1: only the roles of α and γ are reversed.

We next want to see that the subformations $(R/(t^m),R/(t^m),(1,0))$ and $(R/(t^m),(R/(t^m))^{\wedge},(0,1))$ of $(K_q,H_q,(\alpha_q,\gamma_q))$ $(q \in \operatorname{Ass}(H))$ can be split off. This will be carried out more generally below, for subformations where $0 \leq t \leq m$ in the statement of (5.9).

We assume we have completed Steps 1 and 2, so that

$$(5.10) \qquad t^m K_q = 0, \quad q \in \mathrm{Ass}(H), \quad (t) = qA_q .$$

Let $(K_q, H_q, (\alpha_q, \gamma_q)) = (M, N, (\beta, \delta))$. Suppose $m \in M$ has minimal annihilator, (x^t), say. Either $\mathrm{Ann}(\beta(m))$ or $\mathrm{Ann}(\delta(m)) = (x^t)$. If it is $\beta(m)$, we may assume

$$(5.11) \qquad \beta(m) = x^{m-t} b_1 \in N$$

where $\{b_1, \ldots, b_n\}$ is an $R/(x^m)$-basis for N. Let $\delta(m) = \Sigma\, r_i\, b_i^*$, where $\{b_1^*, \ldots, b_n^*\}$ is the dual basis. Since $x^t \delta(m) = 0$, $x^{m-t} | r_i$. Define $r\colon N \to N\hat{}$ by

$$r(b_j) = \begin{cases} \Sigma\, r_i\, x^{t-m}\, b_i^*, & j = 1 \\ 0 & , \ j > 1 . \end{cases}$$

Then if we set $\rho = r - r\hat{}$, ρ is skew-symmetric and $\rho(b_1) = \sum_{i>2} r_i\, x^{t-m} b_i^*$; so after an operation of type (2.11)(b) we get

$$\delta(m) = r_1\, b_1^*$$

By considering the expression $\phi_h(\beta(m) + \delta(m), \beta(m) + \delta(m)) = 0$ we find that $x^t | r_1$.

Let $G = \{m = m_1, m_2, \ldots, m_n\}$ be a minimal set of generators of M. Since $\mathrm{Ann}(m) \subseteq \mathrm{Ann}(m_j)$, the coefficient of b_1 in the expression of $\beta(m_j)$ is divisible by x^{m-t} (or is zero). Hence by adding appropriate multiples of m to the m_j, $j \geq 2$, we may assume

$$(5.12) \qquad \beta(m_j) = \sum_{i \geq 2} s_{ij}\, b_i, \quad j = 2, \ldots, n .$$

Now there is $m' \in M$ such that $\beta(m') = 0$ and $\delta(m') = x^t b_1^*$ because $\phi_h(\beta(m_j) + \delta(m_j), x^t b_1^*) = 0$, for all j by (5.11) and (5.12). So $x^t b_1^* \in [\mathrm{im}(\beta,\delta)]^{\perp} = \mathrm{im}(\beta,\delta)$. Replacing $m = m_1$ with $m - r_1 x^{-t} m'$, we get

$$(5.13) \qquad \begin{aligned} \beta(m - r_1 x^{-t} m') &= x^{m-t} b_1, & \beta(m') &= 0 \\ \delta(m - r_1 x^{-t} m') &= 0, & \delta(m') &= x^t b_1^* . \end{aligned}$$

It is easy to see that we may take m' to be part of a minimal generating set (along with $m - r_1 x^{-t} m'$); say, $m_1 = m - r_1 x^{-t} m'$ and $m_2 = m'$ in G above.

By considering the expression $\phi_h(\beta(m_1) + \delta(m_1), \beta(m_j) + \delta(m_j)) = 0$, $j \geq 3$, we see that the coefficient of b_1^* in the expression for $\delta(m_j)$ is divisible by x^t, $j \geq 3$. Again, by adding appropriate multiples of m_2 to the m_j, $j \geq 3$, we can take

(5.14) $$\delta(m_j) = \sum_{i \geq 2} r_{ij} b_i^*, \quad j \geq 3 .$$

Putting (5.12), (5.13) and (5.14) together, we have shown $(M,N,(\beta,\delta))$ has an orthogonal summand $(R/(x^t) \oplus R/(x^{m-t}), R/(x^m), ((0,x^{m-t}),(x^t,0))$ as desired.

The proof of (5.9) when $\lambda = -1$ is much easier. Given $(M,N,(\beta,\delta)) \in F^{-1}(F(R))$ we know from (1.20) that $\mathrm{im}(\beta,\delta)$ is a summand of $N \oplus N^\wedge \cong (R/(x^m))^{2n}$, so that $M \cong (R/(x^m))^n$. Following the construction of L_1^{-1} in (2. 7), we can thus extend $(M,N,(\beta,\delta))$ to an element

(5.15) $$A = \begin{pmatrix} \beta & \mu \\ \delta & \nu \end{pmatrix} \in Sp_{2n}(R/(x^m))$$

where $Sp_{2n}(-)$ is the symplectic group. Now by [P2,(3.12),(3.14)], there is a symmetric $\rho\colon (R/(x^m))^n \to (R/(x^m))^n$ such that $\beta + \rho\delta$ is invertible. Then after a basis change (in N), we can take $\beta = \mathrm{id}$ in (5.15). Hence an operation of type (1.18)(b) using $-\delta$ gives $\delta = 0$ in (5.15).

§6. Poincaré complexes.

The aim of the next three sections is to prove (1.21). The proof is rather long because in this section we first introduce another definition of $W_1^\lambda(CM_p)$, in terms of chain complexes. This is a natural thing to do, since, for example when $p = 0$ and A is local, (1.21) is a consequence of Sharpe's normal form for elements of the unitary Steinberg group. This in turn is motivated by the geometric description of odd-dimensional surgery theory, so the chain complexes are taking the place of manifolds. The idea of using chain complexes to prove (1.21) comes from Ranicki's work ([R1] and [R2]), on which we will rely heavily. For the convenience of the reader, we recall some of the definitions, making explicit the simplifications that come from the assumption, $1/2 \in A$.

Let (C,d) be an n-dimensional chain complex of finitely-generated projective A-modules, C_i, and let $\phi_r\colon C^r \to C_{n-r}$ be a sequence of homomorphisms, $0 \leq r \leq n$, where $C^r := \mathrm{Hom}(C_r,A)$. Then (C,d,ϕ) is a Poincaré complex if, for $0 \leq r \leq n$,

(6.1)

$$\text{a)}\quad d\phi_r = (-1)^{n-r}\,\phi_{r+1}\,d^*$$

$$\text{b)}\quad \phi_r = \lambda(-1)^{r(n-r)}\,\phi_{n-r}^*$$

$$\text{c)}\quad \phi_r \text{ induces } H^r(C) \overset{\cong}{\to} H_{n-r}(C)$$

where $\phi^* = \mathrm{Hom}(\phi,A)$ and $d^* = \mathrm{Hom}(d,A)$.

It is useful to reformulate this. Given finitely-generated projective A-modules P and Q, there is an isomorphism

$$\begin{aligned} \mathrm{Hom}(P,A) \otimes Q &\overset{\cong}{\to} \mathrm{Hom}(P,Q) \\ f \otimes q &\to \{p \to f(p)q\} \end{aligned}$$

Taking this and $C \cong C^{**}$ as identifications, let $C \otimes C$ be the chain complex with

$$(C \otimes C)_k = \sum_{i+j=k} \mathrm{Hom}(C^i,C_j)\ ,$$

and differential

$$\delta(\phi\colon C^i \to C_j) = (C^i \overset{\phi}{\to} C_j \overset{d}{\to} C_{j-1}) + (-1)^j(C^{i-1} \overset{d^*}{\to} C^i \overset{\phi}{\to} C_j)$$

Thus, the collection of maps $\{\phi_i\}$ satisfying a) and b) above give rise to a "λ-symmetric cycle" in $H_n(C \otimes C)$. In [R1] and [R2] the collection of such maps is denoted ϕ_0, and there is also postulated a sequence of higher chain operators $\phi_k \in (C \otimes C)_{n+k}$, $k \geq 1$, (not the

ϕ_k above!) satisfying certain conditions [R1, p. 104]. All these chain operators together define a cocycle in a certain $\mathbb{Z}/2$-hypercohomology group, $Q^n(C,\lambda)$. Since $1/2 \in A$, this simplifies as follows.

Let $T_\lambda\colon C \otimes C \to C \otimes C$ be the chain map defined, for $f \in \mathrm{Hom}(C^i, C_j) \subseteq (C \otimes C)_{i+j}$, by

(6.2) $$T_\lambda(f) = (-1)^{ij}\lambda f^* \in \mathrm{Hom}(C^j, C_i).$$

Since $1/2 \in A$, the sequence

(6.3) $$\ldots \xrightarrow{(1-T_\lambda)_*} H_n(C \otimes C) \xrightarrow{(1+T_\lambda)_*} H_n(C \otimes C) \xrightarrow{(1-T_\lambda)_*} H_n(C \otimes C) \to \ldots$$

is exact, so $(1 + T_\lambda)_*$ induces

(6.4) $$H_n(C \otimes C)/\mathrm{im}(1 - T_\lambda)_* \xrightarrow{\cong} \mathrm{im}(1 + T_\lambda)_* \subseteq H_n(C \otimes C) .$$

This is the isomorphism $Q_n(C,\lambda) \xrightarrow{\cong} Q^n(C,\lambda)$ of [R1, pp. 102-3] and we here adopt the notation

(6.5) $$Q_n(C,\lambda) := H_n(C \otimes C)/\mathrm{im}(1 - T_\lambda)_* .$$

We view the homomorphisms ϕ_r satisfying (6.1)(a) and (b) as elements of $\mathrm{im}(1 + T_\lambda)_*$, but in practice we work with a sequence of antecedents (under (6.4)) $\{\psi_r\} \in Q_n(C,\lambda)$.

(6.6) Definition. Let A be a commutative ring with $1/2 \in A$. A λ-symmetric chain complex over A is a triple $(C,d,\{\psi\})$ where (C,d) is a chain complex of finitely-generated projective A-modules and $\{\psi\} \in Q_k(C,\lambda)$. It is called n-dimensional if (C,d) is homotopy equivalent to a complex (D,d') with $D_i = 0$, $i > n$ or $i < 0$, and $\{\psi\} \in Q_n(C,\lambda)$. It is λ-Poincaré if, in addition, the maps $\phi_r := \psi_r + (-1)^{r(n-r)}\lambda\psi^*_{n-r}$ induce isomorphisms $H^r(C) \xrightarrow{\cong} H_{n-r}(C)$, $0 \le r \le n$.

Let (C,d_C) and (D,d_D) be chain complexes and $f\colon (C,d_C) \to (D,d_D)$, a chain map. Define the mapping cone of f, $(C(f),d_{C(f)})$ by $C(f)_r = D_r \oplus C_{r-1}$ and

$$(d_{C(f)})_r = \begin{pmatrix} d_D & (-1)^{r-1}f \\ 0 & d_C \end{pmatrix} : D_r \oplus C_{r-1} \to D_{r-1} \oplus C_{r-2} .$$

If C and D are n- and $(n+1)$-dimensional respectively, then $C(f)$ is $(n+1)$-dimensional. Let $(C(f) \hat{\otimes} C(f),\hat{d})$ be the chain complex with

$$(6.7)\quad (C(f)\,\hat{\otimes}\,C(f))_k = \sum_{i+j=k} \mathrm{Hom}(D^i, D_j) \oplus \sum_{r+s=k-1} \mathrm{Hom}(C^r, C_s)$$

and, for $(\delta\psi_i, \psi_r) \in \mathrm{Hom}(D_i, D_j) \oplus \mathrm{Hom}(C^r, C_s)$

$$(6.8)\quad \hat{d}(\delta\psi_i, \psi_r) = d_D\,\delta\psi_i + (-1)^j(\delta\psi)_{i+1} d_D^* + (-1)^n f\,\psi_i\,f^*$$

In analogy with what has been done above, we define

$$(6.9)\quad Q_m(f,\lambda) = H_m(C(f)\,\hat{\otimes}\,C(f))/\mathrm{im}(1 - T_\lambda)_*.$$

(6.10) Definition. Let A be a ring with $1/2 \in A$. A λ-symmetric pair over A is a pair $(f\colon C \to D, \{\delta\psi, \psi\})$ where $f\colon C \to D$ is a chain map of chain complexes of finitely-generated projective A-modules and $\{\delta\psi, \psi\} \in Q_m(f,\lambda)$, for some m. It is (n+1)-dimensional if C and D are n- and (n+1)-dimensional, respectively, and $\{\delta\psi, \psi\} \in Q_{n+1}(f,\lambda)$. Finally, it is relative λ-Poincaré of dimension $n + 1$ if, in addition, the pair $(\delta\phi, \phi) := (\delta\psi + T_\lambda(\delta\psi), \psi + T_\lambda\psi)$ induces isomorphisms,

$$(6.11)\quad \begin{array}{c} H^r(C(f)) \overset{\cong}{\to} H_{n+1-r}(D) \\ (x,y) \to f\phi x + (\delta\phi)y \end{array}$$

where $x \in C^{r-1}$, $y \in D^r$.

In this case the λ-symmetric complex $(C, d_C, \{\psi\})$ is λ-Poincaré of dimension n, and is called the boundary of $(f\colon C \to D, \{\delta\psi, \psi\})$.

Let M_p denote the category of A-modules M such that $\mathrm{ht}\, M \geq p$, together with the zero module. Then

$$P_n^\lambda(M_p)$$

denotes the collection of n-dimensional λ-Poincaré complexes $(C, d, \{\psi\})$ such that $H_*(C) \in M_p$. A cobordism (resp. M_p-cobordism) between $(C, d, \{\psi\})$ and $(C', d', \{\psi'\}) \in P_n^\lambda(M_p)$ is an (n+1)-dimensional λ-Poincaré pair $(f \oplus f'\colon C \oplus C' \to D, \{\delta\psi, \psi \oplus \psi'\})$ with boundary $(C \oplus C', d \oplus d', \{\psi \oplus \psi'\})$ (resp. such that $H_*(D) \in M_p$).

(6.12) Definition-Proposition [R1, (3.2)] M_p-cobordism is an equivalence relation on $P_n^\lambda(M_p)$. The cobordism classes form an abelian group

$$\Omega_n^\lambda(M_p)$$

with addition induced by direct sum of complexes.

(6.13) Next let $(f\colon C \to D, \{\delta\psi, \psi\})$ be an (n+1)-dimensional λ-symmetric pair whose boundary $(C, d, \{\psi\})$ is λ-Poincaré. From this an n-dimensional Poincare complex $(C',d',\{\psi'\})$ is constructed as follows. Let $C'_r = C_r \oplus D_{r+1} \oplus D^{n-r+1}$

$$d_{C'} = \begin{pmatrix} d_C & 0 & (-1)^{n+1}(1+T_\lambda)\psi f^* \\ (-1)^r f & d_D & (-1)^r(1+T_\lambda)\delta\psi \\ 0 & 0 & (-1)^r d_D^* \end{pmatrix},$$

$$\psi' = \begin{pmatrix} \psi & 0 & 0 \\ 0 & 0 & 0 \\ 0 & 1 & 0 \end{pmatrix}.$$

(6.14) Proposition. With the above notation, if $(C,d,\{\psi\}) \in P_n^\lambda(M_p)$ and $H_*(D) \in M_p$ then $(C',d',\{\psi'\}) \in P_n^\lambda(M_p)$ and is M_p-cobordant to $(C,d,\{\psi\})$.

Proof: [R1, (4.1)(ii)] The cobordism is $((g,g')\colon C \oplus C' \to D')$, $\{0,\psi \oplus \psi'\})$ where

$$d_{D'} = \begin{pmatrix} d_C & (-1)^{n+1}(1+T_\lambda)\psi f^* \\ 0 & (-1)^r d_D^* \end{pmatrix} : D'_r = C_r \oplus D^{n-r+1} \to D'_{r-1} = C_{r-1} \oplus D^{n-r+2},$$

$$g = \binom{1}{0}\colon C_r \to D'_r = C_r \oplus D^{n-r+1}$$

and

$$g' = \begin{pmatrix} 1 & 0 & 0 \\ 0 & 0 & 1 \end{pmatrix}\colon C'_r = C_r \oplus D_{r+1} \oplus D^{n-r+1} \to D'_r .$$

We next need to assemble some facts about chain complexes (C,d) for which $H_*(C) \in M_p$. We begin with a quotation of the Acyclicity Lemma of Peskine and Szpiro [PS, Lemme 1.8].

(6.15) Lemma. Let R be local Noetherian and let

$$0 \to L_s \to L_{s-1} \to \dots \to L_1 \to L_0 \to 0$$

be a finite complex of finitely-generated R-modules. Suppose that for every $i > 0$,

1) depth $L_i \geq i$ and

2) depth $H_i(L) = 0$ or $H_i(L) = 0$.

Then $H_i(L_*) = 0$, $i \geq 1$.

Using this result we may restrict the homology of certain complexes.

(6.16) Proposition. Let A be a CM ring and let

$$0 \to C_{p+k} \xrightarrow{d_{p+k}} C_{p+k-1} \to \dots \to C_1 \xrightarrow{d_1} C_0 \to 0$$

be a chain complex of finitely-generated projective A-modules for which $H_*(C) \in M_p$. Then $H_i(C) = 0, \quad i > k$.

<u>Proof</u>: Applying the Acyclicity Lemma to the complex $C_{p+k} \to \dots \to C_{k+1} \to \mathrm{im}\, d_{k+1}$ at all height p primes shows $H_i(C) \in M_{p+1}$ for each $i > k$. Localizing at all height $p+1$ primes shows $H_i(C) \in M_{p+2}$; eventually, $H_i(C) = 0, \quad i > k$.

We also need a kind of universal coefficient theorem.

(6.17) <u>Proposition</u>. Let A be a CM ring and let

$$0 \to C_{p+k} \xrightarrow{d_{p+k}} C_{p+k-1} \to \dots \to C_1 \xrightarrow{d_1} C_0 \to 0$$

be a chain complex of finitely-generated projective A-modules such that $H_*(C) \in M_p$ and $p \geq k$. Then there is an isomorphism

$$H^p(C) \cong \mathrm{Ext}^p(H_0(C), A)$$

<u>Remark</u>: Since $\mathrm{ht}\, H_0(C) \geq p$ the arguments used in (1.6) show $\mathrm{Ext}^p(H_0(C), A) \cong \mathrm{Hom}(H_0(C), V_p)$.

<u>Proof</u>: When $k = 0$, the previous proposition says C_* is a resolution of $H_0(C)$, so the result is immediate from the definition of Ext. Hence we assume $k > 0$.

If $p = 1$ (hence $k = 1$, since $p \geq k > 0$), then $\mathrm{Ext}^1(H_0(C), A) \cong \mathrm{cok}(C^0 \to (\mathrm{im}\, d_1)^*)$. From the exactness of $(\mathrm{im}\, d_1)^* \rightarrowtail C^1 \to (\ker d_1)^*$ and the fact that $(\ker d_1)^* \rightarrowtail C^2$ (dualize the exact sequence $C_2 \to \ker d_1 \twoheadrightarrow H_1(C)$), we find that

$$(\mathrm{im}\, d_1)^* \rightarrowtail C^1 \xrightarrow{d_2^*} C^2$$

is exact. This means $\mathrm{cok}(C^0 \to (\mathrm{im}\, d_1)^*) \cong H^1(C)$ as required.

Finally, if $p \geq 2$, then $\mathrm{Ext}^p(H_0(C), A) \cong \mathrm{Ext}^{p-1}(\mathrm{im}\, d_1, A) \cong \mathrm{Ext}^{p-2}(\ker d_1, A) \cong \dots \cong \mathrm{Ext}^1(\mathrm{im}\, d_{p-1}, A) \cong \mathrm{cok}(C^{p-1} \to (\ker d_{p-1})^*)$. But $(\ker d_{p-1})^* = (\mathrm{im}\, d_p)^*$ because $H^*_{p-1} = 0 = \mathrm{Ext}^1(H_{p-1}, A)$ (the latter because $\mathrm{Ext}^1(H_{p-1}, A) \cong \mathrm{Hom}(H_{p-1}, V_1)$ and $\mathrm{ht}\, H_{p-1} \geq p \geq 2$). As above we find that $\mathrm{cok}(C^{p-1} \to (\mathrm{im}\, d_p)^*) \cong H^p(C)$ so the proof is done.

Finally, we need a spectral sequence describing representatives in $H_n(C \otimes C)$ of the duality maps $\{\psi\} \in Q_n(C, \lambda)$ (cf. (6.5)) where (C, d) is an n-dimensional λ-symmetric complex. Let $F_s(C \otimes C) := \sum_{i \leq s} \mathrm{Hom}(C^i, C)$, a subcomplex of $C \otimes C$. The corresponding spectral sequence has

$$E^2_{s,t} = H_s(\mathrm{Hom}(C^*, H_t(C)))$$

with H_s being computed using the differential on the chain complex $E_* = \mathrm{Hom}(C^*, H_t(C))$ induced from d^* on C^*. It converges to the associated graded groups of the filtration

$$F_s\, H_{s+t}(C \otimes C) := \mathrm{im}[H_{s+t}(F_s(C \otimes C)) \to H_{s+t}(C \otimes C)].$$

Note that in the special case where $H_i(C) = 0$, $i > 0$, and $H_0(C) \in CM_n$, so that $C^0 \overset{d_1^*}{\rightarrowtail} C^1 \to \ldots \to C^n \twoheadrightarrow H^n(C)$ is exact,

(6.18)

$$H_s(\mathrm{Hom}(C, H_t(C)) = \mathrm{Ext}^{n-s}(H^n(C), H_t(C)) = \begin{cases} \mathrm{Ext}^{n-s}(H^n(C), H_0(C)), & t = 0 \\ 0 & , t \neq 0 \end{cases}$$

When $s + t = n$, this is the well-known fact that chain homotopy classes of chain maps $C^* \to C$ are in 1-1 correspondence with the homomorphisms $H^n(C) \to H_0(C)$ they induce.

§7. Poincaré complexes and Witt groups.

In this section it is assumed that A is an n-dimensional regular domain. As always, $1/2 \in A$ and CM_p means $CM_p(A)$. We prove the connection between stable isometry classes in $F^\lambda(CM_p)$ (cf. (1.15)) and homotopy classes in $P^\lambda_{p+1}(CM_p)$, $p \geq 0$. There is an analogous (and more apprent) one-to-one correspondence between isometry classes in $\Omega^\lambda(CM_p)$ and homotopy classes $P^\lambda_p(M_p)$. These one-to-one correspondences give rise to isomorphisms

$$\text{(7.1)}\qquad \begin{aligned} W^\lambda_0(CM_p) &\cong \Omega^\lambda_p(M_p) \\ W^\lambda_1(CM_p) &\cong \Omega^\lambda_p(M_{p+1}) , \end{aligned}$$

which might be thought of as a replacement for the "resolution theorem" in algebraic K-theory. We prove only a connection between $F^\lambda(CM_p)$ and $\Omega^\lambda_p(M_{p+1})$ since it is all that is needed here.

(7.2) Proposition. Let A be a regular Noetherian domain, $1/2 \in A$. For each $p \geq 0$, there is a map of sets

$$A: \left\{\begin{matrix}\text{stable isometry classes}\\ \text{in } F^\lambda(CM_p)\end{matrix}\right\} \to \left\{\begin{matrix}\text{homotopy classes}\\ \text{in } P^\lambda(M_p)\end{matrix}\right\}$$

Proof: Let $(K,H,(\alpha,\gamma))$ be given. Choose a complex R of dimension p such that

$$H_i(R) = \begin{cases} H^\wedge, & i = 0 \\ 0, & i \neq 0 \end{cases}$$

where $(-)^\wedge$ means $\mathrm{Hom}(-,V_p)$. Since A is Cohen-Macaulay, $\mathrm{depth}_{\mathrm{Ann}(M)}(A) = p$ ([K , Thm. 136]), so by [M , p. 103, Prop.] $H_i(R) = 0$, $i < p$, while $H^p(R) = \mathrm{Ext}^p(H^\wedge,A) \cong \mathrm{Hom}(H^\wedge,V_p) = H^{\wedge\wedge} \cong H$. Hence, setting

$$R^t_i = \mathrm{Hom}(R_{p-i},A),$$

R^t is a resolution of H. Set

$$D = R \oplus R^t$$

and define $\psi_i: D^i \to D_{p-i}$ by

$$R^i \oplus (R^t)^i \xrightarrow{\begin{pmatrix}0 & 0\\ 1 & 0\end{pmatrix}} R_{p-i} \oplus (R^t)_{p-i} .$$

Then $(D,\{\psi\}) \in P_p(M_p)$. (The differential in D is excluded here from the notation.)

Next let D' be a resolution of $K^\wedge$; as above, $H^p(D') = K^{\wedge\wedge} \cong K$ and $H^i(D') = 0$, $i \neq p$. By general principles, there is a chain map $f'\colon D \to D'$, unique up to homotopy, such that

$$(\alpha,\gamma) = H^p(f')\colon H^p(D') \to H^p(D).$$

Let $f' = (g,h)$, where $g\colon R \to D'$, $h\colon R^t \to D$. Then (referring to (7.3) below), $f'_\%\{\psi\} := \{f'\psi f'^*\} = \{hg^*\} = \{hg^* - 1/2(1 - T_\lambda)(hg^*)\} = \{1/2(1 + T_\lambda)hg^*\}$, which is trivial in $\Omega_p(D',\lambda)$ since it induces the zero map $\alpha^\wedge\gamma + \lambda\gamma^\wedge\alpha\colon K = H^p(D') \to H_0(D') = K^\wedge$ (cf. (1.11)). Now from [R1,] we have the exact sequence

$$\text{(7.3)} \qquad \Omega_{p+1}(D',\lambda) \to \Omega_{p+1}(f',\lambda) \to \Omega_p(D,\lambda) \xrightarrow{f'_\%} \Omega_p(D',\lambda)$$

But $\Omega_{p+1}(D',\lambda) = 0$ since (6.18) shows $H_{p+1}(D' \otimes D') = 0$. This means there is a unique $\{\delta\psi,\psi\} \in \Omega_{p+1}(f',\lambda)$ giving a λ-symmetric complex

$$\text{(7.4)} \qquad (f'\colon D \to D',\{\delta\psi,\psi\})$$

unique up to a homotopy equivalence inducing the identity on the boundary $(D,\{\psi\})$. It is easily checked to be λ-Poincaré of dimension $p+1$, with $H_*(D) \in M_p$.

Similarly, set $D'' = R^t$ and let $f''\colon D \to D''$ be projection to the second factor, so that $H^p(f'')\colon H^p(D'') \to H^p(D)$ is inclusion $H^\wedge \to H \oplus H^\wedge$ to the second factor. Once again there is a relative λ-Poincaré complex,

$$(f''\colon D \to D'',\{\delta\psi'',\psi\})$$

well-defined up to homotopy equivalence inducing the identity on the boundary $(D,\{\psi\})$.

Referring to [R1, p. 135], let

$$\text{(7.5)} \qquad (C = D' \cup_D D'',d_C,\{\eta\})$$

be the glueing of D' and D'' along D. Evidently, C is a $(p+1)$-dimensional complex with $H_*(C) \in M_p$. Since D' and D'' were relative λ-Poincaré with common boundary D, $(C,d_C,\{\eta\}) \in P^\lambda_{p+1}(M_p)$. This is $A(K,H,(\alpha,\gamma))$.

For the usual reasons, the homotopy type of C (as defined in [R1,p. 140]) is independent of the resolutions R and D' of $H^\wedge$ and $K^\wedge$; as observed above the classes $\{\delta\psi',\psi\} \in \Omega_{p+1}(f',\lambda)$ and $\{\delta\psi'',\psi\} \in$

$Q_{p+1}(f",\lambda)$ are also well-defined. Hence the homotopy type of $(C,d_C,\{\eta\}) \in P^{\lambda}_{p+1}(M_p)$ depends only on $(K,H,(\alpha,\gamma))$.

We next need to show invariance of homotopy type after

1) stabilization of $(K,H,(\alpha,\gamma))$ (1.12(a))

2) isometry of $(K,H,(\alpha,\gamma))$ (1.12(b)).

For 1) let $(K_1,H_1(\alpha_1,\gamma_1))$ be a stabilization of $(K,H,(\alpha,\gamma))$ so that we have, in particular, a commutative diagram

(7.6)
$$\begin{array}{ccc} L & = & L \\ \downarrow & & \downarrow \\ K_1 & \xrightarrow{\alpha_1} & H_1 \\ \downarrow & & \downarrow \\ K & \xrightarrow{\alpha} & H \end{array}$$

Correspondingly, there is a homotopy commutative diagram of chain complexes

$$\begin{array}{ccc} D_1' & \xleftarrow{g_1} & R_1 \\ \uparrow \ell & & \uparrow k \\ D' & \xleftarrow{g} & R \end{array}$$

realizing the bottom square of (7.6) as its H^p. By construction of C,

$$C_r = D_r' \oplus D_{r-1} \oplus D_r", \quad \text{and}$$

$$d_C = \begin{pmatrix} d_{D'} & (-1)^{r-1}f' & 0 \\ 0 & d_D & 0 \\ 0 & (-1)^{r-1}f" & d_{D"} \end{pmatrix} : D_r' \oplus D_{r-1} \oplus D_r" \to D_{r-1}' \oplus D_{r-2} \oplus D_{r-1}".$$

If $C_1 := D_1' \cup_{D_1} D_1"$ is the chain complex constructed for $(K_1,H_1,(\alpha_1,\gamma_1))$ define a chain map $S\colon C \to C_1$ by

$$C_r = D_r' \oplus R_{r-1} \oplus R^t_{r-1} \oplus D_r" \xrightarrow{(\ell\oplus k\oplus 0\oplus 0)} (D_1')_r \oplus (R_1)_{r-1} \oplus (R_1^t)_{r-1} \oplus (D_1")_r.$$

It is not difficult to verify that S is a homotopy equivalence in $P^{\lambda}_p(M_{p+1})$.

For part 2), the isometry of $(H \oplus H^{\wedge},\phi_h)$

$$\begin{pmatrix} 1 & 0 \\ \rho & 1 \end{pmatrix} : H \oplus H^{\wedge} \to H \oplus H^{\wedge}$$

is induced as H^p of a map of the λ-symmetric complex

$$R^t \oplus R \xrightarrow{G=\begin{pmatrix} 1 & 0 \\ k & 1 \end{pmatrix}} R^t \oplus R = D$$

where $k: R^t \to R$ satisfies $(1 + T_\lambda)k = 0$. It is easily verified that

a) the glueing C_1 of D' and D'' along D obtained from $f'G: D \to D'$ and $f'': D \to D''$ is a λ-Poincaré complex constructed from $(K,H,(\alpha,\gamma + \rho\alpha))$ (by A), and

b) C_1 is homotopic to C by a map restricting to the identity on D' and D'', and to K on D.

Finally, it is now easy to show that if the formations $(K,H,(\alpha,\gamma))$ and $(K',H',(\alpha',\gamma'))$ are isomorphic, then their corresponding Poincare complexes are homotopy equivalent (for any choice of the resolutions R and D').

Next we define the inverse to A.

(7.7) <u>Proposition</u>. Let A be a regular noetherian domain containing $1/2$. Then for each $p \geq 0$ there is a map of sets

$$B: \left\{\begin{matrix}\text{homotopy classes} \\ \text{in } P^\lambda_{p+1}(M_p)\end{matrix}\right\} \to \left\{\begin{matrix}\text{stably isometry classes} \\ \text{in } F^\lambda(CM_p)\end{matrix}\right\}$$

<u>Proof:</u> Let $(C,d,\{\eta\}) \in P^\lambda_{p+1}(M_p)$ be given. By (6.16) we have

$$H_i(C) = \begin{cases} \in M_p, & i = 0,1 \\ 0 \ , & i \neq 0,1 \ . \end{cases}$$

We first show how to find a chain map

$$f: C \to D$$

where $D = (D_{p+1} \to \dots \to D_1 \to 0)$ satisfies

$$H^i(D) = \begin{cases} H \in CM_{p+1}, & i = p+1 \\ 0 \ , & i \neq p+1 \ , \end{cases}$$

and

$$H^{p+1}(f): H^{p+1}(D) \to H^{p+1}(C) \tag{7.8}$$

is surjective. To do this we need to define $\bar{f}: H \to H^{p+1}(C)$ and fill in the verticals in

$$\begin{array}{ccccccccc} \dots \to & D^{p-1} & \overset{\delta}{\to} & D^p & \to & D^{p+1} & \twoheadrightarrow & H & \\ & \downarrow f^{p-1} & & \downarrow f^p & & \downarrow f^{p+1} & & \downarrow \bar{f} := H^{p+1}(f) & \\ \dots \to & C^{p-1} & \overset{d^*_p}{\to} & C^p & \overset{d^*_{p+1}}{\to} & C^{p+1} & \twoheadrightarrow & H^{p+1}(C) & . \end{array}$$

For this, let $I = \text{Ann}\, H^*(C)$, an ideal of height $\geq p$ by assumption, and choose a regular sequence $x_1,\ldots,x_p \in I$ (cf. [K,Thm.136]). Let D^* be the associated (sum of) Koszul resolutions of $H:=(A/(x_1,\ldots,x_p))^n$ where $H^{p+1}(C)$ has n generators. Then $\bar{f}$ is defined by a choice of such generators and f^{p+1} is any lifting of $\bar{f}$. Since the bottom sequence is exact at C^{p+1}, f^{p+1} can be lifted to f^p. However, it may not (á priori) be possible to fill in f^{p-1} because C^* is not in general exact at C^p. To remedy this, we change D^* by replacing the regular sequence $\{x_1,\ldots,x_p\}$ by $\{x_1^2,\ldots,x_p^2\}$ (still a regular sequence in I by [K, p.103]), thus multiplying the values of f^p on generators of D^p by elements of I. This keeps commutativity of the second square. Now in any case, $d^*_{p+1} \cdot f^p \cdot \delta = 0$ so that, since $IH^p(C) = 0$, the above change in f^p (and δ) means $\operatorname{im}(f^p\delta) \subseteq \operatorname{im} d^*_p$. Thus, f^{p-1} can be chosen to make the left square commute. Completing the rest of the diagram is immediate because $H^i(C) = 0$, $i < p$.

Consider now the exact sequence from [R1,3.1] (cf. (7.3))

$$(7.9) \qquad Q_{p+2}(D,\lambda) \to Q_{p+2}(f,\lambda) \to Q_{p+1}(C,\lambda) \xrightarrow{f_\%} Q_{p+1}(D,\lambda)$$

According to (6.18), $H_{p+1}(D \otimes D) \cong \operatorname{Ext}^1(H,H^\wedge)$ and the construction and naturality of the spectral sequence giving this result shows that $f_\%\{\eta\}$ is represented by the composition $D^p \xrightarrow{f^p} C^p \to C_1 \xrightarrow{f_1} D_1 \to H_1(D):=H$, viewed as a cycle in the chain complex $\operatorname{Hom}(D^*,H^\wedge)$. Recall, however, that in the construction of f above we arranged that $\operatorname{im} f^p \subseteq I(C_p)$. Hence the cocycle above is actually zero, so $f_\%\{\eta\} = 0$. Thus, we obtain a class $\{\delta\eta,\eta\} \in Q_{p+2}(f,\lambda)$ and hence a λ-symmetric pair

$$(f\colon C \to D,\{\delta\eta,\eta\})$$

with $H_*(D) \in M_p$.

Now do algebraic surgery on f ([R1, p. 145] or (6.14)) getting $(C',d_{C'},\{\eta'\})$, where $d_{C'}$ and η' are described in [R1, p. 145] or (6.14).

Let $C(f)_{+1}$ denote the mapping cone of f, $C(f)$, shifted in degree,

$$(C(f)_{+1})_r = C(f)_{r+1}\ .$$

Let D' be the complex defined by

$$(D')_r := D^{p+2-r}, \quad d_{D'} := d^*_D.$$

Then we have a short exact sequence of complexes

$$(7.10) \qquad C(f)_{+1} \overset{i}{\rightarrowtail} C' \overset{f'}{\twoheadrightarrow} D'$$

where i is the inclusion of the first two factors and f' is the projection to the third. From this we get a canonical chain homotopy equivalence

$$C(f') \overset{\cong}{\to} C(f)$$

which we use to identify $H_*(f)$ with $H_*(f')$.

We are now ready to define the λ-formation $(K,H,(\alpha,\gamma))$ associated to the λ-Poincaré complex $(C,d_C,\{\eta\})$ by $\mathcal{B}$. Namely, set

- $K = H^{p+1}(C(f)) \equiv H^{p+1}(C(f'))$
- $H = H^{p+1}(D)$
- α: $H^{p+1}(C(f)) \to H^{p+1}(D)$ is the map $H^{p+1}(j)$, where $j: D \to C(f)$.
- γ is the composition $H^{p+1}(C(f)) := H^p(C(f)_{+1}) \overset{\delta}{\to} H^{p+1}(D') := H_1(D) \cong H^\wedge$, where δ is the boundary in the cohomology sequence of (7.10).

Specifically, straightforward computations show that α is induced by

$$(a,b) \to b,$$

and γ by

$$(a,b) \to \lambda f\ \phi(a) + \lambda(\delta\phi)b \in D_1$$

where $a \in C^p$, $b \in D^{p+1}$, $\phi = (1 + T_\lambda)\eta$ and $\delta\phi = (1 + T_\lambda)\delta\eta$.

(7.11) Remark. a) The formula for γ induces (up to sign) the duality maps

$$H^r(C(f)) \to H_{p+2-r}(D)$$

for the λ-symmetric complex $(f: C \to D,\{\delta\eta,\eta\})$ (cf. (6.11)). Consequently, the latter is λ-Poincaré if and only if γ is an isomorphism.

b) The formula for γ also shows that $\mathcal{B}$ applied $(C,d,\{-\eta\})$ gives $(K,H,(\alpha,-\gamma))$.

Before we show $(K,H,(\alpha,\gamma))$ above is actually a formation, we sketch how it depends on the choices in its construction. Fix the chain map $f: C \to D$. Then if $\{\delta'\eta,\eta\} \in Q_{p+2}(f)$ is another element mapping to $\{\eta\}$ in (7.9), then there is $\{\tau\} \in Q_{p+2}(D,\lambda)$ such that $\tau = \delta'\eta - \delta\eta: D^* \to D$. If $\rho: H \to H^\wedge$ is the map on homology induced by τ and (α',γ') are the maps produced from the choice $\{\delta'\eta,\eta\}$, then $\rho = -\lambda\rho^\wedge$, $\alpha' = \alpha$ and $\gamma' = \gamma + \rho\alpha$. Thus $(K,H,(\alpha,\gamma))$ and $(K,H,(\alpha',\gamma'))$ are isometric ((1.12b)). If we choose another chain map $g: C \to E$ where E satisfies the conditions D did and $H^{p+1}(g) = H^{p+1}(f)$ then the resultant formation changes by an isomorphism. If another surjection $H' \to H^{p+1}(C)$ is realized by H^{p+1} of a chain map, it is not difficult to show that the resultant formation changes by stabilizations ((1.12(a)) and/or destabilization ((1.14)).

We now show $(K,H,(\alpha,\gamma))$ is a λ-formation, beginning with the injectivity of

$$(\alpha,\gamma): H^{p+1}(C(f)) \to H^{p+1}(D) \oplus H^{p+1}(D') := H^{p+1}(D) \oplus H_1(D)$$

First the following observations:

i) (α,γ) is given on the chain level by

$$C^p \oplus D^{p+1} \to D^{p+1} \oplus D_1$$

$$(a,b) \to (b,\lambda f\phi(a) + \lambda\delta\phi(b))$$

ii) $(a,b) \in C^p \oplus D^{p+1}$ is a cocycle if $d_C^* a + (-1)^{p+1} f^* b = 0$;

iii) (a,b) is a coboundary if, for some $(a',b') \in C^{p-1} \oplus D^p$, $(a,b) = (d_C^* a' + (-1)^p f^* b', d_D^* b')$;

iv) (Co)cycle conditions on D^{p+1} and D_1 on the right side of (i) are automatic;

v) the right side of (i) is a coboundary if there are $x \in D^p$, $y \in D_2$ such that $(d_D^* x, d_D y) = (b,\lambda f\phi(a) + \lambda(\delta\phi)b)$.

Assume then that (in (i)) $(b,\lambda f\phi a + \lambda(\delta\phi)b) = (d_D^* x, d_D y)$. By (iii) it suffices to find a' such that

$$d_C^* a' = (-1)^{p+1} f^* x + a$$

(taking $x = b'$). But $f\phi$ induces an injection $H^p(C) \to H_1(D) = H^\wedge$ (because $H_2(C(f)) = 0$), so it suffices to show $(-1)^{p+1} f\phi f^* x + f\phi a$ is a boundary.

Now from the definition of $(f: C \to D, \{\delta\eta,\eta\})$

$$\begin{aligned}(-1)^{p+1} f\phi f^* x &= -d_D(\delta\phi)x - (-1)(\delta\phi)d_D^* x \\ &\sim (\delta\phi)d_D^* x \\ &= (\delta\phi)b\end{aligned}$$

where $\sim$ means "homologous." By assumption, $d_D y = \lambda f\phi a + \lambda(\delta\phi)b$, so

$$f\phi a \sim -(\delta\phi)b$$

Hence

$$\begin{aligned}&(-1)^{p+1} f\phi f^* x + f\phi a \\ &\sim (\delta\phi)b + f\phi a \\ &\sim (\delta\phi)b - (\delta\phi)b = 0.\end{aligned}$$

Next we show $\mathrm{im}(\alpha,\gamma)$ is totally isotropic. This is equivalent to $\gamma^\wedge\alpha + \lambda\alpha^\wedge\gamma = 0$. We have by (i) above that $\alpha^\wedge\gamma$ is induced by

$$C(f)^{p+1} := C^p \oplus D^{p+1} \to D_1 \to C_0 \oplus D_1 := C(f)_1$$
$$(a,b) \to \lambda f\phi a + \lambda(\delta\phi)b \to (0,\lambda f\phi a + \lambda(\delta\phi)b)$$

and $\gamma^\wedge\alpha$ by

$$C^p \oplus D^{p+1} \to D^{p+1} \to C_0 \oplus D_1$$
$$(a,b) \to b \to ((-1)^{p+1}\phi f^*b,(-\delta\phi)b)$$

Hence $\gamma^\wedge\alpha + \lambda\alpha^\wedge\gamma$ is induced by

$$(a,b) \to ((-1)^{p+1}\phi f^*b, f\phi a).$$

Since (a,b) is a cocycle in $C(f)^{p+1}$,

$$(-1)^{p+1}\phi f^*b = -\phi d_C^*a = -d_C\phi a$$

so

$$d_{C(f)}(-\phi a,0) = (-d_C\phi a, f\phi a) = ((-1)^{p+1}\phi f^*b, f\phi a)$$

Thus $\gamma^\wedge\alpha + \lambda\alpha^\wedge\gamma = 0$, as claimed.

It is finally necessary to show that $\mathrm{im}(\alpha,\gamma) \subseteq H \oplus H^\wedge$ is a sublagrangian. This is in the spirit of the above and is left to the reader.

The next result compares the functions in (7.2) and (7.7).

(7.12) Proposition. The composition $\mathcal{B}\mathcal{A}$ is the identity.

Proof: Start with a formation $(K,H,(\alpha,\gamma)) \in F^\lambda(\mathcal{CM}_p)$ and apply $\mathcal{A}$ to get a $(p+1)$-dimensional λ-Poincaré complex C. We will show there is an obvious choice of the chain map $f: C \to D$ in construction $\mathcal{B}$ so that if $(K',H',(\alpha',\gamma'))$ is the resulting formation, then $K = K'$, $H = H'$, $\alpha = \alpha'$ and $\gamma = \gamma'$.

To begin, let $D = R \oplus R^t$ and D' be chosen as in (7.2) with

$$H^p(D) = H^p(R) \oplus H^p(R^t) = H \oplus H^\wedge$$
$$H^p(D') = K$$

and $f' = (g,h): D = R \oplus R^t \to D'$ such that

$$H^p(f) = (\alpha,\gamma).$$

Let $f'' = pr_2: D = R \oplus R^t \to R^t := D''$ and glue to get (7.5),

$$(C = D' \cup_D D', d_C, \{\eta\}) .$$

Now in the construction of $\mathcal{B}$, define $p: C \to C(f")$ by

$$C_r = D_r' \oplus D_{r-1} \oplus D_r'' \to D_{r-1} \oplus D_r'' := C(f")_r$$

$$(a,b,c) \to (b,c)$$

The following are easily verified:

(i) there are inverse chain homotopy equivalences $R \overset{\mu}{\to} C(f")_{+1}$ and $C(f")_{+1} \overset{\nu}{\to} R$, the first induced from the short exact sequence of chain complexes

$$R \rightarrowtail D \overset{f"}{\twoheadrightarrow} R^t = D"$$

and defined by $x \to (x,0,0)$ in

$$R_r \to (C(f")_{+1})_r = D_r \oplus D_{r+1}'' = R_r \oplus R_r^t \oplus D_{r+1}'' .$$

The second is $(x,y,z) \to x$ in $(C(f")_{+1})_r \to R_r$.

(ii) There is a short exact sequence of chain complexes $D' \to C \overset{p}{\to} C(f")$ (inclusion to the first summand and projection to the last two), hence a chain homotopy equivalence $\kappa: D' \to C(p)_{+1}$ induced by $x \to (x,0,0,0)$ in

$$D_r' \to (C(p)_{+1})_r := C_r \oplus C(f")_{r+1} = D_r' \oplus D_{r-1} \oplus D_r' \oplus C(f")_{r+1}$$

(iii) $(p: C \to C(f"),\{0,\eta\})$ is a λ-symmetric complex.

We now complete the construction in $\mathcal{B}$ by setting $H' = H^{p+1}(C(f"))$, $K' = H^{p+1}(C(p))$, $\alpha' = H^{p+1}(j)$ where j is the canonical map $j: C(f") \to C(p)$ and $\gamma': H^{p+1}(C(p)) \to H_1(C(f"))$ is the map induced by the duality map in the λ-symmetric complex of (iii) above.

Using i) and ii) we identify K with K' by $H^p(\kappa)$ and H with H' by $H^p(\mu)$. It is easily checked that the diagrams

$$\begin{array}{ccc} H^{p+1}(C(p)) & \overset{\alpha'}{\to} & H^{p+1}(C(f")) \\ \downarrow \cong & & \downarrow \cong \\ H^p(D') & \overset{\alpha}{\to} & H^p(R) \end{array}$$

and

$$\begin{array}{ccc} H^{p+1}(C(p)) & \overset{\gamma'}{\to} & H_1(C(f")) \\ \downarrow \cong & & \downarrow \cong \\ H^p(D') & \overset{\gamma}{\to} & H^p(R^t) \equiv H_0(R) \end{array}$$

commute. This completes the proof of (7.12).

(7.13) We next show that the function A above actually induces a homomorphism of abelian groups

$$W_1^\lambda(CM_p) \to \Omega_{p+1}^\lambda(M_p)$$

Since A in (7.2) clearly preserves sum, it is sufficient to show that the Poincare complex (C,d_C,η) obtained from the graph formation $(H\hat{},H,(\alpha,1))$ is M_p-cobordant to one for which $H_* \equiv 0$. By the proof of (7.12) there are natural choices so that $\mathcal{B}(C,d_C,\eta)$ gives γ an isomorphism; but in (7.11)(a) we observed that in the construction of $\mathcal{B}$, $(f: C \to D,\{\delta\eta,\eta\})$ is λ-Poincaré if and only if γ is an isomorphism. Now an easy calculation shows that the result $(C',d',\{\eta'\})$ of algebraic surgery on $(f: C \to D,\{\delta\eta,\eta\})$ yields $H_*(C') \equiv 0$ (cf. (6.14)). Since $(C',d',\{\eta'\})$ is M_p-cobordant to $(C,d_C,\{\eta\})$, the latter represents zero in $\Omega_{p+1}^\lambda(M_p)$.

§8. Two applications of the theory of Poincaré complexes to Witt groups

The first application is the proof of (1.21).

(8.1) Proposition. Let A be a regular domain. Then a λ-formation $(K,H,(\alpha,\gamma)) \in F^{\lambda}(CM_p)$ represents zero in $W_1^{\lambda}(CM_p)$ if and only if it is stably isometric to a graph formation.

Proof: One implication is clear. Conversely, suppose $[K,H,(\alpha,\gamma)] = 0$ in $W_1^{\lambda}(CM_p)$. Then the λ-Poincaré complex $(C,d,\{\eta\})$ associated to it (by (7.2)) is trivial in $\Omega_{p+1}^{\lambda}(M_p)$ by (7.13). If $(g\colon C \to D,\{\delta\eta,\eta\})$ is the null M_p-cobordism (a λ-Poincaré complex), then by (6.16) the only possible non-zero cohomology groups in the exact sequence of $g\colon C \to D$ appear in

$$(8.2)\quad H^p(g) \to H^p(D) \to H^p(C) \to H^{p+1}(g) \to H^{p+1}(D) \to H^{p+1}(C) \to H^{p+2}(g) \to H^{p+2}(D).$$

Now, proceeding as in (7.7), find a λ-symmetric complex $(h\colon D \to E, \{\delta\phi,\delta\eta\})$ such that the homomorphism in degree $p + 2$ is the identity, $H^{p+2}(E) \in CM_p$ and E is a sum of Koszul resolutions (cf. the construction of f at the beginning of the proof of (7.7)). Doing algebraic surgery using h ([R1, p. 145] or (6.14)) yields a λ-symmetric complex $(D',d_{D'},\{\delta'\eta\})$ which has boundary homotopy equivalent to $(C,d_C,\{\eta\})$ (by [R1, (4.1)(i)]) and for which $H^{p+2}(D') = 0 = H_0(D')$. (Compare [R2, pp. 335-6].

Hence (using (7.12)) we can assume in (8.2) that $H^{p+2}(D) = 0 = H_0(D)$. But by (6.17), $H^p \cong \mathrm{Ext}^p(H_0(D),A) = 0$ and by duality $H^{p+2}(g) \cong H_0(D) = 0$. Hence the remaining terms in (8.2) are

$$H^p(C) \rightarrowtail H^{p+1}(g) \to H^{p+1}(D) \twoheadrightarrow H^{p+1}(C) \ .$$

In particular the surjectivity on H^{p+1} induced by g qualifies g for use in the construction of $\mathcal{B}$ in (7.7) applied to C. By (7.11)(a), the resultant λ-formation has γ an isomorphism and by (7.12) it is stably isometric to $(K,H,(\alpha,\gamma))$.

The next result amounts to exactness of (2.1)(b) at $W_1^{\lambda}(CM_1)$. For its proof we need a special case $(p = 1)$ of the result referred to in (7.1),

$$(8.3)\qquad A_*\colon W_1^{\lambda}(CM_1) \xrightarrow{\cong} \Omega_2^{\lambda}(M_1) \ .$$

This is due to Ranicki ([R2, p. 359]).

(8.4) Proposition. Let A be a regular Noetherian domain containing $1/2$. Then every element $[K,H,(\alpha,\gamma)] \in W_1^{\lambda}(CM_1)$ admits a representative for which α is injective.

Proof: Let $(C,d,\{\eta\}) \in P_2^\lambda(M_1)$ be associated to $(K,H,(\alpha,\gamma))$ by (7.2). Suppose it is possible to find a λ-symmetric pair $(f\colon C \to D, \{\delta\eta,\eta\})$ such that

(8.5)
- $H_i(D) = 0, \quad i \neq 2$
- $H^2(f)\colon H^2(D) \to H^2(C)$ is an isomorphism at all height one primes

Then $H^1(C) \xrightarrow{\cong} H^2(f)$, which easily implies $(f\colon C \to D,\{\delta\eta,\eta\})$ is λ-Poincaré. By (7.11)(a), surgery on f (cf. (6.14)) yields a complex $(C',d',\{\eta'\})$ which is M_1-cobordant to $(C,d,\{\eta\})$ and whose associated λ-formation (in (7.7)) $(K',H',(\alpha',\gamma'))$ has γ' an isomorphism at all height one primes.

Now (7.11)(b), together with (8.3) and the fact that $(C',d',\{\eta'\})$ is inverse to $(C',d',\{-\eta'\})$ in $\Omega_1^\lambda(M_1)$, show that $[K',H',(\alpha',-\gamma')]$ is inverse to $[K',H',(\alpha',\gamma')]$ in $W_1^\lambda(CM_1)$. But by (1.16) the latter is also inverse to $[K',H',(\gamma',-\lambda\alpha')]$. Hence in $W_1^\lambda(CM_1)$, $[K',H',(\alpha',\gamma')] = [K',H',(\gamma',\lambda\alpha')]$, so we are done.

It remains to provide (8.5). Let q be a height one prime of A and let $R = A_q$, $(x) = qA_q$. By (5.9), $(K_q,H_q,(\alpha,\gamma)_q)$ is isomorphic to a sum of formations of the form

$$(R/(x^t) \oplus R/(x^{m-t}),R/(x^m),((0,x^t),(x^{m-t},0)))$$

where $0 \le t \le m$. Evidently, each such formation can be destabilized to $(R/(x^t),R/(x^t),(0,1))$. So if $(\overline{C},\overline{d},\overline{\eta}) \in P_2^\lambda(M_1(R))$ is a formation constructed from $(K_q,H_q,(\alpha_q,\gamma_q))$ by (7.2), there is a homotopy equivalence

$$h\colon (C,d,\{\eta\})_q \xrightarrow{\cong} (\overline{C},\overline{d},\{\overline{\eta}\})$$

Moreover, since $\alpha = 0$ in $(R/(x^t),R/(x^t),(0,1))$ we can extract from the construction in (7.7) (cf. also proof of (7.12)) a λ-symmetric complex over R,

$$(\overline{g}\colon \overline{C} \to \overline{D},(\delta\overline{\eta},\overline{\eta}))$$

such that $H^2(\overline{g})\colon H^2(\overline{D}) \xrightarrow{\cong} H^2(\overline{C})$. Setting $\overline{f} = \overline{g}h$, we get a λ-symmetric complex

$$(\overline{f}\colon C_q \to \overline{D},\{\overline{\delta}\eta,\eta_q\})$$

with

$$H^2(\overline{f})\colon H^2(\overline{D}) \xrightarrow{\cong} H^2(C_q) .$$

Now if $H^2(\bar{D}) = (R/(x^t))^m$, let $M = (A/q^{(t)})^m$, where $q^{(t)}$ denotes the t-th symbolic power. Then $M \in CM_1$ because A is locally factorial. Hence we may construct a chain complex of projective A-modules

$$E = (E_2 \to E_1)$$

with

$$H_i(E) = \begin{cases} M, & i = 1 \\ 0, & i = 2 \end{cases}$$

and from the natural inclusion $M \to (R/(x^t))^m$, a commutative diagram,

$$\begin{array}{ccccc} E_2 & \to & E_1 & \to & M \\ \downarrow & & \downarrow & & \downarrow \\ \bar{D}_2 & \to & \bar{D}_1 & \to & (R/(x^t))^m \end{array}$$

Thus the corresponding map of complexes $E \to \bar{D}$ is a homotopy equivalence at q. Finally, multiplying $\bar{f}$ and $\bar{\delta}\eta$ by some element of $A - q$ gives $\bar{f}(C) \subseteq E$ and $\bar{\delta}\eta(E^*) \subseteq E$, hence a λ-symmetric complex

(8.6) $$(f: C \to E, \{\delta\eta, \eta\})$$

where $f = \bar{f}|C$, $\delta\eta = \bar{\delta}\eta|E^*$ and $H^2(f_q): H^2(E)_q \overset{\cong}{\to} H^2(C)_q$. Taking the sum of the complexes (8.6) for each height one q gives the desired λ-symmetric complex in (8.5).

Bibliography

[AB] M. Auslander and M. Bridger, Stable Module Theory, Mem. A.M.S. 94 (1969).

[Ba] J. Barge, Formes Hermitiennes sur les $\mathbb{Z}[X,X^{-1}]$-modules, Bull. Soc. Math. France, Mem. 48 (1976), 7-10.

[B] H. Bass, On the ubiquity of Gorenstein rings, Math. Zeit. 82 (1963), 8-28.

[BE] D. Buchsbaum and D. Eisenbud, Algebra structures on finite free resolutions, Am. J. Math. 99 (1977), 447-485.

[CF] L. Claborn and R. Fossum, Generalizations of the notion of class group, Ill. J. Math. 12 (1968), 228-253.

[Ch] C. Chevalley, Anneaux de Chow et applications, Séminaire C. Chevalley, 2^{e}année, Sec. Math. Paris (1958).

[C1] J.-L. Colliot-Thélène, Formes quadratiques sur les anneaux semi-locaux réguliers, Mém. Soc. Math. France 59 (1979), 13-31.

[C2] ______________, Formes quadratiques multiplicatives et variétés algébriques, Bull. Soc. Math. France 106 (1978), 113-151.

[C3] _______________, Formes quadratiques multiplicatives et variétés algébriques: deux compléments, Bull. Math. Soc. France 108 (1980), 213-227.

[CI] _______________ and F. Ischebeck, L'équivalence rationelle sur les cycles de dimension zéro des variétés algébriques réeles, C. R. Acad. Sc. Paris 292 (1981), 723-725.

[CS] _______________ and J.-J. Sansuc, Fibrés quadratiques et composantes connexes réeles, Math. Ann. 244 (1979), 105-134.

[CRW] T. Craven, A. Rosenberg, and R. Ware, The map of the Witt ring of a domain into the Witt ring of its field of fractions, Proc. A. M.S. 51 (1975), 25-30.

[DK] H. Delfs and M. Knebusch, Semi-algebraic topology over a real closed field I: Paths and components in the set of rational points of analegraic variety, Math. Zeit. 177 (1981), 107-129.

[F] D. Ferrand, Construction des fibrés de rang deux, Asterisque 36, 149-169.

[Fo] R. Fossum, Vector bundles over spheres are algebraic, Inv. Math. 8 (1969), 222-225.

[Fu] W. Fulton, Rational equivalence on singular varieties, Publ. Math. I.H.E.S. 45 (1975), 147-164.

[G] S. Gersten, Some exact sequences in the higher K-theory of rings, Springer Lecture Notes No. 341 (1973), 211-244.

[Gr] A. Grothendieck, Éléments de geometrie algébrique, Pub. Math. I.H.E.S. 32 (1967).

[Ho] M. Hochster, Cohen-Macaulay modules, Lecture Notes in Math. No. 311, Springer Verlag (New York, Heidelberg).

[Hu] D. Husemoller, Fibre Bundles, McGraw-Hill (1966).

[K] I. Kaplansky, Commutative Rings (Rev. Ed.), University of Chicago Press, (1974).

[K1] M. Knebusch, On algebraic curves over real closed fields II, Math. Zeit. 151 (1976), 189-205.

[K2] _______________, Specialization of quadratic and symmetric bilinear forms, and a norm theorem, Acta Arith. 24 (1973), 279-299.

[K3] _______________, Symmetric bilinear forms over algebraic varieties, Conf. on quadratic forms (Kingston 1976), Queens Papers in Pure and Applied Math. 46 (1977), 103-283.

[KOS] M. A. Knus, M. Ojanguren and R. Sridharan, Quadratic forms and Azumaya algebras, Journal fur die reine und angewandte Math. 303/304 (1978), 231-248.

[KM] A. Kustin and M. Miller, Algebra structures on minimal resolutions of Gorenstein rings, In: Commutative algebra: analytic methods, Marcel Dekker (1982), 45-66.

[L] T.-Y. Lam, The algebraic theory of quadratic forms, Benjamin (1973), Reading.

[M] H. Matsumura, Commutative Algebra (Sec. Ed.), Benjamin/Cummings (1980), Reading.

[MH] J. Milnor and D. Husemoller, Symmetric Bilinear Forms, Springer Verlag (1973), New York-Heidelberg.

[N] M. Nagata, Local Rings, Interscience (1962), New York-London.

[O1] M. Ojanguren, A splitting theorem for quadratic forms, Comment. Math. Helv. 57 (1982), 145-157.

[O2] ______________, Quadratic forms over regular rings, J. Indian Math. Soc., to appear.

[P1] W. Pardon, The exact sequence of a localization for Witt groups, Lecture Notes in Math., No. 551 (1976), 336-379.

[P2] ______________, An invariant determining the Witt class of a unitary transformation over a semi-simple ring, J. Alg. 44 (1977), 394-410.

[P3] ______________, A "Gersten conjecture" for Witt groups, Springer Lecture Notes in Math. No. 966 (Proceedings of Conference on Alg. K-Theory Oberwolfach, 1980).

[P4] ______________, A relation between the Witt group of a regular local ring and the Witt groups of its residue class fields, preprint.

[PS] C. Peskine and L. Szpiro, Dimension projective finie et cohomologie locale, Pub. Math. I.H.E.S. 42 (1973), 47-119.

[Q] D. Quillen, Higher algebraic K-Theory I, Springer Lecture Notes in Math. No. 341 (1973), 85-147.

[R1] A. Ranicki, The algebraic theory of surgery I: Foundations, Proc. L.M.S. 40 (1980), 87-192.

[R2] ______________, Exact sequences in the algebraic theory of surgery, Math. Notes, Princeton University Press (1981), Princeton.

[SV] D. W. Sharpe and P. Vamós, Injective Modules, Cambridge University Press (1972), Cambridge.

AN UPPER BOUND FOR ALGEBRAIC K-THEORY

Victor Snaith*

§1: INTRODUCTION

The title of this paper coincides with that of section IV.3 of [Sn 1]. In that section I gave an upper bound, in terms of unitary K-homology, for the algebraic K-theory (mod ℓ^ν) of a ring-*after the latter has been inflicted with Bott periodicity*. In this paper I will improve the results of [Sn 1, §IV.3] in two ways:

(i) We remove the condition that the ring we consider should have ℓ^ν-th roots of unity.

(ii) We show that the indecomposable quotient of the upper bound of [Sn 1] is a (better) upper bound.

The upper bound is sometimes "easier" to compute. For example, when A is a finite ring, the upper bound is determined by the representation theory of GL_nA ($n \geq 1$). See [Sn 1, §IV.3].

Recall [B;G-Q;Q1] that the algebraic K-groups of a ring A, with unit, may be defined as

$$K_iA = [S^i, BGLA^+] = \pi_i(BGLA^+), \quad (i \geq 1)$$

and

$$K_i(A;Z/n) = [S^{i-1} \cup_n e^i, BGLA^+] = \pi_i(BGLA^+;Z/n), \quad (i \geq 2)$$

where $BGLA^+$ is obtained from the classifying space of the infinite general linear group of A by applying the "plus"-construction relative to the commutator subgroup of GLA. One can extend these groups to lower dimensions, although we will not need them, in such a way as to make

$$K_*A = \bigoplus_{i \geq 0} K_iA$$

and

$$K_*(A;Z/n) = \bigoplus_{i \geq 0} K_i(A;Z/n)$$

into graded rings (provided that $16|n$ if n is even and $9|n$ if $3|n$). In fact we will require only a (graded) multiplication to exist on $K_*(A;Z/n)$ for which it

---

1980 Mathematics Subject Classification. 18F25, 55N15, 55Q10, 55P47, 55P42.

*Research was partially supported by grant #A4633 of the Natural Sciences and Engineering Research Council of Canada.

Let ℓ be an *odd* prime, suppose that $A = Z[1/\ell,\xi]$ where ξ is a primitive ℓ-th root of unity. Since $K_1A \cong A^*$ we may choose a "Bott element"

$$\beta_1 \in K_2(A;Z/\ell)$$

such that $\partial(\beta_1) = \xi$. If $\nu \geq 1$ is an integer then (by a Bockstein sequence argument)

$$\beta_1^{\ell^{\nu-1}(\ell-1)} \in K_{2(\ell-1)\ell^{\nu-1}}(A;Z/\ell)$$

is the mod ℓ reduction of

$$\beta_\nu \in K_{2(\ell-1)\ell^{\nu-1}}(A;Z/\ell^\nu)$$

which may be chosen to be invariant under the Galois group, $\mathrm{Gal}(\mathbb{Q}(\xi)/\mathbb{Q}) \cong (Z/\ell)^*$, which acts on A and thence on its K-theory. By a transfer argument (c.f. [B]) there is a (unique) element

$$\beta_\nu \in K_{2(\ell-1)\ell^{\nu-1}}(Z[1/\ell];Z/\ell^\nu)$$

whose image is the former β_ν under the natural map induced by the inclusion of $Z[1/\ell]$ into A.

For any ring R, containing $1/\ell$, there is a pairing

$$K_*(Z[1/\ell];Z/\ell^\nu) \otimes K_*(R;Z/\ell^\nu) \to K_*(R;Z/\ell^\nu).$$

Hence we may define the *Bott periodic algebraic K-theory of R*, denoted by $\mathcal{K}_i(R;Z/\ell^\nu)$ to be the direct limit $(s = 2\ell^{\nu-1}(\ell-1))$

$$K_i(R;Z/\ell^\nu) \xrightarrow{(\beta_\nu\cdot\_)} K_{i+s}(R;Z/\ell^\nu) \xrightarrow{(\beta_\nu\cdot\_)} \cdots$$

over iterated multiplication by β_ν. Clearly $\mathcal{K}_i(R;Z/\ell^\nu) \cong \mathcal{K}_{i+s}(R;Z/\ell^\nu)$. $\mathcal{K}_*(R;Z/\ell^\nu)$ was introduced in [Sn 2, Part IV]. When R is nice these groups are known, in terms of étale cohomology, by the main result of [Th 1]. For further details the reader is referred to [D-F-S-T]; [Sn 1/3], [Th 2] and [D-F].

So far we have no definition of $\mathcal{K}_*(R;Z/2^\nu)$, in order to define this we need to assume R has ℓ^ν-th roots of unity. Suppose that the units of R, R*, contains a central copy of Z/ℓ^a with $a \geq \nu$. Choose such a copy and call its subgroup, Z/ℓ^ν, the ℓ^ν-th roots of unity of R. Since R is a $Z[\xi_{\ell^\nu}]$-module, in a manner depending on the above choice, we obtain a pairing (assuming $\nu \geq 2$ if $\ell = 2$)

$$K_*(Z[\xi_{\ell^\nu}];Z/\ell^\nu) \otimes K_*(R;Z/\ell^\nu) \to K_*(R;Z/\ell^\nu).$$

Here ξ_{ℓ^ν} is a primitive ℓ^ν-th root of unity. In addition $K_*(Z[\xi_{\ell^\nu}];Z/\ell^\nu) \cong K_*(Z[1/\ell,\xi_{\ell^\nu}];Z/\ell^\nu)$ by the localisation sequence [G-Q] and the results of [Q2]. Hence there exists an element $x \in K_2(Z[\xi_{\ell^\nu}];Z/\ell^\nu)$ such that $\partial(x) = \xi_{\ell^\nu}$. In addition the image of x mod ℓ equals the image of β_1 under the homomorphism

$$K_2(Z[\xi];Z/\ell) \to K_2(Z[\xi_{\ell^\nu}];Z/\ell).$$

Hence the image of $\beta_\nu \in K_{2(\ell-1)\ell^{\nu-1}}(A,Z/\ell^\nu)$ in $K_{2(\ell-1)(\ell^{\nu-1})}(Z[\xi_{\ell^\nu}];Z/\ell^\nu)$ may be assumed to equal $x^{(\ell-1)\ell^{\nu-1}}$. Consequently if R contains $1/\ell$ and ξ_{ℓ^ν} we see that $K_i(R;Z/\ell^\nu)$ equal the direct limit of

$$K_i(R;Z/\ell^\nu) \xrightarrow{x^{\ell^{\nu-1}(\ell-1)}} K_{i+s}(R;Z/\ell^\nu) \xrightarrow{x^{\ell^{\nu-1}(\ell-1)}} \dots .$$

Therefore, in general if R contains a central ℓ^ν-th root of unity in the above sense (and if $\nu \geq 2$ when $\ell = 2$) we define $K_*(R;Z/\ell^\nu)$ to be the above direct limit which is isomorphic to the direct limit of

$$K_i(R;Z/\ell^\nu) \xrightarrow{(x.\_)} K_{i+2}(R;Z/\ell^\nu) \xrightarrow{(x.\_)} \dots$$

Notice that if R has ℓ^ν-th roots of unity then $K_i(R;Z/\ell^\nu) \cong K_{i+2}(R;Z/\ell^\nu)$.

Our first result gives an upper bound for $K_*(R;Z/\ell^\nu)$ in the following situation.

<u>1.1</u>: Let ℓ be an *odd* prime and let ξ denote a primitive ℓ-th root of unity. Let R be a ring in which ℓ is invertible then by $R \otimes_Z Z[\xi]$ we will mean, as is usual, the ring $R[t]/\Phi(t)$ where $\Phi(t) = 1 + t + \dots + t^{\ell-1}$. Hence the image of t, denoted by ξ, is a central ℓ-th root of unity in $R \otimes_Z Z[\xi]$. Let $(Z/\ell)^* = G$ act as a Galois group by $v(\xi) = z^v$ invariant subring $(R \otimes_Z Z[\xi])^G = R$. Also, as a (left) R-module $R \otimes_Z Z[\xi]$ is free with normal basis $\{v(\xi);\ v \in G\}$.

We will only use this ring extension when the ring R does not already have convenient ℓ-th roots of unity.

<u>1.2</u>: <u>EXAMPLES</u>

(a) R is a field of characteristic different from ℓ and $S = R(\xi)$, where ξ is a primitive ℓ-th root of unity.

(b) $R = \mathcal{O}[1/\ell]$ where $\mathcal{O}$ is a ring of algebraic integers in a local or global field and $S = \mathcal{O}[1/\ell,\xi]$.

(c) Group rings over the rings in example (b).

Choose t such that $t - 1 = \ell u$ with $(\ell,u) = 1$ and set $j_*(X)$ to be the generalised homology theory defined by the fibre sequence (ψ^t is the Adams operation)

$$\cdots \to j_\alpha(X) \to KU_\alpha(X;Z/\ell) \xrightarrow{\psi^t - 1} KU_\alpha(X;Z/\ell) \to \cdots .$$

With the above notation our first result is the following.

1.3: THEOREM

Let ℓ be an odd prime and let R be a commutative ring in which ℓ is invertible. Set $S = R \otimes_Z Z[\xi]$ and assume the units, S^*, have the form $\langle\xi\rangle \times H$.

Then there is an additive surjection, for each α,

$$J : j_\alpha(BSLS^+ \times BH)^G \twoheadrightarrow K_\alpha(R;Z/\ell).$$

Here $G \cong (Z/\ell)^*$ is Gal(S/R) and $BSLS^+$ is the fibre of det : $BGLS^+ \to BS^*$. Also $(\_)^G$ denotes the subgroup of G-invariants.

Let ν be a positive integer and assume $\nu \geq 2$ if $\ell = 2$. When R is a commutative ring whose units have the form $R^* \cong Z/\ell^a \times G$, $(\nu \leq a \leq \infty)$ define a generalised homology theory, h_*, as follows. If $a = \infty$ then $h_* = KU_*(\_;Z/\ell^\nu)$, mod ℓ^ν unitary K-theory. Otherwise choose a prime s so that $s - 1 = \ell^a u$ with $(u,\ell) = 1$ and define h_* by the fibre sequence

$$\cdots \to h_\alpha(X) \to KU_\alpha(X;Z/\ell^\nu) \xrightarrow{\psi^s - 1} KU_\alpha(X;Z/\ell^\nu) \to \cdots .$$

Under these circumstances it was proved in [Sn 1, §IV.3] that

1.4: THEOREM

With the above notation there is an additive surjection, for each α,

$$J : h_\alpha(BSLR \times BG_1) \twoheadrightarrow K_\alpha(R;Z/\ell^\nu).$$

Our second result improves these upper bounds for K_*. The graded groups $j_*(BSLS)^G$ and $h_*(BSLR)$ become graded rings when endowed with the product induced by direct sum of matrices.

1.5: THEOREM

In Theorems 1.3 and 1.4 the homomorphism, J, annihilates decomposables in the algebras $j_*(BSLS)^G$ and $h_*(BSLR)$ respectively.

These results are proved as follows. The homology theories h_* and j_* may be constructed by localising the mod ℓ^ν stable homotopy functor, $\pi_*^S((\_) \wedge (BZ/\ell^\nu)_+;Z/\ell^\nu)$ ($X_+ \equiv X$ disjoint union a base point), by inverting a "Bott element" in a manner analogous to that used to define K_*.

The epimorphism, J, is induced by the structure map [M1; M2] $D : Q(BGLR^+)_+ \to BGLR^+$ ($QX = \varinjlim_n \Omega^n\Sigma^n X$) of the infinite loopspace $BGLR^+$. The mod ℓ^ν homotopy of $Q(BGLR^+)_+$ is the stable homotopy ring, $\pi_*^S(BGLR^+)_+;Z/\ell^\nu)$, and J is the homomorphism $D_\#[1/\beta_\nu]$ on mod ℓ^ν homotopy. However, we have to use carefully chosen models for the spaces involved in order to obtain a model for $D_\#$ which commutes with multiplication by β_ν. These models are obtained, by means of the group completion theorem of Barratt-Priddy-Quillen, from the classifying spaces of certain matrix groups. The required properties of our models follow from a selection of diagrams of homomorphisms, given in the Appendix, which commute up to inner automorphisms.

Theorem 1.3 is proved in §2 and Theorem 1.5 in §3.

Applications of §§1.3-1.5 and examples are given in §4.

§2: CONSTRUCTION OF THE UPPER BOUND

Let ℓ be a prime.

As explained in [H-S, §3 pp. 22/23] one may form from BGL_nA and $B\Sigma_m\int GL_nA$ the "plussed" spaces $(BGL_nA)^+$ (if $n \geq 3$) and $(B\Sigma_m\int GL_nA)^+$ (if $m \geq 5$).

Let K_n denote $(BGL_nA)^+$ and $S_{m,n}$ denote $(B\Sigma_m\int GL_nA)^+$ for A a fixed commutative ring.

Define

$$K = \varinjlim_{(n,\ell)=1} K_n \quad \text{and}$$

$$S = \varinjlim_{(mn,\ell)=1} S_{m,n}$$

where the limits are taken over the maps induced by the diagonal group inclusions $\Delta_q : GL_n \to GL_{nq}$, $\Delta_q : \Sigma_m\int GL_n \to \Sigma_{mq}\int GL_n$ and $\Sigma_m\int\Delta_q : \Sigma_m\int GL_n \to \Sigma_m\int GL_{nq}$.

The spaces K and S will be our models for $BGLA^+$ and $Q(BGLA \cup pt.)$ respectively. In addition the diagrams A2/3 in the Appendix imply that the inclusion maps

$$\{j_{m,n} : \Sigma_m\int GL_n \to GL_{mn}\}$$

induce a map

$$J : S \to K.$$

2.1: LEMMA

(a) There are isomorphisms ($\nu \geq 1$ and $i \geq 2$)

$$\pi_i(K;Z/\ell^\nu) \cong K_i(A;Z/\ell^\nu)$$

$$\pi_i(S;Z/\ell^\nu) \cong \pi_i^S(BGLA_+;Z/\ell^\nu)$$

(X_+ denotes X disjoint union a base point).

(b) The homomorphism

$$J_\# : \pi_i(S;Z/\ell^\nu) \to \pi_i(K;Z/\ell^\nu)$$

is split surjective.

Proof: (a) Let $QX = \varinjlim_{n} \Omega^n\Sigma^n X$ then $Q(BGLA_+) \simeq QBGLA \times QS^0$ is a space having one component for each integer (corresponding to the degree of a map representing a point in the QS^0 factor). There is a group completion map $t_{m,n} : S_{m,n} \to Q(BGLA_+)$ which lands in the m-component. In addition there are homotopy commutation diagrams

$$\begin{array}{ccc} S_{m,n} & \xrightarrow{t_{m,n}} & Q(BGLA_+) \\ \Delta_q \downarrow & & \downarrow q \\ S_{mq,n} & \xrightarrow{t_{mq,n}} & Q(BGLA_+) \end{array}$$

and

$$\begin{array}{ccc} S_{m,n} & \xrightarrow{t_{m,n}} & Q(BGLA_+) \\ \Sigma_m\int\Delta_q \downarrow & & \downarrow Q(q \cup pt.) \\ S_{m,qn} & \xrightarrow{t_{m,qn}} & Q(BGLA_+). \end{array}$$

In these diagrams q denotes q-fold loop-sum and $Q(q \cup pt.)$ denotes the map induced by q-fold Whitney sum while Δ_q, $\Sigma_m\int\Delta_q$ are induced by the diagonal homomorphisms. Since q and $Q(q \cup pt.)$ induce isomorphisms on mod ℓ^ν homotopy if $(q,\ell) = 1$ these diagrams together with the group completion theorem of Barratt-Priddy-Quillen (c.f. [H-S]) gives isomorphisms

$$\begin{array}{ccc} \pi_i(S;Z/\ell^\nu) & \cong \varinjlim_{(mn,\ell)=1} & \pi_i(S_{m,n};Z/\ell^\nu) \\ \downarrow & & \downarrow \varinjlim(t_{m,n})_\# \\ \pi_i^S(BGLA_+;Z/\ell^\nu) & \cong \varinjlim_{(mn,\ell)=1} & \pi_i^S(BGLA_+;Z/\ell^\nu). \end{array}$$

The identification of $\pi_i(K;Z/\ell^\nu)$ follows in a similar manner from the diagrams

$$\begin{array}{ccc} K_m & \xrightarrow{i_m} & BGLA^+ \\ \Delta_q \downarrow & & \downarrow q \\ K_{qm} & \xrightarrow{i_{qm}} & BGLA^+. \end{array}$$

Diagrams A9/10 in the Appendix imply that $\Delta_q : GL_q \to \Sigma_q\int GL_n$ induces a map $I : K \to S$ such that J.I is a homotopy equivalence at the prime ℓ, which proves part (b).

The tensor product induces two maps, μ, for fixed u

$$(BGL_uA)^+ \times K \xrightarrow{\mu} K$$

and

$$(BGL_uA)^+ \times S \xrightarrow{\mu} S$$

by virtue of diagrams A4, A7/8 in the Appendix.

Since tensor product commutes with the group inclusions $\{j_{m,n}\}$ we obtain the following commutative diagram for each fixed $u \geq 1$.

$$\begin{array}{ccc} BGL_uA^+ \times S & \xrightarrow{\mu} & S \\ {\scriptstyle 1\times J}\downarrow & & \downarrow{\scriptstyle J} \\ BGL_uA^+ \times K & \xrightarrow{\mu} & K \end{array} \tag{2.2}$$

Following [Sn 1, §IV.3] we will use the map, μ, of (2.2) to construct multiplication on $K_*(A;Z/\ell^\nu)$ by the Bott elements of §1. We can then use (2.2) to obtain a map from the localised mod ℓ^ν homotopy of S to that of K-inverting a Bott element in each case. This map will be surjective because $J_\#$ is surjective before localisation, by §2.1(b).

2.3: PRODUCTS INDUCED BY μ

Firstly we consider the K-theory product. Let $i_t : K_t = BGL_tA^+ \to BGLA^+$ be the natural map. In [Lo, Ch. 11] is defined a family of maps (compatible up to homotopy) $\{\gamma_{s,t} : K_s \wedge K_t \to BGLA^+\}$ which induces the algebraic K-theory product.

Fix an integer u in the diagram (2.2). Suppose we have maps $x : P^n(\nu) \to K_u$ and $y : P^n(\nu) \to K_t$ where $P^n(\nu)$ is the Moore space $S^{n-1} \cup_{\ell^\nu} e^n$ $(n \geq 2)$. Let $\chi : P^{n+m}(\nu) \to P^n(\nu) \wedge P^n(\nu)$ denote the dual Moore space pairing [B] then xy is represented by $\gamma_{u,t}(x \wedge y)\chi$ in $[P^{n+m}(\nu), BGLA^+] = K_{n+m}(A;Z/\ell^\nu)$. Let $\pi : X \times Y \to X \wedge Y$ denote the map which collapses $X \vee Y$ to a point. Thus

$$\gamma_{u,t} \cdot \pi \in [K_u \times K_t, BGLA^+]$$

and, by definition [Lo, Ch. 11],

$$\gamma_{u,t} \cdot \pi = i_{ut} \cdot \mu - i_{ut}\Delta_t\pi_1 - i_{ut}\Delta_u\pi_2$$

where π_1, π_2 are the projection maps and the sum is formed using the H-space (direct sum) addition on $BGLA^+$. Since $BGLA^+$ is a loopspace the Hopf construction induces a map

$$H^* : [K_u \times K_t; BGLA^+] \to [K_u \wedge K_t; BGLA^+].$$

Explicitly, the Hopf construction gives $K_u \times K_t \xrightarrow{h} \Omega\Sigma(K_u \wedge K_t)$ and the loop structure gives a retraction $D : \Omega\Sigma BGLA^+ \to BGLA^+$ so $H^*[f] = [D(\Omega\Sigma f)h]$. Since $H^*(g\pi_i) = 0$ $(i = 1,2)$ we have in $[K_u \wedge K_t, BGLA^+]$

$$\gamma_{u,t} = H^*(\gamma_{u,t} \cdot \pi)$$
$$= H^*(i_{ut} \cdot \mu).$$

Therefore in $[P^{n+m}(\nu), BGLA^+]$

$$xy = H^*(i_{ut}\mu(x \times y))\chi.$$

Now let us translate this product over to the model for $K_i(A;Z/\ell^\nu)$ in terms of the space $K = \varinjlim_{(t,\ell)=1} K_t$.

If $P^m(\nu) \xrightarrow{y} K_t \to K$ is a map then the K-theory class represented by it in $K_m(A;Z/\ell^\nu) = [P^m(\nu), BGLA^+]$ is $[i_t y]/t$. Suppose that

$$a \in \pi_n(BGLA^+;Z/\ell^\nu) \text{ and}$$

$$b \in \pi_m(BGLA^+;Z/\ell^\nu) \text{ are represented by}$$

$P^n(\nu) \xrightarrow{x} K_u \to K$ and $P^m(\nu) \xrightarrow{y} K_t \to K$ respectively. Thus $[i_u x] = ua$, $[i_t y] = tb$ in $K_*(A;Z/\ell^\nu) = \pi_*(BGLA^+;Z/\ell^\nu)$. If $B = \varinjlim BGLA^+$, where the limit is taken over all maps $BGLA^+ \xrightarrow{q} BGLA^+$ with $(q,\ell) = 1$ then we have a homotopy equivalence $\varinjlim i_t : K \to B$.

Consider the following commutative diagram in which ϕ is an ℓ-local equivalence.

(2.4)

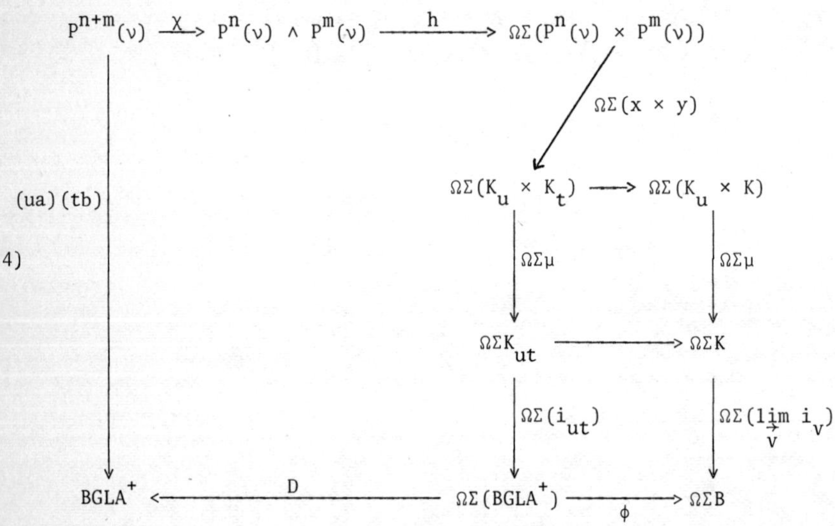

In this diagram, by the previous discussion, the inner composite represents $utab \in K_{n+m}(A;Z/\ell^\nu)$. However the outer composite is given in terms of the space K and the map μ of (2.2).

Suppose now that A is a commutative ring such that either (i) ℓ is invertible in A and

$$P^{2(\ell-1)\ell^{\nu-1}} \xrightarrow{x} K_u \to K$$

represents the Bott element $\beta_\nu \in K_{2(\ell-1)\ell^{\nu-1}}(A;Z/\ell^\nu) \cong \pi_{2(\ell-1)\ell^{\nu-1}}(K;Z/\ell^\nu)$ or (ii) A contains ξ_{ℓ^ν}, and ℓ^ν th root of unity and

$$P^2(\ell^\nu) \xrightarrow{x} K_u \to K$$

represents the Bott element $x \in K_2(A;Z/\ell^\nu) \cong \pi_2(K;Z/\ell^\nu)$. (We may take u = 1 here if we wish.)

Definition: For $y \in \pi_m(K;Z/\ell^\nu)$ define $x \# y \in \pi_{m+\deg(x)}(K;Z/\ell^\nu)$ by the composition (in the notational of (2.4))

$$x \# y = [D\phi^{-1}\Omega\Sigma((\lim_{v} i_v)\mu(x \times y))h\chi].$$

The content of §2.1 and (2.4) is the following.

2.5: PROPOSITION

With the above notation the #-product localisation $\pi_i(K;Z/\ell^\nu)[1/x]$ is isomorphic to $K_i(A;Z/\ell^\nu)$ as defined in §1.

Now set $Q = \varinjlim Q((BGLA^+)_+)$ where the limit is taken over all maps q and $Q(q \cup pt.)$ (as in §2.1(proof)) for $(q,\ell) = 1$. We have maps

$$(\Omega\Sigma\phi'' =)\phi' : \Omega\Sigma Q(BGLA^+)_+ \to \Omega\Sigma Q$$

and $D' : \Omega\Sigma Q(BGLA^+)_+ \to Q(BGLA^+)_+$ where ϕ' is a homotopy equivalence at the prime ℓ and D' comes from the loop-space structure on $QX = \Omega^\infty\Sigma^\infty X = \Omega Q\Sigma X$.

Definition: For $y \in \pi_m(S;Z/\ell^\nu)$ define $x \# y \in \pi_{m+\deg(x)}(S;Z/\ell^\nu)$ by the composition $x \# y = [D'(\phi')^{-1}\Omega\Sigma((\lim_{a,b} t_{a,b})\mu(x \times y))h\chi]$.

2.6: PROPOSITION

With the above notation the homomorphism $J_\#$ of §2.1 satisfies $J_\#(x \# y) = x \# J_\#(y)$ and hence induces an epimorphism

$$J_\# : \pi_*(S;Z/\ell^\nu)[1/x] \to \pi_*(K;Z/\ell^\nu)[1/x] \cong K_*(A;Z/\ell^\nu).$$

Proof: By §2.1(b) and §2.5 we have only to justify the first assertion, which follows at once from (2.2) and the following commutative diagrams.

$$\text{(I)}\qquad \begin{array}{ccccc} Q(BGLA^+)_+ & \xrightarrow{d} & QBGLA^+ & \xrightarrow{e} & BGLA^+ \\ \downarrow{\scriptstyle Q(q\,\cup\,pt.)} & & \downarrow{\scriptstyle Q(q)} & & \downarrow{\scriptstyle q} \\ Q(BGLA^+)_+ & \xrightarrow{d} & QBGLA^+ & \xrightarrow{e} & BGLA^+. \end{array}$$

$$\text{(II)}\qquad \begin{array}{ccc} Q(BGLA^+)_+ & \xrightarrow{ed} & BGLA^+ \\ {\scriptstyle q}\downarrow & & \downarrow{\scriptstyle q} \\ Q(BGLA^+)_+ & \xrightarrow{ed} & BGLA^+. \end{array}$$

$$\text{(III)}\qquad \begin{array}{ccc} \Omega\Sigma Q(BGLA^+)_+ & \xrightarrow{\Omega\Sigma(ed)} & \Omega\Sigma BGLA^+ \\ {\scriptstyle D'}\downarrow & & \downarrow{\scriptstyle D} \\ Q(BGLA^+)_+ & \xrightarrow{ed} & BGLA^+. \end{array}$$

In (I) and (II) e is the structure map of the (Whitney sum) infinite loopspace structure and d is induced by the map $(BGLA^+)_+ \to BGLA^+$ which identifies the disjoint base point with a base point in $BGLA^+$. (I) and (II) commute because q is an infinite loopmap. They are used to show that the following diagram commutes.

$$\begin{array}{ccc} S & \xrightarrow{J} & K \\ \downarrow & & \downarrow \\ \varinjlim Q(BGLA^+)_+ = Q & \xrightarrow[\varinjlim(ed)]{} & B = \varinjlim BGLA^+ \end{array}$$

(III) commutes for the same reason, which completes the proof.

2.7: PROOF OF THEOREM 1.3

We deduce this result from Theorem 1.4, which was proved in [Sn 1 §IV.3]. Thus we fix an odd prime ℓ and commutative rings R, $S = R \otimes Z[\xi]$ with $1/\ell \in R^*$.

We have a commutative diagram

$$\begin{array}{ccc} \pi_*(S(R);Z/\ell)[1/\beta_1] & \xrightarrow{\alpha_1} & \pi_*(S(S);Z/\ell)[1/\beta_1] \\ \downarrow{\scriptstyle J_\#} & & \downarrow{\scriptstyle J_\#} \\ \pi_*(K(R);Z/\ell)[1/\beta_1] & \xrightarrow{\alpha_2} & \pi_*(K(S);Z/\ell)[1/\beta_1] \\ \downarrow{\scriptstyle \cong} & & {\scriptstyle \cong}\downarrow \\ K_*(R;Z/\ell) & \xrightarrow{\alpha_3} & K_*(S;Z/\ell) \end{array} \qquad (2.8)$$

in which $K(R)$, $S(R)$ etc. are the K- and S-spaces associated to the ring R.

If X = S(S) or K(S) diagrams A4/6 of the Appendix give the following commutative diagrams

$$\begin{array}{ccc} K_q(S) \times X & \xrightarrow{\mu} & X \\ \Delta_u \times 1 \downarrow & & \| \\ K_{qu}(S) \times X & \xrightarrow{\mu} & X \end{array}$$

Hence #-multiplication by $\beta_1 \in \pi_{2(\ell-1)}(K_u(S);Z/\ell)$ and by the Bott element $x \in \pi_2(K_1(S);Z/\ell)$ give rise to the same localisation. We have, by definition of β_1 in §1, up to multiplication by units (for $u >> 0$)

$$\Delta_{u_\#}(x^{(\ell-1)}) = q^{\ell-1}\beta_1 + (\text{nilpotents}).$$

Hence

$$\begin{aligned}\pi_*(S(S);Z/\ell)[1/\beta_1] &\cong \pi_*(S(S);Z/\ell)[1/x] \\ &\cong j_*(BSLS \times BH)\end{aligned}$$

by [Sn 1, §IV.3]. The map α_3 has image equal to $K_*(S;Z/\ell)^G$ by a simple transfer argument, since $(|G|,\ell) = 1$, [B; D-F-S-T]. Clearly the image of α_1 consists of G-invariant elements and $J_\#$ commutes with the Galois action of G so the result follows from (2.8).

§3: THE UPPER BOUND HOMOMORPHISM ON DECOMPOSABLES

In this section we prove Theorem 1.5, concerning $J_\#(xy)$ when x,y belong either to $j_*(BSLS)^G$ or $h_*(BSLR)$, in the notation of Theorems 1.3/1.4. Actually, in the second case, when R has ℓ^ν-th roots of unity we will actually give a formula for $J_\#(xy)$ in $K_*(R;Z/\ell^\nu)$. This is stronger than the statement of Theorem 1.5.

Firstly we remark that it is sufficient to treat the case when R contains ℓ^ν-th roots of unity. For in the other case (namely that of §1.3) let $S = R \otimes_Z Z[\xi]$ as in §2.7 and consider diagram (2.8). From [B] α_2 (and α_3) is an isomorphism onto the Galois invariants and α_1 is multiplicative so that $\alpha_2(J_\#(xy)) = 0$ if and only if $J_\#(xy) = 0$ and

$$\alpha_2 J_\#(xy) = J_\#(\alpha_1(xy)) = J_\#((\alpha_1 x)(\alpha_1 y)) = 0.$$

Therefore, for the rest of this section, we will assume that R is a commutative ring having an ℓ^a-th root of unity ($\nu \leq a \leq \infty$) and that $R^* \cong Z/\ell^a \times G_1$ as in §1.4. Since we are dealing with $J_\#$ we continue to use the notation developed in §2. We also use the notation of §A.12 for certain maps.

Suppose that $a,b \in K_i(R;Z/\ell^\nu)$ are represented in terms of the space K by maps

$$q_a : P^i(\nu) \to K_n \to K$$

and

$$q_b : P^i(\nu) \to K_m \to K$$

respectively.

If $(n+m,\ell) = 1$ we have a commutative diagram

$$\begin{array}{ccccc} & & BGLR^+ \times BGLR^+ & \xrightarrow{\oplus} & BGLR^+ \\ & \nearrow^{na \times mb} & \uparrow{\scriptstyle i_n \times i_m} & & \uparrow{\scriptstyle i_{n+m}} \\ P^i(\nu) \xrightarrow{\Delta} P^i(\nu) \times P^i(\nu) & & & & \\ & \searrow_{q_a \times q_b} & & & \\ & & K_n \times K_m & \xrightarrow{\oplus} & K_{n+m} \end{array}$$

which shows that following fact.

(3.1) $\begin{cases} \text{The map } \oplus(q_a \times q_b)\Delta \text{ represents } \dfrac{na+mb}{n+m} \in K_i(R;Z/\ell^\nu) \\ \text{in terms of the space K.} \end{cases}$

Let $z \in \pi_i^S((BGLR)_+;Z/\ell^\nu) \cong \pi_i(S;Z/\ell^\nu)$ and let $w \in \pi_i^S((BGLR)_+;Z/\ell^\nu)$. Let u,v be integers such that $(uv(u+v),\ell) = 1$ and such that, in terms of S, these classes are represented by maps $\hat{f},\hat{g}$ in the homotopy commutative diagrams below.

$$\begin{array}{ccccccc} & & P^i(\nu) & & \xrightarrow{\quad z \quad} & & Q(BGLR_+) \\ & \swarrow^{f} & & \searrow^{\hat{f}} & & & \downarrow{\scriptstyle uQ(u \cup pt.)} \\ BH^+_{u,u} & \xrightarrow{\alpha_u} & BH^+_{u,u} & \to & S_{u,u} & \xrightarrow{t_{u,u}} & Q(BGLR_+) \\ & & & & & \searrow & \downarrow \\ & & & & & S & \xrightarrow[\underset{a,b}{\underrightarrow{\lim}}\, t_{a,b}]{} Q \end{array}$$

$$\begin{array}{ccccccc} & & P^j(\nu) & & \xrightarrow{\quad w \quad} & & Q(BGLR_+) \\ & \swarrow^{g} & & \searrow^{\hat{g}} & & & \downarrow{\scriptstyle vQ(v \cup pt.)} \\ BH^+_{v,v} & \xrightarrow{\alpha_v} & BH^+_{v,v} & \to & S_{v,v} & \xrightarrow{t_{v,v}} & Q(BGLR_+) \\ & & & & & \searrow & \downarrow \\ & & & & & S & \xrightarrow[\underset{a,b}{\underrightarrow{\lim}}\, t_{a,b}]{} Q \end{array}$$

Note that f,g exist because α_u, α_v are ℓ-equivalences as are the maps $BH^+_{u,u} \to S_{u,u}$, $BH^+_{v,v} \to S_{v,v}$.

Since the product, denoted by *, in

$$h_*(BSLR \times BG_1) \cong \pi_*(S;Z/\ell^\nu)[1/x]$$

is induced by direct sum ($\oplus$) on SLR and by the product on R*, these operations give a pairing on $Q(BGLR_+)$ (equivalent at ℓ to $Q((BSLR \times BA^*)_+)$) which fits into the following diagram (c.f. [Ma 2]).

$$\begin{array}{ccc} BH^+_{u,u} \times BH^+_{v,v} & \xrightarrow{t_{u,u} \times t_{v,v}} & Q((BSLR \times BA^*)_+)^2 \\ \downarrow{\scriptstyle \Phi} & & \downarrow \\ B\Sigma_{uv} \int GL_{u+v} R^+ & \xrightarrow{t_{uv,u+v}} & Q(BGLR_+) \end{array} .$$

Therefore $J_\#(z * w)$ is represented (compare the products in §2) by the following composition-taking the left-hand route.

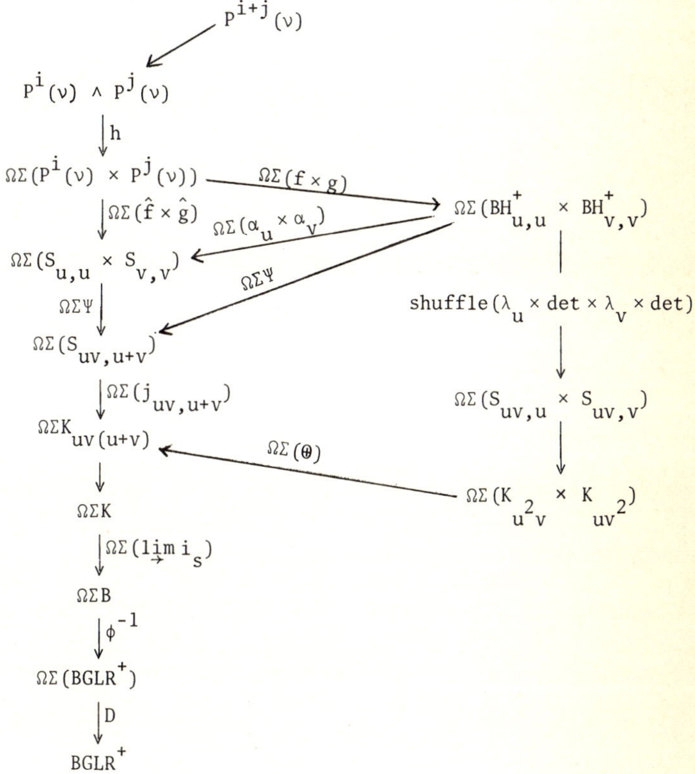

However if det denotes the determinant map $1\int\det : S_{n,m} \to S_{n,1}$ the right-hand route is (in view of previous remarks about $\oplus : K_n \times K_m \to K_{n+m}$) equal to

$$\frac{u^2vJ_{\#}([\lambda_u f] \cdot [\det g]) + (-1)^{ij}uv^2J_{\#}([\det f][\lambda_v g])}{u^2v + uv^2},$$

where $[\det f]$, $[\det g] \in \pi_*(B(\Sigma_\infty\int R^*)^+;Z/\ell^\nu)[1/x] \cong h_*(pt.)$ and $(\_.\_)$ is the $h_*(pt.)$-module structure on $h_*(X)$. Hence we have shown the following.

3.2: PROPOSITION

With the above notation

$$J_{\#}(z * w) = \frac{1}{(u+v)}\{J_{\#}([\lambda_u f][\det g] + (-1)^{ij}J_{\#}([\det f][\lambda_v g])\}.$$

There is a commutative diagram

$$\begin{array}{ccc} h_*(BSLR \times BG_1) & \xrightarrow{\varepsilon} & h_*(pt.) \\ \cong\downarrow & & \cong\downarrow \\ \pi_*(S;Z/\ell^\nu)[1/x] & \xrightarrow[\det_{\#}]{} & \pi_*((B\Sigma_\infty\int R^*)^+;Z/\ell^\nu)[1/x] \end{array}$$

in which ε is the usual augmentation homomorphism. Hence we obtain from §3.2 the following version of Theorem 1.5.

3.3: THEOREM

If $z,w \in \ker\{\varepsilon : h_*(BSLR \times BG_1) \to h_*(pt.)\}$ then $J_{\#}(z * w) = 0$.

Proof: In §3.2 $[\det f]$ and $[\det g]$ represent (up to automorphisms) $\varepsilon(z)$ and $\varepsilon(w)$ respectively.

§4: APPLICATIONS

When R or S in Theorems 1.3/1.4 are regular rings satisfying the conditions of [D-F-S-T; Th. 2] the groups $K_*(R;Z/\ell^\nu)$ and $K_*(S;Z/\ell^\nu)$ are known. For example, in §1.2(b).

$$K_1(\mathcal{O}[1/\ell,\xi];Z/\ell) \cong \begin{cases} (\mathrm{Pic} \otimes Z/\ell) \oplus (\bigoplus_1^k Z/\ell) \\ \\ \mathrm{Tor}(Z/\ell,\mathrm{Pic}) \oplus S^*/(S^*)^\ell \end{cases}$$

where $S = \mathcal{O}[1/\ell,\xi]$, Pic denotes the ideal class group of S and k is the number of primes in $\mathcal{O}[\xi]$ over ℓ.

Hence for such regular rings (a wide class, as explained in [D-F-S-T]) §§1.3-1.5 may be useful in detecting $h_*(BSLR)$ and $j_*(BSLS)$ mod decomposables.

However, when R,S are not regular then §§1.3/1.5 are the best available information on what to expect from $K_*(R;Z/\ell^\nu)$ and $K_*(S;Z/\ell^\nu)$.

4.1: EXAMPLE

When R is a finite commutative ring in which ℓ is invertible then $S = R[\xi]$ is also finite and $j_*(BSLS \times BH)$ of §1.3 is describeable in terms of representations. For j_* is determined by $KU_*(BSLS \times BH;Z/\ell)$ and the Adams operation, ψ^t. However (see [Sn 1, §IV.3.26]) in this case

$$KU_i(BSLS \times BH;Z/\ell) = \begin{cases} \varinjlim_n \mathrm{Hom}_c(R(SL_nS \times H);Z/\ell) & \text{if } i \text{ even} \\ 0 & \text{if } i \text{ odd} \end{cases}$$

where Hom_c denotes those additive homomorphisms on the complex representation rings $R(SL_nS \times H)$ which annihilate some power of the augmentation ideal.

APPENDIX

Let GL_n denote GL_nA throughout this section and let Σ_m denote the m-th symmetric group. $\Sigma_m \int GL_n$ will denote the wreath product obtained by letting Σ_m permute the factors of $(GL_n)^m$. Hence there is an inclusion

$$j_{m,n} : \Sigma_m \int GL_n \to GL_{mn}.$$

The diagrams of group homomorphisms which are listed in this section all commute up to an inner automorphism of the receiving group. Verifications will generally be left to the reader.

Let Δ_q denote the q-fold diagonal $\Delta_q : GL_n \to GL_{nq}$ sending M to

$$\Delta_q(M) = \begin{pmatrix} M & & & & \\ & M & & & \\ & & \cdot & & \\ & & & \cdot & \\ & & & & \cdot \\ & & & & & M \end{pmatrix}.$$

Also Δ_q will denote the restriction of this homomorphism to a subgroup of GL_n.

Let $\otimes$ denote tensor product of matrices.

A1:

$$\begin{array}{ccc}
GL_\ell \times \Sigma_m\int GL_n & \xrightarrow{\;\otimes\;} & \Sigma_m\int GL_{\ell n} \\
\Big\downarrow{\scriptstyle 1 \times j_{m,n}} & & \Big\downarrow{\scriptstyle j_{m,\ell n}} \\
GL_\ell \times GL_{mn} & \xrightarrow{\;\otimes\;} & GL_{\ell mn}
\end{array}$$

A2:

$$\begin{array}{ccc}
\Sigma_m\int GL_n & \xrightarrow{\;\Delta_\ell\;} & \Sigma_{m\ell}\int GL_n \\
\Big\downarrow{\scriptstyle j_{m,n}} & & \Big\downarrow{\scriptstyle j_{m\ell,n}} \\
GL_{mn} & \xrightarrow{\;\Delta_\ell\;} & GL_{\ell mn}
\end{array}$$

A3:

$$\begin{array}{ccc}
\Sigma_m\int GL_n & \xrightarrow{\;\Sigma_m\int\Delta_q\;} & \Sigma_m\int GL_{nq} \\
\Big\downarrow{\scriptstyle j_{m,n}} & & \Big\downarrow{\scriptstyle j_{m,nq}} \\
GL_{mn} & \xrightarrow{\;\Delta_q\;} & GL_{mnq}
\end{array}$$

The inner automorphism which makes this diagram commute is τ, the permutation of the mq $n \times n$ blocks according to the formula $(0 \le j \le q-1;\ 1 \le k \le n)$

$$\tau(jm+k) = (k-1)q + j + 1.$$

A4:

$$\begin{array}{ccc}
GL_n \times GL_m & \xrightarrow{\;\otimes\;} & GL_{mn} \\
\Big\downarrow{\scriptstyle \Delta_q \times 1} & & \Big\downarrow{\scriptstyle \Delta_q} \\
GL_{nq} \times GL_m & \xrightarrow{\;\otimes\;} & GL_{mnq}
\end{array}$$

A5:

$$\begin{array}{ccc}
GL_\ell \times \Sigma_m\int GL_n & \xrightarrow{\;\otimes\;} & \Sigma_m\int GL_{\ell n} \\
\Big\downarrow{\scriptstyle \Delta_q \times 1} & & \Big\downarrow{\scriptstyle \Sigma_m\int\Delta_q} \\
GL_{\ell q} \times \Sigma_m\int GL_n & \xrightarrow{\;\otimes\;} & \Sigma_m\int GL_{\ell nq}
\end{array}$$

A6:

$$\begin{array}{ccc}
GL_\ell \times \Sigma_m\int GL_n & \xrightarrow{\;\otimes\;} & \Sigma_m\int GL_{\ell n} \\
 & & \Big\downarrow{\scriptstyle \Delta_q} \\
\Big\downarrow{\scriptstyle \Delta_q \times \Delta_q} & & \Sigma_{mq}\int GL_{\ell n} \\
 & & \Big\downarrow{\scriptstyle \Sigma_{mq}\int\Delta_q} \\
GL_{\ell q} \times \Sigma_{mq}\int GL_n & \xrightarrow{\;\otimes\;} & \Sigma_{mq}\int GL_{\ell nq}
\end{array}$$

<u>A7</u>:

$$\begin{array}{ccc}
GL_{\ell} \times \Sigma_m \int GL_n & \xrightarrow{\otimes} & \Sigma_m \int GL_{\ell n} \\
{\scriptstyle 1 \times \Delta_q}\downarrow & & \downarrow{\scriptstyle \Delta_q} \\
GL_{\ell} \times \Sigma_{mq} \int GL_n & \xrightarrow{\otimes} & \Sigma_{mq} \int GL_{\ell n}
\end{array}$$

<u>A8</u>:

$$\begin{array}{ccc}
GL_{\ell} \times \Sigma_m \int GL_n & \xrightarrow{\otimes} & \Sigma_m \int GL_{\ell n} \\
{\scriptstyle 1 \times \Sigma_m \int \Delta_q}\downarrow & & \downarrow{\scriptstyle \Sigma_m \int \Delta_q} \\
GL_{\ell} \times \Sigma_m \int GL_{nq} & \xrightarrow{\otimes} & \Sigma_m \int GL_{\ell nq}
\end{array}$$

<u>A9</u>:

$$\begin{array}{ccc}
GL_n = \Sigma_1 \int GL_n & \xrightarrow{\Delta_q} & \Sigma_q \int GL_n \\
{\scriptstyle \Delta_u}\downarrow & & \downarrow{\scriptstyle \Sigma_q \int \Delta_u} \\
GL_{nu} & \xrightarrow{\Delta_q} & \Sigma_q \int GL_{nu}
\end{array}$$

<u>A10</u>:

$$\begin{array}{ccc}
GL_n & \xrightarrow{\Delta_q} & \Sigma_q \int GL_n \\
& {\scriptstyle \Delta_{qu}}\searrow & \downarrow{\scriptstyle \Delta_u} \\
& & \Sigma_{qu} \int GL_n
\end{array}$$

<u>A11</u>: The final diagram is more elaborate and requires the setting up of some notation. Let $\psi : \Sigma_u \times \Sigma_v \to \Sigma_{uv}$ denote the homomorphism induced by letting $\sigma \in \Sigma_u$ and $\tau \in \Sigma_v$ act on the set of (lexicographical ordered) ordered pairs

$$\{(1,1),(1,2),\ldots,(1,v),(2,1),\ldots,(2,v),\ldots,(u,v)\}$$

by $(\sigma,\tau)(i,j) = (\sigma(i),\tau(j))$. Let $\Delta_n A^*$ denote the subgroup of $GL_n A$ consisting of multiples of the identity matrix. Hence $\Delta_n A^* \times SL_n A$ maps homomorphically to $GL_n A$ by sending (xI_n,M) to $x^u M$. Call this homomorphism λ_u. Denote by $H_{u,m}$ the group $\Sigma_u \int \Delta_m A^* \times SL_m A$ and let $\det(xI_m,M) = x^m$. Set $\alpha_u(xI_m,M) = (x^u I_m,M)$ and finally define

$$\Psi : H_{u,m} \times H_{v,n} \to \Sigma_{uv} \int GL_{m+n}$$

by setting

$$\Psi \left\{ \sigma \begin{array}{c} (x_1 I_m, M_1) \\ \vdots \\ (x_u I_m, M_u) \end{array}, \tau, \begin{array}{c} (y_1 I_n, N_1) \\ \vdots \\ (y_v I_n, N_v) \end{array} \right\}$$

equal to

$$\psi(\sigma,\tau)\left\{\begin{matrix} \begin{pmatrix} x_1y_1M_1 & 0 \\ 0 & x_1y_1N_1 \end{pmatrix} & & & & \\ & \cdot & & & \\ & & \cdot & & \\ & & & \cdot & \\ & & & & \begin{pmatrix} x_1y_vM_1 & 0 \\ 0 & x_1y_vN_v \end{pmatrix} \\ & & & & & \cdot \\ & & & & & & \cdot \\ & & & & & & & \cdot \\ & & & & & & & & \cdot \end{matrix}\right\}.$$

A12:

$$\begin{array}{ccc} H_{u,u}\times H_{v,v} & \xrightarrow{\alpha_u\times\alpha_v} & H_{u,u}\times H_{v,v} \\ \Big\downarrow{\scriptstyle (\mathrm{shuffle})(\lambda_u\times 1\int\det\times\lambda_v\times 1\int\det)} & & \Big\downarrow{\scriptstyle\Psi} \\ \Sigma_u\int GL_u\times\Sigma_v\int A^*\times\Sigma_v\int GL_v\times\Sigma_u\int A^* & & \Sigma_{uv}\int GL_{u+v} \\ \Big\downarrow{\scriptstyle\otimes\times\otimes} & & \Big\downarrow{\scriptstyle j_{uv,u+v}} \\ \Sigma_{uv}\int GL_u\times\Sigma_{uv}\int GL_v & \xrightarrow{\oplus} & GL_{uv(u+v)} \\ & {\scriptstyle j\times j}\searrow \qquad \nearrow{\scriptstyle\oplus} & \\ & GL_{u^2v}\times GL_{uv^2} & \end{array}$$

The explanation of this diagram is as follows. Let

$$x = \begin{pmatrix} x_1I & & \\ & x_2I & \\ & & \ddots \end{pmatrix},\quad y = \begin{pmatrix} y_1I & & \\ & y_2I & \\ & & \ddots \end{pmatrix},\quad M = \begin{pmatrix} M_1 & & \\ & \ddots & \\ & & M_u \end{pmatrix} \text{ and } N = \begin{pmatrix} N_1 & & \\ & \ddots & \\ & & N_v \end{pmatrix}.$$

Via the anti-clockwise route (σ,x,M,τ,y,N) goes first to

$$\left(\sigma\begin{pmatrix} x_1^uM_1 & & \\ & x_2^uM_2 & \\ & & \ddots \end{pmatrix},\tau\begin{pmatrix} y_1^vI & & \\ & y_2^vI & \\ & & \ddots \end{pmatrix},\tau\begin{pmatrix} y_1^vN_1 & & \\ & \ddots & \\ & & \ddots \end{pmatrix},\sigma\begin{pmatrix} x_1^uI & & \\ & x_2^uI & \\ & & \ddots \end{pmatrix}\right)$$

and then to

$$\left(\psi(\sigma\tau)\begin{pmatrix} x_1^u y_1^v M_1 & & & & \\ & x_1^u y_2^v M_1 & & & \\ & & \ddots & & \\ & & & x_2^u y_1^v M_2 & & \\ & & & & x_2^u y_2^v M_2 & \\ & & & & & \ddots \end{pmatrix} ; \psi(\sigma\tau)\begin{pmatrix} x_1^u y_1^v N_1 & & \\ & x_1^u y_2^v N_2 & \\ & & \ddots \end{pmatrix}\right)$$

and finally to the direct sum of these two matrices. However, via the clockwise route this element goes to $(\sigma, x^n, M, \tau, y^v, N)$ and then to

$$\left(\psi(\sigma\tau)\begin{pmatrix} x_1^u y_1^v M_1 & 0 & & & \\ 0 & x_1^u y_1^v N_1 & & & \\ & & x_1^u y_2^v M_1 & 0 & \\ & & 0 & x_1^u y_2^v N_2 & \\ & & & & \ddots \end{pmatrix}\right).$$

Then two routes clearly differ by conjugation by a permutation matrix.

REFERENCES

[B] W. Browder, Algebraic K-theory with coefficients Z/p, Springer-Verlag Lecture Notes in Mathematics, #654(1978) 40-84.

[D-F] W.G. Dwyer and E.M. Friedlander, Algebraic K-theory and arithmetic, Bull. Amer. Math. Soc. 6(3), 1982, 453-55.

[D-F-S-T] W. Dwyer, E.M. Friedlander, V. Snaith and R.W. Thomason, Algebraic K-theory eventually surjects onto topological K-theory, to appear in Inventiones Math.

[G-Q] D. Grayson (after D.G. Quillen), Higher algebraic K-theory II, Lecture Notes in Mathematics, #551, 217-240, Springer-Verlag.

[H-S] B. Harris and G.B. Segal, K_i groups of rings of algebraic integers, Annals of Math. (1)101(1975) 20-33.

[Lo] J-L. Loday, K-théorie algébrique et representations de groupes, An. Sci. Ec. Norm. Sup. 4e serie t.9 (1976) 309-377.

[Ma 1] J.P. May, Geometry of iterated loopspaces, Lecture Notes in Mathematics #271, Springer-Verlag.

[Ma 2] J.P. May (with contributions by Frank Quinn, Nigel Ray and Jørgen Tornehave), E_∞-ring spaces and E_∞-ring spectra, Lecture Notes in Mathematics, #577, Springer-Verlag.

[Q] D.G. Quillen, Higher algebraic K-theory I, Lecture Notes in Mathematics, #341, 77-139, Springer-Verlag.

[Q 2] D.G. Quillen, On the cohomology and K-theory of the general linear groups over a finite field, Annals of Math. 96(1972) 552-586.

[Sn 1] V.P. Snaith, Algebraic K-theory and localised stable homotopy, to appear Mem. Amer. Math. Soc.

[Sn 2] V.P. Snaith, Algebraic cobordism and K-theory, Mem. Amer. Math. Soc., #221 (1979).

[Sn 3] V.P. Snaith, A brief survey of Bott periodic algebraic K-theory, to appear in "Current Trends in Algebraic Topology", Can. Math. Soc. Conf. Proceedings Series.

(Sn 4] V.P. Snaith, A topological "proof" of a theorem of Ribet, ibid.

[Th 1] R.W. Thomason, Algebraic K-theory and étale cohomology, preprint, M.I.T. (1980).

[Th 2] R.W. Thomason, The Lichtenbaum-Quillen conjecture for $K/\ell_*[1/\beta]$, to appear in "Current Trends in Algebraic Topology", Can. Math. Soc. Conf. Proceedings Series.

DEPARTMENT OF MATHEMATICS
THE UNIVERSITY OF WESTERN ONTARIO
LONDON, ONTARIO, N6A 5B7
CANADA

TOWARDS THE LICHTENBAUM-QUILLEN CONJECTURE

CONCERNING THE ALGEBRAIC K-THEORY OF SCHEMES

V.P. Snaith
University of Western Ontario
London, Ontario, CANADA

§1: I will sketch the steps in my programme to solve the Lichtenbaum-Quillen conjecture. At the time of the Bielefeld conference I was confident of the completion of the last step, but, as the title reveals, things have not yet gone according to plan. However, as we shall see, my arguments give a reduction to an equivalent problem concerning Hurewicz images in $KU_*(BGLA;Z/\ell)$ or $j_*(BGLA;Z/\ell)$ (where j_*-theory will be defined below).

I will describe a series of commutative diagrams which involve j-theory and K-theory Hurewicz maps. These diagrams - the first of which appeared in [Sn2] and suggested this entire programme - enable us to characterise the kernel of the map from algebraic K-theory to étale K-theory; the map about which the Lichtenbaum-Quillen conjecture speculates.

Invaluable in this work has been the following:

(i) the work of Eric Friedlander, who reformulated the conjecture in terms of his étale K-theory [F] and later, with Bill Dwyer [D-F] extended Soulé's work [Sou], on rings of algebraic integers, in terms of étale K-theory.

(ii) the work of Chuck Weibel [W] which enables one to work with Karoubi-Villamayor K-theory and to use his work with Barry Dayton [D-W] to reduce to dimension two.

(iii) the work of Bob Thomason - the sine qua non - [Th1,Th2] which permits one to replace étale K-theory by "Bott periodic" algebraic K-theory (which I introduced in [Sn] and [Sn2]).

To all of them I would like to express my gratitude for conversations, correspondence and for keeping me informed of their work.

§2: The Lichtenbaum-Quillen conjecture asserts that, if ℓ is a prime, the mod ℓ^ν algebraic K-groups of a ring, A, or scheme, X, satisfying Condition (0.1) of [Th1] are determined by its étale cohomology, at least in high dimensions. Condition (0.1) referred to above is the following: X should be (i) a regular scheme, quasi projective over a noetherian ring, S, which has finite Krull dimension (ii) all residue fields of X should have uniformly bounded étale cohomological dimension for ℓ-torsion sheaves where ℓ is a prime which is invertible in S. Such schemes are quite common place and include the following examples.

2.1: Examples Let ℓ be an odd prime.

(a) Let A denote either a ring of integers in a function field of characteristic q ($\neq$ 0 or ℓ) or $\mathcal{O}[1/\ell]$ where $\mathcal{O}$ is a ring of integers in a number field.

For technical reasons connected with the identification of étale and Bott periodic algebraic K-theory we assume in addition that if $\ell = 3$ A has a 9-th root of unity.

(b) Let A be the quotient field of a ring from example (a).

(c) Let X be a regular, quasi projective scheme over an algebraically closed field containing $1/\ell$.

(d) A separably closed field of characteristic not ℓ or more generally a strictly local Hensel ring containing $1/\ell$.

2.2: The simplest and most specific formulation of the Lichtenbaum-Quillen conjecture asserts that, for schemes, X, as above, the map from algebraic K-theory to étale K-theory [F,§4;Sn1,§1.3]

$$\rho : K_i(X;Z/\ell^\nu) \to K_i^{et}(X;Z/\ell^\nu) \tag{2.3}$$

is an isomorphism when i is sufficiently large.

From [D-F-S-T] and [Th1] we know that (2.3) is eventually onto in the case where $K_i^{et}(X;Z/\ell^\nu)$ are finite for each i. This is true for the examples in §2.1 (a)-(d). Hence the conjecture centres around the injectivity of ρ. To study this we appeal to [Th1] to give a different description of ρ in (2.3). For this we must introduce "Bott periodic" algebraic K-theory [Sn,Ch.IV]. Since $1/\ell \in S$, $K_*(X;Z/\ell^\nu)$ is a module over $K_*(Z[1/\ell];Z/\ell^\nu)$. As described in [D-F-S-T] one may choose a "Bott element" $\beta_\nu \in K_{2\ell^{\nu-1}(\ell-1)}(Z[1/\ell];Z/\ell^\nu)$ and form the localisation, $K_*(X;Z/\ell^\nu)[1/\beta^\nu]$, which will be denoted by $\mathcal{K}_*(X;Z/\ell^\nu)$. The homomorphism of (2.3) induces

$$\rho : \mathcal{K}_*(X;Z/\ell^\nu) \to K_*^{et}(X;Z/\ell^\nu) \tag{2.4}$$

2.5: Theorem [Th1]

For X as above (2.4) is an isomorphism (if $\ell = 3$ we assume S has a 9-th root of unity and if $\ell = 2$ that S has a 16-th root).

Theorem 2.5 identifies (2.3) with the natural map - also denoted by ρ -

$$\rho : K_i(X;Z/\ell^\nu) \to \mathcal{K}_i(X;Z/\ell^\nu). \tag{2.6}$$

2.7: At this point we make two observations.

Firstly, if we are trying to show that (2.6) is an isomorphism for all sufficiently large i then it suffices to study the case $\nu = 1$, by virtue of the coefficient sequences associated to the sequences $Z/\ell^{\nu-1} \to Z/\ell^\nu \to Z/\ell$.

Secondly, by a device of Jouanolou [J] there exists a morphism π : Spec A $\to$ X which induces isomorphisms on $K_*(\_;Z/\ell^\nu)$, $\mathcal{K}_*(\_;Z/\ell^\nu)$ and $K_*^{et}(\_;Z/\ell^\nu)$. Also if X satisfies Condition (0.1) of [Th1] so does Spec A. Hence we may concentrate on the affine case, where we abbreviate $K_*(\text{Spec } A;Z/\ell^\nu)$ to $K_*(A;Z/\ell^\nu)$, and similarly with K_*^{et} and $\mathcal{K}_*$.

Actually we will sometimes assume that A has the property that its units have the form $A^* \cong Z/\ell^a \times H$, (i.e. the ℓ-primary roots of unity split off from A^*). When $a = \infty$ this is automatic but if $1 \le a < \infty$ we will assume that $S^* \cong Z/\ell^a \times H$ and that $X \to S$ is split. These conditions will ensure that the ring A, given by Jouanolou's device, has units of the desired form.

Given a commutative ring, A, whose units have the form $A^* \cong Z/\ell^a \times H$ $(1 \le a \le \infty)$

we choose a prime, t, such that $\nu_\ell(t-1) = a < \infty$ and let $j_*(Y;Z/\ell^\nu)$ denote the J-theory (with mod ℓ^ν coefficients) given by the fibre of $\psi^t-1 : (KUZ/\ell^\nu)_* \to (KUZ/\ell^\nu)_*$. Here ψ^t is the K-theory Adams operation and $(KUZ/\ell^\nu)_*$ is Z/2-graded. Hence for any space Y there is an exact sequence

$$\ldots \to KU_{\alpha+1}(Y;Z/\ell^\nu) \to j_\alpha(Y;Z/\ell^\nu) \to KU_\alpha(Y;Z/\ell^\nu) \xrightarrow{\psi^t-1} \ldots$$

When a = ∞ we replace J-theory by $KU_*(Y;Z/\ell^\nu)$.

For A as above let $BSLA^+ =$ (fibre : $BGLA^+ \to BA^*$), the fibre of the determinant map.

2.8: Theorem Let ℓ be a prime and ν an integer (and if $\ell = 2$ assume $\nu \geq 2$). For A as above let h_* denote $j_*(\_;Z/\ell^\nu)$ if $1 \leq a < \infty$ and $KU_*(\_;Z/\ell^\nu)$ if $a = \infty$. Then there is a commutative diagram in which H is the Hurewicz map, for all $i \geq 2$.

$$\begin{array}{ccc} \pi_i(BSLA^+ \times BH;Z/\ell^\nu) & \xrightarrow{H} & h_i(BSLA^+ \times BH) \\ \downarrow & & \downarrow J \\ K_i(A;Z/\ell^\nu) & \xrightarrow{\rho} & K_i(A;Z/\ell^\nu) \end{array}$$

Theorem 2.8 is proved in [Sn2, §IV.3] using localised stable homotopy to describe $h_*(BSLA^+ \times BH)$. An improvement of that result is to be found in [Sn4].

A simple example of Theorem 2.8 is the case when A is a $\overline{\mathbb{F}}_q$-algebra, where $(q,\ell) = 1$. For then we have an associative pairing $m : BGLA^+ \wedge BGL\overline{\mathbb{F}}_q^+ \to BGLA^+$. Since $BU \simeq BGL\overline{\mathbb{F}}_q^+$ at the prime ℓ, m induces a map from the direct limit, $\varinjlim_n \pi_{i+2n}(BLGA^+ \wedge BGL\overline{\mathbb{F}}_q^+;Z/\ell^\nu)$, which equals $KU_i(BGLA^+;Z/\ell^\nu)$ to $\varinjlim_n \pi_{i+2n}(BGLA^+;Z/\ell^\nu) = K_i(A;Z/\ell^\nu)$.

One may describe multiplication by $\beta_\nu \in K_{2\ell^{\nu-1}(\ell-1)}(Z[1/\ell];Z/\ell^\nu)$ when ℓ is odd by means of the composition with Adams' map

$$A_\nu : P^{m+2\ell^{\nu-1}(\ell-1)}(\nu) \to P^m(\nu)$$

where $P^m(\nu) = S^{m-1} \cup_{\ell^\nu} e^m$. A_ν is a K-theory (and J-theory) isomorphism. Hence if we identify $K_s(A;Z/\ell^\nu)$ with the direct limit

$$K_s(A;Z/\ell^\nu) \xrightarrow{(\Sigma^{s-m}A_\nu)_\#} K_{s+d}(A;Z/\ell^\nu) \xrightarrow{(\Sigma^{s+d-m}A_\nu)_\#} \ldots \quad (2.9)$$

($d = 2\ell^{\nu-1}(\ell-1)$ and $s \geq m$) we obtain the following diagram:

2.10: Theorem Let A be as in §2.8 and let ℓ be an odd prime which is invertible in A. Then for each $\nu \geq 1$ there is a commutative diagram

$$\begin{array}{ccc} K_i(A;Z/\ell^\nu) & \xrightarrow{\rho} & K_i(A;Z/\ell^\nu) \\ & \searrow^{H} & \downarrow I \\ & & h_i(BGLA^+) \end{array}$$

This diagram exists for $i \geq s$ where s depends on ℓ and ν - when $\nu = 1$, $s = 3$ will suffice for all ℓ. h_* is as in §2.8.

I is defined by sending a homology class $f : P^u(\nu) \to BGLA^+$ to its Hurewicz image. This respects the direct limit of (2.9).

<u>2.11: Corollary</u> Let ℓ be an odd prime.

For all i, I is injective in §2.10.

<u>Proof:</u> By §2.10 Ker $\rho \subseteq$ Ker H in dimensions $\geq$ s. Also the diagram of §2.9 shows Ker $\rho \supseteq$ Ker H if $i \geq 3$ so the result follows since $K_i(A;Z/\ell^\nu)$ is generated by the images of the groups $\{K_{i+w\ell^{\nu-1}(\ell-1)}(A;Z/\ell^\nu); w \geq 0\}$ under ρ and there is no loss of generality in choosing i large since both $K_*(A;Z/\ell^\nu)$ and $h_*(BGLA^+)$ have "Bott periodicity" (of period two, in fact).

As we shall see in §3 the analogue of §2.10 in dimension 2 would suffice to solve the Lichtenbaum-Quillen conjecture. However, the following diagram characterises ker ρ in dimension two.

<u>2.12: Theorem</u> Let A be as in §§2.8,2.10 and let ℓ be an odd prime and $\nu = 1$. There is a commutative diagram

$$\begin{array}{ccc} K_2(A;Z/\ell) & \xrightarrow{\rho} & K_2(A;Z/\ell) \\ & \searrow^{\hat{H}} & \downarrow I \\ & & h_0(BGLA^+) \end{array}$$

where $\hat{H}(x) = H(x) - H(x)^\ell$, the product being taken in the Pontrjagin product on $h_*(BGLA^+)$ which comes from the direct sum of matrices. (Here we have identified $h_0(Y) \cong h_2(Y)$.)

<u>Proof:</u> We first construct $\hat{\rho} : K_2(A;Z/\ell) \to K_{2\ell}(A;Z/\ell)$ such that (up to Bott periodicity) $H(\hat{\rho}(x)) = H(x) - H(x)^\ell$. To do this let $P^{2\ell}(1) \to \Omega\Sigma P^2(1)$ be the adjoint of Adams map. Set $\hat{\rho}(f)$ equal to the composite of this and

$$\Omega\Sigma f : \Omega\Sigma P^2(1) \to \Omega\Sigma(BGLA^+)$$

with the retraction $D : \Omega\Sigma(BGLA^+) \to BGLA^+$ coming from the (direct sum) loop structure. One finds that $I(\rho(\hat{\rho}(x)) = H(x) - H(x)^\ell$, if we identify $K_n \cong K_{n+2}$ by Bott periodicity. Set $\rho' = \rho(\hat{\rho}(x)) : K_2(A;Z/\ell) \to K_2(A;Z/\ell)$, then ρ' makes the required diagram commute. We must show $\rho' = \rho$. However this follows by composing with the injections (induced by multiplication with v)

$$v_\# : K_i(A;Z/\ell) \rightarrowtail K_{i+1}(A[v,v^{-1}];Z/\ell)$$

$$v_\# : {}_i(A;Z/\ell) \rightarrowtail K_{i+1}(A[v,v^{-1}];Z/\ell)$$

For
$$\rho(x) = \rho'(x)$$
$$\iff \quad Iv_\#\rho(x) = Iv_\#\rho'(x)$$

but $Iv_\#\rho'(x) = v_\# I\rho'(x) = v_\#(H(x)-H(x)^\ell) = v_\# H(x) = Hv_\#(x) = I\rho v_\#(x) = Iv_\#\rho(x)$, since $v_\#$ annihilates decomposables.

<u>2.13: Corollary</u> For A, ℓ as in §§2.8-2.12

$$\mathrm{Ker}(\rho : K_2(A;Z/\ell) \to K_2(A;Z/\ell)) = \{x \in K_2(A;Z/\ell) \mid H(x) = H(x)^\ell \in h_0(BGLA^+)\}.$$

§3: Throughout this section ℓ will denote an odd prime. A will be a commutative $Z[1/\ell]$-algebra whose units have the form $A^* \cong Z/\ell^a \times H$ $(1 \le a \le \infty)$ - i.e. the ℓ-primary roots of unity split off. In addition we assume that Spec A satisfies Condition (0.1) of [Th1], which was explained at the beginning of §2. When $\ell = 3$ we assume $a \ge 2$.

When considering the Lichtenbaum-Quillen conjecture for A we will encounter some related rings as follows.

For $n \ge 2$ set

$$\dot{\Delta}^n_A = A[x_1,\ldots,x_n]/((\Sigma x_i - 1)x_1 x_2 \ldots x_n) \tag{3.1}$$

$$\dot{\Delta}^1_A = A, \qquad \dot{\Delta}^0_A = A[x,x^{-1}], \qquad \dot{\Delta}^{-1}_A = A[x,y,x^{-1},y^{-1}] .$$

3.2: If $1 \le a < \infty$ let h_* denote $j_*(\_;Z/\ell)$ as defined in §2.7 and if $a = \infty$ let h_* denote $KU_*(\_;Z/\ell)$ - both considered as being $Z/2$-graded.

We will say that A satisfies condition $H(n,\ell)$, for $n \ge -1$, if

$$\underline{H(n,\ell)} : \{x \in h_0(BGL\dot{\Delta}^n_A) \mid x = x^\ell\} = \underline{0} .$$

There is some hope that one may be able to verify $H(n,\ell)$, although at this moment I only know how to verify $H(1,\ell)$ for a finite field - without using the results of Quillen on $K_*(\mathbb{F}_q)$.

3.3: Lemma Let ℓ be any prime and let Y be a two-fold loop space. Then the Hurewicz map

$$H_K : \pi_2(Y;Z/\ell^\nu) \to KU_0(Y;Z/\ell^\nu)$$

has kernel contained in the image under reduction mod ℓ^ν of $\{y \in \pi_2(Y) \mid \ell^N y = 0$ for some $N\}$.

A similar result holds for the Hurewicz homomorphism for $j_*(Y;Z/\ell^\nu)$ by virtue of the fact that (see §2.7) H_K factors through H_J in the cases we will use.

3.4: Theorem If A, ℓ are as above and A satisfies condition $H(n,\ell)$ of §3.2 for $n \ge -1$ then

$$\mathrm{Ker}(\rho : K_{n+1}(A;Z/\ell) \to \mathcal{K}_{n+1}(A;Z/\ell))$$

is contained in the image of the ℓ-primary torsion in $K_{n+1}(A)$.

Proof: If $n \ge 2$ and $\Lambda = Z/\ell^\nu$ or Z it is shown in [D-W;W] that there is a natural splitting

$$K_2(\dot{\Delta}^n_A;\Lambda) \cong K_2(A;\Lambda) \oplus K_{n+1}(A;\Lambda)$$

Similarly for $n = 0,1,-1$ $K_{n+1}(A;\Lambda)$ is a summand in $K_2(\dot{\Delta}^n_A;\Lambda)$, as is well-known. Here we have used the regularity of A.

The effect of $H(n,\ell)$ is to guarantee (by §2.13) that $\mathrm{Ker}(\rho : K_2(\dot{\Delta}^n_A;Z/\ell) \to \mathcal{K}_2(\dot{\Delta}^n_A;Z/\ell))$ is Ker H, the Hurewicz kernel. Thus Lemma 3.3 yields the result.

3.5: Theorem Let A, ℓ be as above. Assume that A satisfies $H(n-1,\ell)$ and that

$$\rho : K_{n+1}(A;Z/\ell^\nu) \to \mathcal{K}_{n+1}(A;Z/\ell^\nu) = K^{et}_{n+1}(A,Z/\ell^\nu)$$

is onto for all $\nu \ge 1$. Then

$$\rho : K_n(A;Z/\ell) \to K_n(A;Z/\ell)$$

is an injective.

Proof: This follows from §3.5 together with a diagram chase around diagrams like the following one. Details are left to the reader (see [Sn3]).

$$\begin{array}{ccccccccc} \cdots & \longrightarrow & K_{n+1}(A;Z/\ell^{N+1}) & \longrightarrow & K_{n+1}(A;Z/\ell^{N}) & \longrightarrow & K_n(A;Z/\ell) & \longrightarrow & \cdots \\ & & \downarrow \rho & & \downarrow \rho & & \downarrow & & \\ \cdots & \longrightarrow & K_{n+1}(A;Z/\ell^{N+1}) & \longrightarrow & K_{n+1}(A;Z/\ell^{N}) & \longrightarrow & K_n(A;Z/\ell) & \longrightarrow & \cdots \end{array}$$

3.6: Remark As I mentioned in §2.2, for the rings A which we considered here, the results of ([D-F-S-T],[Th1]) imply that $\rho : K_{n+1}(A;Z/\ell^{\nu}) \to K_{n+1}(A;Z/\ell^{\nu})$ is onto if n is large enough and if $H^i_{et}(A;Z/\ell^{\nu})$ is finite for each i. However, for Theorem 3.5 we need that ρ is onto for all choices of ν in dimension n+1. This is not yet known, although, by the Gersten-Quillen resolution, it probably suffices to prove such a uniform result for fields.

I will conclude the paper by giving a series of examples for which all such conditions are verified, except for $H(n,\ell)$, and in these examples I will give the explicit value of the K-groups which results.

3.7: Examples In each of the following examples let ℓ be an odd prime and assume that A satisfies Condition $H(n-1,\ell)$ of §3.2.

(a) Suppose A is, as in §2.1(a), a ring of integers in a function field of characteristic not or the localisation away from ℓ of the integers in a number field. In [D-F] it is shown that ρ is onto for all dimensions ≥ 2 so that our assumptions yield $(i \geq 1)$

$$K_{2i}(A;Z/\ell^{\nu}) \cong H^0_{et}(A;Z/\ell^{\nu}(i)) \oplus H^2_{et}(A;Z/\ell^{\nu}(i-1)) \quad \text{if } n = 2i \, ,$$

$$K_{2i-1}(A;Z/\ell^{\nu}) = H^1_{et}(A;Z/\ell^{\nu}(i)) \quad \text{if } n = 2i-1 \, ,$$

where $Z/\ell^{\nu}(i)$ denotes Z/ℓ^{ν} Tate-twisted i times. If $\ell = 3$ we assume that A has a 9-th root of unity.

In order to apply Theorem 3.5 we need that A contains at least ℓ-th roots of unity. If not we apply the results to $A(\zeta_{\ell})$, the integers in the field extended by an ℓ-th root of unity. We recover the K-theory of A as the Galois invariants of the K-theory of $A(\zeta_{\ell})$.

If $A = \mathcal{Q}[1/\ell]$ contains an ℓ^{ν}-th root of unity these groups become $(i \geq 1)$

$$K_{2i}(A;Z/\ell^{\nu}) = (\mathrm{Pic}(A)/\ell^{\nu}) \oplus (\bigoplus_1^k Z/\ell^{\nu})$$

$$K_{2i-1}(A;Z/\ell^{\nu}) \cong A^*/(A^*)^{\ell^{\nu}} \oplus ({}_{\ell^{\nu}}\mathrm{Pic}(A))$$

where Pic(A) is the ideal class group of A, ${}_n\mathrm{Pic}(A) = \{x \in \mathrm{Pic}\,A \mid nx = 0\}$ and k is the number of distinct primes in $\mathcal{Q}$ over ℓ. Let k be the quotient field of A as in this example and let $k(\zeta_{\ell^{\infty}})$ be the field obtained by adjoining all ℓ-primary roots of unity to k. The results of this example apply equally well if A is replaced by the integral closure of $A(\zeta_{\ell^\infty})$ in $k(\zeta_{\ell^\infty})$.

(b) As in §2.1(b), let A be the quotient field of a ring from example (a). Our assumptions give the same formula for $K_n(A;Z/\ell^\nu)$ as in example (a). If the field contains ℓ^ν-th roots of unity these groups become the following ($i \geq 1$)

$$K_{2i}(A;Z/\ell^\nu) = ({}_{\ell^\nu}\mathrm{Br}(A)) \oplus \mu_{\ell^\nu}(A) \qquad \text{if } n = 2i\,,$$

$$K_{2i-1}(A;Z/\ell^\nu) = A^*/(A^*)^{\ell^\nu} \qquad \text{if } n = 2i-1\,.$$

Here Br is the Brauer group and μ_n the group of n-th roots of unity.

(c) Let A be the ring given by Jouanolou's device (see §2.7) when X in §2.1(c) is a smooth, connected curve over an algebraically closed field in which ℓ is invertible. Our assumptions on A, together with the result of Thomason (see [F§3.10]) which states ρ is onto for all ν, in all dimensions, yield

$$K_n(X;Z/\ell^\nu) = \begin{cases} Z/\ell^\nu \oplus Z/\ell^\nu & \text{if } n \text{ is even,} \\ \bigoplus_1^{2g} Z/\ell^\nu & \text{if } n \text{ is odd} \end{cases}$$

where g is the genus of X.

(d) If A is a separably closed field or strictly local Hensel ring in which ℓ is invertible our assumptions on A yield

$$K_*(A;Z/\ell^\nu) = Z/\ell^\nu[\beta]$$

where $\deg \beta = 2$.

In fact from this one sees that

$$H_*(BGLA;Z/\ell^\nu) \cong Z/\ell^\nu[c_1,c_2,\dots]$$

($\deg c_j = 2j$) in this case.

Let $\zeta_\ell = \exp(2\pi i/\ell)$.

<u>3.8: Theorem</u> Let $\ell \neq 2,3$ be a prime. Assume that $Z[1/\ell,\zeta_\ell]$ satisfies $H(2i-3,\ell)$ and $H(2i-2,\ell)$ for $i \geq 2$, even. Then

$$\frac{|K_{2i-2}(Z)|_\ell}{|K_{2i-1}(Z)|_\ell} = |\zeta_Z(1-i)|_\ell = \left|\frac{B_i}{i}\right|_\ell$$

where $|\_|_\ell$ denotes the ℓ-adic valuation, $\zeta_Z(s)$ is the zeta function [Sou] and B_i is the i-th Bernoulli number.

<u>Proof</u>: Set $A = Z[1/\ell]$. The localisation sequence and the fact that $K_i(\mathbb{F}_\ell;Z/\ell) = 0$ for $i \geq 1$ implies that it suffices to compute

$$\frac{|K_{2i-2}(A)|_\ell}{|K_{2i-1}(A)|_\ell}\,.$$

For $\varepsilon = 1,2$ we have exact sequences

$$(K_{2i-\varepsilon}(A))/\ell^\nu \rightarrowtail K_{2i-\varepsilon}(A;Z/\ell^\nu) \twoheadrightarrow {}_{\ell^\nu}K_{2i-1-\varepsilon}(A)\,.$$

The K-groups in question are finite so that, by taking inverse limits, we obtain

$$_{\ell^\infty}K_{2i-\varepsilon}(A) \cong \varprojlim_{\nu} K_{2i-\varepsilon}(A;Z/\ell^\nu).$$

If $\hat{H}^m(i) = \varprojlim_{\nu} H^m_{et}(A;Z/\ell^\nu(i))$ is ℓ-adic étale cohomology, we obtain

$$\frac{|K_{2i-2}(A)|_\ell}{|K_{2i-1}(A)|_\ell} = \frac{|\hat{H}^0(i-1)|_\ell \cdot |\hat{H}^2(i)|_\ell}{|\hat{H}^1(i)|_\ell}$$

which, by work of Mazur-Wiles, Coates and Lichtenbaum (see [Th1]) equals

$$|\zeta_Z(1-i)|_\ell = |\frac{Bi}{i}|_\ell$$

REFERENCES

[D-F] W.G. Dwyer & E.M. Friedlander: Étale K-theory and arithmetic; Bull. Amer. Math. Soc. (6)3(1982) 453-455.

[D-F-S-T] W.G. Dwyer, E.M. Friedlander, V.P. Snaith & R.W. Thomason: Algebraic K-theory eventually surjects onto étale K-theory; Inventiones Math. 66(1982) 481-491.

[D-W] B.H. Dayton & C.A. Weibel: A spectral sequence for higher K-theory of affine glued schemes; Algebraic K-theory, Evanston 1980, Springer-Verlag Lecture Notes in Mathematics #854, pp. 24-92.

[F] E.M. Friedlander: Etale K-theory II; to appear in Ann. Sci. Ec. Norm. Sup. .

[J] J-P. Jouanolou: Une suite exacte de Mayer-Vietoris en K-théorie algébrique; Springer-Verlag Lecture Notes in Mathematics #341, pp. 293-316.

[Sn] V.P. Snaith: Algebraic cobordism and K-theory; Mem. Amer. Math. Soc. #221 (1979).

[Sn1] V.P. Snaith: A brief survey of Bott periodic K-theory; Current Trends in Algebraic Topology, University of Western Ontario 1981, Can. Math. Soc. Conference Proceedings Vol. 2.

[Sn2] V.P. Snaith: Localised stable homotopy and algebraic K-theory; to appear Mem. Amer. Math. Soc. .

[Sn3] V.P. Snaith: Unitary K-homotopy and the Lichtenbaum-Quillen conjecture; to appear.

[Sn4] V.P. Snaith: An upper bound for algebraic K-theory; these proceedings.

[Sou] C. Soulé: K-théorie des anneaux de corps de nombres et cohomologie étale; Inventiones Math. 55(1979) 251-295.

[Th1] R.W. Thomason: The Lichtenbaum-Quillen conjecture for $K/\ell_*[\beta^{-1}]$; Current Trends in Algebraic Topology, University of Western Ontario 1981, Can. Math. Soc. Conference Proceedings Vol. 2.

[Th2] R.W. Thomason: Algebraic K-theory and étale cohomology; preprint M.I.T. (1981).

[W] C.A. Weibel: Mayer-Vietoris sequences and mod p K-theory; Conf. Oberwolfach (1980); Springer-Verlag Lecture Notes in Mathematics, to appear.

HOMOLOGY OF GL_n, CHARACTERISTIC CLASSES AND MILNOR K-THEORY

by

A.A. Suslin

Introduction

The main purpose of the present paper is to study the homomorphism $c_{n,n}: K_n(F) \to H^o(\mathrm{Spec}\ F, \mathcal{K}_n) = K_n(F)$ occurring in the theory of characteristic classes for higher K-groups. It's easy to see that $c_{1,1}$ is the identity map. Using this and the product formula one sees immediately that the restriction of $c_{n,n}$ on the symbolic part of $K_n(F)$ is just multiplication by $(-1)^{n-1}\cdot(n-1)!$ This remark made reasonable to conjecture that $c_{n,n}$ itself coincides (at least up to torsion) with multiplication by $(-1)^{n-1}\cdot(n-1)!$ However recently Soulé using explicit computations of K groups of number fields proved that this conjecture is false and conjectured instead that the image of $c_{n,n}$ is (up to factorials) the symbolic part of $K_n(F)$. We prove below (theorem 5.2) that this is actually true. The main result of the paper (theorem 3.4) states that if F is an infinite field, then $H_n(GL(F)) = H_n(GL_n(F))$ and the natural map $\underbrace{F^*\times\ldots\times F^*}_{n} = H_1(GL_1(F))\times\ldots\times H_1(GL_1(F)) \to H_n(GL_n(F))$ defines an isomorphism $K_n^M(F) \to H_n(GL_n(F))/H_n(GL_{n-1}(F))$ (here $K_n^M(F)$ are Milnor's K-groups of F -see [1], [6]). This theorem gives rise to the homomorphism $\varphi: K_n(F) = \pi_n(BGL(F)^+) \to H_n(BGL(F)^+) = H_n(GL(F)) = H_n(GL_n(F)) \to H_n(GL_n(F))/H_n(GL_{n-1}(F) = K_n^M(F)$ and we prove in § 5 that $c_{n,n}$ coincides with the composition $K_n(F) \xrightarrow{\varphi} K_n^M(F) \to K_n(F)$, where $K_*^M(F) \to K_*(F)$ is the natural ring homomorphism induced by the isomorphism $K_1^M(F) = F^* = K_1(F)$. We show also that if $a_1,\ldots,a_n \in F^*$ and if $\ell(a_i) \in K_1(F)$ is the element corresponding to a_i, then $\varphi(\ell(a_1)\cdot\ldots\cdot\ell(a_n)) = (-1)^{n-1}\cdot(n-1)!\cdot\{a_1,\ldots,a_n\}$ and in particular $\varphi(K_n(F)) \supset (n-1)!\cdot K_n^M(F)$. The map φ has also some other interesting applications: first of all the previous formula shows that the composition $K_n^M(F) \to K_n(F) \to K_n^M(F)$ coincides with multiplication by $(-1)^{n-1}\cdot(n-1)!$ and in particular the kernel of $K_n^M(F) \to K_n(F)$ is annihilated by $(n-1)!$ - compare [18]. We prove also that the image of the map $K_3(F) \to K_3^M(F)$ coincides with the kernel of Milnor's homomorphism $K_3^M(F) \to I^3/I^4$ (see [6]). Thus the study of the map $K_3(F) \to K_3^M(F)$ is closely related to Milnor's conjectures concerning quadratic forms [6].

One of the useful technical results proved in § 1 states that for any infinite field F, the natural embedding

$$GL_n(F) \longrightarrow \begin{pmatrix} 1 & * \\ 0 & GL_n(F) \end{pmatrix}$$

is a homology isomorphism. This result is trivial in characteristic zero, but the proof in positive characteristic is much more delicate.

Everywhere below the word field means (if not specified otherwise) infinite field.

§ 1. Homology of the affine groups

<u>Lemma 1.1</u> <u>Suppose that</u> $\varphi_1,\ldots,\varphi_n : F \hookrightarrow \Omega$ <u>are field embeddings such that for any</u> $x \in F^*$, $\prod_{i=1}^{n} \varphi_i(x) = 1$. <u>Then</u> F <u>is a finite field of</u> p^m <u>elements with</u> $n \geq (p-1)\cdot m$.

<u>Proof:</u> Suppose first that F is finite and $\mathrm{card}(F) = p^m$. Then we may suppose that $\Omega = F$ and $\varphi_i = \sigma^{t_i}$, where σ is the Frobenius automorphism $\sigma(x) = x^p$. For $x \in F^*$ we have the equality: $\prod_{i=1}^{n} x^{p^{t_i}} = 1$. Hence $\sum_{i=1}^{n} p^{t_i} \equiv 0 \bmod (p^m-1)$. Denote by n_j the number of $t_i \equiv j \bmod m$. Then $n = \sum_{j=0}^{m-1} n_j$ and $\sum_{j=0}^{m-1} n_j p^j \equiv 0 \bmod (p^m-1)$.

(1.1.1) Sublemma. If the n_j's are non negative integers, not all of which are zero and $\sum_{j=0}^{m-1} n_j \cdot p^j \equiv 0 \bmod (p^m-1)$, then $\sum_{j=0}^{m-1} n_j \geq (p-1)\cdot m$.

<u>Proof:</u> If $n_j \geq p$ for certain j, then replacing n_j by $n_j - p$ and n_{j+1} by $n_{j+1} - 1$ we'll get a new system of n_j's satisfying the same properties but with a smaller sum. So we may suppose that $0 \leq n_j \leq p-1$ for all j. In this case $0 < \sum_{j=0}^{m-1} n_j \cdot p^j \leq (p-1)(1+p+\ldots+p^{m-1}) = p^m-1$. Hence $\sum_{j=0}^{m-1} n_j \cdot p^j = p^{m-1}$, $n_j = p-1$ for all j and $\sum_{j=0}^{m-1} n_j = m(p-1)$.

(1.1.2) The above arguments show that if F is finite and $\mathrm{card}(F) = p^m$, then $n \geq (p-1)m$. This shows immediately that F cannot be an infinite algebraic extension of a finite field. It's clear also that F cannot be a field of characteristic zero. Suppose finally that F is transcendental over the prime subfield F_p. Let $t \in F$ be transcendental over F_p and put $t_i = \varphi_i(t) \in \Omega$. For an arbitrary non-zero polynomial $f \in F_p[T]$ we have $\prod_{i=1}^{n} f(t_i) = \prod_{i=1}^{n} \varphi_i(f(t)) = 1$. Hence $(t_1,\ldots,t_n)$ is an Ω-valued point of the variety $V \subset \mathbb{A}_{F_p}$ given by the system of equations $\prod_{i=1}^{n} f(x_i) = 1$ for any non-zero $f \in F_p[T]$.

(1.1.3) Sublemma. The variety V is empty.

<u>Proof:</u> Otherwise V would contain an algebraic point x. Choose $f \in F_p[T]$ such that $f(x_1) = 0$. Then we get the contradiction $0 = \prod_{i=1}^{n} f(x_i) = 1$.

<u>Corollary 1.2</u> <u>If</u> $F^{\otimes n}$ <u>has the natural diagonal action of</u> F <u>then</u> $H_0(F^*, F^{\otimes n}) = 0$.

<u>Proof:</u> $H_0(F^*, F^{\otimes n}) = F^{\otimes n}/I$, where I is the ideal generated by elements $(f \otimes \ldots \otimes f) - 1$ $(f \in F^*)$. By lemma 1.1 the ideal I is not contained in any maximal ideal. Hence $I = F^{\otimes n}$.

<u>Proposition 1.3</u> <u>Denote by</u> $V_n = (F^{\otimes n})^{S_n}$ <u>the subring of</u> $F^{\otimes n}$ <u>consisting of symmetric tensors</u>. <u>Then</u> $H_0(F^*, V_n) = 0$.

Proof: Denote by $V_n^{K_1,\dots,K_r} \subset V_n$, where K_i are positive integers with $\sum_{i=1}^{r} K_i = n$, the subspace of V_n generated by the tensors of the following form: $\sum_{\sigma\in S_n/S_{K_1}\times\dots\times S_{K_r}} (\overbrace{f_1\otimes\dots\otimes f_1}^{K_1}\otimes\dots\otimes\overbrace{f_r\otimes\dots\otimes f_r}^{K_r})^{\sigma}$. Then each $V_n^{K_1,\dots,K_r}$ is an F^*-invariant subspace of V_n and $V_n = \sum V_n^{K_1,\dots,K_r}$. Let $V_n^{(i)}$ be equal to $\sum_{\Sigma(K_j-1)\le i} V_n^{K_1,\dots,K_r}$. Then $V_n^{(i)}$ defines an F^*-invariant filtration on V_n. It's easy to see that if $\sum(K_j-1) = i$ then the map $F^r \to V_n^{(i)}/V_n^{(i-1)}$ given by $f_1\times\dots\times f_r \to \sum_{\sigma\in S_n/S_{K_1}\times\dots\times S_{K_r}} (\overbrace{f_1\otimes\dots\otimes f_1}^{K_1}\otimes\dots\otimes\overbrace{f_r\otimes\dots\otimes f_r}^{K_r})^{\sigma} \bmod V_n^{(i-1)}$ is multilinear and hence defines an F^*-equivariant homomorphism $F^{\otimes r} \longrightarrow V_n^{(i)}/V_n^{(i-1)}$. So we have an F-equivariant epimorphism $\coprod_{K_1,\dots,K_r;\Sigma(K_j-1)=i} F^{\otimes r} \longrightarrow V_n^{(i)}/V_n^{(i-1)}$ and from (1.2) we deduce that $H_o(F^*,V_n^{(i)}/V_n^{(i-1)}) = 0$. Now it follows by induction on i, starting with $i=-1$, that $H_o(F^*,V_n^{(i)}) = 0$.

Lemma 1.4 Let $n_1,\dots,n_r$ be positive integers and for any i let $T^{n_i}(F)$ be either the external or symmetric power of F (over $\mathbb{Z}$ or the prime subfield). Put $n = \sum_{i=1}^{r} n_i$ and consider the natural projection $F^{\otimes n} \to T^{n_1}(F)\otimes\dots\otimes T^{n_r}(F)$. The kernel of this projection is a V_n-submodule of $F^{\otimes n}$ and hence $T^{n_1}(F)\otimes\dots\otimes T^{n_r}(F)$ has the natural structure of a V_n-module.

Proof: Denote by K_i the kernel of the map $F^{\otimes n_i} \to T^{n_i}(F)$. Then $K = \sum_{i=1}^{r} F^{\otimes n_1}\otimes\dots\otimes K_i\otimes\dots\otimes F^{\otimes n_r}$ and it's sufficient to check that each summand is V_n-submodule of $F^{\otimes n}$. We'll consider only the case $i=1$. If T^{n_1} is the symmetric power, then $K_1\otimes\dots\otimes F^{\otimes n_r}$ is generated by tensors of the form $v^{(i,j)}-v$, where $1\le i<j\le n_1$ and if $w\in V_n$, then $w\cdot(v^{(i,j)}-v) = (wv)^{(i,j)}-(wv)\in K_1\otimes\dots\otimes F^{\otimes n_r}$. If T^{n_1} is the exterior power, then $K_1\otimes\dots\otimes F^{\otimes n_r}$ is generated by tensors v such that $v^{(i,j)} = v$ for some $1\le i<j\le n_1$. If $w\in V_n$ then $(wv)^{(i,j)} = w^{(i,j)}v^{(i,j)} = wv$ and hence $wv\in K_1\otimes\dots\otimes F^{\otimes n_r}$.

Lemma 1.5 Suppose that A is a commutative ring, G a commutative group and $\varphi: G\to A^*$ a homomorphism. Let M be an A-module and consider on M the induced G-module structure. If $H_o(G,A) = 0$, then $H_p(G,M) = 0$ for any $p\ge 0$.

Proof: $H_o(G,A) = A/I$, where I is the ideal in A generated by elements $\varphi(g)-1$ $(g\in G)$. Since M is an A-module and G acts on M by A-linear transformations the groups $H_p(G,M)$ are naturally A-modules. Since G is abelian it's well known that for any $g\in G$, the map $g: M\to M$ induces identity homomorphisms on all homology groups. So each $H_p(G,M)$ is annihilated by I and hence is zero.

Corollary 1.6 Assuming the conditions of (1.4), $H_p(F^*,T^{n_1}(F) \otimes \ldots \otimes T^{n_2}(F)) = 0$.

Proposition 1.7 Suppose that F is a field of characteristic $p > 0$ and V is a vector space over F. Then for any i and any $j > 0$ $H_i(F^*,H_j(V,F_p)) = 0$.

Proof: It is well known that for an arbitrary vector space V over F_p the graded abelian group $H_*(V,F_p)$ is naturally isomorphic to $S(V) \otimes \Lambda(V)$ if $p \neq 2$ (here generators of the symmetric algebra have degree two and generators of the exterior algebra have degree one) and to $S(V)$ if $p = 2$ (the generators have degree one). (Caution: this is not a ring isomorphism). Hence $H_j(V,F_p)$ is isomorphic as F^*-module to a direct sum of modules of the kind considered in (1.4) and our statement follows from (1.6).

Corollary 1.8 Let F be any field, V a vector space over F and k any prime field. Then for any i and any $j > 0$ $H_i(F^*,H_j(V,k)) = 0$.

Proof: If char $k \neq$ char F, then $H_j(V,k) = 0$. If char k = char $F > 0$, then our statement is proved in 1.7. If char $k = 0 =$ char F, then $H_j(V,k) = \Lambda^j(V)$. For any positive integer n the action of $n \in F^*$ on $H_j(V,k)$ coincides with multiplication by n^j and since F is abelian we deduce that $H_i(F^*,H_j(V,k))$ is annihilated by n^j-1 and hence is zero (since it is a $\mathbb{Q}$-vector space).

Theorem 1.8 Let $G_1 \subset GL_n(F)$, $G_2 \subset GL_m(F)$ be subgroups such that at least one of them contains the group of diagonal matrices. Let $M \subset M_{n,m}(F)$ be a vector subspace such that $G_1M = M = MG_2$. Then the natural embedding

$$i : \begin{pmatrix} G_1 & 0 \\ 0 & G_2 \end{pmatrix} \hookrightarrow \begin{pmatrix} G_1 & M \\ 0 & G_2 \end{pmatrix}$$

is a homology isomorphism.

Proof: It suffices, in view of the Hochschild-Serre spectral sequence, to check that $H_i(G_1 \times G_2,H_j(M,k)) = 0$ if $j > 0$. Suppose, for example, that $G_1 \supset F^*$. Then applying once more the Hochschild-Serre spectral sequence: $H_p(G_1/F^* \times G_2,H_q(F^*,H_j(M,k))) \Rightarrow$ $\Rightarrow H_{p+q}(G_1 \times G_2,H_j(M,k))$ and (1.7), we get the desired result.

§ 2. The spaces of unimodular vectors

If F is a field and V,W finite dimensional vector spaces over F then by $\mathrm{Um}(V,W)$ we'll denote the simplicial set whose nondegenerate p-simplices are sequences $\left(\binom{v_o}{w_o},\ldots,\binom{v_p}{w_p}\right)$ where $v_i \in V$, $w_i \in W$ and $v_o,\ldots,v_p$ are linearly independent, the i-th face of such simplex being obtained by omitting $\binom{v_i}{w_i}$. The following statement is a particular case of the acyclicity theorem of van der Kallen [3]. The proof given below is taken from [3].

Lemma 2.1 The space $\mathrm{Um}(V,W)$ is $(\dim V-2)$-acyclic.

We'll need a slightly more precise result. In the same notations as above, denote by $GP(V,W)$ the simplicial set whose nondegenerate p-simplices are sequences $\left(\binom{v_o}{w_o},\ldots,\binom{v_p}{w_p}\right)$ such that $v_o,\ldots,v_p$ are in general position (i.e. any $\min(p+1, \dim V)$ of them are linearly independent). Then $\mathrm{Um}(V,W)$ is a $(\dim V-1)$-skeleton of $GP(V,W)$ and (2.1) follows from:

Lemma 2.2 The space $GP(V,W)$ is acyclic.

Proof: Suppose that $x = \sum n_i\cdot\left(\binom{v_o^i}{w_o^i},\ldots,\binom{v_p^i}{w_p^i}\right)$ is a p-cycle. Since F is infinite, we can find v such that $v,v_o^i,\ldots,v_p^i$ are in general position for any i. Choose any w and put $y = \sum n_i\left(\binom{v}{w},\binom{v_o^i}{w_o^i},\ldots,\binom{v_p^i}{w_p^i}\right) \in C_{p+1}(GP(V,W))$. Then $dy = x$.

We'll denote by $C_*(V,W)$ the augmented simplicial complex of $\mathrm{Um}(V,W)$ with dimensions shifted by one, i.e. $C_o(V,W) = \mathbb{Z}$, $C_i(V,W) = C_{i-1}(\mathrm{Um}(V,W))$ = the free abelian group with basis $\left(\binom{v_1}{w_1},\ldots,\binom{v_i}{w_i}\right)$ where $v_1,\ldots,v_i$ are linearly independent ($1 \le i \le \dim V$). By (2.1) all the homology groups of $C_*(V,W)$ are zero except for H_n ($n=\dim V$). We'll denote $H_n(C_*(V,W))$ by $H(V,W)$. Denote by $\mathrm{Aff}(V,W)$ the affine group $\begin{pmatrix}\mathrm{Aut}\, V & \mathrm{Hom}(V,W)\\ 0 & 1\end{pmatrix}$. We have the natural action of $\mathrm{Aff}(V,W)$ on $C_*(V,W)$ and the induced action of $\mathrm{Aff}(V,W)$ on $H(V,W)$. We'll denote $H_o(\mathrm{Aff}(V,W),H(V,W)) = H(V,W)_{\mathrm{Aff}(V,W)}$ by $S(V,W)$. If the space W is zero we'll omit it in the above notations. We'll also use following notations: $C_*(F^n,F^m) = (C_*^{n,m}(F), C_*^{n,o}(F) = C_*(F^n) = C_*^n(F)$, $H(F^n,F^m) = H_{n,m}(F)$, $H_{n,o}(F) = H(F^n) = H_n(F)$, $\mathrm{Aff}(F^n,F^m) = A_{n,m}(F)$, $S(F^n,F^m) = S_{n,m}(F)$, $S_{n,o}(F) = H_n(F)_{GL_n(F)} = S_n(F)$.

(2.3) The differential in the simplicial complex of $GP(F^n)$ induces an epimorphism $\delta : C_n(GP(F^n)) \longrightarrow H_n(F)$. Moreover, by (2.2) we have an exact sequence

$$0 \longleftarrow H_n(F) \xleftarrow{\delta} C_n(GP(F^n)) \xleftarrow{d} C_{n+1}(GP(F^n)).$$

We'll denote by $e_1,\ldots,e_n$ the standard basis of F^n. If $a_i \in F^*$, then $e_1,\dots,e_n, \sum a_ie_i$ are in general position and by $<a_1,\dots,a_n>$ we'll denote the class of the cycle $\delta((e_1,\dots,e_n, \sum a_ie_i))$ modulo the action of $GL_n(F)$.

(2.3.1) Proposition. The group $S_n(F)$ is generated by all $<a_1,\dots,a_n>$ and they are subject only to the following relations: if $\lambda_1,\dots,\lambda_n \in F^*$ are all distinct, then

$$<\lambda_1a_1,\dots,\lambda_na_n> - <a_1,\dots,a_n> = \sum_{i=1}^n (-1)^{i+n} <a_1(\lambda_1-\lambda_i),\dots,\widehat{a_i(\lambda_i-\lambda_i)},\dots,a_n(\lambda_n-\lambda_i),\lambda_i>.$$

Proof: From the exact sequence of (2.3) wet get:

$$0 \longleftarrow S_n(F) \xleftarrow{\bar\delta} C_n(GP(F^n))_{GL_n(F)} \xleftarrow{\bar d} C_{n+1}(GP(F^n))_{GL_n(F)} .$$

The action of $GL_n(F)$ on $C_n(GP(F^n))$ and $C_{n+1}(GP(F^n))$ permutes the basis elements and every basis element is equivalent under the action of $GL_n(F)$ to a unique element of the form $(e_1,\dots,e_n,a)$ (resp. $(e_1,\dots,e_n,a,b)$, where corresponding vectors are in general position. The last condition is evidently equivalent to the fact that $a = (a_1,\dots,a_n)^T$ with $a_i \in F^*$ (resp. $a = (a_1,\dots,a_n)^T$, $b = (\lambda_1a_1,\dots,\lambda_na_n)^T$ with $a_i \in F^*$, $\lambda_i \in F^*$ and $\lambda_i \neq \lambda_j$ if $i \neq j$). So the above exact sequence takes the form:

$$0 \longleftarrow S_n(F) \xleftarrow{\bar\delta} \coprod_{(a_1,a_n),a_i\in F^*} \mathbb{Z} \xleftarrow{\bar d} \coprod_{\substack{(a_1,\dots,a_n),a_i\in F^* \\ (\lambda_1,\dots,\lambda_n),\lambda_i\in F^*,\ \lambda_i\neq\lambda_j \text{ if } i\neq j}} \mathbb{Z}$$

Denoting the generator corresponding to (a) (resp. to (a,λ)) by e_a (resp. by $e_{a,\lambda}$) we have: $\bar\delta(e_a) = <a>$, $\bar d(e_{a,\lambda}) = d(e_1,\dots,e_n, a,a\lambda) \bmod GL_n(F)$

$$= \sum_{i=1}^n (-1)^{i-1}(e_1,\dots,\hat e_i,\dots,e_n,a,a\lambda) +$$

$(-1)^n(e_1,\dots,e_n,a\lambda) + (-1)^{n+1}(e_1,\dots,e_n,a))$ mod $GL_n(F)$. We have the following equivalences modulo the action of $GL_n(F)$:

$$(e_1,\dots,\hat e_i,\dots,e_n,a,a\lambda) \sim (e_1,\dots,e_{n-1}, \begin{pmatrix} a_1 \\ \cdots \\ a_{i-1} \\ a_{i+1} \\ \cdots \\ a_n \\ a_i \end{pmatrix}, \begin{pmatrix} \lambda_1a_1 \\ \cdots \\ \lambda_{i-1}a_{i-1} \\ \lambda_{i+1}a_{i+1} \\ \cdots \\ \lambda_n a_n \\ \lambda_i a_i \end{pmatrix}) \sim (e_1,\dots,e_n, \begin{pmatrix} a_1(\lambda_1-\lambda_i) \\ \cdots \\ a_{i-1}(\lambda_{i-1}-\lambda_i) \\ a_{i+1}(\lambda_{i+1}-\lambda_i) \\ \cdots \\ a_n(\lambda_n-\lambda_i) \\ \lambda_i \end{pmatrix})$$

and hence $\bar d(e_{a,\lambda}) = \sum_{i=1}^n (-1)^i e_{(a_1(\lambda_1-\lambda_i),\dots,\widehat{a_i(\lambda_i-\lambda_i)},\dots,a_n(\lambda_n-\lambda_i),\lambda_i)} +$ $(-1)^n e_{(a_1\lambda_1,\dots,a_n\lambda_n)} + (-1)^{n+1} e_{(a_1,\dots,a_n)}$.

(2.4) Proposition. Suppose that $a_i, \lambda_i \in F$ and $\lambda_i \neq \lambda_j$ if $i \neq j$. Then we have the following identity in $K_n^M(F)$:

$$\{\lambda_1a_1,\dots,\lambda_na_n\} - \{a_1,\dots,a_n\} = \sum_{i=1}^n (-1)^{i+n}\{a_1(\lambda_1-\lambda_i),\dots,\widehat{a_i(\lambda_i-\lambda_i)},\dots,a_n(\lambda_n-\lambda_i),\lambda_i\}.$$

Proof: We'll use induction on n. When $n = 1$ everything is trivial. In the general case we have $\sum_{i=1}^{n-1} (-1)^{i+n} \{a_1(\lambda_1-\lambda_i),\ldots,a_n(\lambda_n-\lambda_i),\lambda_i\} =$

$$= \sum_{i=1}^{n-1} (-1)^{i+n} [\{a_1(\lambda_1-\lambda_i),\ldots,a_n\lambda_n,\lambda_i\} + \{a_1(\lambda_1-\lambda_i),\ldots,1-\frac{\lambda_i}{\lambda_n},\lambda_n\}]$$

$$= \sum_{i=1}^{n-1} (-1)^{i+n-1}\{a_1(\lambda_1-\lambda_i),\ldots,\widehat{a_i(\lambda_i-\lambda_i)},\ldots,a_{n-1}(\lambda_{n-1}-\lambda_i),\lambda_i,a_n\lambda_n\}$$

$$+ \sum_{i=1}^{n-1} (-1)^{i+n} \{a_1(\lambda_1-\lambda_i),\ldots,\widehat{a_i(\lambda_i-\lambda_i)},\ldots,a_{n-1}(\lambda_{n-1}-\lambda_i),\lambda_i-\lambda_n,\lambda_n\}.$$

Applying the induction hypothesis to $a_1,\ldots,a_{n-1},\lambda_1,\ldots,\lambda_{n-1}$ and also to $a_1,\ldots,a_{n-1},\lambda_1-\lambda_n,\ldots,\lambda_{n-1}-\lambda_n$ we see that the last expression is equal to
$\{a_1\lambda_1,\ldots,a_{n-1}\lambda_{n-1},a_n\lambda_n\} - \{a_1,\ldots,a_{n-1},a_n\lambda_n\} -$
$\{a_1(\lambda_1-\lambda_n),\ldots,a_{n-1}(\lambda_{n-1}-\lambda_n),\lambda_n\} + \{a_1,\ldots,a_{n-1},\lambda_n\} =$
$\{a_1\lambda_1,\ldots,a_n\lambda_n\} - \{a_1,\ldots,a_n\} - \{a_1(\lambda_1-\lambda_n),\ldots,a_{n-1}(\lambda_{n-1}-\lambda_n),\lambda_n\}$.

(2.4.1) Corollary. There exists a unique homomorphism $S_n(F) \longrightarrow K_n^M(F)$ taking $\langle a_1,\ldots,a_n\rangle$ to $\{a_1,\ldots,a_n\}$.

(2.5) The pairings.

Let V be a vector space over F. We'll say that $x = (v_o,\ldots,v_{p-1}) \in C_{p-1}(GP(V))$ and $y = (w_o,\ldots,w_{q-1}) \in C_{q-1}(GP(V))$ are in general position if the vectors $v_o,\ldots,v_{p-1},w_o,\ldots,w_{q-1}$ are in general position, in which case define $x \cdot y \in C_{p+q-1}(GP(V))$ to be the simplex $(v_o,\ldots,v_{p-1},w_o,\ldots,w_{q-1})$. We extend this definition, by linearity, to the case of arbitrary chains (two chains are in general position if any simplex of the first chain is in general position with any simplex of the second chain). We always take C_{-1} to be $\mathbb{Z}$ and extend the previous notions in the obvious way to (-1) - dimensional chains (which are in general position with all other chains). We define $d: C_o \longrightarrow C_{-1}$ to be the augmentation. One sees easily that if x and y are in general position then dx and y and also x and dy are in general position and moreover $d(x \cdot y) = dx \cdot y + (-1)^{\deg x+1} x \cdot dy$ (compare with proof of (2.2)).

(2.5.1) Suppose that $V = V_1 \oplus V_2$. We have evidently embeddings $Um(V_1,V_2) \hookrightarrow (Um(V_1 \oplus V_2) \hookleftarrow Um(V_2)$. Moreover, if $x \in C_p(V_1,V_2) = C_{p+1}(Um(V_1,V_2))$ and $y \in C_q(V_2) = C_{q+1}(Um(V_2))$ then x and y are in general position and $x \circ y \in C_{p+q}(V)$. In this way we get a pairing $C_*(V_1,V_2) \otimes C_*(V_2) \to C_*(V)$ of complexes and hence a pairing $H(V_1,V_2) \otimes H(V_2) \to H(V)$ of homology groups. To distinguish these everywhere defined pairings from the partially defined ones above we'll use the sign $\times$ to denote the latter pairings. If $x \in C_p(V_1,V_2)$, $y \in C_q(V_2)$, $\alpha \in Aff(V_1,V_2)$ then $\alpha x \times y = \alpha(x \times y)$ and hence the previous pairing defines a map $S(V_1,V_2) \otimes H(V_2) \to S(V)$. Finally, if $V = V_1' \oplus V_2'$ and $\dim V_1 = \dim V_1'$, $\dim V_2 = \dim V_2'$, then the images of the maps $S(V_1,V_2) \otimes H(V_2) \to S(V)$ and $S(V_1',V_2') \otimes H(V_2') \to S(V)$ evidently coincide.

(2.5.2) In the same notations as above we have pairings $C_*(V_1) \otimes C_*(V_2) \to C_*(V)$, $H(V_1) \otimes H(V_2) \to H(V)$. If $x \in C_p(V_1)$, $y \in C_q(V_2)$, $\alpha \in \mathrm{Aut}(V_1)$, $\beta \in \mathrm{Aut}(V_2)$ then $\alpha x \times \beta y = \mathrm{diag}(\alpha,\beta) \cdot (x \times y)$ and hence we get an induced pairing $S(V_1) \otimes S(V_2) \to S(V)$. These pairings are associative in the obvious sence and in particular $H_*(F) = \amalg H_n(F) = \amalg H(F^n)$ and $S_*(F) = \amalg S_n(F) = \amalg S(F^n)$ are naturally associative rings.

(2.5.3) Lemma. For any $a_1,\dots,a_p \in F^*$ we have:

$$\langle a_1\rangle \times \dots \times \langle a_p\rangle = \sum_{1\le i_1<\dots<i_K\le p} (-1)^K \cdot \langle a_1,\dots,\underset{i_1}{1},\dots,\underset{i_K}{1},\dots,a_p\rangle$$

Proof: $\langle a\rangle$ is the class modulo the action of $GL_1(F)$ of the cycle $(a)-(1)$ and hence $\langle a_1\rangle \times \dots \times \langle a_p\rangle$ is the class modulo the action of $GL_p(F)$ of the cycle

$$((a_1)-(1)) \times \dots \times ((a_p)-(1)) = \sum_{1\le i_1<\dots<i_K\le p} (-1)^{p-K}(e_1,\dots,a_{i_1}e_{i_1},\dots,a_{i_K}e_{i_K},\dots,e_p)$$

$$= \delta\Big(\sum_{1\le i_1<\dots<i_K\le p} (-1)^K(e_1,\dots,a_{i_1}e_{i_1},\dots,a_{i_K}e_{i_K},\dots,e_p,a_1e_1+\dots+a_pe_p)$$

$$\equiv \delta\Big(\sum_{1\le i_1<\dots<i_K\le p} (-1)^K(e_1,\dots,e_p,a_1e_1+\dots+e_{i_1}+\dots+e_{i_K}+\dots+a_pe_p)$$ modulo the action of $GL_p(F)$.

(2.6) The long exact sequence

$$0 \leftarrow \mathbb{Z} = C_o(F^n) \leftarrow C_1(F^n) \leftarrow \dots \leftarrow C_n(F^n) \leftarrow H_n(F) \leftarrow 0$$

of $GL_n(F)$-modules gives rise to the iterated connecting homomorphism

$$\varepsilon_n : H_n(GL_n(F),\mathbb{Z}) \longrightarrow H_o(GL_n(F),H_n(F)) = S_n(F).$$

(2.6.1) Lemma. a) The restriction of ε_n on $H_n(GL_{n-1}(F),\mathbb{Z})$ is zero.
b) For any $n,m \ge 0$ the following diagram commutes:

$$\begin{array}{ccc} H_n(GL_n(F)) \times H_m(GL_m(F)) & \longrightarrow & H_{n+m}(GL_{n+m}(F)) \\ \downarrow & & \downarrow \\ S_n(F) \times S_m(F) & \longrightarrow & S_{n+m}(F). \end{array}$$

Proof: a) follows from the fact that $C_1(F^n) \to C_o(F_n)$ has a $GL_{n-1}(F)$-linear splitting given by $r \to r \cdot e_n$. b) follows immediately from the definitions of direct sum operations on $H_i(GL_i(F))$ and S_i.

(2.7) For any n we have the natural map induced by the direct sum operation

$$\theta : \underbrace{F^* \times \dots \times F^*}_{n} = H_1(GL_1(F)) \times \dots \times H_1(GL_1(F)) \to H_n(GL_n(F)) \to$$
$$\to H_n(GL_n(F))/H_n(GL_{n-1}(F)).$$

(2.7.1) Lemma. The map θ is multilinear. Moreover, if $a_i + a_j = 1$ for some $i \neq j$, then $\theta(a_1,\dots,a_n) = 0$.

Proof: The first statement is evident. To prove the second, it suffices to check that $a \times (1-a) \in H_2(GL_1(F))$ for any $a \neq 0,1$. If α,β are commuting elements of the group G, then we'll denote by $(\alpha,\beta) \in H_2(G,\mathbb{Z})$ the homology class of the cycle $[\alpha,\beta]-[\beta,\alpha]$. With these notations, $a \times (1-a) = (\mathrm{diag}(a,1),\mathrm{diag}(1,1-a)) \equiv (\mathrm{diag}(a,a^{-1}),\mathrm{diag}(1,1-a))$ mod $H_2GL_1(F)$. From the Hochschild-Serre spectral sequence corresponding to the group extension $1 \to SL_n(F) \to GL_n(F) \to F^* \to 1$, we get the following exact sequence:

$$0 \to H_2(SL_n(F))_{F^*} \to H_2(GL_n(F)) \to H_2(F^*) \to 0.$$

If n is at least three then ([7],[8]) the group $H_2(SL_n(F))$ is equal to $K_2(F)$ and the action of F^* is trivial. On the other hand, the group $H_2(SL_2(F)) = K_2Sp(F)$ is generated by symbols $\{\{a,b\}\}$, $a,b \in F^*$, which are subject to well known relations ([8]). The action of F^* on $K_2Sp(F)$ is given by the formula $c \cdot \{\{a,b\}\} = \{\{a,bc\}\} - \{\{a,c\}\}$. This shows that $H_2(SL_2(F))_{F^*}$ also coincides with $K_2(F)$ and hence $H_2(GL_2(F)) \xrightarrow{\sim} H_2(GL(F))$. But, the image of $(\mathrm{diag}(a,a^{-1}),\mathrm{diag}(1,1-a))$ in $H_2(GL(F))$ coincides with $(\mathrm{diag}(a,a^{-1},1),\mathrm{diag}(1,1-a,(1-a)^{-1})) = \{a,1-a\} = 0$.

(2.7.2) <u>Corollary.</u> <u>θ defines a homomorphism (which we'll also denote by θ)</u> $K_n^M(F) \to H_n(GL_n(F))/H_n(GL_{n-1}(F))$.

(2.7.3) <u>Lemma.</u> <u>The composition</u> $K_n^M(F) \to H_n(GL_n(F))/H_n(GL_{n-1}(F)) \to S_n(F)$ <u>sends</u> $\{a_1,\dots,a_n\}$ to $\sum_{1\le i_1<\dots<i_K} (-1)^K \langle a_1,\dots,\underset{i_1}{1},\dots,\underset{i_K}{1},\dots,a_n\rangle$. This follows immediately from (2.5.3) and (2.6.1).

(2.7.4) <u>Corollary.</u> <u>The composition</u> $K_n^M(F) \to H_n(GL_n(F))/H_n(GL_{n-1}(F)) \to S_n(F) \to K_n^M(F)$ <u>is the identity map.</u>

§ 3. The main theorem

The natural action of the group $A_{n,m}(F)$ on the complex $C_*^{n,m}(F)$ gives rise to a spectral sequence whose E^1 term is equal to $E^1_{p,q} = H_p(A_{n,m}(F), C_q^{n,m}(F))$ and whose limit is equal to the graded object $H_*(A_{n,m}(F), H_{n,m}(F))$ with dimensions shifted by n. For any $0 \le q \le n$ the action of $A_{n,m}(F)$ permutes transitively the basis elements of $C_q^{n,m}(F)$ and the stabilizer of $(e_{n-q+1},\dots,e_n)$ coincides with $A_{n-q,m+q}(F)$. This shows that $E^1_{p,q} = H_p(A_{n-q,m+q}(F),\mathbb{Z}) = H_p(GL_{n-q}(F),\mathbb{Z})$. We'll denote this spectral sequence by $\varepsilon^{n,m}(F)$.

(3.1) Proposition. The natural embedding $H_n(F) \hookrightarrow H_{n,m}(F)$ induces isomorphisms $H_i(GL_n(F), H_n(F)) \xrightarrow{\sim} H_i(A_{n,m}(F), H_{n,m}(F))$, in particular $S_n(F) \xrightarrow{\sim} S_{n,m}(F)$.

Proof: The natural embedding $C_*^n(F) \hookrightarrow C_*^{n,m}(F)$ of complexes respects the action of $GL_n(F)$ and hence gives rise to a spectral sequence homomorphism $\varepsilon^n(F) \to \varepsilon^{n,m}(F)$. By the above remarks, this homomorphism is an isomorphism on E^1 terms and hence induces an isomorphism on limits.

(3.1.1) Corollary. If $V = V_1 \oplus V_2$ is a vector space over F, then the image of the pairing $H(V_1,V_2) \otimes H(V_2) \to H(V) \to S(V)$ coincides with the image of the pairing $S(V_1) \otimes S(V_2) \to S(V)$.

(3.2.1) It's easy to see that the differential $d^1_{p,q} : E^1_{p,q} = H_p(GL_{n-q}(F)) \to E^1_{p,q-1} = H_p(GL_{n-q+1}(F))$ is induced by the embedding $GL_{n-q}(F) \hookrightarrow GL_{n-q+1}(F)$ when q is odd and is zero when q is even.

(3.2.2) The edge homomorphism $H_n(GL_n(F)) = E^1_{n,o} \to E^\infty_{n,o} \to S_n(F)$ coincides evidently with the map ε_n from (2.6).

(3.3) We'll denote by $S_n^o \subset S_n$ the subgroup of decomposable elements, i.e. the image of $S_1 \otimes S_{n-1} \oplus \dots \oplus S_{n-1} \otimes S_1 \to S_n$. In view of (3.1.1) and (2.5.1), for any decomposition $F^n = V_1 \oplus V_2$ the image of $H(V_1,V_2) \otimes H(V_2) \to S(F^n) = S_n(F)$ lies in $S_n^o(F)$.

(3.3.1) Proposition. For any $\lambda, a_1,\dots,a_n \in F$ and any i the element $\langle a_1,\dots,\lambda a_i,\dots,a_n\rangle - \langle a_1,\dots,a_n\rangle$ lies in $S_n^o(F)$.

Proof: $\langle a_1,\dots,\lambda a_i,\dots,a_n\rangle - \langle a_1,\dots,a_n\rangle = \delta((e_1,\dots,\lambda^{-1}e_i,\dots,e_n,a) - (e_1,\dots,e_i,\dots,e_n,a)) = \delta((e_1,\dots,e_{i-1})\cdot((\lambda^{-1}e_i)-(e_i))\cdot(e_{i+1},\dots,e_n,a)) =$ the class modulo the action of $GL_n(F)$ of the cycle $d(e_1,\dots,e_{i-1})\cdot((\lambda^{-1}e_i)-(e_i))\cdot(e_{i+1},\dots,e_n,a) + (-1)^i(e_1,\dots,e_{i-1})\cdot((\lambda^{-1}e_i)-(e_i))\cdot d(e_{i+1},\dots,e_n,a)$. Put $u = a_1e_1+\dots+a_ie_i = a-a_{i+1}e_{i+1}-\dots-a_ne_n$. It's easy to see that for any $1 \le K \le i-1$, $i+1 \le \ell \le n$, the following vectors are linearly independent: $e_1,\dots,\hat{e}_K,\dots,\hat{e}_\ell,\dots,e_n,a,u; e_1,\dots,\hat{e}_K,\dots,e_n,u$. This shows that all the chains below are in general position and hence the computation makes sense. We have:

$(e_{i+1},\dots,e_n,a) = d(u,e_{i+1},\dots,e_n,a) + u\cdot d(e_{i+1},\dots,e_n,a), (e_1,\dots,e_{i-1}) =$
$= (-1)^{i-1}(d(e_1,\dots,e_{i-1},u) - d(e_1,\dots,e_{i-1})\cdot u)$ and the above cycle is equal to

$$d(e_1,\dots,e_{i-1})\cdot((\lambda^{-1}e_i) - (e_i))\cdot d(u,e_{i+1},\dots,e_n,a)$$
$$- d(e_1,\dots,e_{i-1},u)\cdot((\lambda^{-1}e_i) - (e_i))\cdot d(e_{i+1},\dots,e_n,a)$$
$$+ d(e_1,\dots,e_{i-1})\cdot[((\lambda^{-1}e_i) - (e_i))\cdot u + u\cdot((\lambda^{-1}e_i) - (e_i))]\cdot d(e_{i+1},\dots,e_n,a).$$

It suffices now to check that everyone of these three cycles is decomposable. Set V_2 equal to the subspace generated by $u,e_{i+1},\dots,e_n$ and V_1 to the subspace generated by $e_1,\dots,e_i$. Then $\dim V_2 = n-i+1$, $\dim V_1 = i$. The cycles $d(e_1,\dots,e_{i-1},u)\cdot((\lambda^{-1}e_i) - (e_i))$ and $d(e_1,\dots,e_{i-1})\cdot[((\lambda^{-1}e_i) - (e_i))\cdot u + u\cdot((\lambda^{-1}e_i) - (e_i))]$ lie in $H(V_1)$ and since the cycle $d(e_{i+1},\dots,e_n,a)$ is in general position with V_1 we can find V_2' such that $F^n = V_1 \oplus V_2'$ and $d(e_{i+1},\dots,e_n,a)$ lies in $H(V_2',V_1)$. This shows that the last two terms lie in the image of $H(V_1)\otimes H(V_2',V_1) \to S_n(F)$ and hence are decomposable. In the same manner, we show that the first term lies in the image of $H(V_1',V_2)\otimes H(V_2) \to S_n(F)$ for some V_1' such that $V_1'\oplus V_2 = F^n$ and hence is again decomposable.

(3.3.2) If $i = 1$ or n then the above argument needs slight modification. However, these cases are actually easier than the general case.

(3.3.3) Corollary. *For any* $a_1,\dots,a_n \in F^*$ *the element* $\langle a_1,\dots,a_n\rangle - \langle 1,\dots,1\rangle$ *is decomposable.*

(3.3.4) Proposition. a) $S_0(F) = \mathbb{Z}, S_1(F) = F^*$, b) $S_2(F) = K_2(F)\oplus \mathbb{Z}\cdot\langle 1,1\rangle$ c) *If* $n \geq 3$ *then* $S_n(F) = S_n^o(F)$.

Proof: a) is evident. b) From (2.7.4) we know that $K_2(F)$ is a direct summand of $S_2(F)$. On the other hand, we have a surjective mapping $S_2(F) \to \mathbb{Z} : \langle a,b\rangle \mapsto 1$ which is zero on $K_2(F)$. This shows that $S_2(F) = K_2(F)\oplus\mathbb{Z}\cdot\langle 1,1\rangle \oplus ?$. Finally, $? = 0$ by (3.3.3) since $S_2^o(F) = K_2(F)$. c) We have to show that the element $\langle 1,\dots,1\rangle$ is decomposable if $n \geq 3$. If n is odd then it follows immediately from the main relation on symbols $\langle a_1,\dots,a_n\rangle$ (see (2.3.1), and (3.3.3)). If n is even, we'll consider the element $\overbrace{\langle 1,\dots,1\rangle}^{n-2}\times\langle 1,1\rangle$. This cycle is the sum of an odd number of simplices. Moreover, if we count every simplex with its multiplicity, we get one. Using the standard procedure to present a cycle as a sum of generators (partial homotopy operators of (2.2)), we see that $\langle 1,\dots,1\rangle\times\langle 1,1\rangle = \sum n_i\langle a_1^i,\dots,a_n^i\rangle$ with $\sum n_i = 1$ and hence is congruent to $\langle 1,\dots,1\rangle$ modulo $S_n^o(F)$.

(3.3.5) Corollary. For any $n \geq 2$ *we have:* $S_n(F) = K_n^M(F)\oplus\langle 1,1\rangle\times S_{n-2}(F)$.

Proof: $K_n^M(F)$ is a direct summand of $S_n(F)$ by (2.7.4). It's easy to check that $\langle 1,1\rangle\times S_{n-2}(F)$ lies in the kernel of the map $S_n(F)\to K_n^M(F)$ and hence

$S_n(F) = K_n^M(F) \oplus \langle 1,1\rangle \times S_{n-2}(F) \oplus ?$. To finish the proof it suffices (in view of (3.3.4) and induction) to check the following formula:

(3.3.6) Lemma. For any $a \in F^*$, $\langle a\rangle \times \langle 1,1\rangle = \langle 1,1\rangle \times \langle a\rangle$.

Proof: $\langle a\rangle \times \langle 1,1\rangle - \langle 1,1\rangle \times \langle a\rangle = [(a) \times \langle 1,1\rangle - \langle 1,1\rangle \times (a)] - [(1) \times \langle 1,1\rangle - \langle 1,1\rangle \times (1)]$. The first cycle is obtained from the second by the action of the matrix $\mathrm{diag}(a,1,1)$ and hence their difference is zero in $S_3(F)$.

(3.4) Theorem. a) The differentials $d^2, d^3, \ldots$ in the spectral sequence $\varepsilon^n(F)$ are zero.

b) The map $S_{n-2}(F) \to S_n(F)$ given by $x \to \langle 1,1\rangle \times x$ is injective.

c) The map $H_i(GL_{n-1}(F)) \to H_i(GL_n(F))$ is an isomorphism if $i \leq n-1$

d) $\theta : K_n^M(F) \to H_n(GL_n(F))/H_n(GL_{n-1}(F))$ is an isomorphism.

Proof: We use induction on n. If $n = 1$ or 2, then everything is trivial. In the general case, we choose any cycle $\sigma \in H_2(F)$ representing $\langle 1,1\rangle$ and consider the map $C_*^{n-2}(F)(2) \to C_*^n(F)$ of complexes given by multiplication by $\sigma : x \mapsto \sigma \times x$. This map respects the action of $GL_{n-2}(F)$ and hence induces a map $\varepsilon'^{n-2}(F) \to \varepsilon^n(F)$ of the corresponding spectral sequences where $\varepsilon'^{n-2}(F)$ is $\varepsilon^{n-2}(F)$ shifted by two in the q direction. This map induces isomorphisms $E'^1_{p,q} \to E^1_{p,q}$ if $q \geq 2$. On the other hand $E'^1_{p,q} = 0$ if $q < 2$. This and the induction hypothesis prove immediately a). The induced map $S_{n-2}(F) \to S_n(F)$ on the limits coincides with multiplication by $\langle 1,1\rangle$. The filtration on $S_{n-2}(F)$ induced from ε' has the following form: $S^o_{n-2} = S^1_{n-2} = 0$ and the induced map $S^{i+1/i}_{n-2} \to S^{i+1/i}_n$ is an isomorphism if $i \geq 1$. This shows that $S_{n-2}(F) \to S_n(F) \to S_n(F)/S_n(F)^1$ is an isomorphism and hence $S_{n-2}(F) \hookrightarrow S_n(F)$. Moreover, this shows that $S_n(F)^1 \cap \langle 1,1\rangle \times S_{n-2}(F) = 0$ and since $S_n(F)^o = \varepsilon((H_n(GL_n(F))/H_n(GL_{n-1}(F))) \supset K_n^M(F)$, we see from (3.3.5) that $H_n(GL_n(F))/H_n(GL_{n-1}(F)) = E^2_{n,o} = E^\infty_{n,o} = S_n(F)^o$ coincides with $K_n^M(F)$ and $\ker(H_{n-1}(GL_{n-1}(F)) \to H_{n-1}(GL_n(F))) = E^2_{n-1,1} = E^\infty_{n-1,1} = S_n(F)^{1/o} = 0$. Finally, the terms $E^2_{p,q} = E^\infty_{p,q}$ with $p+q \leq n-1$ are zero, which proves c).

(3.4.1) Corollary. $S_n(F) = K_n^M(F) \oplus K_{n-2}^M(F) \oplus \ldots \oplus K_{n-2r}^M(F)$ where r is the largest integer such that $2r \leq n$.

§ 4. The map from Quillen K-theory to Milnor K-theory

Using theorem 3.4 we can define the homomorphism $H_n(GL(F)) \to K_n^M(F)$ as follows: $H_n(GL(F)) = H_n(GL_n(F)) \to H_n(GL_n(F))/H_n(GL_{n-1}(F)) = K_n^M(F)$ and it follows immediately from (2.6.1) that the resulting map $H_*(GL(F)) \to K_*^M(F)$ is a ring homomorphism. Combining this homomorphism with the Hurewicz map $K_n(F) = \pi_n(BGL(F)^+) \to H_n(BGL(F)^+) = H_n(GL(F))$ we get a map $K_n(F) \to K_n^M(F)$. It's very probable that this map coincides with the map constructed in [18] in terms of Mennicke symbols, but I don't know whether this is true. The map $K_*(F) \xrightarrow{\varphi} K_*^M(F)$ is not a ring homomorphism. It's very probable that it satisfies the following product formula:

(4.1) If $x \in K_n(F)$, $y \in K_m(F)$, then $\varphi(xy) = \frac{-(m+n-1)!}{(m-1)!(n-1)!} \cdot \varphi(x)\,\varphi(y)$. Below we'll prove this formula in the case $m=1$ (the case $m=2$ follows immediately since K_2 is generated by products of elements of K_1).

Products in K-theory are defined as follows (see [4]): The Tensor product on matrices defines a homomorphism $GL_n(F) \times GL_m(F) \xrightarrow{\psi_{n,m}} GL_{n,m}(F)$ (defined up to conjugation by a permutation matrix) and hence gives a well defined (up to homotopy) map $f_{n,m} : BGL_n(F)^+ \times BGL_m(F)^+ \to BGL_{mn}(F)^+ \to BGL(F)^+$. Now we can define, using the H-group structure on $BGL(F)^+$, a map $\gamma_{n,m}$ by the formula: $\gamma_{n,m} = f_{n,m} - f'_{n,m} - f''_{n,m}$, where $f'_{n,m}(x,y) = f_{n,m}(x_o,y), f''_{n,m}(x,y) = f_{n,m}(x,y_o)$ and x_o, y_o are distinguished points in respectively $BGL_n(F)^+$ and $BGL_m(F)^+$. The map $\gamma_{n,m}$ is homotopically trivial on $BGL_n(F)^+ \vee BGL_m(F)^+$ and hence defines (again since $BGL(F)^+$ is an H-group) a unique up to homotopy map $\hat{\gamma}_{n,m} : BGL_n(F)^+ \wedge BGL_m(F)^+ \to BGL(F)^+$ such that the diagram

$$
\begin{array}{ccc}
BGL_n(F)^+ \times BGL_m(F)^+ & \xrightarrow{\gamma_{n,m}} & \\
\downarrow & & BGL(F)^+ \\
BGL_n(F)^+ \wedge BGL_m(F)^+ & \xrightarrow{\hat{\gamma}_{n,m}} &
\end{array}
$$

commutes up to homotopy. The maps $\hat{\gamma}_{n,m}$ agree with one another (up to homotopy) and hence define a (unique up to weak homotopy) map $\hat{\gamma} : BGL(F)^+ \wedge BGL(F)^+ \to BGL(F)^+$ such that $\hat{\gamma}$ restricted to $BGL_n(F)^+ \wedge BGL_m(F)^+$ is homotopic to $\hat{\gamma}_{n,m}$. The pairing $\hat{\gamma}$ defines pairings on the homotopy groups of $BGL(F)^+$, i.e. on the K-groups of F.

(4.2) <u>Lemma.</u> <u>Suppose that</u> $x \in K_n(F)$, $y \in K_m(F)$. <u>Let</u> $\lambda : K_*(F) \to H_*(GL(F))$ <u>denote the Hurewicz homomorphism. Then</u> $\lambda(x \cdot y) = \psi_*(\lambda(x) \times \lambda(y)) - mn\lambda(x)\cdot\lambda(y)$.

<u>Proof:</u> Let $\lambda(x) = \sigma \in H_n(GL(F)), \lambda(y) = \tau \in H_m(GL(F))$. These are primitive homology classes, i.e. if $\Delta : GL \to GL \times \ldots \times GL$ is the diagonal embedding, then $\Delta_*(\sigma) = \sigma \times 1 \times \ldots \times 1 + \ldots + 1 \times 1 \times \ldots \times \sigma$ and the same for τ. We can suppose that $\sigma \in H_n(GL_n(F))$, $\tau \in H_m(GL_m(F))$. From the description of products in K-theory, given above, it follows immediately that $\lambda(x \cdot y) = (\gamma_{n,m})_*(\sigma \times \tau)$. We can decompose $\gamma_{n,m}$ as follows:

$$BGL_n^+ \times BGL_m^+ \xrightarrow{\Delta} (BGL_n^+ \times BGL_m^+) \times (BGL_n^+ \times BGL_m^+) \times (BGL_n^+ \times BGL_m^+)$$
$$\xrightarrow{f_{n,m} \times (-f'_{n,m}) \times (-f''_{n,m})} BGL^+ \times BGL^+ \times BGL^+ \xrightarrow{\Sigma} BGL^+ .$$

Since σ and τ are primitive we see that $\Delta(\sigma\times\tau) = (\sigma\times\tau)\times(1\times 1)\times(1\times 1)+$ $(\sigma\times 1)\times(1\times\tau)\times(1\times 1)+(\sigma\times 1)\times(1\times 1)\times(1\times\tau)+(-1)^{mn}(1\times\tau)\times(\sigma\times 1)\times(1\times 1)$ $(1\times 1)\times(\sigma\times\tau)\times(1\times 1)+(1\times 1)\times(\sigma\times 1)\times(1\times\tau)+(-1)^{mn}(1\times\tau)\times(1\times 1)\times(\sigma\times 1)+$ $(-1)^{mn}(1\times 1)\times(1\times\tau)\times(\sigma\times 1)+(1\times 1)\times(1\times 1)\times(\sigma\times\tau)$.

Since the restriction of the homomorphism ψ to $GL_n\times 1$ coincides with $\Delta : GL_n \to \underbrace{GL_n\times\ldots\times GL_n}_{m} \longrightarrow GL_{mn}$ and since σ is primitive, we deduce that $f_*(\sigma\times 1) = f''_*(\sigma\times 1) = m\sigma$, while $f'_*(\sigma\times 1)=0$. Similarly, $f_*(1\times\tau) = f'_*(1\times\tau) = n\tau$, $f''_*(1\times\tau) = 0$ and $f'_*(\sigma\times\tau) = f''_*(\sigma\times\tau) = 0$. Finally, the map $(-)_*$ coincides with the change of sign on primitive elements and hence, applying $(f\times(-f')\times(-f''))_*$ to the above sum, we get $f_*(\sigma\times\tau)\times 1\times 1 + m\sigma\times(-n\tau)\times 1 + (-1)^{mn}n\tau\times 1\times(-m\sigma) + (-1)^{mn}1\times(-n\tau)\times(-m\sigma)$. Now, applying Σ_* we get: $f_*(\sigma\times\tau) - mn\sigma\cdot\tau - (-1)^{mn}mn\tau\sigma + (-1)^{mn}mn\tau\sigma = f_*(\sigma\cdot\tau) - mn\sigma\cdot\tau$.

(4.3) Corollary. *If* $m=1$ *or* $n=1$ *then* $\varphi(xy) = -mn\varphi(x)\varphi(y)$.

Proof: In this case, $\psi_*(\lambda(x)\times\lambda(y)) \in H_{m+n}(GL_{mn}(F))$ goes to zero under the map $H_{m+n}(GL(F)) \to K^M_{m+n}(F)$.

(4.4) Corollary. *If* $a_1,\ldots,a_n \in F^*$ *and* $\ell(a_i)\in K_1(F) = F^*$ *are the corresponding elements, then* $\varphi(\ell(a_1)\cdot\ldots\cdot\ell(a_n)) = (-1)^{n-1}\cdot(n-1)!\cdot\{a_1,\ldots,a_n\}$.

The corollary shows that the image of $K_n(F)\to K^M_n(F)$ contains $(n-1)!\,K^M_n(F)$. It may be conjectured that this image actually coincides with $(n-1)!\,K^M_n(F)$, but this seems to be a very delicate question, as one sees for example from the following result.

(4.5) Proposition. *The image of the map* $K_3(F)\to K^M_3(F)$ *coincides with the kernel of the homomorphism* $K^M_3(F)\to I^3(F)/I^4(F)$ *defined by Milnor [6].*

This shows that for $n=3$ the above conjecture is equivalent to Milnor's conjecture [6] that $K^M_3(F)/2K^M_3(F) \xrightarrow{\sim} I^3(F)/I^4(F)$.

Proof of (4.5). From the Hochschild-Serre spectral sequence associated with the group extension $1\to K_2(F)\to St(F)\to SL(F)\to 1$, one sees easily that the homomorphism $K_3(F) = H_3(St(F))\to H_3(SL(F))$ is always surjective (moreover, the kernel of this map is always two-torsion and is zero if F is quadratically closed). From this, we see that $K^M_3(F)/im(K_3(F)) = H_3(GL(F))/H_3(SL(F)) + H_3(GL_2(F))$. Consider the Hochschild-Serre spectral sequence associated with the group extension $1\to SL_n(F)\to GL_n(F)\to$ $\to F^*\to 1$. It looks as follows:

$H_3(F^*)$	0	*	*
$H_2(F^*)$	0	*	*
$H_1(F^*)$	0	$H_1(F^*,H_2(SL_n(F))$	*
$H_o(F^*)$	0	$H_o(F^*,H_2(SL_n(F))$	$H_3(SL_n(F))_{F^*}$

Since the map $GL_n(F) \to F^*$ has a section, we see that all differentials starting in the first column are zero and hence $H_3(GL_n(F))/H_3(SL_n(F)) + H_3(GL_1(F)) = H_1(F^*, H_2(SL_n(F))$. This shows that $K_3^M(F)/\mathrm{im}(K_3(F)) = \mathrm{coker}(H_1(F^*, H_2(SL_2(F))) \to H_1(F^*, H_2(SL(F))))$. Consider the exact sequence of F^*-modules $0 \to R(F) \to K_2\,Sp(F) \to K_2(F) \to 0$ where $R(F) = \ker(K_2\,Sp(F) \to K_2(F))$. The exact homology sequence takes the following form

$$0 \to K_3^M(F)/\mathrm{im}(K_3(F)) \to R(F)_{F^*} \to K_2\,Sp\,(F)_{F^*} \to K_2(F) \to 0.$$

It was mentioned above that $K_2\,Sp\,(F)_{F^*} = K_2(F)$. Hence $K_3^M(F)/\mathrm{im}(K_3(F)) = R(F)_{F^*}$ and our result follows from:

(4.6) <u>Proposition.</u> $R(F) \simeq I^3(F)$ <u>as</u> F^*<u>-modules, where the action of</u> F^* <u>on</u> $I^3(F) \subset W(F)$ <u>is induced by the natural action of</u> F^* <u>on</u> $W(F)$ <u>and hence is given by the formula:</u>

$$a \cdot \langle\langle x,y,z\rangle\rangle = \langle\langle x,y,z\rangle\rangle + \langle\langle x,y,z,a\rangle\rangle .$$

<u>Proof:</u> One knows ([16], [17]) that the defining relations on the two-fold Pfister forms $\langle\langle x,y\rangle\rangle$ generating $I^2(F)$ are the same as the defining relations on the symbols $\{\{x,y\}\}$ generating $K_2\,Sp\,(F)$, except for the following additional relation on Pfister forms: $\langle\langle x^2,y\rangle\rangle = 0$. This gives an epimorphism $K_2\,Sp\,(F) \xrightarrow{\theta} I^2(F)$ sending $R(F)$ surjectively on $I^3(F)$ (compare [16], [17]). To finish the proof we have to show that $\ker\theta \cap R(F) = 0$. We have the homomorphism $K_2(F) \xrightarrow{\sigma} K_2\,Sp\,(F)$ given by the formula $\{x,y\} \to \{\{x^2,y\}\}$ and its image coincides with $\ker\theta$. Since the composition $K_2(F) \to K_2\,Sp\,(F) \to K_2(F)$ is just multiplication by two, we deduce that $(\ker\theta) \cap R(F) = \sigma({}_2K_2(F))$. But one knows from [5] that ${}_2K_2(F) = \{-1,F^*\}$ if $\mathrm{char}\,F \neq 2$ (${}_2K_2(F) = 0$ if $\mathrm{char}\,F = 2$) and hence $\sigma({}_2K_2(F)) = 0$.

§ 5. The characteristic class $c_{n,n} : K_n(F) \to K_n(F)$

We'll recall briefly the construction of the characteristic classes $c_{n,i} : K_n(X) \to H^i(X, \mathcal{K}_{n+i})$, ([11],[12],[13]).

(5.1.1) For any scheme X, we'll denote by $\mathcal{K}_n$ the sheaf in the Zariski topology on X, associated with the presheaf $\mathcal{U} \to K_n(\Gamma(\mathcal{U}, O_x))$. It's evident from the definition that for any $x \in X$, the stalk $(\mathcal{K}_n)_x$ coincides with $K_n(O_x)$.

(5.1.2) If G is a discrete group acting on a topological space X and $\mathcal{F}$ is an abelian sheaf on X, endowed with a linear G-action, then by $H^n(X,G;\mathcal{F})$ we'll denote the corresponding equivariant cohomology groups [2]. As usual, $H^1(X,G;O_X^*)$ classifies line bundles on X, endowed with a linear G-action compatible with projection on X.

(5.1.3) Suppose that X is a smooth variety over a field and G is a discrete group acting on X. Let E/X be a vector bundle of constant rank n endowed with a linear G-action compatible with projection $E \to X$. Consider the projectivization $P = \mathbb{P}(E)$ and the canonical line bundle $L = O_P(-1)$ on P. We have natural actions of G on P and L and by $\xi \in H^1(P,G;O_P^*) = H^1(P,G;\mathcal{K}_1)$ we'll denote the cohomology class corresponding to L. The following result is the principal tool for constructing characteristic classes with values in $\mathcal{K}$-cohomology groups. The proof of the weaker fact corresponding to the case of the trivial group G may be found in [14], [15].

(5.1.4) <u>Theorem</u> (Shechtman [13]). <u>The bigraded abelian group</u> $H^*(P,G;\mathcal{K}_*)$ <u>is isomorphic to</u> $H^*(X,G;\mathcal{K}_*) \oplus \xi \cdot H^*(X,G;\mathcal{K}_*) \oplus \ldots \oplus \xi^{n-1} \cdot H^*(X,G;\mathcal{K}_*)$, <u>i.e.</u> $H^i(P,G;\mathcal{K}_j) = H^i(X,G;\mathcal{K}_j) \oplus \xi \cdot H^{i-1}(X,G;\mathcal{K}_{j-1}) \oplus \ldots \oplus \xi^{n-1} \cdot H^{i-n+1}(X,G;\mathcal{K}_{j-n+1})$.

(5.1.5) Now we can act in the usual manner an construct equivariant characteristic classes $c_i(E) \in H^i(X,G;\mathcal{K}_i)$:

Express ξ^n in the form $-c_1\xi^{n-1} - \ldots - c_n$ and take $c_i(E) = c_i$. By definition $c_0(E) = 1 \in H^0(X,G;\mathcal{K}_0)$. The total characteristic class $c(E) = \sum_{i=0}^{n} c_i(E)$ has the usual additive property: if $0 \to E' \to E \to E'' \to 0$ is an exact sequence of G-bundles then $c(E) = c(E') \cdot c(E'')$.

(5.1.6) When the action of G on X is trivial, one can compute the equivariant cohomology of any trivial G sheaf as follows (see [11]): Let I^* be an injective resolution of $\mathcal{F}$ and $C_*(G)$ be a G-projective resolution of $\mathbb{Z}$ (for example the standard complex). Then the total complex of the bicomplex $\mathrm{Hom}(C_*(G), I^*)$ will be a G-injective resolution of $\mathcal{F}$ and hence $H^i(X,G;\mathcal{F})$ coincides with the homology of $\Gamma^G(\mathrm{Hom}(C_*(G),I^*)) = \mathrm{Hom}(C_*(G),\Gamma(I^*))^G = \mathrm{Hom}(C_*(G)_G, \Gamma(I^*))$. The natural map $H_*(\mathrm{Hom}(C_*(G)_G, \Gamma(I^*)) \to \mathrm{Hom}(H_*(C_*(G)_G), H_*(\Gamma(I^*)))$ defines homomorphisms $H^p(X,G;\mathcal{F}) \to \coprod_{i=0}^{p} \mathrm{Hom}(H_{p-i}(G;\mathbf{Z}), H^i(X,\mathcal{F}))$.

(5.1.7) Suppose that G acts trivially on X and E is a G-vector bundle of rank n over X. Then equivariant characteristic classes $c_p(E)$ give rise (in view of (5.1.6)) to homomorphisms $c_{k,p}(E) : H_k(G,\mathbb{Z}) \to H^{p-k}(X,\mathcal{K}_p)$ and clearly $c_{k,p}(E) = 0$ if $p > n = \operatorname{rank} E$.

(5.1.8) Suppose that $X = \operatorname{Spec} A$ is a smooth affine variety and take $G = GL_n(A)$. Then we have the natural action of G on the trivial vector bundle A^n and hence we get homomorphisms

$$c_{k,p}(A^n) : H_k(GL_n(A)) \to H^{p-k}(X,\mathcal{K}_p).$$

It's easy to see that $c_{k,p}(A^{n+1})$ restricted to $H_k(GL_n(A))$ is $c_{k,p}(A^n)$ and taking limits we get the canonical homomorphism $c_{k,p}(A^\infty) : H_k(GL(A)) \to H^{p-k}(X,\mathcal{K}_p)$. Combining these homomorphisms with the Hurewicz maps $K_k(A) \to H_k(GL(A))$, we get the desired characteristic classes $c_{k,p} : K_k(A) = K_k(X) \to H^{p-k}(X,\mathcal{K}_p)$.

(5.1.9) Suppose that X is smooth and irreducible. Denote by F the function field on X. Then $H^0(X,\mathcal{K}_p)$ may be identified with a certain subgroup of $K_p(F)$ ([9]) and the functorial character of the above constructions shows that the following diagram commutes:

$$\begin{array}{ccc} K_p(X) & \xrightarrow{c_{p,p}} & H^0(X,\mathcal{K}_p) \\ \downarrow & & \downarrow \\ K_p(F) & \xrightarrow{c_{p,p}} & H^0(\operatorname{Spec} F,\mathcal{K}_p) = K_p(F). \end{array}$$

This shows that to study the homomorphism $c_{p,p}$ for arbitrary X it's sufficient to study the case of a field, and in the latter case this homomorphism has the form $c_{p,p} : K_p(F) \to K_p(F)$. It's easy to see that $c_{1,1}$ is the identity map and from the product formula ([13]), which in this case takes the form $c_{m+n,m+n}(x,y) = -\frac{(m+n-1)!}{(m-1)!(n-1)!} \cdot c_{m,m}(x) \cdot c_{n,n}(y)$, where $x \in K_m(F)$, $y \in K_n(F)$, it follows that on decomposable elements, $c_{p,p}$ coincides with multiplication by $(-1)^{p-1}(p-1)!$. The following result is the main purpose of the paper.

(5.2) <u>Theorem.</u> <u>The homomorphism</u> $c_{p,p} : K_p(F) \to K_p(F)$ <u>coincides with the composition</u> $K_p(F) \xrightarrow{\varphi} K_p^M(F) \to K_p(F)$ <u>where the second map is the p-component of the natural ring homomorphism</u> $K_*^M(F) \to K_*(F)$ <u>induced by the identity map</u> $K_1^M(F) = F^* = K_1(F)$.

<u>Proof:</u> The homomorphism $c_{p,p}(F^{p-1}) : H_p(GL_{p-1}(F)) \to K_p(F)$ is zero by (5.1.7), and from the main theorem (3.4) we deduce that $c_{p,p}(F^\infty) : H_p(GL(F)) = H_p(GL_p(F)) \to K_p(F)$ defines a homomorphism $\lambda : K_p^M(F) = H_p(GL_p(F))/H_p(GL_{p-1}(F)) \to K_p(F)$. From the definition of $c_{p,p}$, we see immediately that $c_{p,p} = \lambda \circ \varphi$ and we have only to compute λ. For this purpose, since the isomorphism $K_p^M(F) \xrightarrow{\sim} H_p(GL_p)/H_p(GL_{p-1})$ is induced by the homomorphism $F^* \times \ldots \times F^* \to H_p(GL_1 \times \ldots \times GL_1) \to H_p(GL_p)$, we have to compute the

equivariant characteristic class $c_p(F^p) \in H^p(\mathrm{Spec}\ F, G; K_p) = H^p(G, K_p(F))$, where $G = GL_1(F) \times \ldots \times GL_1(F)$ is the group of diagonal matrices. The G-module F^p is the direct sum of one dimensional modules $F_{(i)}$, where $F_{(i)}$ is simply F, but the action of G on $F_{(i)}$ is given by the i-th projection $G \to GL_1$. From (5.1.5), we see that $c_p(F^p)$ is the direct product $c_1(F) \times \ldots \times c_1(F)$, where $c_1(F) \in H^1(GL_1(F), K_1(F)) = \mathrm{Hom}(F^*, F^*)$ is the identity map. Hence $\langle a_1 \times \ldots \times a_p, c_p(F^p)\rangle = \ell(a_1) \cdot \ldots \cdot \ell(a_p)$, where $\ell(a) \in K_1(F) = F^*$ is the element corresponding to a. This shows that $K_p^M(F) \to K_p(F)$ takes $\{a_1, \ldots, a_p\}$ to $\ell(a_1) \cdot \ldots \cdot \ell(a_p)$ and finishes the proof.

(5.3) Corollary. *If F is an algebraic extension of a finite field and $n \geq 2$, then the homomorphism $c_{n,n}$ is zero.*

Proof: If F is infinite, our statement follows from (5.2) and the fact that $K_n^M(F) = 0$. If F is finite, we have the commutative diagram

$$\begin{array}{ccc} K_n(F) & \xrightarrow{c_{n,n}} & K_n(F) \\ \downarrow & & \downarrow \\ K_n(\overline{F}) & \xrightarrow{c_{n,n}} & K_n(\overline{F}) \end{array}$$

where $\overline{F}$ is the algebraic closure of F and it's sufficient to remark that $K_n(F) \to K_n(\overline{F})$ is injective - see [10].

References

[1] Bass, H., Tate, J.: The Milnor ring of a global field. Lect. Notes Math., 1973, V. 342, p. 349-446

[2] Grothendieck, A.: Sur quelques points de l'algebre homologique. Tohoku Math. J., 1957, V. 9, N 2, 3

[3] Kallen, van der W.: Homology stability for general linear groups. Invent. Math., 1980, V. 60, N 3, p. 269-295

[4] Loday, J.L.: K-theorie algebrique et representation de groupes. Ann. Sci. Ecole Norm. Supér., 1976, V. 9, p. 309-337

[5] Mercurjev, A.S., Suslin, A.A.: K-cohomology of Severi-Brauer varieties and norm residue homomorphism. - Preprint LOMI, 1982, 27 p.

[6] Milnor, J.: Algebraic K-theory and quadratic forms. Invent. Math., 1970, V. 9, N 4, p. 318-344

[7] Milnor, J.: Introduction to algebraic K-theory. Ann. Math. Stud., 1971, N 72, 184 p.

[8] Moore, C.C.: Group extensions of p-adic and adelic linear groups. Publ. Math. IHES, 1969, N 35, p. 5-70

[9] Quillen, D.: Higher K-theory I. Lect. Notes Math., 1973, V. 341, p. 85-147

[10] Quillen, D.: On the cohomology and K-theory of the general linear group over a finite field. - Ann. Math., 1972, V. 96, N 3, p. 552-586

[11] Soulé, C.: K-théorie des anneaux d'entiers de corps de nombres et cohomologie étale. - Invent. Math., 1979, V. 55, N 3, p. 251-295

[12] Shechtman, V.V.: Algebraic K-theory and characteristic classes. Uspekhi Mat. Nauk, 1978, V. 33, N 6, p. 239-240

[13] Shechtman, V.V.: Riemann-Roch theorem in algebraic K-theory and the Atiyah-Hirzebruch spectral sequence. Uspekhi Mat. Nauk, 1980, V. 35, N 6, p. 179-180

[14] Sherman, C.: Some theorems on the K-theory of coherent sheaves. Commun. Algebra, 1979, V. 7, N 14, p. 1489-1508

[15] Sherman, C.: K-cohomology of regular schemes. Commun. Algebra, 1979, V. 7, N 10.

[16] Suslin, A.A.: On the cancellation problem for projective modules. Preprint LOMI, 1977

[17] Suslin, A.A.: Cancellation problem for projective modules and related topics. Lect. Notes Math., 1979, V. 734, p. 323-338

[18] Suslin, A.A.: Mennicke symbols and their applications in the K-theory of fields. Preprint LOMI, 1981, 24 p.

Generating the cokernel of $K_3\mathbf{Z} \to K_3\mathbf{F}_7$

by

Michael R. Stein*

Let R be a commutative local ring, and I an ideal of R such that R/I is finite. In [St] I presented a method for computing $K = \mathrm{cok}(K_3 R \to K_3 R/I)$, along with some general theorems which result when R/I is a finite field and $R^* \to (R/I)^*$ is a <u>split</u> surjection. The general method is algorithmic and actually produces a (usually very large) set of generators for K , but even when the order of K is known in advance, it is a difficult and tedious task to reduce the set of generators to a small and manageable one. Indeed, the only non-split cases I was able to do by hand were $R/I = \mathbf{F}_5$ and $R/I = \mathbf{Z}/4$ [St,Propositions 3.7 and 3.8].

A great deal of the tedium in the method lies in creating the list of generators in the first place; this step is also a great source of computational errors. Thus I was led to write a computer program which would set up the list of generators for K , perform some of the easier reductions, and eliminate redundancies before human intervention was required. This note is an account of some new results obtained in this way.

---

*This work was partially supported by the National Science Foundation.

§1. The method.

We will be concerned here only with the case where $I = \underline{m}$, the maximal ideal of R . Since R is local, the groups $K_2(R)$, $K_2(R,\underline{m})$ have presentations with generators $\langle a , b\rangle$, $a,b \in R$, $1-ab \in R^*$; and certain relations discussed below [K;M-S]. (For $K_2(R,\underline{m})$ we require a or $b \in \underline{m}$ as well.) We let $F(R,I)$ ($I=R$ or $\underline{m}$) be the free abelian group on generators

$$[a , b] , (a , b) \in R \times I \cup I \times R , 1-ab \in R^* ,$$

and $S(R,I)$ the subgroup generated by the elements

$$(D2^{op})_I(a,b,c) = [b,a] [c,a] [b+c-abc,a]^{-1}$$
$$(D3)_I(a,b,c) = [a,bc] [ab,c]^{-1} [ac,b]^{-1}$$
$$(D4^{op})_I(a) = [1,a] .$$

Then I showed in [St] that $F(R,I)/S(R,I) \approx D(R,I)$ ($\approx K_2(R,I)$ by [K]) .

Consider the diagram

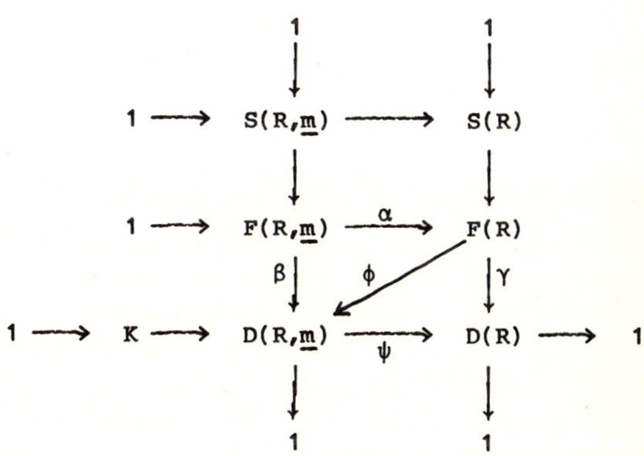

where ϕ is any homomorphism such that $\phi\alpha = \beta$ and $\psi\phi = \gamma$.

Then

$$\begin{aligned}\phi(S(R)) &= K \\ &= \ker(K_2(R,\underline{m}) \longrightarrow K_2(R)) \\ &= \operatorname{cok}(K_3 R \longrightarrow K_3 R/\underline{m}) \ .\end{aligned}$$

A method for constructing such a ϕ was given in [St]:

Let $s\colon R/\underline{m} \longrightarrow R$ be a fixed choice of representatives for the residue classes mod $\underline{m}$, with $s(\overline{0}) = 0$, $s(\overline{1}) = 1$. Then define

$$\phi[a,b] = \phi[sa,sb] \left\langle \frac{a-sa}{1-sa.sb}, sb \right\rangle \left\langle a, \frac{b-sb}{1-a.sb} \right\rangle ,$$

where $\phi[0,x] = \phi[x,0] = 1$, and the other elements $\phi[sa,sb]$ remain to be chosen. Then [St,1.10-1.12] the finite set of elements

$$\begin{aligned}&\phi(D2^{op}(sa,sb,sc)) \\ (1) \qquad &\phi(D3(sa,sb,sc)) \\ &\phi(D4^{op}(sa)) = \phi([1,sa])\end{aligned}$$

generate K.

How do we define $\phi[sa,sb]$ in general? Clearly it is desirable to choose these elements to trivialize as many of the relations (1) as possible in $D(R,\underline{m})$. In particular, we set $\phi[1,x] = \phi[x,1] = 1$. We then write out the list of elements

$$\phi(D2^{op}(sa,sb,sc))$$

$$\phi(D3(sa,sb,sc)$$

and solve for the unknown elements $\phi[sa,sb]$ wherever they occur alone; substituting these values allows us to solve for other of the $\phi[sa,sb]$ and so on, until we are reduced to a list of elements of $D(R,\underline{m})$ which generate K.

Schematically, this reduction process goes as follows:

(A) List the generators $\phi(D2^{op}(sa,sb,sc))$ and $\phi(D3(sa,sb,sc))$ of K.

(B) Find a generator on the list in which only one unknown term $\phi[sa,sb]$ appears; equate that generator to 1 and solve for $\phi[sa,sb]$ as a product of elements of $D(R,\underline{m})$. Delete this generator from the list. Repeat (B) until no such generators remain on the list.

(C) Substitute the solved-for values of the $\phi[sa,sb]$ from (B) into the remaining elements on the list.

(D) Go to (B).

We continue in this way until no more substitutions are possible (usually because all the $\phi[sa,sb]$ have been solved for). The remaining elements on the list generate K.

To the above sketch we may add 2 more steps:

(E) Simplify the remaining list as much as possible by using the relations in $D(R,\underline{m})$.

(F) Prove that the remaining list is a minimal set of generators for K.

For example, take $R = Z_{(5)}$, $I = (5)$ and let $s(\overline{2}) = 2$, $s(\overline{3}) = 3$, $s(\overline{4}) = -1$. We must define $\phi[2,2]$, $\phi[2,-1]$, $\phi[-1,2]$, $\phi[3,3]$, $\phi[3,-1]$, $\phi[-1,3]$ (since $1 - 2.3 \notin R^*$ there is no generator [2,3]!). The list of generators for K will include

$$\phi(D2^{op}(sa,sb,sc)) \quad \text{when} \quad 1 - sa.sb\ ,\ 1 - sa.sc \in R^*$$

$$\phi(D3^{op}(sa,sb,sc)) \quad \text{when} \quad 1 - sa.sb.sc \in R^*\ .$$

Since these are symmetric in sb,sc , it suffices to consider the elements for which sb < sc . Still, there are initially 24 generators for K in this case, and it takes a good half-hour of hand computation to prove they are trivial in $D(\mathbb{Z}_{(5)},5)$, i.e. that $K_3(\mathbb{Z}_{(5)}) \longrightarrow K_3\mathbb{F}_5$ is onto.

It is easy to see that if $R/\underline{m}$ is a finite field with q elements, this method will produce an initial list of about q^3 generators for K .

§2. The computations.

Since steps (A)-(D) above are algorithmic in nature, I wrote a PASCAL program to produce the list of generators for K and do the needed solving and substituting. The program will produce such a list for $R = \mathbb{Z}_{(p)}$ and $\underline{m} = (p^k)$ (any $k > 1$) ; the programmer must specify a choice of representatives $s: \mathbb{F}_p \longrightarrow \mathbb{Z}_{(p)}$. Some of the more mechanical parts of (E) are also written into the program (e.g cancelling x with x^{-1}).

As I showed in [St], the closer the choice of representatives is to being multiplicative, the smaller the number of generators produced for K . Thus for $p = 7$, it was advantageous to choose the representatives $0, \pm 1, \pm 2, \pm \frac{1}{2}$. This resulted in a list of 124 generators, 57 of type D2 and 67 of type D3. A few (nine) of these were eliminated by the program, either by cancelling inverses or by routine use of relation D2. The rest required extensive hand calculation, in the course of which the following identities were discovered.

Theorem 1. The following elements of $D(\mathbb{Z}_{(7)},(7))$ are trivial:

$$\langle 1/2 , 14/9\rangle , \langle 1/2 , -7\rangle , \langle -98/81 , -1/2\rangle , \langle 14 , -1/2\rangle ,$$
$$\langle 14 , -1\rangle , \langle -7/4 , -1/2\rangle , \langle 2 , -7/4\rangle , \langle -1 , 7/5\rangle$$
$$\langle -1 , 14/15\rangle , \langle -1 , 7/12\rangle , \langle 14/25 , 2\rangle .$$

Theorem 2. Each of the following elements equals $\langle 7, -1\rangle$ in $D(\mathbf{Z}_{(7)},(7))$:

$\langle -1/2, 14/9\rangle$, $\langle 1/2, -14/9\rangle$, $\langle -1/2, -14/9\rangle$, $\langle -1/2, 7\rangle$,

$\langle -14, 1/2\rangle$, $\langle -1/2, -7/8\rangle$, $\langle 7/8, 1/2\rangle$, $\langle 7/4, 1/2\rangle$,

$\langle 7/4, 5/8\rangle$, $\langle -2, 7/4\rangle$, $\langle -1, -7/9\rangle$, $\langle -1, 7/2\rangle$,

$\langle -1, 7/3\rangle$, $\langle -1, 7/8\rangle$, $\langle -1, -7/10\rangle$, $\langle 4, 7\rangle$.

Theorem 3. $K = \ker(K_2(\mathbf{Z}_{(7)},7)) \to K_2(\mathbf{Z}_{(7)})$ is cyclic of order 2, generated by $\langle 7, -1\rangle = \{8, -1\}$.

Theorem 4. The following relations hold in $D(\mathbf{Z}_{(7)},(7))$:

$$\langle -1, 7\rangle = \langle 2, 14\rangle^2$$
$$\langle -1, 7\rangle = \langle 1/2, 7/2\rangle^2$$
$$\langle -1, 7\rangle = \langle 1/2, -7/4\rangle^4 .$$

Hence the exact sequence

$$1 \longrightarrow K \longrightarrow D(\mathbf{Z}_{(7)},(7)) \longrightarrow D(\mathbf{Z}_{(7)}) \longrightarrow 1$$

is not split.

Proof. $\langle 2, 14\rangle = \{2, -27\}$, and the tame symbol at 3 maps this element to the residue class of $\dfrac{2^3}{(-27)^0} = 8$ in $\mathbf{F}_3^*$. Hence $\langle 2, 14\rangle$ is a non-trivial element of $K_2(\mathbf{Z}_{(7)},(7))$. If $K \xrightarrow{i} D(\mathbf{Z}_{(7)}(7))$ were split by a homomorphism s, we would have $\langle 7, -1\rangle = si\langle 7, -1\rangle = s(\langle 2, 14\rangle^2)$ $= (s(\langle 2, 14\rangle))^2 = 1$, since $|K| = 2$. (Weibel is responsible for observing that this sequence doesn't split.) □

Theorems 1 and 2 are proved by direct computation. For example,

$$\langle 1/2\ ,\ 14/9\rangle\ \langle 9/4\ ,\ 14/9\rangle = \langle 1\ ,\ 14/9\rangle = 1$$

$$\langle 9/4\ ,\ 14/9\rangle\ \langle 3/2\ ,\ 3/2 \cdot 14/9\rangle^2 = \langle 3/2\ ,\ 7/3\rangle^2$$

$$\langle 3/2\ ,\ 7/3\rangle\ \langle 3/5\ ,\ 7/3\rangle = \langle 0\ ,\ 7/3\rangle = 1$$

$$\langle -1\ ,\ 7/3\rangle\ \langle 3/5\ ,\ 7/3\rangle = \langle 1\ ,\ 7/3\rangle = 1$$

Hence:

$$\begin{aligned}\langle \tfrac{1}{2}\ ,\ \tfrac{14}{9}\rangle &= \langle \tfrac{9}{4}\ ,\ \tfrac{14}{9}\rangle^{-1}\\ &= \langle \tfrac{3}{2}\ ,\ \tfrac{7}{3}\rangle^{-2}\\ &= \langle \tfrac{3}{5}\ ,\ \tfrac{7}{3}\rangle^{2}\\ &= \langle -1\ ,\ \tfrac{7}{3}\rangle^{-2}\\ &= \langle \tfrac{7}{3}\ ,\ -1\rangle^{2}\\ &= 1\ .\end{aligned}$$

Also

$$\begin{aligned}\langle -1\ ,\ 7\rangle\ \langle 4\ ,\ 7\rangle &= \langle 31\ ,\ 7\rangle\\ &= \langle 1\ ,\ 7\rangle\ \langle -5\ ,\ 7\rangle\\ &= \langle 1\ ,\ 7\rangle\ \langle 1\ ,\ 7\rangle\ \langle 1\ ,\ 7\rangle\\ &= 1\ .\end{aligned}$$

Using the computations given in Theorems 1 and 2, we can proceed to simplify the list of generators of K given by our procedure. Here are some examples of those generators:

$\phi(D2^{op}(-1 , -1/2 , -1/2)) = \langle 7 , -1\rangle$

$\phi(D2^{op}(-1/2 , 2 , 2)) = \langle -14/9 , -1/2\rangle^{-3} \langle -7/4 , -1/2\rangle$

$\phi(D2^{op}(-1 , 2 , 2)) = \langle -1/2 , 14/25\rangle^{-2} \langle 7/5 , -1\rangle^{2} \langle -7/9 , -1\rangle$

$\phi(D2^{op}(1/2 , 1 , 1/2)) = \langle 1/2 , 14/3\rangle^{-1} \langle -1/2 , 14/5\rangle \langle -14/3 , 1/2\rangle$

$\phi(D2^{op}(-2 , -2 , -2)) = \langle -14/27 , -2\rangle^{-2}$

$\phi(D3(1 , -1/2 , -1)) = \langle -14/9 , -1/2\rangle^{-1}$

$\phi(D3(-1/2 , -1/2 , -2)) = \langle -7/15 , -2\rangle \langle 7/6 , -2\rangle$

$\phi(D3(-1 , 2 , 2)) = \langle -1/2 , 14/5\rangle^{-1} \langle -14/5 , 1/2\rangle \langle -2 , -7/20\rangle^{2} \langle -1 , 7/3\rangle$

$\phi(D3(2 , -1 , 1/2)) = \langle -14/9 , -1/2\rangle^{-1} \langle -14 , 1/2\rangle^{-1}$

$\phi(D3(-2 , 1/2 , 1/2)) = \langle -1/2 , 14/5\rangle^{2} \langle -14/5 , -1/2\rangle^{-2}$

When this simplification is done, we find that each generator on our list is either trivial, or equals $\langle 7 , -1\rangle$. Since Browder [Br] has shown that $|\mathrm{cok}(K_3\mathbf{Z} \longrightarrow K_3\mathbf{F}_7)| = 2$, we see that $\langle 7 , -1\rangle$ must be non-trivial in K (because $K_3\mathbf{Z} \longrightarrow K_3\mathbf{Z}_{(7)}$ is onto by the localization sequence, $\mathrm{cok}(K_3\mathbf{Z}_{(7)} \longrightarrow K_3\mathbf{F}_7) \approx \mathrm{cok}(K_3\mathbf{Z} \longrightarrow K_3\mathbf{F}_7))$.

§3. Some questions.

As noted above, the localization sequence shows that

$$\mathrm{cok}(K_3\mathbf{Z} \to K_3\mathbf{F}_7) \xrightarrow[\approx]{} \mathrm{cok}(K_3\mathbf{Z}_{(7)} \to K_3\mathbf{F}_7)$$

and the latter is cyclic of order 2 , generated by $\langle 7 , -1\rangle$. Now $\langle 7 , -1\rangle$ is an element of $K_2(\mathbf{Z}_{(7)},(7))$, but <u>not</u> of $K_2(\mathbf{Z} ,(7))$, because $1 - (-1)7 = 8 \notin \mathbf{Z}^*$. Indeed it is clear that the inverse image of $\langle 7 , -1\rangle$ in $K_2(\mathbf{Z},7)$ cannot be a Steinberg or Dennis-Stein symbol.

Question 1: Find an expression for $\langle 7 , -1\rangle$ as an element of $K_2(\mathbb{Z},7)$.

Question 2: Find generators for $\mathrm{cok}(K_3R \to K_3R/\underline{m})$ in other cases where $|R/\underline{m}|$ is small.

Hand computations are under way for $\mathbb{Z}/8$ and $\mathbb{Z}/9$, but little beyond that seems approachable now. Further progress will require theoretical improvements which reduce the initial set of generators for K substantially. It will probably also be necessary to implement an interactive version of the computer program so that the effect of various relations on the generators for K can be tried out easily.

Question 3: Find an explicit generator for $K_3(\mathbb{F}_7)$ using our knowledge that K is generated by $\langle 7 , -1\rangle$.

Loday [unpublished] has discussed some ideas of how to go about this by "lifting" the proof that $\langle 7 , -1\rangle$ is trivial in $K_2(\mathbb{Z}_{(7)})$:

$$\begin{aligned}\langle 7 , -1\rangle &= \langle 1 , -1\rangle \langle 3 , -1\rangle \\ &= \langle 1 , -1\rangle \langle 1 , -1\rangle \langle 1 , -1\rangle \\ &= 1 .\end{aligned}$$

His method depends on an explicit use of the isomorphism $K_3(R) \approx H_3(St(R),\mathbb{Z})$ using an exact sequence of Gersten's.

References

[Br] W. Browder, Letter to R.K. Dennis (May 26, 1978).

[K] F. Keune, The relativization of K_2 , J. Algebra 54 (1978) 159-177.

[M-S] H. Maazen and J. Stienstra, A presentation for K_2 of split radical pairs, J. Pure Appl. Algebra 10 (1977) 271-294.

[St] M.R. Stein, Maps of rings which induce surjections on K_3 , J. Pure Appl. Algebra 21 (1981) 23-49.

UNE NOUVELLE FAMILLE DE GROUPES EN L-THEORIE ALGEBRIQUE

Pierre VOGEL

§0.- INTRODUCTION

Les groupes de chirurgie $L_n(A)$ ont été définis par Wall [12], du moins lorsque A est un anneau de groupe. Si n est pair, les éléments de $L_n(A)$ sont représentés par des formes quadratiques et, si n est impair, $L_n(A)$ est un quotient du groupe unitaire U(A). En fait, il y a plusieurs définitions possibles suivant que l'on considère les groupes L_n^h, L_n^s ou L_n^p .

Dans [12], Wall suggère une nouvelle définition des groupes $L_n(A)$ en considérant des formes quadratiques de degré n sur des complexes algébriques et cette idée a été reprise et développée considérablement par Ranicki [8].

Soient $\mathcal{A} \supset \mathcal{B}$ deux classes de A-modules différentiels gradués, projectifs de type fini, stables par équivalence d'homotopie, par extension et par dualité. On se propose de définir ici des groupes $L_n(\mathcal{A},\mathcal{B})$, en considérant des formes quadratiques sur des complexes de $\mathcal{A}$, le défaut de dégénérescence étant dans $\mathcal{B}$. Ces groupes sont des généralisations des groupes de chirurgie classiques et apparaissent dans de nombreuses situations, comme le montrent les exemples suivants:

Soit $K \subset S^{n+2}$ un noeud. Si M est le complémentaire d'un voisinage tubulaire ouvert de K, on a une application normale de M dans $B^{n+1} \times S^1$ qui est un isomorphisme sur le bord. Comme l'application de M dans $B^{n+1} \times S^1$ est une équivalence d'homologie, la construction standard donne une forme quadratique non dégénérée de degré n+2 sur un $Z[Z]$-complexe Z-acyclique.

Ainsi, si $\mathcal{A}$ est la classe des $Z[Z]$-complexes C tels que $C \otimes_{Z[Z]} Z$ est acyclique, et si $\mathcal{B}$ est la classe des $Z[Z]$-complexes acycliques, on obtient un élément du groupe $L_{n+2}(\mathcal{A},\mathcal{B})$, et on vérifie que cela donne un isomorphisme du groupe de noeuds P.L C_n sur le groupe $L_{n+2}(\mathcal{A},\mathcal{B})$.

Si maintenant le noeud $K \subset S^{n+2}$ vérifie la conjecture de Smith modulo p, c'est-à-dire que le revêtement de S^{n+2} à p feuillets, ramifié le long de K, est une sphère d'homologie, alors l'application de M dans $B^{n+1} \times S^1$ est une équivalence de $\mathbb{Z}[\mathbb{Z}/p]$-homologie, et l'on obtient une forme quadratique non dégénérée de degré n+2 sur un $\mathbb{Z}[\mathbb{Z}]$-complexe $\mathbb{Z}[\mathbb{Z}/p]$-acyclique.

Soit alors $\mathcal{A}_p$ la classe des $\mathbb{Z}[\mathbb{Z}]$-complexes C tels que $C \otimes_{\mathbb{Z}[\mathbb{Z}]} \mathbb{Z}[\mathbb{Z}/p]$ est acyclique. On obtient un isomorphisme du groupe de Cappell et Shaneson [2] de noeuds vérifiant la conjecture de Smith modulo p, dans le groupe $L_{n+2}(\mathcal{A}_p, \mathcal{B})$

Soit $A \to \Lambda$ une localisation injective d'anneaux avec involution. Soient $\mathcal{A}$ la classe des A-complexes, $\mathcal{B}$ la classe des A-complexes Λ-acycliques et $\mathcal{C}$ la classe des A-complexes acycliques. Alors la théorie de localisation en L-théorie peut s'interpréter de la façon suivante:

$$L_n(\mathcal{A},\mathcal{C}) = L_n(A)$$
$$L_n(\mathcal{A},\mathcal{B}) = L_n(\Lambda)$$
$$L_n(\mathcal{B},\mathcal{C}) = L_{n+1}(\Lambda,A)$$

et l'on a une longue suite exacte:

$$\ldots \to L_n(\mathcal{B},\mathcal{C}) \to L_n(\mathcal{A},\mathcal{C}) \to L_n(\mathcal{A},\mathcal{B}) \to L_{n-1}(\mathcal{B},\mathcal{C}) \to \ldots$$

Les groupes $L_n(\mathcal{A},\mathcal{B})$ possèdent un certain nombre de bonnes propriétés qui peuvent se résumer de la façon suivante: les groupes L_n se comportent vis-à-vis des paires de classes $\mathcal{A} \supset \mathcal{B}$ comme les groupes d'homologie H_n vis-à-vis des paires d'espaces topologiques. En particulier, si $\mathcal{A} \supset \mathcal{B} \supset \mathcal{C}$ sont trois classes, on a une longue suite exacte comme ci-dessus; on a également un théorème d'excision et une suite exacte de Mayer-Vietoris.

Si A est commutatif noetherien régulier, les calculs peuvent se poursuivre beaucoup plus loin, et l'on obtient, entre autres, une suite spectrale convergeant vers $L_*(A)$ analogue à la suite spectrale de Quillen convergeant vers la K-théorie de A. On obtient ainsi des résultats assez comparables à ceux de W. Pardon [5], [6]. Les techniques utilisées sont cependant assez différentes car Pardon travaille essentiellement avec des groupes de Witt sur certaines catégories de modules.

Cet article est consacré à l'étude complète des groupes $L_n(\mathcal{A},\mathcal{B})$ dans le cas général. Le cas où A est commutatif est traité en collaboration avec J. Barge et J.J. Sansuc dans [1].

Table des matières

§1. Enoncé des principaux résultats

§2. Quelques points d'algèbre homologique

§3. Formes quadratiques

§4. Les groupes $L_n(\mathcal{A},\mathcal{B})$

§5. La longue suite exacte

§6. Transversalité et excision

§7. Localisation

§1- ENONCE DES PRINCIPAUX RESULTATS

Soient A un anneau unitaire muni d'une involution: $a \to \bar{a}$, et ε un élément du centre de A tel que $\varepsilon\bar{\varepsilon} = 1$. Si e est l'un des symboles s, h, p ou p', on désignera par $C^e(A)$ la catégorie des A-modules (à droite) différentiels $\mathbb{Z}$-gradués:

$$\ldots \to C_n \to C_{n-1} \to \ldots$$

où chaque module C_n est un module de type fini, projectif si $e = p$ ou p', libre si $e = h$, et libre muni d'une base si $e = s$, et où de plus, dans le cas $e \neq p'$, les modules C_n sont tous nuls sauf pour un nombre fini d'entre eux.

Les objets de $C^e(A)$ seront appelés des complexes.

Soit $0 \to C \to C' \to C'' \to 0$ une suite de complexes de $C^e(A)$. On dira que cette suite est e-exacte si elle est exacte et si, dans le cas $e = s$, elle induit, pour chaque dimension, une suite exacte courte de torsion nulle dans le groupe $\tilde{K}_1(A)/\varepsilon$, $\tilde{K}_1(A)$ étant le groupe de K-théorie relative du morphisme d'anneaux $\mathbb{Z} \to A$ [4], [7], [9]

Si C est un complexe de $C^e(A)$, on désignera par $\hat{C}$ le complexe dual défini par:

$$\hat{C}_n = \mathrm{Hom}\,(C_{-n}, A)$$

la différentielle étant choisie de façon que l'évaluation de $\hat{C} \otimes C$ dans A soit un cocycle.

En remarquant que l'involution sur A induit sur chaque A-module à gauche une structure de A-module à droite (et réciproquement), on vérifie que $\hat{C}$ est un complexe de $C^e(A)$ (si $e = s$, il faut choisir sur $\hat{C}$ les bases duales).

<u>DEFINITION</u> 1.1. Une classe $\mathcal{A} \subset C^e(A)$ sera dite exacte si $\mathcal{A}$ contient tous les complexes ayant le type d'homotopie (simple, si $e = s$) du complexe 0, et si, pour chaque suite e-exacte : $0 \to C \to C' \to C'' \to 0$ de complexes de $C^e(A)$, on a la propriété suivante:

Si deux des complexes C, C' et C" appartiennent à $\mathcal{A}$, le troisième appartient également à $\mathcal{A}$.

La plus grande classe exacte est $C^e(A)$. la plus petite classe exacte, formée des complexes ayant le type d'homotopie (simple, si $e = s$) de 0 sera notée $\mathcal{O}$.

<u>DEFINITION</u> 1.2. Soit $\mathcal{A}$ une classe exacte de $C^e(A)$. On désignera par $\hat{\mathcal{A}}$ la classe des complexes $C \in C^e(A)$ tels que $\hat{C}$ appartient à $\mathcal{A}$. La classe $\hat{\mathcal{A}}$ est exacte.

La classe $\mathcal{A}$ sera dite symétrique si $\mathcal{A} = \hat{\mathcal{A}}$.

<u>DEFINITION</u> 1.3. Soit $\mathcal{A}$ une classe exacte de $C^e(A)$. On dira qu'un morphisme $f \in C^e(A)$ est une $\mathcal{A}$-équivalence si le cône de f appartient à $\mathcal{A}$.

<u>EXEMPLE</u> 1.4. Soit M un A-module à gauche (resp. à droite). Alors la classe $\mathcal{A}$ des complexes C tels que $H_*(C,M)$ (resp. $H^*(C,M)$) est nul est une classe exacte. De plus, les $\mathcal{A}$-équivalences sont les équivalences de M-homologie (resp. de M-cohomologie).

<u>EXEMPLE</u> 1.5. Soit $A \to B$ un morphisme d'anneaux et β un sous-groupe de $\tilde{K}_1(B)$. Soit $\mathcal{A}$ la classe des complexes $C \in C^s(A)$ tels que $C \otimes B$ soit acyclique avec torsion dans β. Alors $\mathcal{A}$ est exacte et les $\mathcal{A}$-équivalences sont les équivalences de B-homologie avec torsion dans β.

Si, de plus, $A \to B$ est un morphisme d'anneaux à involution et si β est stable par involution, $\mathcal{A}$ est symétrique.

Soient C et C' deux complexes de $C^e(A)$. On désignera par $Hom(C,C')_n$ l'ensemble des homorphismes A-linéaires de degré n de C dans C'; Hom(C,C') est muni d'une structure de $\mathbf{Z}$-module différentiel gradué en posant, pour tout $f \in Hom(C,C')$ et tout $x \in C$:

$$\partial^\circ(f(x)) = \partial^\circ f + \partial^\circ x$$

$$d(f(x)) = (df)(x) + (-1)^{\partial^\circ f} f(dx) .$$

Si l'on pose, de plus, pour tout $f \in Hom(C,C')$ est $u \in \hat{C}'$:

$$\hat{f}(u) = (-1)^{\partial^\circ f \, \partial^\circ u} u \circ f, ,$$

on obtient un morphisme $f \to \hat{f}$ de Hom(C,C') dans $Hom(\hat{C}',\hat{C})$ qui respecte les degrés et les différentielles.

<u>NOTATION</u> 1.6. Soit C un complexe de $C^e(A)$. On désignera par B(C) le $\mathbf{Z}$-module différentiel gradué $Hom(C,\hat{C})$. L'application composée :
$Hom(C,\hat{C}) \to Hom(\hat{\hat{C}}, \hat{C}) \simeq Hom(C,\hat{C})$ est une involution sur B(C) (toujours notée $f \to \hat{f}$) et B(C) est un $\mathbf{Z}[\mathbf{Z}/2]$-module différentiel gradué.

On désignera par $B(C)^\varepsilon$ le module différentiel gradué B(C) muni de la nouvelle involution $f \to \varepsilon\hat{f}$.

<u>DEFINITION</u> 1.7. Soit C un complexe de $C^e(A)$. On notera $Q_n^\varepsilon(C)$ le groupe d'hyper-homologie $H_n(\mathbf{Z}/2,B(C)^\varepsilon)$. On appellera n-forme ε-quadratique sur C un élément de $Q_n^\varepsilon(C)$. Un n-complexe ε- quadratique sera un couple (C,q), $q \in Q_n^\varepsilon(C)$.

Soit $C \to C'$ un épimorphisme de degré 0 entre deux complexes de $C^e(A)$. On notera $Q_n^\varepsilon(C \to C')$ le groupe $H_n(\mathbf{Z}/2, B(C)^\varepsilon/B(C')^\varepsilon)$. On appellera n-forme ε-quadratique sur $C \to C'$ un élément de $Q_n^\varepsilon(C \to C')$. Une n-paire ε-quadratique sera un couple $(C \to C',q)$, $q \in Q_n^\varepsilon(C \to C')$.

<u>DEFINITION</u> 1.8. Soit $\mathcal{A}$ une classe exacte symétrique dans $C^e(A)$.

Soit (C,q) un n-complexe ε-quadratique. On dira que q ou (C,q) est $\mathcal{A}$-non dégénérée si l'image de q par l'application composée:

$$H_n(\mathbf{Z}/2, B(C)^\varepsilon) \xrightarrow{\text{transfert}} H_n(1, B(C)^\varepsilon) \approx H_n(B(C))$$

est représentée par une $\mathcal{A}$-équivalence de C dans $\hat{C}$.

Soit $(C \to C',q)$ une n-paire ε-quadratique. Soit K le noyau de $C \to C'$. On dira que q ou $(C \to C', q)$ est $\mathcal{A}$-non dégénérée si l'image de q par l'application composée:

$$H_n(\mathbf{Z}/2,\ B(C)^{\varepsilon}/B(C')^{\varepsilon}) \xrightarrow{\text{transfert}} H_n(B(C)/B(C')) \longrightarrow H_n(\mathrm{Hom}(K,\hat{C}))$$

est représentée par une $\mathcal{A}$-équivalence de K dans $\hat{C}$.

<u>DEFINITION</u> 1.9. Soient $\mathcal{A} \supset \mathcal{B}$ deux classes exactes symétriques de $\mathcal{C}^e(A)$. On désignera par ${}_{\varepsilon}L_n^e(\mathcal{A},\mathcal{B})$ l'ensemble des n-complexes ε-quadratiques $\mathcal{A}$-non dégénérés (C,q), $C \in \mathcal{A}$, soumis à la relation de cobordisme suivante:

(C,q) et (C',q') sont cobordants s'il existe une n+1-paire ε-quadratique $\mathcal{B}$-non dégénérée $(\Sigma \to C \oplus C', u)$ telle que Σ appartienne à $\mathcal{A}$ et $\partial u = q \oplus -q'$.

Les groupes ${}_{\varepsilon}L_n^e(\mathcal{A},\mathcal{B})$ ainsi définis sont des généralisations des groupes de chirurgie classiques. Ainsi le groupe ${}_{\varepsilon}L_n^e(A)$ (pour $e = s$, h ou p) est isomorphe à ${}_{\varepsilon}L_n^e(\mathcal{C}^e(A),0)$ et ${}_{\varepsilon}L_n^{p'}(\mathcal{C}^{p'}(A),0)$ est isomorphe à ${}_{\varepsilon}L_n^p(A)$.

En outre, si α est un sous-groupe de $\tilde{K}_0(A)$ stable par involution, ${}_{\varepsilon}L_n^{\alpha}(A)$ est le groupe ${}_{\varepsilon}L_n^p(\mathcal{A},0)$, $\mathcal{A}$ étant la classe des complexes de $\mathcal{C}^p(A)$ dont la classe est dans α.

Si α est un sous-groupe de $\tilde{K}_1(A)/\varepsilon$ stable par involution, ${}_{\varepsilon}L_n^{\alpha}(A)$ est le groupe ${}_{\varepsilon}L_n^s(\mathcal{C}^s(A),\mathcal{B})$, $\mathcal{B}$ étant la classe des complexes acycliques de $\mathcal{C}^s(A)$ avec torsion dans α.

Ceci est démontré par Ranicki [8] mais dans un cadre un peu différent: tout d'abord il utilise un point de vue dual du point de vue adopté ici; de plus, les complexes qu'il considère pour définir les groupes ${}_{\varepsilon}L_n$ sont -1-connexes et de dimension n, alors qu'ici on ne prend aucune restriction de ce genre.

<u>PROPOSITION</u> 1.10. <u>Soient</u> $\mathcal{A} \supset \mathcal{B}$ <u>deux classes exactes symétriques</u>. <u>Pour tout</u> $n \in \mathbf{Z}$, <u>on a des isomorphismes canoniques</u> :

$${}_{\varepsilon}L_{n+2}^e(\mathcal{A},\mathcal{B}) \xrightarrow{\sim} {}_{-\varepsilon}L_n^e(\mathcal{A},\mathcal{B})$$

<u>PROPOSITION</u> 1.11. <u>Soient</u> $\mathcal{A} \supset \mathcal{B}$ <u>deux classes exactes symétriques de</u> $\mathcal{C}^e(A)$. <u>Soit</u> a <u>une unité du centre de</u> A. <u>Alors on a des isomorphismes</u> :

$${}_{\varepsilon}L_n^e(\mathcal{A},\mathcal{B}) \xrightarrow{\sim} {}_{\varepsilon'}L_n^e(\mathcal{A},\mathcal{B})\ ,\quad \text{avec}\quad \varepsilon' = \varepsilon\frac{a}{\bar{a}}$$

<u>THEOREME</u> 1.12. <u>Soient</u> $\mathcal{A} \supset \mathcal{B} \supset \mathcal{C}$ <u>trois classes exactes symétriques de</u> $\mathcal{C}^e(A)$. <u>Alors on a une longue suite exacte</u> :

$$\dots \to {}_\varepsilon L_n^e(\mathcal{B},\mathcal{C}) \to {}_\varepsilon L_n^e(\mathcal{A},\mathcal{C}) \to {}_\varepsilon L_n^e(\mathcal{A},\mathcal{B}) \to {}_\varepsilon L_{n-1}^e(\mathcal{B},\mathcal{C}) \to \dots$$

Ainsi ${}_\varepsilon L_n^e$ se comporte vis-à-vis d'une paire de classes exactes symétriques comme une théorie homologique vis-à-vis d'une paire d'espaces topologiques. On en déduit, par exemple, que si $\mathcal{C}^e(A)$ est filtré par des classes exactes symétriques $\mathcal{A}_i$ $\mathcal{O} = \mathcal{A}_0 \subset \mathcal{A}_1 \subset \dots \subset \mathcal{A}_n = \mathcal{C}^e(A)$, on a une suite spectrale dont le terme E^1 est :

$$E^1_{pq} = {}_\varepsilon L^e_{p+q}(\mathcal{A}_p,\mathcal{A}_{p-1})$$

et qui converge vers ${}_\varepsilon L^e_{p+q}(A) = {}_\varepsilon L^e_{p+q}(\mathcal{C}^e(A),\mathcal{O})$.

<u>DEFINITION</u> 1.13. Soient C un complexe de $\mathcal{C}^e(A)$ et $\mathcal{A}$ une classe exacte. On dira que C est inversible modulo $\mathcal{A}$ s'il existe un complexe $C' \in \mathcal{C}^e(A)$ tel que $C \oplus C'$ appartienne à $\mathcal{A}$.

Soient $\mathcal{A} \supset \mathcal{B}$ deux classes exactes de $\mathcal{C}^e(A)$ telles que tout complexe $C \in \mathcal{A}$ est inversible modulo $\mathcal{B}$. Alors la relation

$$C \sim C' \Leftrightarrow C \oplus \Sigma C' \in \mathcal{B}$$

est une relation d'équivalence sur $\mathcal{A}$ et l'ensemble quotient est un groupe abélien que l'on notera $K(\mathcal{A},\mathcal{B})$.

<u>EXEMPLE</u> 1.14. Si $\mathcal{A}$ est la classe des complexes acycliques de $\mathcal{C}^s(A)$, alors $K(\mathcal{A},\mathcal{O})$ est égal à $\tilde{K}_1(A)/\varepsilon$.

Si $\mathcal{A}$ est égal à $\mathcal{C}^p(A)$ et si $\mathcal{B}$ est la classe des complexes $C \in \mathcal{C}^p(A)$ tels que $[C]$ soit nul dans $\tilde{K}_0(A)$, alors $K(\mathcal{A},\mathcal{B})$ est égal à $\tilde{K}_0(A)$.

<u>THEOREME</u> 1.15. <u>Soient</u> $\mathcal{A} \supset \mathcal{B}$ <u>deux classes exactes symétriques telles que tout complexe</u> $C \in \mathcal{A}$ <u>soit inversible modulo</u> $\mathcal{B}$. <u>Alors on a un isomorphisme canonique</u> :

$${}_\varepsilon L_n^e(\mathcal{A},\mathcal{B}) \simeq \hat{H}^n(Z/2,K(\mathcal{A},\mathcal{B}))$$

Les théorèmes 1.12 et 1.15 permettent de retrouver les suites exactes de Ranicki-Rothenberg. Par exemple, la suite exacte :

$$\ldots \to {}_{\varepsilon}L_n^h(A) \to {}_{\varepsilon}L_n^p(A) \to \hat{H}^n(\mathbb{Z}/2, \tilde{K}_0(A)) \to {}_{\varepsilon}L_{n-1}^h(A) \to \ldots$$

s'obtient en choisissant (dans 1.12) e = p et les classes suivantes :

$\mathcal{A} = \mathcal{C}^p(A)$

$\mathcal{B}$ est la classe des complexes $C \in \mathcal{C}^p(A)$ tels que $[C]$ soit nul dans $\tilde{K}_0(A)$

$\mathcal{C}$ est la classe nulle $\mathcal{O}$.

De même, la suite exacte reliant ${}_{\varepsilon}L_n^s(A)$ et ${}_{\varepsilon}L_n^h(A)$ s'obtient en prenant e = s, et les classes suivantes: la classe $\mathcal{C}^s(A)$, la classe des complexes acycliques et la classe nulle .

Les groupes ${}_{\varepsilon}L_n$ possèdent en outre la propriété d'excision. Cette propriété n' est cependant valable que pour les classes transverses au sens suivant:

<u>DEFINITION</u> 1.16. Soient $\mathcal{A}$ et $\mathcal{B}$ deux classes exactes de $\mathcal{C}^e(A)$. On dira que $\mathcal{A}$ est transverse à $\mathcal{B}$ et l'on écrira $\mathcal{A} \pitchfork \mathcal{B}$, si, pour tout $C \in \mathcal{A}$ et tout $C' \in \mathcal{B}$, tout morphisme de C dans C' se factorise à travers un complexe de $\mathcal{A} \cap \mathcal{B}$.

Si $\mathcal{A}$ et $\mathcal{B}$ sont deux classes exactes de $\mathcal{C}^e(A)$, on notera $\mathcal{A} + \mathcal{B}$ la classe exacte engendrée par $\mathcal{A} \cup \mathcal{B}$.

<u>THEOREME</u> 1.17 (<u>excision</u>). <u>Soient</u> $\mathcal{A}$ <u>et</u> $\mathcal{B}$ <u>deux classes exactes symétriques</u>, $\mathcal{A} \pitchfork \mathcal{B}$. <u>Alors on a un isomorphisme canonique</u> :

$${}_{\varepsilon}L_n^e(\mathcal{A}, \mathcal{A} \cap \mathcal{B}) \xrightarrow{\sim} {}_{\varepsilon}L_n^e(\mathcal{A} + \mathcal{B}, \mathcal{B})$$

Des théorèmes 1.12 et 1.17, on déduit immédiatement le théorème suivant :

<u>THEOREME</u> 1.18 (<u>Mayer-Vietoris</u>). <u>Soient</u> $\mathcal{A}$ <u>et</u> $\mathcal{B}$ <u>deux classes exactes symétriques</u>, $\mathcal{A} \pitchfork \mathcal{B}$. <u>Soit</u> $\mathcal{C}$ <u>une classe exacte symétrique contenue dans</u> $\mathcal{A} \cap \mathcal{B}$. <u>On a alors une longue suite exacte</u> :

$$\ldots \to {}_{\varepsilon}L_n^e(\mathcal{A} \cap \mathcal{B}, \mathcal{C}) \to {}_{\varepsilon}L_n^e(\mathcal{A}, \mathcal{C}) \oplus {}_{\varepsilon}L_n^e(\mathcal{B}, \mathcal{C}) \to {}_{\varepsilon}L_n^e(\mathcal{A} + \mathcal{B}, \mathcal{C}) \to \ldots$$

<u>THEOREME</u> 1.19. <u>Soit</u> $\mathcal{A}$ <u>une classe exacte de</u> $\mathcal{C}^e(A)$. <u>On suppose que</u> $\mathcal{A}$ <u>est transverse à</u> $\hat{\mathcal{A}}$. <u>Alors les groupes</u> ${}_{\varepsilon}L_n^e(\mathcal{A} + \hat{\mathcal{A}}, \mathcal{A} \cap \hat{\mathcal{A}})$ <u>sont nuls</u>.

Si A est commutatif, les classes exactes sont en général transverses, comme le montre l'exemple suivant:

EXEMPLE 1.20. On suppose A commutatif noetherien. Si C est un complexe de $C^e(A)$ on note Supp(C) (support de C) l'ensemble des idéaux premiers p de A tels que $C_{(p)}$ ne soit pas acyclique. Le support de C est un fermé du spectre de A. Soit S une union de fermés de Spec A. Alors la classe A_S des complexes de $C^e(A)$ dont le support est dans S est une classe exacte.

THEOREME 1.21 [1]. *Soient S et S' deux unions de fermés de Spec A. Alors les classes A_S et $A_{S'}$ sont transverses et l'on a :*

$$A_S \cap A_{S'} = A_{S\cap S'} \quad \text{et} \quad A_S + A_{S'} \subset A_{S\cup S'}$$

De plus, si A est régulier et e = p, on a : $A_{S\cup S'} = A_S + A_{S'}$

Si A n'est pas commutatif, on peut assez facilement trouver des classes non transverses :

EXEMPLE 1.22. Soit F un groupe libre non commutatif. Soit A l'anneau K[F], K étant un corps. Soit A la classe des complexes C de $C^e(A)$ tels que $H_*(C)$ soit de type fini sur K. Alors A et $\hat{A}$ ne sont pas transverses.

D'autres propriétés des groupes ${}_\varepsilon L_n$ concernent la localisation.

Soit $A \to \Lambda$ un morphisme d'anneaux. On dira que $A \to \Lambda$ est une localisation de A si Λ est obtenu par inversion formelle d'un système de matrices à coefficients dans A [3].

THEOREME 1.23. *Soit $A \to \Lambda$ une localisation de A qui soit un morphisme d'anneaux avec involution. Soit α un sous-groupe de $\tilde{K}_1(\Lambda)/\varepsilon$ stable par involution. Soit A la classe des complexes $C \in C^s(A)$ tels que $C \otimes \Lambda$ soit acyclique avec torsion dans α. Alors, on a un isomorphisme :*

$${}_\varepsilon L_n^s(C^s(A),A) \xrightarrow{\sim} {}_\varepsilon L_n^\alpha(\Lambda)$$

THEOREME 1.24. *Soit $A \to \Lambda$ une localisation de A qui soit un morphisme d'anneaux avec involution. On suppose que les groupes $\mathrm{Tor}_i^A(\Lambda,\Lambda)$ sont nuls pour $i \geq 2$.*

Soient $A \supset B$ deux classes symétriques de $C^e(A)$ ($e \neq p'$) contenant tous les complexes $C \in C^e(A)$ tels que $C \otimes \Lambda$ soit dans la classe nulle de $C^e(\Lambda)$. Soit $A \otimes \Lambda$ (resp. $B \otimes \Lambda$) la classe exacte de $C^e(\Lambda)$ engendrée par les complexes $C \otimes \Lambda$, $C \in A$

(<u>resp</u>. $C \in \mathcal{B}$).

<u>Alors, la tensorisation par Λ induit un isomorphisme</u> :

$$ {}_{\varepsilon}L^{e}_{n}(A,B) \xrightarrow{\sim} {}_{\varepsilon}L^{e}_{n}(A \otimes \Lambda, B \otimes \Lambda) $$

<u>REMARQUE</u> 1.25. Si $\mathcal{W}$ est la classe des complexes $C \in \mathcal{C}^{e}(A)$ (e = s ou h) tels que $C \otimes \Lambda$ appartienne à la classe nulle de $\mathcal{C}^{e}(A)$, on déduit des théorèmes 1.12 et 1.23 une longue suite exacte :

$$ \ldots \xrightarrow{\partial} {}_{\varepsilon}L^{e}_{n}(\mathcal{W},\mathcal{O}) \to {}_{\varepsilon}L^{e}_{n}(A) \to {}_{\varepsilon}L^{e}_{n}(\Lambda) \xrightarrow{\partial} {}_{\varepsilon}L^{e}_{n-1}(\mathcal{W},\mathcal{O}) \to \ldots $$

C'est la longue suite exacte de localisation démontrée dans [9].

Si e est égal à p, on a une suite analogue, α étant l'image de $\tilde{K}_0(A) \to \tilde{K}_0(\Lambda)$:

$$ \ldots \xrightarrow{\partial} {}_{\varepsilon}L^{p}_{n}(\mathcal{W},\mathcal{O}) \to {}_{\varepsilon}L^{p}_{n}(A) \to {}_{\varepsilon}L^{\alpha}_{n}(\Lambda) \xrightarrow{\partial} {}_{\varepsilon}L^{p}_{n-1}(\mathcal{W},\mathcal{O}) \to \ldots $$

§2 - <u>QUELQUES POINTS D'ALGEBRE HOMOLOGIQUE</u>

2.1 Soient C et C' deux complexes de $\mathcal{C}^{e}(A)$. On appellera aplication de degré n de C dans C' un élément de $\mathrm{Hom}(C,C')_n = \prod_{p\in\mathbb{Z}} \mathrm{Hom}(C_p,C'_{n+p})$. Les applications de C dans C' forment un $\mathbb{Z}$-module différentiel gradué Hom(C,C'). On appellera morphisme de degré n de C dans C' une application f de $\mathrm{Hom}(C,C')_n$ telle que df = 0. On appellera homotopie de degré n de C dans C' une application de la forme dg, $g \in \mathrm{Hom}(C,C')_{n+1}$.

2.2 <u>Le cône d'un morphisme</u>

Soient C et C' deux complexes de $\mathcal{C}^{e}(A)$ et f un morphisme de C dans C'. Soient M(f) le module $C \oplus C'$, i et i' les inclusions canoniques de C et C' dans M(f) et p et p' les projections canoniques de M(f) sur C et C'. Soit n un entier; on peut alors choisir sur M(f) la graduation et la différentielle de façon que l'on ait :

$$ \partial^\circ i' = n \quad \partial^\circ p' = -n \quad \partial^\circ i = 1+n+\partial^\circ f \quad \partial^\circ p = -1-n-\partial^\circ f $$

$$ di' = 0 \quad dp' = fp \quad di = -(-1)^n i'f \quad dp = 0 $$

M(f) sera appelé le (n-)cône du morphisme f; M(f) est un complexe de $\mathcal{C}^{e}(A)$.

$$ 0 \longrightarrow C' \underset{i'}{\overset{p'}{\rightleftarrows}} M(f) \underset{p}{\overset{i}{\rightleftarrows}} C \longrightarrow 0 $$

Soit $\mathcal{A}$ une classe exacte de $C^e(A)$. On dira qu'un morphisme f est une $\mathcal{A}$-équivalence si M(f) (bien défini à suspension près) appartient à $\mathcal{A}$.

LEMME 2.3. *Soient C, C', et C" trois complexes de* $C^e(A)$, $f: C \to C'$ *et* $g: C' \to C''$ *deux* $\mathcal{A}$-*équivalences. Alors* gf *est une* $\mathcal{A}$-*équivalence.*

Démonstration. Il suffit de remarquer que l'on a une suite e-exacte :

$$0 \to M(f) \to M(gf) \oplus M(1_{C'}) \to M(g) \to 0$$

LEMME 2.4. *Considérons le diagramme commutatif suivant dans* $C^e(A)$

$$\begin{array}{ccccccccc} 0 & \longrightarrow & C & \longrightarrow & C' & \longrightarrow & C'' & \longrightarrow & 0 \\ & & {\scriptstyle f}\downarrow & & {\scriptstyle g}\downarrow & & {\scriptstyle h}\downarrow & & \\ 0 & \longrightarrow & C_1 & \longrightarrow & C'_1 & \longrightarrow & C''_1 & \longrightarrow & 0 \end{array}$$

où toutes les flèches sont des morphismes et où les lignes sont e-exactes. Alors, si deux des trois morphismes f, g et h sont des $\mathcal{A}$-*équivalences, le troisième l'est.*

Démonstration. On a la suite e-exacte :

$$0 \to M(f) \to M(g) \to M(h) \to 0$$

LEMME 2.5. *Considérons le diagramme commutatif suivant dans* $C^e(A)$:

$$\begin{array}{ccccccccc} 0 & \longrightarrow & C & \xrightarrow{\alpha} & C' & \xrightarrow{\alpha'} & C'' & \longrightarrow & 0 \\ & & {\scriptstyle f}\downarrow & & {\scriptstyle g}\downarrow & & & & \\ 0 & \longrightarrow & C_1 & \xrightarrow{\beta} & C'_1 & \xrightarrow{\beta'} & C''_1 & \longrightarrow & 0 \end{array}$$

où toutes les flèches sont des morphismes et où les lignes sont e-exactes. Soit λ *une application de* C' *dans* C'_1 *qui rende commutatif le diagramme :*

$$\begin{array}{ccc} C & \xrightarrow{\alpha} & C' \\ {\scriptstyle f}\downarrow & \swarrow{\scriptstyle \lambda} & \downarrow{\scriptstyle g} \\ C'_1 & \xrightarrow{\beta'} & C''_1 \end{array}$$

Soit h *le morphisme de* C" *dans* C_1 *tel que* : $d\lambda = \beta h \alpha'$. *Alors, si deux des trois morphismes* f, g *et* h *sont des* $\mathcal{A}$-*équivalences, le troisième l'est.*

Démonstration. Soient $M(\beta)$ le 0-cône de β, et i, i', p et p' les applications définies dans 2.2. On vérifie que $g' = i'\lambda + ih\alpha'$ est un morphisme de C' dans $M(\beta)$, et le diagramme suivant est commutatif, à lignes e-exactes :

$$\begin{array}{ccccccccc} 0 & \longrightarrow & C & \xrightarrow{\alpha} & C' & \xrightarrow{\alpha'} & C'' & \longrightarrow & 0 \\ & & \downarrow f & & \downarrow g' & & \downarrow h & & \\ 0 & \longrightarrow & C'_1 & \xrightarrow{i'} & M(\beta) & \xrightarrow{p} & C_1 & \longrightarrow & 0 \end{array}$$

D'après le lemme 2.4, si deux des trois morphismes f, g' et h sont des A-équivalences, le troisième l'est.

Or g est égal à $\beta'p'g'$ et $\beta'p'$ est une équivalence d'homotopie (simple si e = s) On en déduit donc le lemme.

§3 -FORMES QUADRATIQUES

Soit W la résolution suivante du $\mathbb{Z}[\mathbb{Z}/2]$-module $\mathbb{Z}$:

$$\mathbb{Z}[\mathbb{Z}/2]e_0 \longleftarrow \mathbb{Z}[\mathbb{Z}/2]e_1 \longleftarrow \ \ldots$$

avec, pour tout $i \geqslant 1$: $\quad de_i = (1+(-1)^i t)e_{i-1}$

t étant le générateur du groupe $\mathbb{Z}/2$.

Si C est un complexe de $C^e(A)$, $B(C)^\varepsilon$ est le $\mathbb{Z}[\mathbb{Z}/2]$-module différentiel gradué $\mathrm{Hom}(C,\hat{C})$, l'involution étant définie par : $f \to \varepsilon\hat{f}$.

On désignera par $\mathcal{Q}^\varepsilon(C)$ le $\mathbb{Z}$-module différentiel gradué suivant:

$$\mathcal{Q}^\varepsilon(C) = W \underset{\mathbb{Z}/2}{\otimes} B(C)^\varepsilon$$

Ainsi, une n-forme ε-quadratique sur C est un élément de $Q_n^\varepsilon(C) = H_n(\mathcal{Q}^\varepsilon(C))$.

Un élément de $\mathcal{Q}^\varepsilon(C)$ est une somme finie de la forme :

$$u = \Sigma\, e_i \otimes u_i \ , \text{ avec } \quad \partial^0 u = i+\partial^0 u_i \ , \ u_i \in B(C) \ ,$$

et l'on a :

$$d(\Sigma\, e_i \otimes u_i) = \Sigma e_i \otimes (u_{i+1} - (-1)^i \varepsilon\hat{u}_{i+1} + (-1)^i du_i) \ .$$

<u>NOTATION</u> 3.2. Si $u = \Sigma\, e_i \otimes u_i$ est un élément de $\mathcal{Q}^\varepsilon(C)$, on posera :

$$\tilde{u} = u_0 + \varepsilon\hat{u}_0 \in B(C)$$

$$S(u) = \Sigma e_{i+1} \otimes (-1)^i \varepsilon\hat{u}_i \in \mathcal{Q}^{-\varepsilon}(C)$$

On vérifie aisément la formule suivante :

<u>FORMULE</u> 3.3. Pour tout $u \in \mathcal{Q}^\varepsilon(C)$, on a : $d(Su) + S(du) = e_0 \otimes \tilde{u}$.

NOTATION 3.4. Soient C et C' deux complexes de $C^e(A)$ et α et β deux applications de C dans C' de degrés respectifs p et q.

Si $u = \Sigma e_i \otimes u_i$ est un élément de $\mathcal{Q}^{\varepsilon}(C')$, on désignera par $\hat{\alpha}u\beta$ l'élément de $\mathcal{Q}^{\varepsilon'}(C)$ $(\varepsilon' = (-1)^p\varepsilon)$ défini par :

$$\hat{\alpha}u\beta = \Sigma e_i \otimes (-1)^{pi}\hat{\alpha}u_i\beta$$

LEMME 3.5. Soient C et C' deux complexes de $C^e(A)$, α et β deux applications de C dans C' de degré p et u un élément de $\mathcal{Q}^{\varepsilon}(C')$ de degré n. Alors, on a les formules :

$$\hat{\alpha}\, S(u)\, \beta = (-1)^p S(\hat{\beta}u\alpha)$$

$$d\left[\hat{\alpha}u\alpha - (-1)^p\, d\hat{\alpha}S(u)\alpha\right] = (-1)^p\, \hat{\alpha}du\alpha + d\hat{\alpha}d(Su)\alpha - (-1)^n d\hat{\alpha}S(u)d\alpha$$

§4 - LES GROUPES $L_n(\mathcal{A},\mathcal{B})$

Dans ce paragraphe, $\mathcal{A} \supset \mathcal{B}$ désigneront deux classes exactes symétriques de $C^e(A)$.

DEFINITION 4.1. Soient (C,q) et (C',q') deux n-complexes ε-quadratiques $\mathcal{B}$-non dégénérés, C et C' appartenant à $\mathcal{A}$.

On dira que (C,q) et (C',q') sont en relation (par la relation $\mathcal{R}$) s'il existe une n+1-paire ε-quadratique $\mathcal{B}$-non dégénérée $(\Sigma \to C \oplus C', u)$ telle que Σ appartienne à $\mathcal{A}$ et $\partial u = q \oplus -q'$

LEMME 4.2 Soient $q = [\phi]$ et $q' = [\phi']$ deux n-formes ε-quadratiques $\mathcal{B}$-non dégénérées sur des complexes C et C' de $\mathcal{A}$. Alors, (C,q) et (C',q') sont en relation si et seulement s'il existe deux suites e-exactes:

$$0 \longrightarrow K \xrightarrow{\beta} \Sigma \xrightarrow{\alpha} C \longrightarrow 0 \qquad \partial^0\alpha = \partial^0\beta = 0$$

$$0 \longrightarrow K' \xrightarrow{\beta'} \Sigma \xrightarrow{\alpha'} C' \longrightarrow 0 \qquad \partial^0\alpha' = \partial^0\beta' = 0$$

et un élément $\psi \in \mathcal{Q}^{\varepsilon}(\Sigma)$ de degré n+1 tels que: $\hat{\alpha}\phi\alpha - \hat{\alpha}'\phi'\alpha' = d\psi$ et $\hat{\beta}'\tilde{\psi}\beta$ est une $\mathcal{B}$-équivalence.

Démonstration. Supposons que (C,q) et (C',q') soient en relation. Alors il existe une suite e-exacte :

$$0 \longrightarrow \Sigma' \xrightarrow{\gamma} \Sigma \xrightarrow{\alpha \oplus \alpha'} C \oplus C' \longrightarrow 0$$

et un élément $\psi \in \mathcal{Q}^{\varepsilon}(\Sigma)$ de degré n+1 tels que :

$$\hat{\alpha}\phi\alpha - \hat{\alpha}'\phi'\alpha' = d\psi$$

$\tilde{\psi}\gamma$ est une $\mathcal{B}$-équivalence.

Soient K (resp. K') le noyau de α (resp. α') et β (resp. β') l'inclusion canonique de K (resp. K') dans Σ. On a le diagramme commutatif suivant:

$$\text{(I)}\qquad \begin{array}{ccccccccc} & & 0 & \longrightarrow & \Sigma' & \longrightarrow & K & \xrightarrow{\alpha'\beta} & C' \longrightarrow 0 \\ & & & & \tilde{\psi}\gamma\downarrow & \overset{\tilde{\psi}\beta}{\swarrow} & \downarrow\hat{\beta}'\tilde{\psi}\beta & & \\ 0 & \longrightarrow & \hat{C}' & \xrightarrow{\hat{\alpha}'} & \hat{\Sigma} & \longrightarrow & \hat{K}' & \longrightarrow & 0 \end{array}$$

et l'on a : $\qquad d(\tilde{\psi}\beta) = -\hat{\alpha}'\tilde{\phi}'\alpha'\beta$

Or $\tilde{\psi}\gamma$ et $\tilde{\phi}'$ sont des $\mathcal{B}$-équivalences. D'après le lemme 2.5, $\hat{\beta}'\tilde{\psi}\beta$ est une $\mathcal{B}$-équivalence.

Réciproquement, supposons qu'il existe deux suites e-exactes:

$$0 \longrightarrow K \xrightarrow{\beta} \Sigma \xrightarrow{\alpha} C \longrightarrow 0$$

$$0 \longrightarrow K' \xrightarrow{\beta'} \Sigma \xrightarrow{\alpha'} C' \longrightarrow 0$$

et un élément $\psi \in \mathcal{Q}^{\varepsilon}(\Sigma)$ vérifiant les conditions du lemme.

A type d'homotopie (simple) près, on peut supposer que $\alpha \oplus \alpha'$ est surjectif, de noyau $\Sigma' \in C^{e}(A)$.

On a alors le diagramme (I) et $\tilde{\phi}'$ et $\hat{\beta}'\tilde{\psi}\beta$ sont des $\mathcal{B}$-équivalences. D'après le lemme 2.5, $\tilde{\psi}\gamma$ est une $\mathcal{B}$-équivalence.

<u>LEMME</u> 4.3. <u>La relation</u> $\mathcal{R}$ <u>est une relation d'équivalence compatible avec la somme directe</u>.

<u>Démonstration</u>. D'après le lemme 4.2, la relation $\mathcal{R}$ est réflexive et symétrique. Elle est de plus clairement compatible avec la somme directe.

Montrons que $\mathcal{R}$ est transitive.

Soient q, q' et q" des n-formes ε-quadratiques $\mathcal{B}$-non dégénérées sur des complexes C, C' et C" de $\mathcal{A}$, représentées par des éléments ϕ, ϕ' et ϕ'' de $\mathcal{Q}^{\varepsilon}(C)$, $\mathcal{Q}^{\varepsilon}(C')$ et $\mathcal{Q}^{\varepsilon}(C'')$.

Supposons que (C,q) et (C',q') soient en relation, ainsi que (C',q') et (C",q").

Il existe alors des suites e-exactes:

$$0 \longrightarrow K \longrightarrow \Sigma \xrightarrow{\alpha} C \longrightarrow 0$$

$$0 \longrightarrow K' \longrightarrow \Sigma \xrightarrow{\alpha'} C' \longrightarrow 0$$

$$0 \longrightarrow K_1 \longrightarrow \Sigma_1 \xrightarrow{\alpha_1} C' \longrightarrow 0$$

$$0 \longrightarrow K_1' \longrightarrow \Sigma_1 \xrightarrow{\alpha_1'} C'' \longrightarrow 0$$

où toutes les flèches sont des morphismes de degré 0, et deux éléments $\psi \in \mathcal{Q}^{\varepsilon}(\Sigma)$ et $\psi_1 \in \mathcal{Q}^{\varepsilon}(\Sigma_1)$ tels que:

$$\hat{\alpha}\phi\alpha - \hat{\alpha}'\phi'\alpha' = d\psi$$

$$\hat{\alpha}_1\phi'\alpha_1 - \hat{\alpha}_1'\phi''\alpha_1' = d\psi_1$$

et où $\tilde{\psi}$ et $\tilde{\psi}_1$ induisent des $\mathcal{B}$-équivalences de K dans $\hat{K}'$ et de K_1 dans $\hat{K}_1'$.

Soit X le produit fibré de Σ et Σ_1 au-dessus de C'. On a le diagramme commutatif suivant :

$$\begin{array}{ccccc} X & \xrightarrow{\gamma} & \Sigma & \xrightarrow{\alpha} & C \\ \gamma_1 \downarrow & & \downarrow \alpha' & & \\ \Sigma_1 & \xrightarrow{\alpha_1} & C' & & \\ \alpha_1' \downarrow & & & & \\ C'' & & & & \end{array}$$

où toutes les flèches sont des morphismes de degré 0. On a:

$$\widehat{\alpha\gamma}\phi\alpha\gamma - \widehat{\alpha_1'\gamma_1}\phi''\alpha_1'\gamma_1 = d(\hat{\gamma}\psi\gamma + \hat{\gamma}_1\psi_1\gamma_1)$$

Considérons alors le diagramme commutatif suivant :

$$\begin{array}{ccccccccc} 0 & \longrightarrow & \operatorname{Ker}\gamma & \longrightarrow & \operatorname{Ker}\alpha\gamma & \longrightarrow & \operatorname{Ker}\alpha & \longrightarrow & 0 \\ & & \downarrow & & \downarrow & & \downarrow & & \\ 0 & \longrightarrow & \widehat{\operatorname{Ker}\alpha_1'} & \longrightarrow & \widehat{\operatorname{Ker}\alpha_1'\gamma_1} & \longrightarrow & \widehat{\operatorname{Ker}\gamma_1} & \longrightarrow & 0 \end{array}$$

où les flèches verticales sont des morphismes induits par $\hat{\gamma}\tilde{\psi}\gamma + \hat{\gamma}_1\tilde{\psi}_1\gamma_1$. Les flèches verticales extrêmes sont des $\mathcal{B}$-équivalences. Il en résulte que $\hat{\gamma}\tilde{\psi}\gamma + \hat{\gamma}_1\tilde{\psi}_1\gamma_1$ induit une $\mathcal{B}$-équivalence de Ker $\alpha\gamma$ sur le dual de Ker $\alpha_1'\gamma_1$ et (C,q) et (C",q") sont en relation.

Ainsi la relation $\mathcal{R}$ est une relation d'équivalence (on l'appellera dorénavant cobordisme) et l'ensemble quotient ${}_{\varepsilon}L_n^e(\mathcal{A},\mathcal{B})$ est un groupe abélien pour la somme directe. De plus, (C,q) et (C,-q) représentent des éléments opposés dans ${}_{\varepsilon}L_n^e(\mathcal{A},\mathcal{B})$.

<u>LEMME</u> 4.4. <u>Soient</u> f <u>une</u> $\mathcal{B}$-<u>équivalence</u> <u>de</u> <u>degré</u> 0 <u>d'un</u> <u>complexe</u> $C \in \mathcal{A}$ <u>dans</u> <u>un</u> <u>complexe</u> $C' \in \mathcal{A}$, <u>et</u> q <u>une</u> n-<u>forme</u> ε-<u>quadratique</u> $\mathcal{B}$-<u>nondégénérée</u> <u>sur</u> C'. <u>Alors</u> (C',q) <u>et</u>

(C,f^*(q)) sont cobordants.

Démonstration. Comme f est une $\mathcal{B}$-équivalence, on peut trouver deux suites e-exactes dans $C^e(A)$:

$$0 \longrightarrow K \longrightarrow \Sigma \xrightarrow{\alpha} C \longrightarrow 0$$

$$0 \longrightarrow K' \longrightarrow \Sigma \xrightarrow{\alpha'} C' \longrightarrow 0$$

où les flèches sont des morphismes de degré 0, α est une équivalence d'homotopie (simple) et où $f\alpha$ est homotope à α' par une homotopie $d\beta$.

Soit $\phi \in \mathcal{Q}^{\varepsilon}(C')$ un représentant de q. On a :

$$\hat{\alpha}\hat{f}\phi f\alpha - \hat{\alpha}'\phi\alpha' = \hat{\alpha}'\phi d\beta + d\hat{\beta}\phi\alpha' + d\hat{\beta}\phi d\beta$$

Posons $\phi = \Sigma e_i \otimes \phi_i$. On vérifie aisément la formule suivante :

$$\hat{\alpha}'\phi d\beta + d\hat{\beta}\phi\alpha' = d\left[\Sigma e_i \otimes ((-1)^n\hat{\alpha}'\phi_i\beta + (-1)^i\hat{\beta}\phi_i\alpha')\right]$$

Soit ϕ' l'élément de $\mathcal{Q}^{-\varepsilon}(C')$ défini par : $\phi = S\phi' + e_o \otimes \phi_o$. On a :

$$0 = d(S\phi') + e_o \otimes d\phi_o = -S(d\phi') + e_o \otimes \tilde{\phi}' + e_o \otimes d\phi_o$$

et $d\phi'$ est nul.

En appliquant la formule 3.5, on obtient:

$$d(\hat{\beta}\phi'\beta + d\hat{\beta}\phi\beta) = d(e_o \otimes d\hat{\beta}\phi_o\beta) + d\hat{\beta}d(S\phi')\beta + (-1^n d\hat{\beta}S\phi' d\beta$$

d'où : $d(\hat{\beta}\phi'\beta + d\hat{\beta}\phi\beta) = (-1)^n d\hat{\beta}\phi d\beta$

Ilen résulte que $\hat{\alpha}\hat{f}\phi f\alpha - \hat{\alpha}'\phi\alpha'$ est un bord $d\psi$, $\psi \in \mathcal{Q}^{\varepsilon}(\Sigma)$. De plus, comme K et K' appartiennent à $\mathcal{B}$, $\tilde{\psi}$ induit une $\mathcal{B}$-équivalence de K dans $\hat{K}'$. On en déduit, d'après le lemme 4.2, que (C',q) est cobordant à (C,f^*q).

Démonstration de la proposition 1.10

Il s'agit de montrer l'isomorphisme entre les groupes ${}_{\varepsilon}L^e_{n+2}(A,\mathcal{B})$ et ${}_{-\varepsilon}L^e_n(A,\mathcal{B})$.

Soit C un complexe de A. Soit ΣC la suspension de C ; ΣC est le complexe C, la graduation et la différentielle ayant été modifiées pour que l'identité de C dans ΣC soit un morphisme de degré 1.

Le morphisme de C dans ΣC induit un isomorphisme équivariant de degré 2 de $B(\Sigma C)^{-\varepsilon}$ sur $B(C)^{\varepsilon}$, d'où un isomorphisme de degré 2 de $\mathcal{Q}^{-\varepsilon}(\Sigma C)$ sur $\mathcal{Q}^{\varepsilon}(C)$ et un isomorphisme de ${}_{-\varepsilon}L^e_n(A,\mathcal{B})$ sur ${}_{\varepsilon}L^e_{n+2}(A,\mathcal{B})$.

<u>Démonstration de la proposition</u> 1.11

Soit a une unité du centre de A. Soit ε' l'élément $\frac{a}{\bar{a}}\varepsilon$ de A. Si C est un complexe de $\mathcal{C}^e(A)$, la multiplication par a induit un isomorphisme de degré 0 de $B(C)^{\varepsilon}$ sur $B(C)^{\varepsilon'}$, d'où un isomorphisme de degré 0 de $\mathcal{Q}^{\varepsilon}(C)$ sur $\mathcal{Q}^{\varepsilon'}(C)$ et un isomorphisme de ${}_{\varepsilon}L^e_n(A,\mathcal{B})$ sur ${}_{\varepsilon'}L^e_n(A,\mathcal{B})$.

<u>LEMME DE CHIRURGIE</u> 4.5. <u>Soit</u> $0 \longrightarrow K \xrightarrow{\alpha} \Sigma \xrightarrow{\beta} C \longrightarrow 0$ <u>une suite</u> e-<u>exacte dans</u> A, <u>les morphismes</u> α <u>et</u> β <u>étant de degré</u> 0. <u>Soit</u> q <u>une</u> n-<u>forme</u> ε-<u>quadratique</u> $\mathcal{B}$-<u>non dégénérée sur</u> C <u>représentée par un élément</u> ϕ <u>de</u> $\mathcal{Q}^{\varepsilon}(C)$. <u>On suppose que</u> $\beta^*\phi$ <u>est un bord</u> $d\psi$.

<u>Alors, il existe une</u> n-<u>forme</u> ε-<u>quadratique</u> $\mathcal{B}$-<u>non dégénérée</u>, q' <u>sur le</u> (-n-1)-<u>cône</u> C' <u>du morphisme</u> $\tilde{\psi}\alpha$ <u>et</u> (C',q') <u>est cobordant à</u> (C,q).

<u>Démonstration</u>. Soit M le 0-cône de α. Notons γ et p les projections de M sur K et Σ . On a :

$$\partial^o p = 0 \quad , \qquad \partial^o \gamma = -1$$

$$dp = \alpha\gamma \quad , \qquad d\gamma = 0$$

De plus, βp est une équivalence d'homotopie (simple) de degré 0 de M sur C. Il en résulte que $(\beta p)^* q$ est une n-forme ε-quadratique $\mathcal{B}$-non dégénérée sur M et (C,q) et $(M,(\beta p)^* q)$ sont cobordants.

Soit C' (resp. X) le (-n-1)-cône du morphisme $\tilde{\psi}\alpha$ (resp $\tilde{\psi}\alpha\gamma$). On a le diagramme commutatif suivant dans A, où les lignes et les colonnes sont e-exactes:

$$\begin{array}{ccccccccc}
 & & & & 0 & & 0 & & \\
 & & & & \downarrow & & \downarrow & & \\
 & & & & \hat{\Sigma} & \xrightarrow{1} & \hat{\Sigma} & & \\
 & & & & \downarrow{\scriptstyle\lambda} & & \downarrow{\scriptstyle\lambda'} & & \\
0 & \longrightarrow & \Sigma & \xrightarrow{f} & X & \xrightarrow{g} & C' & \longrightarrow & 0 \\
 & & \downarrow{\scriptstyle 1} & & \downarrow{\scriptstyle\mu} & & \downarrow{\scriptstyle\mu'} & & \\
0 & \longrightarrow & \Sigma & \longrightarrow & M & \xrightarrow{\gamma} & K & \longrightarrow & 0 \\
 & & & & \downarrow & & \downarrow & & \\
 & & & & 0 & & 0 & &
\end{array}$$

De plus, toutes les flèches du diagramme sont de degré 0 sauf γ et μ' qui sont de degré -1 et λ et λ' qui sont de degré -n-1 .

Soient ρ et ρ' les projections de X et C' sur $\hat{\Sigma}$. On a :

$$\partial^o\rho = \partial^o\rho' = n+1$$

$$\rho = \rho' g$$

$$d\rho = \tilde{\psi}\alpha\gamma\mu \qquad d\rho' = \tilde{\psi}\alpha\mu'$$

D'après le lemme 3.5, on a :

$$d\left[\hat{\mu}\hat{\rho}\psi p\mu - \hat{\mu}\hat{\gamma}\hat{\alpha}S(\psi)p\mu\right] = \hat{\mu}\hat{\rho}d\psi p\mu + \hat{\mu}\hat{\gamma}\hat{\alpha}d(S\psi)p\mu + (-1)^n\hat{\mu}\hat{\gamma}\hat{\alpha}S(\psi)\alpha\gamma\mu$$

Or, on a :

$$\hat{\mu}\hat{\gamma}\hat{\alpha}d(S\psi)p\mu = -\hat{\mu}\hat{\gamma}\hat{\alpha}S(\hat{\beta}\phi\beta)p\mu + e_o \otimes \hat{\mu}\hat{\gamma}\hat{\alpha}\tilde{\psi}p\mu = e_o \otimes \hat{\mu}\hat{\gamma}\hat{\alpha}\tilde{\psi}p\mu$$

Posons :

$$\omega = \hat{\mu}\hat{\rho}\psi p\mu - \hat{\mu}\hat{\gamma}\hat{\alpha}S(\psi)p\mu + (-1)^n\, e_o \otimes \varepsilon\hat{\rho}p\mu$$

On a alors :

$$d\omega = (\beta p\mu)^*\phi - g^*(e_o \otimes \varepsilon\hat{\rho}'\alpha\mu' - (-1)^n\hat{\mu}'\alpha S(\psi)\alpha\mu')$$

Comme $(\beta p)^* q$ est $\mathcal{B}$-non dégénérée et que $\widehat{f\tilde{\omega}\lambda}$ est un isomorphisme, $e_o \otimes \varepsilon\hat{\rho}'\alpha\mu' - (-1)^n\hat{\mu}'\hat{\alpha}S(\psi)\alpha\mu'$ représente une n-forme ε-quadratique $\mathcal{B}$-non dégénérée q' sur C' et (C',q') est cobordant à (C,q).

THEOREME 4.6. *Si n est un entier positif ou nul et si e est égal à s, h ou p, le groupe ${}_\varepsilon L^e_n(C^e(A),\mathcal{O})$ est isomorphe au groupe de chirurgie ${}_\varepsilon L^e_n(A)$.*

Démonstration. Pour démontrer le théorème, il est possible d'utiliser le lemme de chirurgie et d'exprimer le groupe ${}_\varepsilon L^e_n(C^e(A),\mathcal{O})$ en terme de générateurs et de relations avec des complexes de longueur 1 ou 2. Il est alors facile de reconnaître la définition classique du groupe ${}_\varepsilon L^e_n(A)$.

On peut également utiliser la définition de Ranicki des groupes ${}_\varepsilon L^e_n(A)$ [6] : Si (C,q) est un complexe de Poincaré ε-quadratique de dimension n au sens de Ranicki on en déduit une n-forme ε-quadratique non dégénérée sur $\hat{C}$. Cette correspondance induit un morphisme de groupe de ${}_\varepsilon L^e_n(A)$ dans ${}_\varepsilon L^e_n(C^e(A),\mathcal{O})$, compatible avec la périodicité.

Soit u un élément de $L^e_n(C^e(A),\mathcal{O})$ représenté par un couple (C,q). Si k est un entier assez grand, le stabilisé de (C,q) par la périodicité $L_n \xrightarrow{\sim} L_{n+4k}$ sera le couple $(\Sigma^{2k}C,q')$ et le dual de $\Sigma^{-2k}C$ sera un complexe positif de dimension inférieure ou égale à n+4k . Dans ce cas, la classe de $(\Sigma^{-2k}C,q')$ est dans l'image du morphisme :

$$ {}_{\varepsilon}L^{e}_{n+4k}(A) \longrightarrow {}_{\varepsilon}L^{e}_{n+4k}(\mathcal{C}^{e}(A),\mathcal{O}) $$

Il en résulte, par périodicité, que la classe de (C,q) appartient à l'image de ${}_{\varepsilon}L^{e}_{n}(A)$ et le morphisme de ${}_{\varepsilon}L^{e}_{n}(A)$ dans ${}_{\varepsilon}L^{e}_{n}(\mathcal{C}^{e}(A),\mathcal{O})$ est surjectif.

Pour l'injectivité du morphisme, on procède de même; on stabilise pour s'assurer que les complexes utilisés satisfont les conditions dimensionnelles de Ranicki.

On montrera plus loin que ${}_{\varepsilon}L^{p'}_{n}(\mathcal{C}^{p'}(A),\mathcal{O})$ est isomorphe au groupe ${}_{\varepsilon}L^{p}_{n}(A)$.

<u>DEFINITION</u> 4.7. Soit $\mathcal{A}$ une classe exacte de $\mathcal{C}^{e}(A)$. On dira qu'un complexe $C \in \mathcal{C}^{e}(A)$ est inversible modulo $\mathcal{A}$ s'il existe un complexe $C' \in \mathcal{C}^{e}(A)$ tel que $C \oplus C'$ appartienne à $\mathcal{A}$.

<u>LEMME</u> 4.8. <u>Soient $\mathcal{A} \supset \mathcal{B}$ deux classes exactes de $\mathcal{C}^{e}(A)$. On suppose que tout complexe de $\mathcal{A}$ est inversible modulo $\mathcal{B}$. Alors, la relation</u> :

$$ C \sim C' \iff C \oplus \Sigma C' \in \mathcal{B} $$

<u>est une relation d'équivalence sur $\mathcal{A}$ compatible avec la somme directe</u>.

Le groupe quotient sera noté $K(\mathcal{A},\mathcal{B})$.

<u>Démonstration</u>. Soit C un complexe de $\mathcal{A}$. Il existe $C' \in \mathcal{C}^{e}(A)$ tel que $C \oplus C'$ appartienne à $\mathcal{B}$ et C' est un complexe de $\mathcal{A}$. Or $C \oplus C'$ est le 0-cône de l'application nulle de $\Sigma^{-1}C'$ dans C. Par suite, l'application nulle de C' dans ΣC est une $\mathcal{B}$-équivalence et il existe une $\mathcal{B}$-équivalence de $C \oplus C'$ dans $C \oplus \Sigma C$. Il en résulte que $C \oplus \Sigma C$ appartient à $\mathcal{B}$.

On vérifie alors immédiatement que la relation $\sim$ est une relation d'équivalence sur $\mathcal{A}$ compatible avec la somme directe et que l'ensemble quotient $K(\mathcal{A},\mathcal{B})$ est un groupe abélien.

<u>LEMME</u> 4.9. <u>Soit θ l'application quotient de $\mathcal{A}$ dans $K(\mathcal{A},\mathcal{B})$. Alors, si</u> $0 \longrightarrow C \xrightarrow{\alpha} C' \xrightarrow{\beta} C'' \longrightarrow 0$ <u>est une suite e-exacte dans $\mathcal{A}$, α et β étant des morphismes de degré 0, on a</u> :

$$ \theta(C') = \theta(C) + \theta(C'') $$

Démonstration. On a une suite e-exacte :

$$0 \longrightarrow C \oplus \Sigma C \longrightarrow C' \oplus \Sigma C \oplus \Sigma C'' \longrightarrow C'' \oplus \Sigma C'' \longrightarrow 0$$

et $C' \oplus \Sigma C \oplus \Sigma C''$ appartient à $\mathcal{B}$.

Démonstration du théorème 1.15.

Soient $\mathcal{A} \supset \mathcal{B}$ deux classes exactes symétriques de $C^e(A)$. On suppose que tout complexe de $\mathcal{A}$ est inversible modulo $\mathcal{B}$. Alors, $K(\mathcal{A},\mathcal{B}))$ est un groupe abélien muni d'une involution induite par la dualité.

On va montrer que θ induit un isomorphisme de ${}_{\varepsilon}L^e_n(\mathcal{A},\mathcal{B})$ sur $\hat{H}^n(\mathbf{Z}/2,K(\mathcal{A},\mathcal{B}))$.

Soit (C,q) un n-complexe ε-quadratique $\mathcal{B}$-non dégénéré, C appartenant à $\mathcal{A}$. Comme q induit une $\mathcal{B}$-équivalence de degré n de C dans $\hat{C}$, on a :

$$\theta(C) = (-1)^n \theta(\hat{C})$$

D'autre part, si (C,q) est cobordant à 0, il existe une suite e-exacte dans $\mathcal{A}$:

$$0 \longrightarrow K \longrightarrow \Sigma \longrightarrow C \longrightarrow 0$$

et une $\mathcal{B}$-équivalence de degré n+1 de K dans $\hat{\Sigma}$. On a alors, dans ce cas:

$$\theta(C) = \theta(\Sigma) - \theta(K) = \theta(\Sigma) + (-1)^n\theta(\hat{\Sigma})$$

On en déduit, par passage au quotient, un morphisme de groupe θ' de ${}_{\varepsilon}L^e_n(\mathcal{A},\mathcal{B})$ dans $\hat{H}^n(\mathbf{Z}/2,K(\mathcal{A},\mathcal{B}))$. Le morphisme θ' fait correspondre à la classe de (C,q) la classe de $\theta(C)$.

Montrons que θ' est surjectif. Soit $u \in \hat{H}^n(\mathbf{Z}/2,K(\mathcal{A},\mathcal{B}))$ un élément représenté par $\theta(C)$, $C \in \mathcal{A}$.

On a : $\theta(C) = (-1)^n \theta(\hat{C})$

On en déduit que $(C,0)$ est un n-complexe ε-quadratique $\mathcal{B}$-non dégénéré qui donne u dans $\hat{H}^n(\mathbf{Z}/2,K(\mathcal{A},\mathcal{B}))$; θ' est donc surjectif.

Montrons que θ' est injectif.

Soit (C,q) un n-complexe ε-quadratique, C appartenant à $\mathcal{A}$, tel que (C,q) donne 0 dans $\hat{H}^n(\mathbf{Z}/2,K(\mathcal{A},\mathcal{B}))$.

Alors on a : $\theta(C) = \theta(\Sigma) + (-1)^n \theta(\hat{\Sigma})$

Sans changer le type d'homotopie (simple) de Σ, on peut supposer qu'il existe un épimorphisme homotope à 0 de Σ sur C, de noyau $K \in \mathcal{A}$. Choisissons un élément

$u \in \mathcal{Q}^{\varepsilon}_{n+1}(\Sigma \overset{f}{\to} C)$ de bord q. Alors, u est $\mathcal{B}$-non dégénéré et (C,q) est cobordant à 0.

§5- LA LONGUE SUITE EXACTE (Théorème 1.12)

Soient $\mathcal{A} \supset \mathcal{B} \supset \mathcal{C}$ trois classes exactes symétriques de $C^e(A)$. On va construire une longue suite exacte :

$$\ldots \xrightarrow{\partial} {}_{\varepsilon}L^e_n(\mathcal{B},\mathcal{C}) \xrightarrow{i} {}_{\varepsilon}L^e_n(\mathcal{A},\mathcal{C}) \xrightarrow{j} {}_{\varepsilon}L^e_n(\mathcal{A},\mathcal{B}) \xrightarrow{\partial} {}_{\varepsilon}L^e_{n-1}(\mathcal{B},\mathcal{C}) \longrightarrow \ldots$$

où i et j résultent des inclusions $\mathcal{C} \subset \mathcal{B}$ et $\mathcal{B} \subset \mathcal{A}$.

5.1 Construction du bord ∂.

Soient C un complexe de $\mathcal{A}$ et q une n-forme ε-quadratique sur C représentée par un élément ϕ de $\mathcal{Q}^{\varepsilon}(C)$. Soient M le -n-cône du morphisme $\tilde{\phi}$ de C dans $\hat{C}$, α l'injection canonique de $\hat{C}$ dans M, β et ρ les projections canoniques de M sur C et $\hat{C}$.

$$0 \longrightarrow \hat{C} \underset{\alpha}{\overset{\rho}{\rightleftarrows}} M \xrightarrow[\beta]{} C \longrightarrow 0$$

On a : $\partial°\alpha = -n \qquad \partial°\rho = n \qquad \partial°\beta = -1$

$d\alpha = 0 \qquad d\beta = 0 \qquad d\rho = \tilde{\phi}\beta$

Posons : $\psi = e_0 \otimes \hat{\beta}\rho - \hat{\beta}S(\phi)\beta$

On a alors : $d\psi = -e_0 \otimes \hat{\beta}\tilde{\phi}\beta + \hat{\beta}d(S\phi)\beta = -\hat{\beta}S(d\phi)\beta = 0$

et ψ représente une n-1-forme ε-quadratique sur M.

On vérifie que le diagramme suivant est commutatif :

$$\begin{array}{ccccccccc} 0 & \longrightarrow & \hat{C} & \xrightarrow{\alpha} & M & \xrightarrow{\beta} & C & \longrightarrow & 0 \\ & & \downarrow 1 & & \downarrow \tilde{\psi} & & \downarrow \varepsilon & & \\ 0 & \longrightarrow & \hat{C} & \xrightarrow{\hat{\beta}} & \hat{M} & \xrightarrow{\hat{\alpha}} & C & \longrightarrow & 0 \end{array}$$

et la classe de ψ est non dégénérée.

Si la forme ε-quadratique q est $\mathcal{B}$-non dégénérée, M appartient à $\mathcal{B}$ et $(M,[\psi])$ représente un élément $\partial(C,\phi)$ de ${}_{\varepsilon}L^e_{n-1}(\mathcal{B},\mathcal{C})$.

Supposons que $(C,[\phi])$ soit cobordant à 0. Alors il existe une suite e-exacte dans $\mathcal{A}$: $\qquad 0 \longrightarrow K \xrightarrow{\lambda} \Sigma \xrightarrow{\mu} C \longrightarrow 0$

λ et μ étant des morphismes de degré 0, et un élément u de $\mathcal{Q}^{\varepsilon}(\Sigma)$ tel que :

$\hat{\mu}\phi\mu = du$ et $\tilde{u}\lambda$ est une $\mathcal{B}$-équivalence

Soit $\hat{K}'$ le -1-cône du morphisme $\hat{\mu} : \hat{C} \to \hat{\Sigma}$. Soient f et s les injections canoniques de $\hat{\Sigma}$ et $\hat{C}$ dans $\hat{K}'$, et g et t les projections canoniques de $\hat{K}'$ sur $\hat{C}$ et $\hat{\Sigma}$.

$$0 \longrightarrow \hat{\Sigma} \underset{f}{\overset{t}{\rightleftarrows}} \hat{K}' \underset{g}{\overset{s}{\rightleftarrows}} \hat{C} \longrightarrow 0$$

On a : $\partial^\circ f = -1$ $\quad \partial^\circ t = 1$ $\quad \partial^\circ g = \partial^\circ s = 0$

$df = dg = 0 \quad dt = \hat{\mu}g \quad ds = f\hat{\mu}$

On vérifie que $\tilde{u}\lambda$ et $s\tilde{\phi}\mu + f\tilde{u}$ sont des morphismes et le diagramme suivant est commutatif :

$$\begin{array}{ccc}
K & \xrightarrow{\tilde{u}\lambda} & \hat{\Sigma} \\
\downarrow{\scriptstyle\lambda} & & \downarrow{\scriptstyle f} \\
\Sigma & \xrightarrow{s\tilde{\phi}\mu+f\tilde{u}} & \hat{K}' \\
\downarrow{\scriptstyle\mu} & & \downarrow{\scriptstyle g} \\
C & \xrightarrow{\tilde{\phi}} & \hat{C}
\end{array}$$

Soient M' le -n-cône de $s\tilde{\phi}\mu+f\tilde{u}$ et M" le -n-1-cône de $\tilde{u}\lambda$. On a un diagramme commutatif à lignes et colonnes e-exactes :

$$\begin{array}{ccccccccc}
 & & 0 & & 0 & & 0 & & \\
 & & \downarrow & & \downarrow & & \downarrow & & \\
0 & \longrightarrow & \hat{\Sigma} & \xrightarrow{\alpha''} & M'' & \xrightarrow{\beta''} & K & \longrightarrow & 0 \\
 & & \downarrow{\scriptstyle f} & & \downarrow{\scriptstyle h} & & \downarrow{\scriptstyle \lambda} & & \\
0 & \longrightarrow & \hat{K}' & \xrightarrow{\alpha'} & M' & \xrightarrow{\beta'} & \Sigma & \longrightarrow & 0 \\
 & & \downarrow{\scriptstyle g} & & \downarrow{\scriptstyle k} & & \downarrow{\scriptstyle \mu} & & \\
0 & \longrightarrow & \hat{C} & \xrightarrow{\alpha} & M & \xrightarrow{\beta} & C & \longrightarrow & 0 \\
 & & \downarrow & & \downarrow & & \downarrow & & \\
 & & 0 & & 0 & & 0 & &
\end{array}$$

On notera ρ' et ρ'' les rétractions canoniques de α' et α''. On a donc :

$$d\rho' = s\tilde{\phi}\mu\beta' + f\tilde{u}\beta' \qquad d\rho'' = \tilde{u}\lambda\beta''$$

On a : $\hat{k}\psi k = e_o \otimes \hat{\beta}'\hat{\mu}g\rho' - \hat{\beta}'\hat{\mu}S(\phi)\mu\beta'$

On vérifie alors que, si l'on pose : $v = -e_o \otimes \hat{\beta}'t\rho' - \hat{\beta}'S(u)\beta'$, on a :

$$\hat{k}\psi k = dv$$

Pour montrer que $(M,[\psi])$ est cobordant à 0, il reste à montrer que M'et M" appartiennent à $\mathcal{B}$ et que $\tilde{v}h$ est une $\mathcal{C}$-équivalence. Or M" est le cône de $\tilde{u}\lambda$ qui est une $\mathcal{B}$-équivalence. Il en résulte que M' et M" appartiennent à $\mathcal{B}$.

D'autre part, on a le diagramme commutatif suivant :

$$\begin{array}{ccccccccc}
0 & \longrightarrow & \hat{\Sigma} & \xrightarrow{\alpha''} & M'' & \xrightarrow{\beta''} & K & \longrightarrow & 0 \\
 & & \downarrow{\scriptstyle -1} & & \downarrow{\scriptstyle \tilde{\nabla}h} & & \downarrow{\scriptstyle (-1)^n \varepsilon \hat{t\lambda}} & & \\
0 & \longrightarrow & \hat{\Sigma} & \xrightarrow{\hat{\beta}'} & \hat{M}' & \xrightarrow{\hat{\alpha}'} & K' & \longrightarrow & 0
\end{array}$$

Or, il est facile de vérifier que $\hat{t\lambda}$ est une équivalence d'homotopie (simple) de K vers le dual K' de $\hat{K}'$. On en déduit que $\tilde{\nabla}h$ est une $\mathcal{C}$-équivalence et $(M,[\psi])$ est cobordant à 0.

Ainsi, si $(C,[\phi])$ est cobordant à 0, $\partial(C,\phi)$ est nul.

Soient, maintenant, C et C' deux complexes de $\mathcal{A}$ et $[\phi]$ et $[\phi']$ des n-formes ε-quadratiques $\mathcal{B}$-non dégénérées sur C et C'. Si $(C,[\phi])$ est cobordant à $(C',[\phi'])$, $(C \oplus C',[\phi \oplus -\phi'])$ est cobordant à 0 et $\partial(C \oplus C',\phi \oplus -\phi') = \partial(C,\phi) + \partial(C',-\phi')$ est nul. Cela a lieu en particulier si $C = C'$ et $\phi = \phi'$. On en déduit :

$$\partial(C',\phi') + \partial(C',-\phi') = 0$$

Il en résulte que, si $(C,[\phi])$ est cobordant à $(C',[\phi'])$, $\partial(C,\phi)$ est égal à $\partial(C',\phi')$.

Ainsi l'application ∂ est bien définie de ${}_{\varepsilon}L_n^e(\mathcal{A},\mathcal{B})$ dans ${}_{\varepsilon}L_{n-1}^e(\mathcal{B},\mathcal{C})$.

5.2. <u>$ji = 0$</u> . Soit q une n-forme ε-quadratique $\mathcal{C}$-non dégénérée sur un complexe $C \in \mathcal{B}$. L'application nulle de 0 dans C est une $\mathcal{B}$-équivalence.

D'après le lemme 4.4, (C,q) représente 0 dans le groupe ${}_{\varepsilon}L_n^e(\mathcal{A},\mathcal{B})$. L'application ji est donc nulle.

5.3. <u>$\partial j = 0$</u> . Soit $[\phi]$ une n-forme ε-quadratique $\mathcal{C}$-non dégénérée sur un complexe $C \in \mathcal{A}$. Alors, $\tilde{\phi}$ est une $\mathcal{C}$-équivalence de C sur $\hat{C}$ et $\partial(C,\phi)$ est représenté par un couple (M,q) , $M \in \mathcal{C}$. D'après le lemme 4.4, (M,q) est cobordant à 0 et $\partial(C,\phi)$ est nul.

L'application ∂j est donc nulle.

5.4. <u>$i\partial = 0$</u> . Soit $[\phi]$ une n-forme ε-quadratique $\mathcal{B}$-non dégénérée sur un complexe $C \in \mathcal{A}$.

En reprenant les notations de 5.1, $\partial(C,\phi)$ est représenté par $(M,[\psi])$.

Soient M' le O-cône du morphisme β de M dans C, λ l'inclusion canonique de C dans M' et t et μ les projections canoniques de M' sur C et M :

$$0 \longrightarrow C \underset{\lambda}{\overset{t}{\longleftrightarrow}} M' \underset{\mu}{\longrightarrow} M \longrightarrow 0$$

Les applications λ, μ et t sont de degré 0 et l'on a :

$$d\lambda = d\mu = 0 \qquad dt = \beta\mu$$

On a : $\hat{\mu}\psi\mu = e_o \otimes \hat{\mu}\hat{\beta}\rho\mu - \hat{\mu}\hat{\beta}S(\phi)\beta\mu = e_o \otimes \hat{dt}\rho\mu - \hat{dt}S(\phi)dt$

D'après le lemme 3.5, on a également :

$$(-1)^n d(\hat{t}\phi t - \hat{dt}S(\phi)t - e_o \otimes \hat{t}\tilde{\phi}t) = -e_o \otimes \hat{t}\phi dt - \hat{dt}S(\phi)dt$$

On en déduit , si l'on pose :

$$u = e_o \otimes \hat{t}\rho\mu + (-1)^n\hat{t}\phi t - (-1)^n\hat{dt}S(\phi)t - (-1)^n e_o \otimes \hat{t}\tilde{\phi}t$$

que l'on a : $\hat{\mu}\psi\mu = du$

D'autre part, on vérifie que $\tilde{u}\lambda$ est égal à $\varepsilon\hat{\hat{\mu}}\rho - (-1)^n\hat{t}\tilde{\phi}$, et si f est l'équivalence d'homotopie (simple) canonique de $\hat{C}$ dans M', $\hat{f}\tilde{u}\lambda$ est égal à $(-1)^n\varepsilon$. On en déduit que $\tilde{u}\lambda$ est une $\mathcal{C}$-équivalence. Comme, de plus, C et M' appartiennent à $\mathcal{A}$, $\partial(C,\phi)$ est annulé par i.

L'application $i\partial$ est donc nulle.

5.5. Exactitude en ${}_{\varepsilon}L^{e}_{n}(\mathcal{A},\mathcal{C})$.

Soit $[\phi]$ une n-forme ε-quadratique $\mathcal{C}$-non dégénérée sur un complexe $C \in \mathcal{A}$, tel que $(C,[\phi])$ soit annulé par j. Il existe alors une suite e-exacte dans $\mathcal{A}$:

$$0 \longrightarrow K \xrightarrow{\ \alpha\ } \Sigma \xrightarrow{\ \beta\ } C \longrightarrow 0$$

et un élément $\psi \in \mathcal{Q}^{\varepsilon}(\Sigma)$ tels que : $\partial^o\alpha = \partial^o\beta = 0$

$\hat{\beta}\phi\beta = d\psi$

$\tilde{\psi}\alpha$ est une $\mathcal{B}$-équivalence.

D'après le lemme de chirurgie 4.5, $(C,[\phi])$ est cobordant à (C',q'), C' étant un cône de $\tilde{\psi}\alpha$. Ainsi, C' appartient à $\mathcal{B}$ et la classe de $(C,[\phi])$ est dans l'image de i.

5.6. Exactitude en ${}_{\varepsilon}L^{e}_{n}(\mathcal{A},\mathcal{B})$.

Soit $[\phi]$ une n-forme ε-quadratique $\mathcal{B}$-non dégénérée sur un complexe $C \in \mathcal{A}$, tel que $\partial(C,\phi)$ soit nul. Soient M le -n-cône de $\tilde{\phi}$, α et σ les injections canoniques de $\hat{C}$ et C dans M et β et ρ les projections canoniques de M sur C et $\hat{C}$. Si $\partial(C,\phi)$ est nul, il existe une suite e-exacte dans $\mathcal{B}$:

$$0 \longrightarrow K \xrightarrow{\lambda} \Sigma \xrightarrow{\mu} M \longrightarrow 0$$

et un élément u de $\mathcal{Q}^{\varepsilon}(\Sigma)$ tels que :

$$\partial^{\circ}\lambda = \partial^{\circ}\mu = 0 \qquad \partial^{\circ}u = n$$

$$\hat{\mu}\psi\mu = \hat{\mu}(e_0 \otimes \hat{\beta}\rho - \hat{\beta}S(\phi)\beta)\mu = du$$

$\tilde{u}\lambda$ est une $\mathcal{C}$-équivalence

Soient s une section de μ et C' le produit fibré de $\hat{C}$ et Σ au dessus de M, le morphisme $C' \to \Sigma$ étant de degré 0. On a le diagramme suivant :

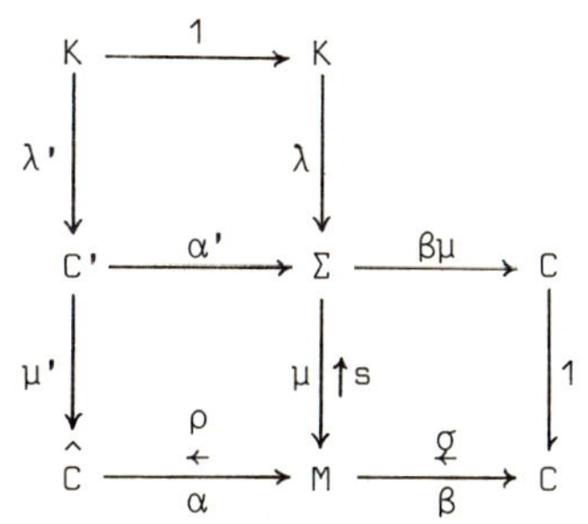

Comme $s\sigma$ est une section (de degré 1) de $\beta\mu$, il existe un morphisme f de degré 0 de C dans C' tel que : $\quad d(s\sigma) = -(-1)^n\alpha' f$

On vérifie aisément que $\mu' f$ est égal à $\tilde{\phi}$.

Soit v l'élément de $\mathcal{Q}^{-\varepsilon}(\Sigma)$ défini par : $\quad u = e_0 \otimes u_0 + S(v)$

On a :

$$du = e_0 \otimes du_0 - S(dv) + e_0 \otimes \tilde{v} = e_0 \otimes \hat{\mu}\hat{\beta}\rho\mu - \hat{\mu}\hat{\beta}S(\phi)\beta\mu$$

d'où :

$$S(dv) = \hat{\mu}\hat{\beta}S(\phi)\beta\mu = -S(\hat{\mu}\hat{\beta}\phi\beta\mu)$$

$$\Rightarrow dv = -\hat{\mu}\hat{\beta}\phi\beta\mu$$

D'après le lemme 3.5, on a :

$$d\left[\widehat{s\sigma}vs\sigma + d\widehat{s\sigma}S(v)s\sigma\right] = -\widehat{s\sigma}dvs\sigma + d\widehat{s\sigma}d(Sv)s\sigma + (-1)^n d\widehat{s\sigma}Svd(s\sigma)$$

Or on a les formules suivantes :

$$-\widehat{s\sigma}dvs\sigma = \widehat{s\sigma}\widehat{\beta\mu}\phi\beta\mu s\sigma = -\phi$$

$$d\widehat{s\sigma}d(Sv)s\sigma = -(-1)^n\hat{f}\hat{\alpha}'dus\sigma - e_0 \otimes d\widehat{s\sigma}du_0 s\sigma = -e_0 \otimes d\widehat{s\sigma}du_0 s\sigma$$

$$(-1)^n d\widehat{s\sigma}Svd(s\sigma) = (-1)^n\hat{f}\hat{\alpha}'u\alpha' f - (-1)^n e_0 \otimes d\widehat{s\sigma}u_0 d(s\sigma)$$

On en déduit :

$$d\left[\hat{s}\hat{\sigma}vs\sigma + d\hat{s}\hat{\sigma}S(v)s\sigma\right] = -\phi + (-1)^n\hat{f}\hat{\alpha}'u\alpha'f - d(e_0 \otimes d\hat{s}\hat{\sigma}u_0 s\sigma)$$

Ainsi ϕ est homologue à $(-1)^n\hat{f}\hat{\alpha}'u\alpha'f$. Comme f est une $\mathcal{B}$-équivalence, $(C,[\phi])$ est cobordant à $(C',[(-1)^n\hat{\alpha}'u\alpha'])$.

Considérons maintenant le diagramme commutatif suivant :

$$\begin{array}{ccccccccc}
 & & 0 & \longrightarrow & K & \xrightarrow{\lambda'} & C' & \xrightarrow{\mu'} & \hat{C} & \longrightarrow & 0 \\
 & & & & \downarrow{\scriptstyle \tilde{u}\lambda} & \swarrow{\scriptstyle \tilde{u}\alpha'} & \downarrow{\scriptstyle \hat{\alpha}'\tilde{u}\alpha'} & & & & \\
0 & \longrightarrow & \hat{C} & \xrightarrow{\hat{\mu}\hat{\beta}} & \hat{\Sigma} & \xrightarrow{\hat{\alpha}'} & \hat{C}' & \longrightarrow & 0 & &
\end{array}$$

Les flèches horizontales et verticales sont des morphismes et l'on a :

$$d(\tilde{u}\alpha') = (\hat{\mu}\hat{\beta}\rho\mu + \varepsilon\hat{\mu}\hat{\rho}\beta\mu)\alpha' = \hat{\mu}\hat{\beta}\mu'$$

D'après le lemme 2.5, comme $\tilde{u}\lambda$ est une $\mathcal{C}$-équivalence, $\hat{\alpha}'\tilde{u}\alpha'$ est une $\mathcal{C}$-équivalence et la classe de $(C',(-1)^n[\hat{\alpha}'u\alpha'])$ est dans l'image de j.

5.7. <u>Exactitude en</u> ${}_{\varepsilon}L^{e}_{n-1}(\mathcal{B},\mathcal{C})$

Soit $[\phi]$ une n-1-forme ε-quadratique $\mathcal{C}$-non dégénérée sur un complexe C de $\mathcal{B}$, telle que $(C,[\phi])$ s'annule dans ${}_{\varepsilon}L^{e}_{n-1}(\mathcal{A},\mathcal{C})$. Il existe une suite e-exacte dans $\mathcal{A}$:

$$0 \longrightarrow K \xrightarrow{\alpha} \Sigma \xrightarrow{\beta} C \longrightarrow 0$$

et un élément ψ de $\mathcal{Q}^{\varepsilon}(\Sigma)$ tels que : $\partial°\alpha = \partial°\beta = 0$, $\partial°\psi = n$

$$d\psi = \hat{\beta}\phi\beta$$

$\tilde{\psi}\alpha$ est une $\mathcal{C}$-équivalence

Considérons alors le couple $(K,\hat{\alpha}\psi\alpha)$. Soient M le -n-cône de $\hat{\alpha}\tilde{\psi}\alpha$, λ et σ les injections canoniques de $\hat{K}$ et K dans M, et μ et ρ les projections canoniques de M sur K et $\hat{K}$. Posons : $u = e_0 \otimes \hat{\mu}\rho - \hat{\mu}S(\hat{\alpha}\psi\alpha)\mu$. Alors $(M,[u])$ représente $\partial(K,\hat{\alpha}\psi\alpha)$.

Soit s une section de β. Il existe un morphisme γ de C dans K défini par :

$$ds = \alpha\gamma$$

Comme le morphisme $\hat{\alpha}\tilde{\psi}$ de Σ dans $\hat{K}$ est une $\mathcal{C}$-équivalence, il induit une $\mathcal{C}$-équivalence du cône de α sur M. Or le cône de α a le type d'homotopie (simple) de C. On en déduit une $\mathcal{C}$-équivalence f de C dans M. On trouve la formule :

$$f = \lambda\hat{\alpha}\tilde{\psi}s + (-1)^n\sigma\gamma$$

$$\begin{array}{ccccccccccc}
0 & \longrightarrow & K & \xrightarrow{\alpha} & \Sigma & \underset{\beta}{\overset{s}{\rightleftarrows}} & C & \longrightarrow & 0 & & \\
 & & & & \downarrow{\scriptstyle \hat{\alpha}\hat{\psi}} & & \downarrow{\scriptstyle f} & & & & \\
 & & 0 & \longrightarrow & \hat{K} & \underset{\lambda}{\overset{\rho}{\rightleftarrows}} & M & \underset{\mu}{\overset{\varrho}{\rightleftarrows}} & K & \longrightarrow & 0
\end{array}$$

D'après le lemme 3.5, on a :

$$\begin{aligned} d(\hat{s}\psi s - d\hat{s}S(\psi)s) &= \hat{s}d\psi s + d\hat{s}d(S\psi)s - (-1)^n d\hat{s}\hat{S}\psi ds \\ &= \phi - d\hat{s}\hat{S}(\beta\phi\beta)s + e_0 \otimes d\hat{s}\hat{\psi}s - (-1)^n\hat{\gamma}\hat{\alpha}S\psi\alpha\gamma \\ &= \phi + e_0 \otimes \hat{\gamma}\hat{\alpha}\hat{\psi}s - (-1)^n\hat{\gamma}\hat{\alpha}S\psi\alpha\gamma \end{aligned}$$

D'autre part, on a :

$$\hat{f}(e_0 \otimes \hat{\mu}\rho - \hat{\mu}S(\hat{\alpha}\psi\alpha)\mu)f = -(-1)^n e_0 \otimes \hat{\gamma}\hat{\alpha}\hat{\psi}s - \hat{\gamma}\hat{\alpha}S\psi\alpha\gamma$$

On en déduit que ϕ est homologue à $(-1)^n f^*(e_0 \otimes \hat{\mu}\rho - \hat{\mu}S(\hat{\alpha}\psi\alpha)\mu)$, et par conséquent la classe de $(C,(-1)^n[\phi])$ est égale à $\partial(K,\hat{\alpha}\psi\alpha)$.

Il en résulte que la classe de $(C,[\phi])$ est dans l'image de ∂.

§6- TRANSVERSALITE ET EXCISION

Soient $\mathcal{A}$ et $\mathcal{B}$ deux classes exactes de $C^e(A)$. On dira que $\mathcal{A}$ est transverse à $\mathcal{B}$ ($\mathcal{A} \pitchfork \mathcal{B}$) si tout morphisme d'un complexe de $\mathcal{A}$ dans un complexe de $\mathcal{B}$ se factorise par un complexe de $\mathcal{A} \cap \mathcal{B}$.

Si $\mathcal{A}$ et $\mathcal{B}$ sont deux classes exactes, on notera $\mathcal{A} + \mathcal{B}$ la classe engendrée par $\mathcal{A}$ et $\mathcal{B}$.

LEMME 6.1. *Soient $\mathcal{A}$ et $\mathcal{B}$ deux classes exactes, $\mathcal{A} \pitchfork \mathcal{B}$. Alors, un complexe C appartient à $\mathcal{A} + \mathcal{B}$ si et seulement s'il existe une $\mathcal{B}$-équivalence d'un complexe de $\mathcal{A}$ dans C.*

Démonstration. Soit $\mathcal{C}$ la classe des complexes $C \in C^e(A)$ tels qu'il existe une $\mathcal{B}$-équivalence d'un complexe de $\mathcal{A}$ dans C. Il est clair que $\mathcal{C}$ est contenue dans $\mathcal{A} + \mathcal{B}$ et contient $\mathcal{A}$ et $\mathcal{B}$. Il suffit donc de montrer que $\mathcal{C}$ est exacte.

Soit $0 \longrightarrow C \xrightarrow{\alpha} C' \longrightarrow C'' \longrightarrow 0$ une suite e-exacte, C et C' appartenant à $\mathcal{C}$.

Il existe deux complexes K et K' dans $\mathcal{A}$ et des $\mathcal{B}$-équivalences $f: K \to C$ et

$f': K' \to C'$, que l'on peut supposer surjectives.

Soit M' le cône de f'. Comme $\mathcal{A}$ est transverse à $\mathcal{B}$, le morphisme composé $K \to C \to C' \to M'$ se factorise par un complexe $X \in \mathcal{A} \cap \mathcal{B}$. Soit M le 1-cône du morphisme $K \to X$.

Comme X appartient à $\mathcal{A} \cap \mathcal{B}$, M appartient à $\mathcal{A}$ et le morphisme composé $M \to K \to C$ est une $\mathcal{B}$-équivalence. De plus, le morphisme composé $M \to K \to M'$ est homotope à 0 et le morphisme de M vers C' se factorise par K'.

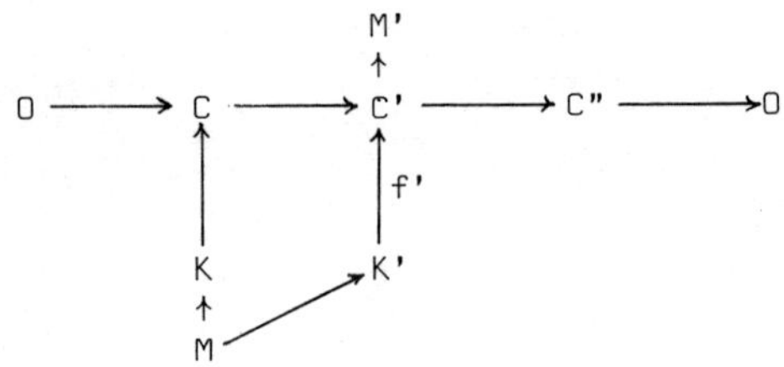

Soit K'' le cône du morphisme $M \to K'$; K'' appartient à $\mathcal{A}$ et s'envoie dans C'' par une $\mathcal{B}$-équivalence. Donc C'' appartient à $\mathcal{C}$.

Comme $\mathcal{C}$ est stable par équivalence d'homotopie (simple) et par suspension, on en déduit que $\mathcal{C}$ est exacte, et par conséquent, égale à $\mathcal{A} + \mathcal{B}$.

6.2 <u>Démonstration du théorème d'excision</u> 1.17.

Soient $\mathcal{A}$ et $\mathcal{B}$ deux classes exactes symétriques, $\mathcal{A} \pitchfork \mathcal{B}$. On a un morphisme canonique : $F : {}_{\varepsilon}L^{e}_{n}(\mathcal{A}, \mathcal{A} \cap \mathcal{B}) \longrightarrow {}_{\varepsilon}L^{e}_{n}(\mathcal{A} + \mathcal{B}, \mathcal{B})$

On va montrer que F est bijective.

Soit (C,q) un n-complexe ε-quadratique $\mathcal{B}$-non dégénéré, $C \in \mathcal{A} + \mathcal{B}$.

D'après 6.1, il existe une $\mathcal{B}$-équivalence f de K dans C , $K \in \mathcal{A}$. D'après le lemme 4.4, (C,q) est cobordant à $(K, f^{*}q)$ et F est surjective.

Soit (C,q) un n-complexe ε-quadratique $\mathcal{A} \cap \mathcal{B}$-non dégénéré, $C \in \mathcal{A}$. On suppose que (C,q) s'annule par F. Alors, il existe une suite exacte dans $\mathcal{A} + \mathcal{B}$:

$$0 \longrightarrow K \xrightarrow{\alpha} \Sigma \xrightarrow{\beta} C \longrightarrow 0$$

et un élément $\psi \in \mathcal{Q}^{\varepsilon}(\Sigma)$ tels que : $\partial^{o}\alpha = \partial^{o}\beta = 0 \qquad \partial^{o}\psi = n+1$

$$d\psi = \hat{\beta}\phi\beta$$

$\tilde{\psi}\alpha$ est une $\mathcal{B}$-équivalence.

D'après le lemme 6.1, on peut trouver un diagramme commutatif :

$$\begin{array}{ccccccccc}
0 & \longrightarrow & K & \xrightarrow{\alpha} & \Sigma & \xrightarrow{\beta} & C & \longrightarrow & 0 \\
 & & \uparrow f & & \uparrow g & & \uparrow 1 & & \\
0 & \longrightarrow & K' & \xrightarrow{\alpha'} & \Sigma' & \xrightarrow{\beta'} & C & \longrightarrow & 0
\end{array}$$

où les lignes sont e-exactes, où K' et Σ' sont dans $\mathcal{A}$ et où f et g sont des $\mathcal{B}$-équivalences. On a alors : $d(\hat{g}\psi g) = \hat{\beta}'\phi\beta'$

$\hat{g}\psi g\alpha'$ est une $\mathcal{A} \cap \mathcal{B}$-équivalence.

Ainsi, (C,q) est cobordant à 0 et F est injective.

6.3. Démonstration du théorème 1.19.

Soit $\mathcal{A}$ une classe exacte, $\mathcal{A} \pitchfork \hat{\mathcal{A}}$. On va montrer la nullité des groupes $_{\varepsilon}L_n^e(\mathcal{A} + \hat{\mathcal{A}}, \mathcal{A} \cap \hat{\mathcal{A}})$.

Soit (C,q) une n-forme ε-quadratique $\mathcal{A} \cap \hat{\mathcal{A}}$-non dégénérée, $C \in \mathcal{A} + \hat{\mathcal{A}}$.

Si j est un entier, on désignera par (P_j) la propriété suivante :

" Il existe un complexe $\Sigma \in \mathcal{A}$ et une $\hat{\mathcal{A}}$-équivalence f de Σ dans C tels que $f^*(q)$ soit représenté par un élément de $\mathcal{Q}^{\varepsilon}(\Sigma)$ de la forme $\sum_{i=0}^{j} e_i \otimes \phi_i$."

D'après le lemme 6.1, la propriété (P_j) est vérifiée pour j assez grand. Montrons par récurrence que (P_j) est vérifiée pour tout $j \geqslant -1$.

Supposons donc vérifiée la propriété (P_j) , $j \geqslant 0$. Il existe un complexe $\Sigma \in \mathcal{A}$ et un $\hat{\mathcal{A}}$-équivalence f de Σ dans C tels que $f^*(q)$ soit représenté par $e_0 \otimes \phi_0 + \ldots + e_j \otimes \phi_j$.

Comme cet élément est un cycle de $\mathcal{Q}^{\varepsilon}(\Sigma)$, on a : $d\phi_j = 0$ et ϕ_j est un morphisme de Σ dans $\hat{\Sigma}$. Comme $\mathcal{A}$ est transverse à $\hat{\mathcal{A}}$, ϕ_j se factorise par un complexe $X \in \mathcal{A} \cap \hat{\mathcal{A}}$. Soient Σ' le -1-cône de $\Sigma \to X$ et α le morphisme de Σ' dans Σ . Il est clair que Σ' appartient à $\mathcal{A}$ et que $f\alpha$ est une $\hat{\mathcal{A}}$-équivalence. De plus $\hat{\alpha}\phi_j\alpha$ est un bord $d\psi$. Il est alors facile de vérifier que $(f\alpha)^*(q)$ est représenté par un élément de la forme $e_0 \otimes \phi'_0 + \ldots + e_{j-1} \otimes \phi'_{j-1}$

Ainsi, par récurrence, on vérifie toutes les propriétés (P_j) et il existe un complexe $\Sigma \in \mathcal{A}$ et une $\hat{\mathcal{A}}$-équivalence f de Σ dans C tels que $f^*(q)$ soit nul. Sans changer le type d'homotopie (simple) de Σ, on pourra supposer que f est surjective de noyau $K \in \hat{\mathcal{A}}$: $0 \to K \xrightarrow{\alpha} \Sigma \xrightarrow{f} C \longrightarrow 0$.

Soient $\phi \in \mathcal{Q}^{\varepsilon}(C)$ un représentant de q et $\psi \in \mathcal{Q}^{\varepsilon}(\Sigma)$ un élément tel que $d\psi = \hat{f}\phi f$.

Considérons le diagramme commutatif suivant :

$$\begin{array}{ccccccccc}
0 & \longrightarrow & K & \xrightarrow{\alpha} & \Sigma & \xrightarrow{f} & C & \longrightarrow & 0 \\
 & & \downarrow{\scriptstyle \tilde{\psi}\alpha} & \swarrow{\scriptstyle \tilde{\psi}} & & & \downarrow{\scriptstyle \hat{\alpha}\tilde{\psi}} & & \\
0 & \longrightarrow & \hat{C} & \xrightarrow{\hat{f}} & \hat{\Sigma} & \xrightarrow{\hat{\alpha}} & \hat{K} & \longrightarrow & 0
\end{array}$$

Comme Σ et $\hat{K}$ appartiennent à A, $\hat{\alpha}\tilde{\psi}$ est une A-équivalence. De plus, on a : $d\tilde{\psi} = \hat{f}\tilde{\phi}f$ et $\tilde{\phi}$ est une A-équivalence. On en déduit, d'après le lemme 2.5, que $\tilde{\psi}\alpha$ est une A-équivalence.

D'autre part, K et $\hat{\Sigma}$ appartiennent à $\hat{A}$ et $\tilde{\psi}\alpha$ est une $A \cap \hat{A}$-équivalence. Il en résulte que (C,q) est cobordant à 0 .

<u>COROLLAIRE</u> 6.4. <u>Le groupe</u> ${}_{\varepsilon}L_n^{p'}(C^{p'}(A),0)$ <u>est isomorphe à</u> ${}_{\varepsilon}L_n^{p}(A)$.

<u>Démonstration</u>. Soit A la classe exacte de $C^{p'}(A)$ formée des complexes ayant le type d'homotopie de complexes de $C^p(A)$. Comme les catégories homotopiques $C^p(A)$ et A sont équivalentes, ${}_{\varepsilon}L_n^{p'}(A,0)$ est isomorphe à ${}_{\varepsilon}L_n^{p}(C^p(A),0) = {}_{\varepsilon}L_n^{p}(A)$.

Soit B la classe exacte de $C^{p'}(A)$ formée des complexes ayant le type d'homotopie de complexes de dimension finie. Alors, $\hat{B}$ est la classe des complexes ayant le type d'homotopie de complexes minorés. On vérifie que B est transverse à $\hat{B}$, que $B \cap \hat{B}$ est égal à A et que $B + \hat{B}$ est égal à $C^{p'}(A)$.

D'après le théorème 1.19, on a : ${}_{\varepsilon}L_n^{p'}(C^{p'}(A),A) = 0$

En utilisant la longue suite exacte du théorème 1.12, on en déduit que l'application canonique de ${}_{\varepsilon}L_n^{p'}(A,0)$ dans ${}_{\varepsilon}L_n^{p}(C^p(A),0)$ est un isomorphisme et ${}_{\varepsilon}L_n^{p'}(C^{p'}(A),0)$ est isomorphe à ${}_{\varepsilon}L_n^{p}(A)$.

§7. <u>LOCALISATION</u>

Dans ce paragraphe, on suppose que $A \to \Lambda$ est une localisation d'anneaux au sens de Cohn [3], qui soit, de plus, un morphisme d'anneaux avec involution.

7.1. <u>Démonstration du théorème</u> 1.23.

Soient α un sous-groupe de $\tilde{K}_1(\Lambda)_{\not{\varepsilon}}$ stable par involution, A(resp. A') la classe des complexes C de $C^s(A)$ (resp. $C^s(\Lambda)$) tels que $C \otimes \Lambda$ (resp. C) soit acyclique avec torsion dans α.

La tensorisation par Λ induit un morphisme :

$$ {}_{\varepsilon}L^{s}_{n}(C^{s}(A),A) \longrightarrow {}_{\varepsilon}L^{s}_{n}(C^{s}(\Lambda),A') = {}_{\varepsilon}L^{\alpha}_{n}(\Lambda) $$

On va montrer que c'est un isomorphisme.

Soit $\mathcal{B}$ (resp. $\mathcal{B}'$) la classe des complexes C de $C^{s}(\Lambda)$ (resp. $C^{s}(\Lambda)$) tels que $C \otimes \Lambda$ (resp. C) soit acyclique. On a un diagramme commutatif :

$$ \begin{array}{ccccccccc} \cdots \longrightarrow & {}_{\varepsilon}L^{s}_{n}(\mathcal{B},A) & \longrightarrow & {}_{\varepsilon}L^{s}_{n}(C^{s}(A),A) & \longrightarrow & {}_{\varepsilon}L^{s}_{n}(C^{s}(A),\mathcal{B}) & \longrightarrow & {}_{\varepsilon}L^{s}_{n-1}(\mathcal{B},A) & \longrightarrow \cdots \\ & \downarrow & & \downarrow & & \downarrow & & \downarrow & \\ \cdots \longrightarrow & {}_{\varepsilon}L^{s}_{n}(\mathcal{B}',A') & \longrightarrow & {}_{\varepsilon}L^{s}_{n}(C^{s}(\Lambda),A') & \longrightarrow & {}_{\varepsilon}L^{s}_{n}(C^{s}(\Lambda),\mathcal{B}') & \longrightarrow & {}_{\varepsilon}L^{s}_{n-1}(\mathcal{B}',A') & \longrightarrow \cdots \end{array} $$

Or, ${}_{\varepsilon}L^{s}_{n}(C^{s}(\Lambda),\mathcal{B}')$ est égal à ${}_{\varepsilon}L^{h}_{n}(\Lambda)$, et il est démontré dans [10] que le morphisme de ${}_{\varepsilon}L^{s}_{n}(C^{s}(A),\mathcal{B})$ dans ${}_{\varepsilon}L^{h}_{n}(\Lambda)$ est bijectif, du moins si ε est égal à 1, mais la démonstration s'applique pour tout ε.

D'autre part, tout complexe de $\mathcal{B}$(resp. $\mathcal{B}'$) est inversible modulo A (resp. A') et l'on a [10] :

$$ K(\mathcal{B},A) = K(\mathcal{B}',A') = \tilde{K}_{1}(\Lambda)/\alpha $$

On en déduit que l'application de ${}_{\varepsilon}L^{s}_{n}(\mathcal{B},A)$ dans ${}_{\varepsilon}L^{s}_{n}(\mathcal{B}',A')$ est bijective, et, d'après le lemme des cinq, l'application de ${}_{\varepsilon}L^{s}_{n}(C^{s}(A),A)$ dans ${}_{\varepsilon}L^{\alpha}_{n}(\Lambda)$ est bijective

7.2. <u>Localisation des complexes</u>

Soit $\overline{C}(A)$ la catégorie des A-modules différentiels $\mathbb{Z}$-gradués projectifs et minorés. Si e est différent de p', on a une fonction canonique de $C^{e}(A)$ dans $\overline{C}(A)$ qui est une injection si e est différent de s.

Soient $\overline{\mathcal{W}} \subset \overline{C}(A)$ la classe des complexes C de $\overline{C}(A)$ tels que $C \otimes \Lambda$ soit acyclique, et $\mathcal{W}$ la classe exacte des complexes C de $C^{e}(A)$ tels que $C \otimes \Lambda$ ait le type d'homotopie (simple) de 0.

<u>LEMME</u> 7.3 [10]. <u>Soient</u> K <u>un complexe de</u> $C^{e}(A)$ <u>et</u> C <u>un complexe de</u> $\overline{\mathcal{W}}$. <u>Alors tout morphisme de</u> K <u>dans</u> C <u>se factorise par un complexe de</u> $\mathcal{W}$.

<u>DEFINITION</u> 7.4. Un complexe C de $\overline{C}(A)$ est local si tout morphisme d'un complexe de $\overline{\mathcal{W}}$ dans C est homotope à 0.

Voici quelques résultats démontrés dans [10] concernant la localisation des complexes.

THEOREME 7.5. *Il existe un foncteur E de $\overline{C}(A)$ dans lui-même et un morphisme η de Id dans E tels que :*

Pour tout $C \in \overline{C}(A)$, E(C) est local

Pour tout $C \in \overline{C}(A)$, le cône de η : $C \to E(C)$ appartient à $\overline{W}$.

LEMME 7.6. *Si $C \in \overline{C}(A)$ est local, l'application canonique de $H_i(C)$ dans $H_i(C)\otimes\Lambda$ est bijective.*

LEMME 7.7. *On a les formules suivantes :*

$$\Lambda \underset{A}{\otimes} \Lambda \simeq \Lambda \qquad \mathrm{Tor}_1^A(\Lambda,\Lambda) = 0$$

On supposera dorénavant que $\mathrm{Tor}_i^A(\Lambda,\Lambda)$ est nul pour tout $i \geqslant 2$.

LEMME 7.8. *Si C est un complexe local de $\overline{C}(A)$, l'application canonique de $H_i(C)$ dans $H_i(C \otimes \Lambda)$ est bijective.*

Démonstration. Soit $\mathcal{E}$ la catégorie des A-modules à droite M tels que $M \to M \otimes \Lambda$ soit bijectif. Soit M un module de $\mathcal{E}$. Il existe une suite exacte :

$$0 \longrightarrow N \longrightarrow L \longrightarrow M \longrightarrow 0$$

où L est un Λ-module libre et N appartient à $\mathcal{E}$. Comme L est libre, $\mathrm{Tor}_i(L,\Lambda)$ est nul pour $i \geqslant 1$, et $\mathrm{Tor}_{i+1}(M,\Lambda)$ est isomorphe à $\mathrm{Tor}_i(N,\Lambda)$ pour $i \geqslant 1$. On a de plus la suite exacte suivante:

$$0 \to \mathrm{Tor}_1(M,\Lambda) \to N \otimes \Lambda \to L \otimes \Lambda \to M \otimes \Lambda \to 0$$

On en déduit que $\mathrm{Tor}_1(M,\Lambda)$ est nul pour tout M de $\mathcal{E}$ et par suite, $\mathrm{Tor}_i(M,\Lambda)$ est nul pour tout $i \geqslant 1$ et tout module M de $\mathcal{E}$.

D'autre part, il existe une suite spectrale de terme $E^2_{pq} = \mathrm{Tor}_p(H_q(C),\Lambda)$ convergeant vers $H_{p+q}(C \otimes \Lambda)$. Comme C est local, $H_q(C)$ appartient à $\mathcal{E}$ et E^2_{pq} est nul pour $p \geqslant 1$. De plus, E^2_{po} est égal à $H_p(C) \otimes \Lambda$ c'est-à-dire à $H_p(C)$.

On en déduit que l'application $C \to C \otimes \Lambda$ induit un isomorphisme en homologie.

LEMME 7.9. *Soient* C *et* C' *deux complexes de* $C^e(A)$ *et* f *un morphisme de* $C \otimes \Lambda$ *dans* $C' \otimes \Lambda$. *Alors il existe une* $\mathcal{W}$-*équivalence* $\alpha : K \to C$ *et un morphisme* $g : K \to C'$ *tels que* $g \otimes \Lambda$ *soit homotope à* $f \circ \alpha \otimes \Lambda$

Démonstration. Soit E(C') le localisé de C'. On a le diagramme suivant :

$$\begin{array}{ccc} C' & \longrightarrow & E(C') \\ \downarrow & & \downarrow \\ C' \otimes \Lambda & \longrightarrow & E(C') \otimes \Lambda \end{array}$$

et, d'après le lemme 7.8, le morphisme $E(C') \to E(C') \otimes \Lambda$ induit un isomorphisme en homologie. Il n'y a donc pas d'obstruction à factoriser le morphisme composé $C \to C \otimes \Lambda \to C' \otimes \Lambda \to E(C') \otimes \Lambda$ à travers E(C').

Soit g' le morphisme obtenu de C dans E(C') et M le cône de $C' \to E(C')$. Comme M appartient à $\overline{\mathcal{W}}$ et que C appartient à $C^e(A)$, l'application composée $C \to E(C') \to M$ se factorise à travers un complexe X de $\mathcal{W}$. Soient K le -1-cône du morphisme obtenu de C dans X et α le morphisme de K dans C . Par construction, $g'\alpha$ se factorise à travers C' par un morphisme g . On a donc un diagramme homotopiquement commutatif :

$$\begin{array}{ccccc} C' & \longrightarrow & E(C') & \longrightarrow & M \\ {\scriptstyle g}\uparrow & & {\scriptstyle g'}\uparrow & & \uparrow \\ K & \xrightarrow{\alpha} & C & \longrightarrow & X \end{array}$$

Il est clair que α est une $\mathcal{W}$-équivalence. Pour vérifier que $g \otimes \Lambda$ est homotope à $f \circ \alpha \otimes \Lambda$, il suffit de remarquer que le morphisme $C' \otimes \Lambda \to E(C') \otimes \Lambda$ est une équivalence d'homotopie.

LEMME 7.10. *Soient* C *et* C' *deux complexes de* $C^e(A)$ *et* f *un morphisme de* C *dans* C'. *Alors, si* $f \otimes \Lambda$ *est homotope à* 0, *il existe une* $\mathcal{W}$-*équivalence* $\alpha: K \to C$ *telle que* $f\alpha$ *soit homotope à* 0.

Démonstration. Soit E(C') le localisé de C'. Comme $E(C') \to E(C') \otimes \Lambda$ induit un isomorphisme en homologie, l'application composée $C \to C' \to E(C')$ est homotope à 0.

Soit M' le -1-cône de $C' \to E(C')$. Alors f se factorise à travers M' par une application g' :

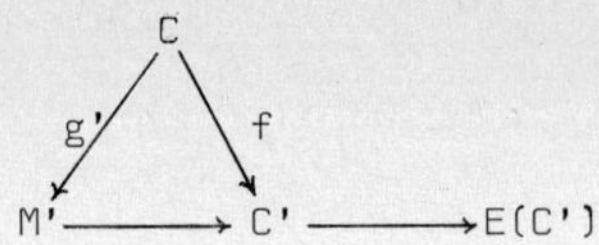

Comme M' appartient à $\overline{\mathcal{W}}$ et que C appartient à $\mathcal{C}^e(A)$, g' se factorise à travers un complexe X de $\mathcal{W}$ par une application g :

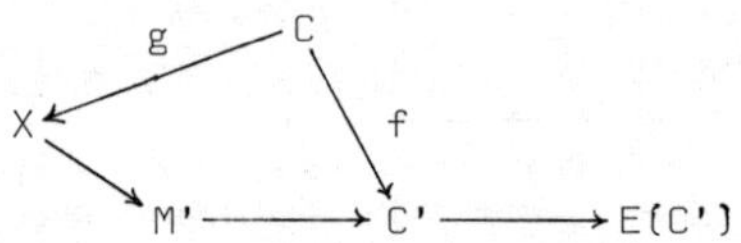

Soient alors K le -1-cône de g et $\alpha : K \longrightarrow C$ la projection canonique. Il est clair que α est une $\mathcal{W}$-équivalence et que $f\alpha$ est homotope à 0.

<u>LEMME</u> 7.11. <u>Soit $\mathcal{A}$ une classe exacte de $\mathcal{C}^e(A)$ contenant $\mathcal{W}$. Soit C' un complexe de $\mathcal{A} \otimes \Lambda$. Alors, il existe un complexe $C \in \mathcal{A}$ et une équivalence d'homotopie (simple) de $C \otimes \Lambda$ dans C'.</u>

<u>Démonstration</u>. Soit $\mathcal{C}$ la classe de $\mathcal{C}^e(\Lambda)$ formée des complexes ayant le type d'homotopie (simple) de $C \otimes \Lambda$, $C \in \mathcal{A}$.

Pour montrer que $\mathcal{C}$ est égal à $\mathcal{A} \otimes \Lambda$, il suffit de montrer que $\mathcal{C}$ est exacte.

Soit $0 \to C \overset{\alpha}{\to} C' \overset{\beta}{\to} C'' \to 0$ une suite e-exacte dans $\mathcal{C}^e(\Lambda)$, C et C' appartenant à $\mathcal{C}$. Il existe deux complexes K et K' de $\mathcal{A}$ et des équivalences d'homotopie (simple) $u : K \otimes \Lambda \to C$ et $u' : K' \otimes \Lambda \to C'$. On a donc un diagramme homotopiquement commutatif :

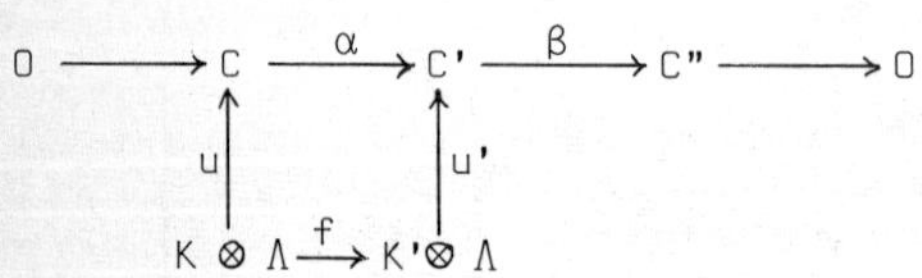

D'après le lemme 7.9, on peut, quitte à modifier K, supposer que f est de la forme $g \otimes \Lambda$. Soit K" le cône de g. Alors, C" a le type d'homotopie (simple) de $K'' \otimes \Lambda$.

Comme $\mathcal{C}$ est stable par équivalence d'homotopie (simple) et par suspension, $\mathcal{C}$ est exacte et, par suite, égale à $\mathcal{A} \otimes \Lambda$ •

7.12. <u>Démonstration du théorème</u> 1.24.

On suppose que $\mathcal{A} \supset \mathcal{B}$ sont deux classes exactes symétriques de $\mathcal{C}^e(A)$ ($e \neq p'$) contenant $\mathcal{W}$. Soit F le morphisme canonique :

$$F : {}_\varepsilon L_n^e(\mathcal{A},\mathcal{B}) \longrightarrow {}_\varepsilon L_n^e(\mathcal{A} \otimes \Lambda, \mathcal{B} \otimes \Lambda)$$

a) <u>Surjectivité de</u> F. Soit v un élément de ${}_\varepsilon L_n^e(\mathcal{A} \otimes \Lambda, \mathcal{B} \otimes \Lambda)$ représenté par un couple (C',q').

Soit j un entier positif ou nul. Considérons la propriété P_j suivante :

" Il existe un complexe $C \in \mathcal{A}$, une équivalence d'homotopie (simple) f de $C \otimes \Lambda$ dans C', un cycle $\phi' = \Sigma\, e_i \otimes \phi'_i$ de $\mathcal{Q}^\varepsilon(C \otimes \Lambda)$ et des éléments ϕ_i , $i \geqslant j$, de $B(C)$ tels que : $f^* q'$ est représenté par ϕ'

pout tout $i \geqslant j$, $\phi'_i = \phi_i \otimes \Lambda$

pour tout $i \geqslant j$, $(-1)^i d\phi_i + \phi_{i+1} - (-1)^i \varepsilon \hat{\phi}_{i+1} = 0$ "

D'après le lemme 7.11, la propriété P_j est vérifiéc pour j assez grand. Soit $j > 0$ et supposons P_j vérifiée. Comme ϕ' est un cycle, on a :

$$(-1)^{j-1} d\phi'_{j-1} + \phi'_j + (-1)^j \varepsilon \hat{\phi}'_j = 0$$

On vérifie alors que $\phi_j + (-1)^j \varepsilon \hat{\phi}_j$ est un cycle et que $(\phi_j + (-1)^j \varepsilon \hat{\phi}_j) \otimes \Lambda$ est un bord. D'après le lemme 7.10, on peut, quitte à modifier C, supposer que $\phi_j + (-1)^j \varepsilon \hat{\phi}_j$ est un bord $d\alpha$. On a alors :

$$(-1)^{j-1} d\phi'_{j-1} + d(\alpha \otimes \Lambda) = 0$$

D'après le lemme 7.9, on peut, quitte à modifier C, supposer que l'on a :

$$(-1)^{j-1} \phi'_{j-1} + \alpha \otimes \Lambda = \alpha' \otimes \Lambda + d\beta \qquad \text{avec } d\alpha' = 0$$

On pose alors : $\phi_{j-1} = (-1)^j (\alpha - \alpha')$

et l'on a :

$$\phi' - d(e_{j-1} \otimes \beta) = \Sigma\, e_i \otimes \phi''_i$$

et pour tout $i \geqslant j-1$, $\phi''_i = \phi_i \otimes \Lambda$ et $(-1)^i d\phi_i + \phi_{i+1} - (-1)^i \varepsilon \hat{\phi}_{i+1} = 0$

On a ainsi montré la propriété P_{j-1} .

En particulier, on a la propriété P_o, ce qui implique que F est surjective.

b) <u>Injectivité de</u> F. Soit u un élément de ${}_\varepsilon L_n^e(\mathcal{A},\mathcal{B})$ annulé par F et représenté par un n-complexe ε-quadratique $\mathcal{B}$-non dégénéré (C,q), $C \in \mathcal{A}$.

Comme F(u) est nul, il existe une n+1-paire ε-quadratique $\mathcal{B} \otimes \Lambda$-non dégénérée

$(\Sigma' \xrightarrow{f'} C \otimes \Lambda, q')$ telle que Σ' appartienne à $A \otimes \Lambda$ et $\partial q'$ soit égal à $q \otimes \Lambda$.

D'après les lemmes 7.11 et 7.9, on peut supposer que Σ' est de la forme $\Sigma \otimes \Lambda$, $\Sigma \in A$ et f' de la forme $f \otimes \Lambda$, f étant un épimorphisme.

Soit $\phi = \Sigma\, e_i \otimes \phi_i$ un représentant de q. Soit j un entier positif ou nul. Considérons la propriété P_j suivante :

" Il existe un complexe $\Sigma_1 \in A$, une $\mathcal{W}$-équivalence surjective g de Σ_1 sur Σ, un représentant $\psi' = \Sigma\, e_i \otimes \psi'_i$ de $(g \otimes \Lambda)^* q'$ et des éléments ψ_i , $i \geqslant j$ de $B(\Sigma_1)$ tels qu que :

$$\text{pour tout } i \geqslant j \text{ , } \psi'_i = \psi_i \otimes \Lambda$$

$$\text{pour tout } i \geqslant j \text{ , } (-1)^i d\psi_i + \psi_{i+1} - (-1)^i \varepsilon\hat{\psi}_{i+1} = \hat{g}\hat{f}\phi_i fg \quad . \text{"}$$

Cette propriété a lieu pour j assez grand. Soit $j > 0$ et supposons P_j vérifiée. On vérifie que $\psi_j + (-1)^j \varepsilon\hat{\psi}_j - \hat{g}\hat{f}\phi_{j-1} fg$ est un cycle et que $(\psi_j + (-1)^j \varepsilon\hat{\psi}_j - \hat{g}\hat{f}\phi_{j-1} fg) \otimes \Lambda$ est un bord.

D'après le lemme 7.10, on peut, quitte à modifier Σ_1, supposer que $\psi_j + (-1)^j \varepsilon\hat{\psi}_j - \hat{g}\hat{f}\phi_{j-1} fg$ est un bord $d\alpha$. On a alors :

$$(-1)^{j-1} d\psi'_{j-1} + d\alpha \otimes \Lambda = 0$$

D'après le lemme 7.9, on peut, quitte à modifier Σ_1, supposer que l'on a :

$$(-1)^{j-1} \psi'_{j-1} + \alpha \otimes \Lambda = \alpha' \otimes \Lambda + d\beta \qquad \text{avec} \quad d\alpha' = 0$$

On pose alors :

$$\psi_{j-1} = (-1)^j (\alpha - \alpha')$$

et l'on a : $\psi' - d(e_{j-1} \otimes \beta) = \Sigma\, e_i \otimes \psi''_i$

$$\text{pour tout } i \geqslant j-1 \text{ , } \psi''_i = \psi_i \otimes \Lambda$$

$$\text{pour tout } i \geqslant j-1 \text{ , } (-1)^i d\psi_i + \psi_{i+1} - (-1)^i \varepsilon\hat{\psi}_{i+1} = \hat{g}\hat{f}\phi_i fg$$

On a montré ainsi la propriété P_{j-1}. En particulier, on a la propriété P_o. On vérifie alors que $(\Sigma_1 \xrightarrow{gf} C, [\psi])$ est une $n+1$-paire ε-quadratique $\mathcal{B}$-non dégénérée, que Σ_1 appartient à A et que $\partial[\psi]$ est égal à q. En conséquence, F est injective.

BIBLIOGRAPHIE

[1] J.BARGE, J.J.SANSUC et P.VOGEL, Une suite spectrale pour les formes quadratiques sur un anneau commutatif noetherien régulier,à paraître.

[2] S.CAPPELL and J.SHANESON, The codimension two placement problem and homology equivalent manifolds, Ann. of Math. (2) 99 (1974), p. 277-348.

[3] P.M.COHN, Inversive localization in noetherian rings, Comm. Pure Appl. Math. 26 (1973), p. 679-691.

[4] J.W.MILNOR, Whitehead torsion, Bull. Amer. Math. Soc. 72 (1966), p. 358-426.

[5] W.L.PARDON, A "Gerstein Conjecture" for Witt groups, à paraître aux Proceedings de K-théorie algébrique, Oberwolfach, 1980.

[6] W.L.PARDON, A relation between the Witt group of a regular local ring and the Witt group of its residue class fields, preprint.

[7] A.A.RANICKI, Algebraic L-theory III Twisted Laurent extensions, Algebraic K-theory III, Battelle Institute conference 1972, Lecture Notes in Math. 343, Springer-Verlag (1973), p. 412-463.

[8] A.A.RANICKI, The algebraic theory of surgery I foundations, Proc. London Math. Soc. (3) 40 (1980), p. 87-192.

[9] P.VOGEL, Torsion de Whitehead généralisée, C.R.A.S. 290 (1980) p. 491-493.

[10] P.VOGEL,On the obstruction group in homology surgery, Publ. Math. de l'I.H.E.S. 55 (1982), p. 165-206.

[11] P.VOGEL, Localization in algabraic L-theory, Lecture Notes in Math. 788, Springer-Verlag (1980), p. 482-495.

[12] C.T.C.WALL, Surgery on compact manifolds, Academic Press,New-York and London, 1970.

INSTITUT DE MATHEMATIQUES

Université de NANTES

2, rue de la Houssinière

F- 44072 NANTES Cedex

A Survey on the K-theory of polynomial extensions.

by Ton Vorst

§1. Introduction.

In this paper we will try to give a survey of the K-theory of polynomial extensions. Furthermore we will state some open problems.

Approximately ten years ago Bass([Ba2]) devoted a section of his list of problems in "classical" algebraic K-theory to the K-theory of polynomial extensions. In the same Battelle proceedings Gersten ([Ge]) in his problems about higher K-functors also had two problems on this subject.

Since that time two developments had their influence on the K-theory of polynomial rings. First there was the more profound study which was made of higher K-theory. (The seminal paper of Quillen ([Qu1]) using the Q-construction for defining K-theory was published in the proceedings of the same conference). The second development was the solution of the Serre problem by Quillen and Suslin. Nowadays a lot of the problems posed in the papers of Bass and Gersten are solved and new problems have come into the picture. These are the reasons why I have written this survey.

All rings A in this paper are assumed to be commutative and noetherian, unless stated otherwise. By studying the K-theory of polynomial extensions we mean the study of the groups

$$N_pK_n(A) = \mathrm{Ker}\{K_n(A[T_1,\dots,T_p]) \to K_n(A)\}.$$

where the map is the map induced by the A-algebra homomorphism sending each T_i to zero.

We say that A is K_n-regular if $N_pK_n(A) = 0$ for all $p \geq 1$. One often meets the notation $N^pK_n(A)$. By this is meant $N_1(N_1(N_1 \dots N_1(K_n(A))\dots))$, where we apply p times N_1. It is straightforward to show that $N^pK_n(A) = 0$ for all $p \geq 1$ if and only if $N_pK_n(A) = 0$ for all $p \geq 1$. One often writes $NK_n(A)$ in stead of $N_1K_n(A)$.

The organisation of this paper is as follows. In § 2 we give some general results and define the concept of the K-regularity index. Module structures on $N^pK_n(A)$ will be described in § 3. In § 4 we calculate this K-regularity index for rings of dimension one while § 5 is devoted to the study of K_n-regularity for rings of higher dimension. A proposition of this section will be proved in the appendix. The relation of the K-theory of polynomial extensions to other K-theories can be found in § 6, while § 7 is devoted to unstable problems of which the Serre problem is an example. Finally, in the last section we will state some open problems. I like to thank Wilberd van der Kallen and Chuck Weibel for giving comments on an earlier version of this paper.

§ 2. General results and the K-regularity index.

The following theorem makes it possible to restrict the study of the question of K_n-regularity to local rings.

2.1. Theorem. ([Vo2] Cor. 1.9). Let A be a commutative ring. Then for $p \geq 1$ and $n \in \mathbb{Z}$ we have:

(i) If $S \subset A$ is a multiplicatively closed set of non-zero divisors, then $N_pK_n(A) = 0$ implies $N_pK_n(S^{-1}A) = 0$. If moreover A is reduced we have:

(ii) $N_pK_n(A) = 0$ implies $N_pK_n(A_{\underline{p}}) = 0$ for every prime ideal $\underline{p}$ of A.

(iii) The map $N_pK_n(A) \to \prod_{\underline{m} \subset A} N_pK_n(A_{\underline{m}})$ is injective, where the product is over the maximal ideals of A and the map is induced by the natural maps $A \to A_{\underline{m}}$.

2.2. Remark. In [Vo2] only the cases $p = 1$ and $n \geq 0$ are proved, but the obvious generalization of the proof also works in the other cases. One only needs a localization sequence for negative K-theory, but this can be found in [Ca].

Consider the fundamental sequence

$$0 \to N_pK_n(A) \to N_pK_n(A[T]) \oplus N_pK_n(A[T^{-1}]) \to N_pK_n(A[T,T^{-1}]) \to N_pK_n(A) \to 0,$$

which one gets by taking kernels of the map between the fundamental sequences for $A[T_1,\ldots,T_p]$ and A. If A is K_n-regular then A[T] is

K_n-regular and hence by the above theorem $A[T,T^{-1}]$ is K_n-regular. But this implies that $N_pK_{n-1}(A) = 0$ for every $p \geq 1$. Hence we have:

2.3. Corollary. (van der Kallen [Vo2]). K_n-regularity implies K_{n-1} regularity.

This partially solves problem III in [Ba2] which asks: Does K_1-regularity imply K_o-regularity? And more specifically Bass asks: does $NK_1(A) = 0$ imply $NK_o(A) = 0$. But we don't know yet the answer to the question whether $NK_n(A) = 0$ implies $NK_{n-1}(A) = 0$. By what we said before it is clear that this problem is related to the following problem posed by Sharma and Strooker.

2.4. Problem. ([Ba2] Prob. VI). Does $NK_n(A) = 0$ imply that A is K_n-regular?

On the other hand we have the following result.

2.5. Proposition. (Weibel, see [An] lemma 5.7). If $N^qK_n(A) = 0$ then also $N^pK_n(A) = 0$ for all $1 \leq p \leq q$.

2.6. Remarks. (i) In [Ba2] a ring is called Laurent K_n-regular if $A[T_1,T_1^{-1},\dots,T_m,T_m^{-1}]$ is K_n-regular for every $m \geq 0$. Again from 2.1.(i) it is clear that Laurent K_n-regularity is equivalent with K_n-regularity. In [Ba2] it is said that Laurent K_n-regularity is stronger but the example which Bass gives, doesn't work (see also [DW1] 4.4).
(ii). The condition in 2.1 that A is reduced isn't very serious. If J is a nilpotent ideal of A then
$K_o(A) \cong K_o(A/J)$ and by the fundamental sequence one sees from this that $K_n(A) \cong K_n(A/J)$ for all $n \leq 0$. Hence
$N_pK_n(A) \cong N_pK_n(A/J)$ for all $p \geq 1$ and $n \leq 0$. So to check K_n-regularity for $n \leq 0$ one may always divide out the nilradical. On the other hand if $a \in A$ is a nilpotent element then
$(1+aT_1) \in GL_1(A[T_1,\dots,T_p]) \in GL(A[T_1,\dots,T_p])$ yields a nonzero element in every $N_pK_1(A)$ and hence if A is not reduced it will not be K_1-regular and therefore not K_n-regular for $n \geq 1$.
Knowing this it is obvious that it suffices to study K_n-regularity for reduced rings and then 2.1 says that K_n-regularity may be studied locally.

From 2.3 it follows that the following definition makes sense.

2.7. <u>Definition</u>. Let A be a ring. We define the K-regularity index, K-reg(A) as follows

$$\text{K-reg}(A) = \begin{cases} -\infty & \text{if there is no } n \in Z \text{ for which } A \text{ is } K_n\text{-regular} \\ n & \text{if } A \text{ is } K_n\text{-regular but not } K_{n+1}\text{-regular} \\ +\infty & \text{if } A \text{ is } K_n\text{-regular for every } n \in Z. \end{cases}$$

2.8. <u>Theorem</u>. (Quillen [Q1]). If A is regular then K-reg(A) = + ∞.

K_0-regularity for regular rings has been proved by Grothendieck, while K_1-regularity has been proved by Bass, Heller and Swan ([BHS]).

If dim A = 0 there are two possibilities for K-reg(A). If A is regular then K-reg(A) = + ∞ and if A is not regular it must have a non-zero nilradical J and by remark 2.6(ii) we know that A is K_0-regular, since A/J is a product of fields, but not K_1-regular and hence K-reg(A) = 0.

§ 3. <u>Module structures on NK-theory</u>.

Most of the results which we mention in this section can be found in a more extensive form (with proofs) in a paper of Weibel ([We1]). Let W(A), where A is a commutative ring with 1, be the ring of Witt vectors. I.e. $W(A) = 1 + tA[[t]] \subset A[[t]]$ where the additive structure of W(A) comes from the multiplicative structure on A[[t]], while the multiplication on W(A) is the unique continuous functorial operation * with $(1-at)*(1-bt) = (1-abt)$. An introduction to the theory of Witt rings suited for K-theory can be found in [Bl1]. The following result is more or less implicit in the work of Bloch ([Bl2]).

3.1. <u>Theorem</u>. Let A be a commutative ring with 1. $N^pK_n(A)$ has a W(A)-module structure for all $p \geq 1$ and $n \in Z$.

A proof of this theorem (for $n \geq 1$) can be found in [St]. The multiplication by an element of the form (1-at) is the map $\phi_a^*: N^pK_n(A) \to N^pK_n(A)$ which can be described as follows: Let

$\phi_a: A[T] \to A[T]$ be the A-algebra homomorphism with $\phi_a(T) = aT$. Then ϕ_a^* is the restriction to $N^pK_n(A)$ of the following map

$$N^{p-1}K_n(\phi_a): N^{p-1}K_n(A[T]) \to N^{p-1}K_n(A[T]).$$

Using this description of the multiplication by (1-at) one has the following result

3.2. <u>Theorem</u>. ([Vo2] and [Wel]). Let A be a commutative ring with 1 and $S \subset A$ a multiplicatively closed set. If either A is reduced or S doesn't contain zero divisors we have an isomorphism

$$\{1-st \mid s \in S\}^{-1}W(A) \otimes_{W(A)} N^pK_n(A) \xrightarrow{\sim} N^pK_n(S^{-1}A).$$

Theorems 2.1 (ii) and (iii) are an immediate consequence of this result, but historically 3.2 is a consequence of the proof of 2.1.

Let A be $S^{-1}\mathbb{Z}, \hat{\mathbb{Z}}_p$ or a $\mathbb{Q}$-algebra then the following expression is a power series which defines an element of W(A).

$$(1-t)^a = 1 - at + \binom{a}{2}t^2 - \binom{a}{3}t^3 + \ldots$$

3.3. <u>Proposition</u>. ([Wel]). If A is $S^{-1}\mathbb{Z}, \hat{\mathbb{Z}}_p$ or a $\mathbb{Q}$-algebra, then the map $\lambda_t: a \to (1-t)^a$ defines a ring injection from A to W(A).

3.4. <u>Corollary</u>. ([Wel]). If A is a $S^{-1}\mathbb{Z}$, $\mathbb{Q}$ or $\hat{\mathbb{Z}}_p$-algebra then $N^pK_n(A)$ is resp. a $S^{-1}\mathbb{Z}$, A or $\hat{\mathbb{Z}}_p$-module.

Combining 3.2 and 3.3 we get the following corollary

3.5. <u>Corollary</u>. ([Bal]). Let A be a reduced commutative ring and $m \in \mathbb{Z}$ such that $A[\frac{1}{m}]$ is regular, then $NK_n(A)$ is an m-torsion abelian group.

Another result which can be proved using the W(A)-module structure is the following, which was proved by Chase [Ch] for n = 0,1 and of which a proof for general n can be found in [Wei]:

3.6. <u>Proposition</u>. If $p^m = 0$ in A, then $N^rK_n(A)$ is a p-group.

This result answers problem 18 of [Ge].
We already introduced the map ϕ_a^*: $N^pK_n(A) \to N^pK_n(A)$ which is the multiplication by $(1-at) \in W(A)$. There are two other important kinds of operations on $N^pK_n(A)$. For every $m \in \mathbb{N}$ define the A-algebra homomorphism
Θ_m: $A[T] \to A[T]$ by Θ_m: $T \to T^m$. The restriction of

$$N^{p-1}K_n(\Theta_m): N^{p-1}K_n(A[T]) \to N^{p-1}K_n(A[T])$$

to $N^pK_n(A)$ has been called the Verschiebung V_m. Since Θ_m makes A[T] a free module of rank m over itself, we also have a contravariant transfer map (the Frobenius map):

$$F_m: N^pK_n(A) \to N^pK_n(A).$$

By the usual norm argument we have that $F_m \circ V_m = mI$. By a remark which will be explained in § 6, there exists for every $\alpha \in N^pK_n(A)$ an $M \in \mathbb{N}$ s.t. $m \geq M$ implies $F_m(\alpha) = 0$. From this one easily deduces the following result.

3.7. Proposition. (Farrell, see [vdK] p.310). Every group $N^pK_n(A)$ is zero or not finitely generated.

§ 4. Rings of dimension one.

In this section we only consider reduced rings A such that $\dim(A) = 1$. Weibel ([We2], prop. 2.8) has shown that A is always K_{-1}-regular in this case and hence $K\text{-reg}(A) \geq -1$. We denote by $\bar{A}$ the integral closure of A in its total quotient ring Q(A). We have the following theorem:

4.1. Theorem. Let A be a noetherian ring as above such that $\bar{A}$ is a finite A-module. Then
(i) A is K_0-regular iff A is seminormal.
(ii) A is K_1-regular iff A is seminormal and for every maximal ideal $\underline{m}' \subset \bar{A}$, $\bar{A}/\underline{m}'$ is a separable extension of $A/\underline{m}$, where $\underline{m} = A \cap \underline{m}'$.

4.2. Remarks. A commutative ring A is called seminormal if it is reduced and whenever $b, c \in A$ satisfy $b^3 = c^2$, there is an $a \in A$ with $a^2 = b$, $a^3 = c$. This definition recently has been given by Swan ([Sw]), who shows that if Q(A) is a product of fields A is seminormal if and only if

$$A = \{x \in \bar{A} | x \in A_{\underline{p}} + \text{Rad } \bar{A}_{\underline{p}}, \ \underline{p} \in \text{Spec}A\}$$

Before Swan's paper this last condition was mostly used for the definition of seminormal. The new definition is a slight modification of an older definition of Hamann ([Ha]). Swan's paper is a good reference concerning seminormality.

The concept of seminormality arose from a classification problem of algebraic varieties. Bombieri ([Bo]) proved that if A is the coordinate ring of an affine algebraic curve over an algebraically closed field then R is seminormal if and only if at the singular points the curve analytically consists of smooth branches with linearly independent tangents. A generalization of this result for arbitrary ground fields has been obtained by Davis ([Da]).

Part (i) of 4.1 has been proved by Bass and Murthy ([BM]). They first showed, that for rings as in the theorem, one has $\tilde{K}_o(A[T_1,\ldots,T_n]) \cong \text{Pic}(A[T_1,\ldots,T_n])$ and next they proved that Pic-regularity is equivalent with seminormality. This last result has been extended to rings of arbitrary dimension fulfilling the other conditions by Traverso ([Tr]).

Theorem 4.1.(i) is also true if one deletes the condition that $\bar{A}$ is a finite A-module (see e.g. [BC2]).

The proof of theorem 4.1.(ii) can be found in [Vo1].For $n = 2$ we have the following theorem.

4.3. <u>Theorem</u>. ([Vo2]). Let A be a one-dimensional k-algebra of essentially finite type over a field k. Then A is K_2-regular iff A is regular.

In the proof of this theorem one needs the description of a seminormal curve over a field as given by Davis (see preceeding remark). We don't know yet how to prove the theorem if we don't have a ground field. By corollary 2.3 and theorem 2.8 we may replace K_2-regular by K_n-regular for $n \geq 2$ in 4.3.

Since the rings of theorem 4.3 fulfil the conditions of theorem 4.1 we can state the following theorem.

4.4. Theorem. Let A be a reduced algebra which is of essentially finite type over a field k such that dim(A) = 1. Then

$$\text{K-reg}(A) = \begin{cases} -1 & \text{iff A is not seminormal} \\ 0 & \text{iff A is seminormal but there exists a maximal ideal } \underline{m}' \text{ of } \bar{A} \text{ such that } \bar{A}/\underline{m}' \text{ is an inseparable extension of } A/m, \text{ where } \underline{m} = A \cap \underline{m}' \\ 1 & \text{iff A is not regular but seminormal and maximal ideals } \underline{m}' \text{ of } \bar{A} \text{ as above don't exist} \\ \infty & \text{iff A is regular} \end{cases}$$

By this result and remark 2.8 one is led to the following problem.

4.5. Problem. Let A be a non-regular ring. Do the following inequalities always hold

$$-\dim(A) \leq \text{K-reg}(A) \leq \dim(A)?$$

The question of the left-hand inequality was first posed by Weibel ([We2]), while the question of the right hand inequality was posed in [Vo2]. It can be proved that the left-hand inequality can not be improved, but for the right-hand inequality one even has the following question.

4.6. Question. Is there a non-regular ring A for which K-reg(A) $\geq$ 2?

§ 5. Rings of higher dimensions.

For rings of dimension greater than one we don't have such a nice description of the K-regularity index as we had in the preceeding section. For years, one of the main problems was the following question of Murthy ([Ba2], problem IV):

5.1. Problem. Does there exist a noetherian integral domain A which is factorial (or even only normal) for which $NK_0(A) \neq 0$?

This problem has been solved affirmatively by Weibel and Swan. Weibel ([We2]) showed that Nagata's example 6 of "Examples of bad noetherian rings" in [Na] is a two-dimensional local normal domain

for which $NK_o(A) \neq 0$. This ring can be described as follows. Let K be a field of characteristic $p \neq 0$ with an infinite set $\{b_1, c_1, b_2, \ldots\}$ of p-independent elements. Then take $R = K^p[[x,y]] \otimes_{K^p} K$, $d = \Sigma(b_i x^i + c_i y^i)$ and $e = d^p \in R$. Then $A = R[Z]/(Z^p - e)$ is a local normal 2-dimensional ring for which $NK_o(A) \neq 0$.
Furthermore in theorem 4.6 of [We2] a method of Swan is described to find UFD's for which $NK_o \neq 0$. One of the examples one finds, is $k[x,y,s,t]/(x^3 - y^2 - st)$, where k is a field.

Later Bloch and Murthy (see [We4]) showed that also the normal domain $A = k[x,y,z]/(x^2 - y^3 - z^7)$, with k a field, is not K_o-regular (and hence K-reg(A) < 0). Srinivas ([Sr]) used the techniques which were required for the proof of this last result to get some more examples of rings which are not K_o-regular. His results were generalized by Coombes and Srinivas to the following.

5.2. <u>Proposition</u>. ([CS]). Let $C \subset P^n_k$ be a projectively normal curve over an algebraically closed field k. Let Y be the affine cone over C. Assume that $H^1(C, \mathcal{O}_C(1)) \neq 0$. Then Y is not K_o-regular.

Although one doesn't have a general theory of K_n-regularity for rings of higher dimension there are a lot of examples known. In the first place we have the following result of Murthy and Pedrini.

5.3. <u>Proposition</u>. ([MP]). Let A be one of the following types of rings where k is a field:

(i) $A = k[x,y,z]/(x^n - yz)$.

(ii) A is the homogeneous coordinate ring of an arithmetically normal embedding of P^1_k in P^n_k.

(iii) k is algebraically closed and A is the coordinate ring of a surface X birationally equivalent to a ruled surface of genus > 0, and such that X has only rational singularities.

Then A is K_o-regular.

Pedrini considered what happens if one glues together some regular varieties. For the definition of the glueing process we refer to [Pe]. By his techniques he can show that e.g. $k[x,y,z]/(y^3 + z^2 - xyz)$ and $k[x,y,z]/(xy^2 - z^2)$, where k is a field, are K_o-regular.

Next we consider the following class of rings A. Let k be a field (or more generally assume that k is K_1-regular) and $R = k[x_o,x_1,\ldots,x_n]$. Let I be a non-trivial ideal of R generated by square-free monomials. Set A = R/I (a ring of this type is called a square free discrete Hodge algebra over k ([CEP])). Then we have the following result.

5.4. Proposition. Let A be as above. K-reg(A) = 1 if and only if I is not generated by a subset of $\{x_o,x_1,\ldots,x_n\}$, in which case A is regular and hence K-reg(A) = + ∞.

The proof of this result can be found in the appendix. A ring of this type is the coordinate ring of a union of intersections of the coördinate hyperplanes in k^{n+1}. The K_1-regularity of this kind of rings has been proved in a more general setting by Dayton and Weibel ([DW2]), but their result can be reduced to this proposition by 2.1. One can also give an alternative description of these rings as follows. Let Σ be a simplicial subcomplex of the standard n-simplex Δ_n with vertices {0,1,...,n}. Let I be the ideal of $k[x_o,x_1,\ldots,x_n]$ generated by the monomials $x_{i_o}x_{i_1}\ldots x_{i_r}$ $(0\leq i_o<i_1<\ldots<i_r\leq n)$ for which the simplex with the vertices $\{i_o,i_1,\ldots,i_r\}$ is not a subsimplex of Σ. Then I is an ideal generated by square-free monomials and we have a bijection from the simplicial subcomplexes of Δ_n to those ideals. Reisner [Re] showed that the Cohen-Macaulyness of the ring $k[x_o,x_1,\ldots,x_n]/I$ is related with the topological structure of Σ. Our result shows that the K-regularity doesn't depend on the topology of Σ.

§6. Connections with other K-theories.

For a ring A let Nil(A) be the category whose objects are pairs (P,v) where P is a finitely generated projective A-module and v: P → P a nilpotent endomorphism. The category P(A) of finitely generated projective A-modules is a retract of Nil(A) by: (P,v) → P and P → (P,o). Hence $K_n(\mathrm{Nil}(A)) = K_n(A) \oplus \mathrm{Nil}_n(A)$ where $\mathrm{Nil}_n(A)$ is the kernel of $K_n(\mathrm{Nil}(A)) \to K_n(A)$.
Problem 3 of [Ge] states the question whether $\mathrm{Nil}_{n-1}(A) \cong NK_n(A)$. This result was at that time known for n = 1, see [Ba1]. One has now

the following result

6.1. Theorem. (Quillen [GQ]). Suppose A is a ring, then $NK_n(A) \cong Nil_{n-1}(A)$ for all n.

In fact Gersten's problem also had the question whether $K_n(A[T,T^{-1}]) = K_n(A) \oplus K_{n-1}(A) \oplus Nil_{n-1}(A) \oplus Nil_{n-1}(A)$. This is also solved in [GQ].
Grayson extended theorem by considering more general classes of endomorphisms as follows. Let $T \subset A[t]$ be a multiplicatively closed set of monic polynomials, containing t. Let End(A;T) be the category of pairs (P,v) where P is a finitely generated projective A-module and $v: P \to P$ an endomorphism such that $\tilde{g}(v) = 0$ for some $\tilde{g}(x) \in \tilde{T}$. For $g(t) = t^n + a_{n-1}t^{n-1}+\dots+a_o \in T$ define $\tilde{g}(x) = 1 + a_{n-1}x+\dots+a_o x^n$ and let $\tilde{T} = \{\tilde{g}(x) \in A[x]/g(u) \in T\}$. Then Grayson ([Gr]) showed that

$$Ker[(K_j(End(A;T)) \to K_j(A)] \cong coker[K_{j+1}(A) \to K_{j+1}(\tilde{T}^{-1}A[x])].$$

Stienstra ([St]) describes operations on these K-theories of endomorphisms and relations between these operations. Examples are the operations F_m and V_m (see section 3). In fact one can show that the operation F_m on $NK_n(A)$ comes from the operation on Nil(A) which sends $(P,v) \to (P,v^m)$. This and the nilpotency of v^m are the reason for the fact that for every $\alpha \in NK_n(A)$ there is an M such that $m \geq M$ implies $F_m(\alpha) = 0$.

The K-theory of polynomial extensions is also related to the Karoubi-Villamayer KV-theory [KV] as follows:

6.2. Theorem. (Anderson and Gersten [We3]). If A is a ring with unit, there is a first quadrant spectral sequence (defined for $p \geq 0$, $q \geq 1$)

$$E^1_{pq} = N^pK_q(A) \to KV_{p+q}(A).$$

§7. The unstable problems.

Let A be a commutative ring. We say that a finitely generated projective $A[T_1,\dots,T_n]$ is extended if there is a finitely generated projective A-module P_o such that

$$P_o \otimes_A A[T_1,\dots,T_n] \cong P.$$

Since $K_o(A)$ is the Grothendieck group of finitely generated projective A-modules it is clear that K_o-regularity is related to the following problem.

7.1. <u>Problem</u>. Which rings A have the property that for every n, every finitely generated projective $A[T_1,\dots,T_n]$-module is extended?

Only rings which are K_o-regular can have this property and if a ring is K_o-regular it is possible for every finitely generated projective $A[T_1,\dots,T_n]$-module P to find a free $A[T_1,\dots,T_n]$-module Q such that $P \oplus Q$ is extended.Since regular rings are K_o-regular we can ask, whether they have the property. This is the so-called Bass-Quillen problem ([Ba2], problem IX or [Qu2]). If A is a field we get the Serre-problem, which was solved independently by Quillen ([Qu2]) and Suslin ([Su1]). In fact they showed that regular rings A with $\dim A \leq 2$ have the property described in 7.1. An excellent survey on the Serre problem has been given by Lam ([La]). The state of the art of solving 7.1 for regular rings up to now is:

7.2. <u>Theorem</u>. (i) (Lindel [Li2]). Let k be a field and A a regular k-algebra of essentially finite type over k. Then every finitely generated projective $A[T_1,\dots,T_n]$-module is extended.
(ii) (Lindel [Li1]). Let A be a power series ring over a complete regular local ring of dimension ≤ 2. Then every finitely generated projective $A[T_1,\dots,T_n]$-module is extended.

The second result was independently proved for a power series ring over a field by Mohan Kumar ([Ku]) and Lindel and Lütkebohmert ([LL]).
In the non-noetherian case we have the following result which was independently proved by Lesquain and Simis, Brewer and Costa.
7.3. <u>Theorem</u>. ([LS] and [BC1]). If A is a one-dimensional Bezout domain, then finitely generated projective $A[T_1,\dots,T_n]$-modules are free. If A is a one-dimensional Prüfer domain, then finitely generated projective $A[T_1,\dots,T_n]$-modules are extended.

Now we return to the noetherian case again, and consider non-regular rings. Since by 4.1 we know that a one-dimensional ring is K_o-regular if and only if A is seminormal we may ask whether for seminormal rings A of dimension one every finitely generated projective $A[T_1,\ldots,T_n]$-module is extended. One has the following theorem of Brewer and Costa.

7.4. Theorem. ([BC2]). Let A have dimension one. If A is noetherian, then every finitely generated projective $A[T_1,\ldots,T_n]$-module is extended if and only if A is seminormal. The condition that A is noetherian may be replaced by the condition that $\bar{A}$ is a Prüfer domain.

The following result was inspired by 5.4.

7.5. Theorem. ([Vo4]). Let k be such that for every n, every finitely generated projective $k[T_1,\ldots,T_n]$-module is extended then the same holds for k replaced by $A = k[x_o,x_1,\ldots,x_p]/I$, where I can be generated by square free monomials. Furthermore one can even show that every finitely generated projective A-module P is extended from k (i.e. $P \cong A \otimes_k P'$ for some finitely generated projective k-module P').

Also for K_1-regularity one may state an unstable problem as follows:

7.6. Problem. Which rings A have the property that every $\alpha \in GL_r(A[T_1,\ldots,T_n])$ is a product $\alpha = \beta\gamma$ with $\beta \in GL_r(A)$ and $\gamma \in E_r(A[T_1,\ldots,T_n])$. (Such an α will be called extended).

Again one only has to study this problem for K_1-regular rings, but Cohn ([Co]) has shown that for r = n = 2 and A a field, the matrix

$$\begin{pmatrix} 1 + T_1T_2 & T_1^2 \\ - T_2^2 & 1 - T_1T_2 \end{pmatrix} \in GL_2(A[T_1,T_2])$$

is not extended. Hence it only makes sense to consider the problem for $r \geq 3$. We have the following theorem of Suslin.

7.7. Theorem. ([Su2]). If k is a field and $r \geq 3$ then every

$\alpha \in GL_r(k[T_1,\ldots,T_n])$ is extended.

Again this result can be extended as follows.

7.8. Theorem. ([Vo3]). Let k be a field and A a regular k-algebra of essentially finite type over k, or let A be a power series ring over k. Then every $\alpha \in GL_r(k[T_1,\ldots,T_n])$ is extended.

And in the non-regular case we have the following:

7.9. Theorem. ([Vo4]). Let k be such that for all n, every $\alpha \in GL_r(k[T_1,\ldots,T_n])$ is extended, then the same holds for k replaced by A as in 7.5. Furthermore every $\alpha \in GL_r(A)$ is extended from $GL_r(k)$ (i.e. $\alpha = \beta\gamma$ with $\beta \in GL_r(k)$ and $\alpha \in E_r(A)$).

§ 8. Open problems.

In this section we list for the reader's convenience the open problems we gave in the preceeding sections and some other problems.

The open problems we already stated are 2.4; 4.5 and 4.6. Further we had problems 7.1 and 7.6, with the last one only for $r \geq 3$. Since it only makes sense to state these problems for K_0-resp. K_1-regular rings we ask more specifically.

8.1. Problem. (Bass-Quillen). Do all regular rings have the property of 7.1.

8.2. Problem. Do the following rings have the property of 7.6 for $r \geq 3$: all regular rings or one-dimensional seminormal rings with separable residue class field extensions to $\bar{A}$ (see 4.1(ii))?

Related to 2.3 we ask:

8.3. Problem. If A has the property of 7.6 does this imply that A has the property of 7.1. (See also 2.3)

Another class of problems is the following: Let f: A → B be a map and assume that A is K_n-regular. Which properties of f imply that B is K_n-regular. We know that if B is a localisation of A it is K_n-regular. More specifically we like to ask:

8.4. <u>Problem</u>. If f: A → B is a smooth extension and A is K_n-regular, does this imply that B is K_n-regular.

One can easily show that it is enough to solve the problem for étale extensions. Of course one can also state the following problem:

8.5. <u>Problem</u>. Let f: A → B and B be K_n-regular. Under which conditions on f does one know that A is K_n-regular.

In this direction we may remark that, if B is a rank n free A-module then $N_pK_n(A)$ is n torsion by the usual transfer argument and so A is K_n-regular if for example A is a Q-algebra (see § 3).

References.

[An] Anderson, D.F., Projective modules over subrings of k[X,Y] generated by monomials, Pacific J. Math. 79, 5-17, 1978.

[Ba1] Bass, H., Algebraic K-theory, New York, W.A. Benjamin, 1968.

[Ba2] Bass, H., Some problems in "classical" algebraic K-theory Lecture Notes in Math. 342, New York, Springer Verlag, 3-73, 1973.

[BHS] Bass, H., Heller, A. and Swan, R.G., The Whitehead group of a polynomial extension. Publ. Math. I.H.E.S. 22, 61-80, 1964.

[BC1] Brewer, J. and Costa, D.L., Projective modules over some nonnoetherian polynomial rings, J. Pure Appl. Algebra 13, 157-163, 1978.

[BC2] Brewer, J. and Costa, D.L., Seminormality and projective modules over polynomial rings, J. of Algebra, 58, 208-216, 1979.

[Bl1] Bloch, S. Algebraic K-theory and crystalline cohomology, Publ. Math. I.H.E.S. 47, 188-268, 1978.

[Bl2] Bloch, S. Some formulas pertaining to the K-theory of commutative group schemes, J. of Algebra, 53, 304-326, 1978.

[BM] Bass, H. and Murthy, M.P., Grothendieck groups and Picard groups of abelian group rings, Ann. of Math. 86, 16-73, 1967.

[Bo] Bombieri, E., Seminormalita e singolarita ordinaire, Symposia

Mathematica XI, New York, Academic Press, 205-210, 1972.

[Ca] Carter, D., Localization in lower K-theory, Comm. Algebra 8, 603-622, 1980.

[CEP] De Concini, C., Eisenbud, D and Procesi, C., Hodge algebras, Astérisque 91, Soc. Math. de France, 1982.

[Co] Cohn, P.M., On the structure of GL_2 of a ring. Publ. Math. I.H.E.S., 30, 365-413, 1966.

[CS] Coomber, K.R. and Srinivas, V., Relative K-theory and vector bundles on certain singular varieties, Inv. Math. 70, 1-12, 1982.

[Da] Davis, E., On the geometric interpretation of seminormality, Proc. Amer. Math. Soc. 68, 1-5, 1978.

[DK] Dennis, R.K. and Krusemeyer, M., $K_2(A[X,Y]/XY)$, a problem of Swan, and related computations, J. Pure and Appl. Algebra, 15, 125-148, 1979.

[DW1] Dayton, B. and Weibel, C., K-theory of hyperplanes, Trans Amer. Math. Soc. 257, 119-141, 1980.

[DW2] Dayton, B. and Weibel, C., A spectral sequence for the K-theory of affine glued schemes, Lecture Notes in Math. 854, 24-92, New York, Springer Verlag, 24-92, 1981.

[Ge] Gersten, S., Problems about higher K-functors, Lecture Notes in Math. 341, New York, Springer Verlag, 43-56, 1973.

[GQ] Grayson, D., Higher algebraic K-theory II (after Quillen), Lecture Notes in Math. 551, New York, Springer Verlag, 217-240, 1976.

[Gr] Grayson, D., The K-theory of endomorphisms, J. of Algebra, 48, 439-446, 1977.

[Ha] Hamann, E., The R-invariance of R[X], J. of Algebra, 35, 1-16, 1975.Math. Studies no. 72, Princeton U.P., 1971.

[KU] Mohan Kumar, N., On a question of Bass and Quillen, preprint.

[KV] Karoubi, M. and Villamayor, O., K-theorie algêbrique et K-theorie topologique, Math. Scand. 28, 265-307, 1971.

[La] Lam, T.Y., Serre's conjecture, Lecture Notes in Math. 635, New York, Springer Verlag, 1978.

[Li1] Lindel, H., Projektive Moduln über Polynomring $A[T_1,...,T_m]$ mit einem regulären Grundring A, Manuscripta Math. 23, 143-154, 1978.

[Li2] Lindel, H., On the Bass-Quillen conjecture concerning projective modules over polynomial rings. Inv. Math. 65, 319-323, 1981.

[LL] Lindel, H. and Lütkebohmert, W., Projektive Moduln über polynomialen Erweiterungen von Potentreihenalgebren, Archiv der Math. 28, 51-54, 1977.

[LS] Lequain, Y. and Simis, A., Projective modules over $R[X_1,...,X_n]$, R a Prüfer domain, J. Pure Appl. Algebra, 18, 165-171, 1980.

[Mi] Milnor, J., Introduction to algebraic K-theory, Annals of Math. Studies no. 72, Princeton U.P., 1971.

[MP] Murthy, P. and Pedrini, C., K_o and K_1 of polynomial rings, Lecture Notes in Math. 342, New York, Springer Verlag, 109-121, 1973.

[Na] Nagata, M., Local rings, New York, Interscience, 1962.

[Pe] Pedrini, C., On the K_o of certain polynomial extensions, Lecture Notes in Math. 342, New York, Springer Verlag, 91-108,1973.

[Qu1] Quillen, D., Higher algebraic K-theory I, Lecture Notes in Math. 341, New York, Springer Verlag, 85-147, 1973.

[Qu2] Quillen, D., Projective modules over polynomial rings. Invent. Math. 36, 167-171, 1976.

[Re] Reisner, G., Cohen-Macauly quotients of polynomial rings, Adv. Math. 21, 30-49, 1976.

[St] Stienstra, J., Operations in the higher K-theory of endomorphisms, In Current Trends in Algebraic Topology, Conf. Proc. Can. Math. Soc. Vol.2, 1982.

[Su1] Suslin, A.A., Projective modules over polynomial rings are free, Dokl. Akad. Nauk. 229, 221-238, 1976.

[Su2] Suslin, A.A., On the structure of the special linear group over polynomial rings, Math. USSR Izvestija, 11, 221-238, 1977.

[Sr] Srinivas, V., Vector bundles on the cone over a curve, Comp. Math. 47, 249-269, 1982.

[Sw] Swan, R.G., On seminormality, J. Algebra 67, 210-229, 1980.

[Tr] Traverso, C., Seminormality and Picard group, Ann. Scuola Norm. Sup. Pisa, XXIV, 585-595, 1970.

[vdK] Van der Kallen, W., Generators and relations in algebraic K-theory, Proc. I.C.M. Helsinki, 305-310, 1978.

[Vo1] Vorst, T., Polynomial extensions and excision for K_1, Math. Ann. 244, 193-204, 1979.

[Vo2] Vorst, T., Localization of the K-theory of polynomial extensions, Math. Ann., 244, 33-53, 1979.

[Vo3] Vorst, T., The general linear group of polynomial rings over regular rings, Comm. Alg. 9, 499-509, 1981.

[Vo4] Vorst, T. The Serre problem for discrete Hodge algebras, preprint, 1983.

[We1] Weibel, C., Mayer-Vietoris sequences and module structures on NK_*, Lecture Notes in Math. 854, New York, Springer Verlag, 466-493, 1981.

[We2] Weibel, C., K-theory and analytic isomorphisms, Inv. Math. 61, 177-197, 1980.

[We3] Weibel, C., KV-theory of categories, Trans. Am. Math. Soc., 267, 621-635, 1981.

[We4] Weibel, C., Complete intersection points on affine surfaces, preprint, 1981.

Appendix.

In this appendix we want to prove proposition 5.4. Let Σ be a simplicial subcomplex of Δ_n, the standard simplex on vertices $\{0,1,...,n\}$. Let $I(\Sigma)$ be the ideal of $k[x_o,x_1,...,x_n]$ (k a field) generated by the monomials $x_{i_o} x_{i_1} \ldots x_{i_r}$ $(i_o<i_1<\ldots<i_r)$

where $\{i_o,i_r,...,i_r\}$ is not a simplex of Σ. We denote $k[x_1,...,x_n]/I(\Sigma)$ by $k(\Sigma)$. Let $C(\Sigma)$ be the cone on Σ then $k(C(\Sigma)) = k(\Sigma)[x_{n+1}]$ if $n+1$ is the top vertex of the cone. Hence K_n-regularity for $k(\Sigma)$ means that K_n doesn't change under taking the cone on Σ several times. We have to prove that $k(\Sigma)$ is K_1-regular for every Σ and $k(\Sigma)$ is not K_2-regular if Σ is not just a single simplex.

To prove the K_1-regularity we proceed by induction on n. If $n = 0$ we have $K(\Sigma) = k$ or $k(\Sigma) = k[x_o]$, in both cases $k(\Sigma)$ is K_1-regular. Let $\dim\Sigma = m$. If $m = n$ then $\Sigma = \Delta_n$ and $k(\Delta)$ is regular. Hence suppose $m < n$. Let $\Delta \subset \Sigma$ be an m-dimensional simplex. If $\Delta = \Sigma$ then $k(\Sigma)$ is regular again. If $\Delta \neq \Sigma$ we can find an $i \in \{0,1,...,n\}$ such that i is a vertex of $\Sigma\setminus\Delta$. Suppose $i = n$. Now take $\Sigma_1 = \Delta_{n-1} \cap \Sigma$ where $\Delta_{n-1} = \{0,1,...,n-1\}$. Let $S = \{\sigma \in \Sigma | n \text{ is a vertex of } \sigma\}$ and Σ_3 the simplicial subcomplex of Δ_n which one gets by taking S and all boundaries of elements of S. Take $\Sigma_2 = \Sigma_3 \cap \Delta_{n-1}$, hence $\Sigma_3 = C(\Sigma_2)$ if we take n as the top vertex of $C(\Sigma_2)$. Then $\Sigma = \Sigma_3 \cup \Sigma_1$ and $\Sigma_2 = \Sigma_3 \cap \Sigma_1$ which implies $I(\Sigma) = I(\Sigma_3) \cap I(\Sigma_1)$ and $I(\Sigma_2) = I(\Sigma_3) + I(\Sigma_1)$. This means that we have a cartesian square.

$$\begin{array}{ccc} k(\Sigma) & \longrightarrow & k(\Sigma_1) \\ \downarrow & & \downarrow \\ k(C(\Sigma_2)) & \xrightarrow[i]{} & k(\Sigma_2) \end{array}$$

where the map i splits.

Now we have a long exact Mayer-Vietoris sequence for K-theory (see [Mi] 6.4) and by taking kernels we have such a sequence for NK-theory

$$N_pK_2(k(\Sigma)) \to N_pK_2(k(\Sigma_1)) \oplus N_pK_2(k(\Sigma_2)[x_n]) \xrightarrow{\mu} N_pK_2(k(\Sigma_2)) \to$$

$$N_pK_1(k(\Sigma)) \to N_p(K_1(k(\Sigma_1)) \oplus N_pK_1(k(\Sigma_2)[x_1]) \to N_pK_1(k(\Sigma_2)).$$

Now we know that Σ_1 and Σ_2 are simplicial subcomplexes of Δ_{n-1} and hence by induction we now that $k(\Sigma_1)$ and $k(\Sigma_2)[x_1]$ are K_1-regular. Furthermore we know that μ is surjective since i splits. Hence $N_pK_1(k(\Sigma)) = 0$ for every $p \geq 1$ i.e. $k(\Sigma)$ is K_1-regular. Finally we have to prove that $k(\Sigma)$ is not K_2-regular if Σ is not just a simplex. Suppose Σ has vertices $\{0,1,\ldots,r\}$. Let Δ be a lowest dimensional subsimplex of Δ_r which is not in Σ. We may suppose that $\Delta = \{0,1,\ldots,s\}$. Now we consider the Dennis-Stein symbol

$$\langle x_o, x_1x_2\ldots x_sT\rangle \in NK_2(k(\Sigma))$$

Let $\phi\colon k(\Sigma) \to k[x_o,x_1]/x_ox_1$ be the k-algebra map such that $\phi(x_o) = x_o$, $\phi(x_1) = x_1$, $\phi(x_j) = 1(2\leq j\leq s)$ and $\phi(x_j) = 0(s+1\leq j<r)$. Then

$$NK_2(\phi)(\langle x_o,x_1,\ldots,x_sT\rangle) = \langle x_o,x_1T\rangle \in NK_2(k[x_o,x_1]/x_1x_2)$$

but the right hand side is unequal to zero by the work of Dennis and Krusemeyer ([DK]). Hence $k(\Sigma)$ is not K_2-regular. This finishes the proof of proposition 5.4.

Although the proof of the K_1-regularity can be found for a more general case in [DW2] we think that our proof which uses essentially the same idea is easier to understand and the more general case can be reduced to this case by theorem 2.1. Finally we remark that the result remains the same if we only know that k is a K_1-regular ring.

Ton Vorst
Econometric Institute
P.O. Box 1738,
3000 DR Rotterdam
The Netherlands

Algebraic K-theory and the Adams e-Invariant

Charles A. Weibel*
Mathematics Department
Rutgers University
New Brunswick, NJ
08903 USA

Abstract: The Adams e-invariant used in algebraic topology has an analogue in algebraic K-theory. The target spaces of the e-invariant are infinite loop spaces $F\underline{R}$ and $F\underline{C}$ in the real and complex cases, respectively. We show that these spaces are actually retracts of $BG\ell^{+}(\underline{R})$ and $BG\ell^{+}(\underline{C})$. This accounts for all known torsion in $K_*(\underline{R})$ and $K_*(\underline{C})$.

Define the topological spaces $F\underline{R}$ and $F\underline{C}$ to be the homotopy fibers of the rational localization maps of BO and BU:

$$F\underline{R} \to BO \to BO\otimes\underline{Q} = \Pi K(\underline{Q},4n)$$

$$F\underline{C} \to BU \to BU\otimes\underline{Q} = \Pi K(\underline{Q},2n)$$

We can construct the following table of homotopy groups, which is periodic of order 8:

i(mod 8)	0	1	2	3	4	5	6	7
$\pi_i(F\underline{R})$	0	$\underline{Z}/2$	$\underline{Z}/2$	$\underline{Q}/\underline{Z}$	0	0	0	$\underline{Q}/\underline{Z}$
$\downarrow$	0	inclusion	0	mult. by 2	0	0	0	iso.
$\pi_i(F\underline{C})$	0	$\underline{Q}/\underline{Z}$	0	$\underline{Q}/\underline{Z}$	0	$\underline{Q}/\underline{Z}$	0	$\underline{Q}/\underline{Z}$

The relationship to algebraic K-theory is given by the following result:

*Supported by NSF grant MCS81-02753.

Theorem 1: The space $F\underline{R}$ is a retract of $BG\ell^+(\underline{R})$, and the space $F\underline{C}$ is a retract of $BG\ell^+(\underline{C})$. Consequently, the above table gives direct summands of $K_i(\underline{R})$ and $K_i(\underline{C})$.

This theorem accounts for all known torsion in $K_*(\underline{R})$ and $K_*(\underline{C})$. For example, the $\underline{Q}/\underline{Z}$ in $K_1(\underline{C}) = \underline{C}^*$ is the subgroup of roots of unity, and $K_3(\underline{Z}) \cong \underline{Z}/48$ is a subgroup of the $\underline{Q}/\underline{Z}$ in $K_3(\underline{R})$. The Lichtenbaum Conjecture implies that $\pi_*(F\underline{C})$ is precisely the torsion subgroup of $K_*(\underline{C})$. Similarly, one may conjecture that $\pi_*(F\underline{R})$ is precisely the torsion subgroup of $K_*(\underline{R})$; this is a consequence of a conjecture of Milnor on $H_*(BG\ell_n(\underline{R})_{discrete};\underline{Q}/\underline{Z})$.

The construction of the maps $BG\ell^+(k) \to Fk$ $(k=\underline{R},\underline{C})$ and the relationship to the Adams e-invariant $e:\Omega^\infty\Sigma^\infty \to F\underline{R}$ are given by the following result. (We give $\underline{R},\underline{C}$ the discrete topology in order to construct the spaces $BG\ell^+(\underline{R})$ and $BG\ell^+(\underline{C})$; we use the metric topology to construct the spaces $BO \simeq BG\ell^+(\underline{R}_{top})$ and $BU \simeq BG\ell^+(\underline{C}_{top})$.)

Proposition 2 (Quillen [Q 72]). The compositions

$$BG\ell^+(\underline{R}) \to BO \to BO\otimes\underline{Q}$$

$$BG\ell^+(\underline{C}) \to BU \to BU\otimes\underline{Q}$$

are null-homotopic, and the lifts to the homotopy fibers may be chosen to make the following diagram commute:

$$\begin{array}{ccc} BG\ell^+(\underline{R}) & \longrightarrow & F\underline{R} \\ \downarrow & & \downarrow \\ BG\ell^+(\underline{C}) & \longrightarrow & F\underline{C} \end{array}$$

The composition $\Omega^\infty\Sigma^\infty \to BG\ell^+(\underline{Z}) \to BG\ell^+(\underline{R}) \to F\underline{R}$ is $-e$.

Remark. The maps in Proposition 2 are infinite loop space maps.

Proof. Let k denote either $\underline{\underline{R}}$ or $\underline{\underline{C}}$, with the discrete topology. The universal bundles on $BG\ell^+(k)$ have flat connections, since the structure groups $G\ell_n(k)$ are discrete. Hence the map

$$\begin{array}{ccc} H^*(\Pi K(\underline{\underline{Q}},n);\underline{\underline{Q}}) & \xrightarrow{\cong} & H^*(BG\ell(k_{top});\underline{\underline{Q}}) \\ & & \downarrow \\ H^*(BG\ell^+(k);\underline{\underline{Q}}) & \xrightarrow{\cong} & H^*(BG\ell(k);\underline{\underline{Q}}) \end{array}$$

is zero (a proof of this fact may be found on p. 308 of [Milnor 74]). By obstruction theory, $BG\ell^+(k) \to \Pi K(\underline{\underline{Q}},n)$ is null-homotopic. The choice of a null-homotopy gives a lift $BG\ell^+(k) \to Fk$. By naturality (or by obstruction theory), we can choose compatible nullhomotopies for $k=\underline{\underline{R}}$ and $\underline{\underline{C}}$; this gives compatible lifts, i.e., makes the diagram commute. The fact that the composition is $-e$ was proven explicitly in [Q 72].

To construct the maps $Fk \to BG\ell^+(k)$, we use the infinite cyclic group $C_\infty \cong \underline{\underline{Q}}/\underline{\underline{Z}}$ and its twofold cover, the infinite dihedral group $D_\infty = C_\infty \rtimes \Sigma_2$. It is a theorem of Leonardo da Vinci (see [Weyl, p.66, 99]) that D_∞ is the torsion subgroup of O_2, while C_∞ is the torsion subgroup of $U_1 \cong SO_2$.

Proposition 3. The following sequences are fibrations:

$$\Omega^\infty\Sigma^\infty BC_\infty \to \Omega^\infty\Sigma^\infty BU_1 \to (\Omega^\infty\Sigma^\infty BU_1)\otimes\underline{\underline{Q}}$$

$$\Omega^\infty\Sigma^\infty BD_\infty \to \Omega^\infty\Sigma^\infty BO_2 \to (\Omega^\infty\Sigma^\infty BO_2)\otimes\underline{\underline{Q}}$$

Proof. First, note that $BC_\infty \to BU_1$ is an isomorphism on homology with finite coefficients. This is an easy calculation using [Cartan]. Alternatively, let U_1^δ denote U_1 with the discrete topology. Since $0 \to C_\infty \to U_1^\delta \to U_1^\delta\otimes\underline{\underline{Q}} \to 0$ is an exact sequence of discrete groups, $BC_\infty \to BU_1^\delta \to B(U_1^\delta\otimes\underline{\underline{Q}})$ is a fibration. Hence $BC_\infty \to BU_1^\delta$ is an isomorphism on homology with finite coefficients. But $BU_1^\delta \to BU_1$ is an isomorphism on homology with finite coefficients by [Milnor 81], whence the claim.

To see that $BD_\infty \to BO_2$ is an isomorphism on homology with finite coefficients, apply the comparison theorem to the map of fibrations

$$\begin{array}{ccccc} BC_\infty & \to & BD_\infty & \to & B\Sigma_2 \\ \downarrow & & \downarrow & & \| \\ BU_1 & \to & BO_2 & \to & B\Sigma_2. \end{array}$$

As $H_*(\Omega^\infty\Sigma^\infty(-);\underline{Z}/\ell)$ is a functor of $H_*(-;\underline{Z}/\ell)$ on connected spaces, it follows that $\Omega^\infty\Sigma^\infty BC_\infty \to \Omega^\infty\Sigma^\infty BU_1$ and $\Omega^\infty\Sigma^\infty BD_\infty \to \Omega^\infty\Sigma^\infty BO_2$ are isomorphisms on homology with finite coefficients. By [BK, p.186] these maps are p-adic equivalences for every prime p. As the homotopy groups of $\Omega^\infty\Sigma^\infty BC_\infty$ and $\Omega^\infty\Sigma^\infty D_\infty$ are torsion, the proposition follows.

We now invoke a result of Snaith [Snaith,(2.1) and (3.2)]. There are transfer maps

$$\tau_O : BO \to \Omega^\infty\Sigma^\infty B(\Sigma_\infty \int O_2) \to \Omega^\infty\Sigma^\infty BO_2$$

$$\tau_U : BU \to \Omega^\infty\Sigma^\infty B(\Sigma_\infty \int U_1) \to \Omega^\infty\Sigma^\infty BU_1$$

which are split by the maps $\Omega^\infty\Sigma^\infty BO_2 \to BO$ and $\Omega^\infty\Sigma^\infty BU_1 \to BU$. (The composites are the identity on BO and BU.) We can now fill in the leftmost arrows in following diagrams (whose rows are fibrations):

$$\begin{array}{ccccc} F\underline{R} & \longrightarrow & BO & \longrightarrow & BO\otimes\underline{Q} \\ \downarrow & & \downarrow{\scriptstyle \tau_O} & & \downarrow{\scriptstyle \tau\otimes Q} \\ \Omega^\infty\Sigma^\infty BD_\infty & \to & \Omega^\infty\Sigma^\infty BO_2 & \to & (\Omega^\infty\Sigma^\infty BO_2)\otimes\underline{Q} \\ \downarrow & & \downarrow & & \downarrow \\ F\underline{R} & \longrightarrow & BO & \longrightarrow & BO\otimes\underline{Q} \end{array}$$

$$\begin{array}{ccccc} F\underline{C} & \longrightarrow & BU & \longrightarrow & BU\otimes\underline{Q} \\ \downarrow & & \downarrow{\scriptstyle \tau_U} & & \downarrow{\scriptstyle \tau\otimes Q} \\ \Omega^\infty\Sigma^\infty BC_\infty & \to & \Omega^\infty\Sigma^\infty BU_1 & \to & (\Omega^\infty\Sigma^\infty BU_1)\otimes\underline{Q} \\ \downarrow & & \downarrow & & \downarrow \\ F\underline{C} & \longrightarrow & BU & \longrightarrow & BU\otimes\underline{Q} \end{array}$$

Proof of Theorem 1. The group maps $D_\infty \to O_2$, $C_\infty \to U_1$ factor through the discrete groups $G\ell_2(\underline{R})$, $G\ell_1(\underline{C})$. Hence we have a commutative diagram

$$\begin{array}{ccc} F\underline{R} \to \Omega^\infty\Sigma^\infty BD_\infty \to \Omega^\infty\Sigma^\infty BG\ell_2(\underline{R}) & \longrightarrow & \Omega^\infty\Sigma^\infty BO_2 \\ \downarrow & & \downarrow \\ \Omega^\infty\Sigma^\infty BG\ell(\underline{R}) & \longrightarrow & \Omega^\infty\Sigma^\infty BO \\ \downarrow & & \downarrow \\ BG\ell^+(\underline{R}) & \longrightarrow & F\underline{R} \to BO \end{array}$$

and a similar diagram for C_∞, $BG\ell^+(\underline{C})$, and BU. The upshot is that the composites

$$F\underline{R} \to \Omega^\infty\Sigma^\infty BD_\infty \to BG\ell^+(\underline{R}) \to F\underline{R}$$

$$F\underline{C} \to \Omega^\infty\Sigma^\infty BC_\infty \to BG\ell^+(\underline{C}) \to F\underline{C}$$

are homotopy equivalent to the identity maps.

Naturality. From proposition 2, it is clear that the map $K_*(\underline{R}) \to K_*(\underline{C})$ is compatible with $\pi_*(F\underline{R}) \to \pi_*(F\underline{C})$. It is also the case that the transfer map $K_*(\underline{C}) \to K_*(\underline{R})$ is compatible with the transfer $\pi_*(F\underline{C}) \to \pi_*(F\underline{R})$, since $U \to O$ induces $C_\infty \to D_\infty$. The transfer map $\pi_*(F\underline{C}) \to \pi_r(F\underline{R})$ is an isomorphism if $r \equiv 3 \pmod 8$, multiplication by 2 if $r \equiv 7 \pmod 8$, and zero otherwise.

Product Structure. The spaces Fk ($k=\underline{R},\underline{C}$) are more than just retracts of $BG\ell(k)^+$; they are direct factors, i.e., there is a homotopy equivalence $BG\ell(k)^+ \simeq (Fk)\times(Gk)$. This is a general fact true for all retracts of H-spaces. Of course, $F\underline{R} \to BG\ell^+(\underline{R})$ is not an H-space map (see [Snaith]), so this product structure is relatively uninteresting.

Connection with mod ℓ homotopy. Let ℓ be a positive integer, not congruent to 2 mod 4. Write β_ℓ for that element of $\pi_2(BC_\infty;\underline{Z}/\ell) \cong \underline{Z}/\ell$ corresponding to $1/\ell \in \underline{Q}/\underline{Z}$ under a fixed isomorphism $\underline{Q}/\underline{Z} \cong C_\infty$. This yields a map of the polynomial ring $\underline{Z}/\ell[\beta_\ell]$ into the graded rings $\pi_*(\Omega^\infty\Sigma^\infty BC_\infty;\underline{Z}/\ell)$ and $K_*(\underline{C};\underline{Z}/\ell)$. The image of β_ℓ in $\pi_*(F\underline{C};\underline{Z}/\ell) \cong \pi_*(BU;\underline{Z}/\ell)$ is the reduction of a Bott element $\beta \in \pi_2(BU)$, and $\pi_*(F\underline{C};\underline{Z}/\ell) \cong \underline{Z}/\ell[\beta_\ell]$. This was proven in [Browder 77]. In summary:

Corollary 3. The retraction $BG\ell^+(\underline{C}) \to F\underline{C}$ induces a ring retraction of $K_*(\underline{C};\underline{Z}/\ell)$ onto the polynomial ring $\underline{Z}/\ell[\beta_\ell]$. (The Lichtenbaum conjecture implies that this map is a ring isomorphism.)

When we vary ℓ, the multiplication changes. For example, the map $\pi_{2n}(F\underline{C};\underline{Z}/\ell) \to \pi_{2n}(F\underline{C};\underline{Z}/\ell m)$ induced by $\underline{Z}/\ell \to \underline{Z}/\ell m$ sends $(\beta_\ell)^n$ to m times $(\beta_{\ell m})^n$.

Similarly, let γ_ℓ denote the image $\iota_*(\beta_\ell{}^2)$ of $(\beta_\ell)^2$ in $\pi_4(F\underline{R};\underline{Z}/\ell) \cong \underline{Z}/\ell$ under the transfer map ι_*. The element γ_ℓ generates $\pi_4(F\underline{R};\underline{Z}/\ell)$, but $(\gamma_\ell)^n$ does not generate $\pi_{4n}(F\underline{R};\underline{Z}/\ell)$ since $(\gamma_\ell)^n = 2^{n-1}\cdot\iota_*(\beta_\ell{}^{2n})$. However, it is still the case that the images of the $(\gamma_\ell)^n$ generate the group $\pi_{4n-1}(F\underline{R}) \cong \underline{Q}/\underline{Z}$ as ℓ varies. Thus this $(\underline{Q}/\underline{Z})$-summand of $K_3(\underline{R})$ reproduces itself periodically in $K_{4n+3}(\underline{R})$ (cf. [Browder 77]).

Connections with stable homotopy of spheres. There is a sequence of spaces

$$0 \xrightarrow{J} \Omega^\infty\Sigma^\infty \to BG\ell^+(\underline{Z}) \to BG\ell^+(\underline{R}) \to F\underline{R}$$

which detect the "image of J" by [Adams 65]. The groups $J(\pi_r O)$ are zero unless $r\equiv 0,1,3,7 \pmod 8$. When $r\equiv 0,1 \pmod 8$ and $r\neq 0$, $J(\pi_r O)$ is cyclic of order 2. Waldhausen has shown in [Wald, Cor. 3.6] that every element of $J(\pi_r O)$, $r\equiv 0,1 \pmod 8$ vanishes when mapped to $K_r(\underline{Z})$, save the nonzero element η of $J(\pi_1 O) \cong \pi_1^s$. The image of η in $K_1(\underline{Z})$ is the unit '-1' in $\underline{Z}^*=K_1(\underline{Z})$, and maps nontrivially into $\pi_1(F\underline{R})$.

We are left with $J(\pi_r 0)$ for $r \equiv 3,7 \pmod 8$. Write $r=4n-1$. In this case, the group $J(\pi_{4n-1}0)$ injects into $\pi_{4n-1}(F\underline{R}) \cong \underline{Q}/\underline{Z}$, and hence into $K_{4n-1}(\underline{Z})$ and $K_{4n-1}(\underline{R})$ (see [Adams 65],[Q 72]). The smallest values of 4n-1 deserve special mention: $J(\pi_3 0) \cong \underline{Z}/24$ injects into $K_3(Z) \cong \underline{Z}/48$, and η^3 maps to $\{-1,-1,-1\}$. The group $J(\pi_7 0)$ is $\underline{Z}/240$.

In general, $J(\pi_{4n-1}0)$ is a cyclic group whose order equals the denominator of $B_n/4n$, where B_n is the nth Bernoulli number. The 2-primary part of the order of $J(\pi_{4n-1}0)$ is 2^{i+3}, where $n \equiv 2^i \pmod{2^{i+1}}$; in particular, the order is always divisible by 8. These facts may be found in [Adams 65], [Mahowald 70], and in Appendix B of [Milnor 74].

In order to describe the image of J in $\pi_{8n+3}(\underline{C})$, we need a related family of elements of π_*^s. These are the elements μ_{8n+1} of π_{8n+1}^s which are constructed in [Adams 65]. They have order 2, and μ_1 is η. Gersten showed that the μ and $\eta\mu$ map to generators of $\pi_r(F\underline{R}) \cong \pi_r(BO)$ for $r \equiv 1,2 \pmod 8$. This fact was pointed out by Quillen in [Q 72], and is well-known for r=1,2. The element $\eta^2\mu$ of π_{8n+3}^s is mapped by the e-invariant to $1/2 \in \underline{Q}/\underline{Z} = \pi_{8n+3}(F\underline{R})$. In fact, $\eta^2\mu$ is the unique element of order 2 in the in the subgroup $J(\pi_{8n+3}0)$ of π_{8n+3}^s. We have now recovered the following result of Browder:

<u>Corollary 4</u> (cf. [Browder 77, (4.7)]).

- If $r \equiv 1 \pmod 8$, the element μ of π_r^s generates a $(\underline{Z}/2)$-summand of $K_r(\underline{Z})$ and $K_r(\underline{R})$. This $(\underline{Z}/2)$-summand injects into a $(\underline{Q}/\underline{Z})$-summand of $K_r(\underline{C})$.
- If $r \equiv 2 \pmod 8$, the element $\eta\mu$ generates a $(\underline{Z}/2)$-summand of $K_{r+1}(\underline{Z})$ and $K_{r+1}(\underline{R})$, but is zero in $K_{r+1}(\underline{C})$.
- If $r \equiv 3 \pmod 8$, the element $\eta^2\mu$ is the element of order 2 in the $(\underline{Q}/\underline{Z})$-summand $\pi_{r+2}(F\underline{R})$ of $K_{r+2}(\underline{R})$, and $\eta^2\mu$ is the only element in this $(\underline{Q}/\underline{Z})$-summand mapping to zero in $K_{r+2}(\underline{C})$.

<u>Connections with Arithmetic.</u> Let E be a number field with r_1 real completions and r_2 complex completions. Then the maps

$$BG\ell^+(\underline{Z}) \to BG\ell^+(E) \to \Pi BG\ell^+(\underline{R})\times\Pi BG\ell^+(\underline{C}) \to \Pi F\underline{R} \times \Pi F\underline{C}$$

induce maps

$$K_r(\underline{Z}) \to K_r(E) \to \Pi\pi_r(F\underline{R})\times\Pi\pi_r(F\underline{C}).$$

For example, if $r\equiv 2 \pmod 8$ then $K_r(E)$ contains as summand $(\underline{Z}/2)\times\cdots\times(\underline{Z}/2)$ (r_1 times). To see this, let $e_\nu \in E$ be negative in the νth real embedding, and positive in all other real embeddings. The elements $\{e_\nu,\mu\}$ of $K_r(E)$ yield the r_1 district $(\underline{Z}/2)$-summands.

Similarly, the $\{e_\nu,e_\nu,\mu_{8n+1}\}$ in $K_{8n+3}(E)$ map to the $(0,\ldots,0,1/2,0,\cdots,0)$ in $\Pi\underline{Q}/\underline{Z} = \Pi\pi_{8n+3}(F\underline{R})$. Thus $K_{8n+3}(E)$ contains at least r_1 cyclic subgroups of even order. I do not know how large these subgroups are, but galois theory suggests that they are all of the same size. Shapiro [Sh 81] claimed that all the r_1 summands in $K_3(E)$ have order divisible by 4, but there is a mistake in in his argument.

Similarly, if E has r_2 complex embeddings, then there is a map from $K_{2n+1}(E)$ to the summand $\Pi\underline{Q}/\underline{Z} = \Pi\pi_{2n+1}(F\underline{C})$ of $K_{2n+1}(\underline{C})$. This should often yield r_2 more cyclic subgroups of $K_{2n+1}(E)$. Their size can be estimated by transferring to K_{2n+1} of the maximal totally real subfield of E. For example, Shapiro's claim (if true) would imply that $K_3(\underline{Q}(e^{2\pi i/p}))$ contains $(p-1)/2$ cyclic summands of even order.

ACKNOWLEDGMENTS

The author wishes to thank Vic Snaith for his help in converting a mod p homotopy calculation into a topological theorem, and in particular his help with Proposition 3.

REFERENCES

[Adams 65] J.F. Adams, On the groups J(X)-IV, Topology 5(1966), 21-71.

[BK] A. Bousfield and D. Kan, Homotopy Limits, Completions and Localizations, Lecture Notes in Math. No. 304, Springer-Verlag, 1972.

[Browder 77] W. Browder, Algebraic K-theory with coefficients Z/p, Lecture Notes in Math. No. 657, Springer-Verlag, 1976.

[Cartan] H. Cartan, Algèbre d'Eilenberg-MacLane et homotopie, Seminaire Henri Cartan 1954/1955, Ecole Normale Superieure, Paris, 1956.

[Mahowald 70] M. Mahowald, The order of the image of the J-homomorphism, Bull. AMS 76(1970), 1310-1313.

[Milnor 74] J. Milnor, Characteristic Classes, Ann. of Math Studies No. 76, Princeton Univ. Press, 1974.

[Milnor 81] J. Milnor, On the homology of Lie groups made discrete, preprint (1981).

[Q 72] D. Quillen, Letter from Quillen to Milnor on $\mathrm{Im}(\pi_i O \to \pi_i^s \to K_i\underline{Z})$, Lecture Notes in Math No. 551, Springer-Verlag, 1976.

[Sh 81] J. Shapiro, A modified Stiefel-Whitney invariant and a calculation in K_3 of an algebraic number field, J. Pure Appl. Alg. 24 (1982), 321-330.

[Snaith] V. Snaith, Algebraic cobordism and K-theory, Mem. AMS 221, A.M.S., Providence, 1979.

[Wald] F. Waldhausen, Algebraic K-theory of spaces, a manifold approach, preprint (1981).

[Weyl] H. Weyl, Symmetry, Princeton Univ. Press, Princeton, 1952.

Mennicke-type Symbols for Relative K_2

Charles A. Weibel
Rutgers University

Let I be a radical ideal in a commutative domain A. In his Irvine notes (q.v. p. 57), Bloch gave a Mennicke-type symbol for $K_2(A,I)$, and used it to detect elements in the image of $K_3(A/I) \to K_2(A,I)$. It seems to me that these symbols are a potentially powerful tool for getting our hands on elements of K_3.

The purpose of this essay is to outline some basic properties of such symbols. Denoting the universal Mennicke-type group by $MT_2(A,I)$, our primary result is this:

<u>Theorem</u> (2.5),(3.1). Let A be a semilocal PID with $I=\Pi p_i^{e_i} A$, the p_i being prime. elements. Then there is an exact sequence

$$\prod_i A/p_i^{e_i-1} \xrightarrow{\sigma} K_2(A,I) \xrightarrow{\psi} MT_2(A,I) \xrightarrow{\Pi v_i} \prod_{p_i | e_i} I/p_i I \to 0.$$

In particular, $K_2(A,I) \cong MT_2(A,I)$ if $I = \Pi p_i A$. The image of σ lies in the image of $K_3(A/I) \to K_2(A,I)$.

Here are some instances in which these Mennicke-type symbols are useful:

(0.1) $A = k[t]_{(t)\cup(t-1)}$, $I = (t^2-t)A$, where k is an algebraically closed field.

In this case, the image of $K_3(A/I) \to K_2(A,I)$ is the indecomposable part of $K_3(k)$. Bloch's "-*" map of [Bloch, (5.37)] is a Mennicke-type symbol with values in $H_3(S\ell_2(k),k^*;\underline{Z})$. When $k = \underline{C}$, composition with a volume form gives Bloch's dilogarithm regulator $K_3(\underline{C}) \to \underline{R}$; see [Bloch, §6] and [DS] for details. This Mennicke-type symbol is further analyzed in [W-C].

(0.2) $A = \underline{F}_p[t]_{(t)}$, $I = t^eA$, where $p>2e$ or $p=e$.

In this case, the p-primary subgroup $K_3(A/I,t)$ of $K_3(A/I)$ injects into $K_2(A,I)$. In §2 we will identify a summand of $K_3(A/I)$ called "wild-$K_2(A,I)$" which is isomorphic to $\underline{F}_p[t]/(t^{e-1})$. All the rest of $K_3(A/I,t)$ is detected by Mennicke-type symbols. In future work, I intend to use Mennicke-type symbols to compute $K_3(A/I)$. When e=2, this program should recover the results of Evens and Friedlander [EF]: $K_3(\underline{F}_p[t]/(t^2),t) = \underline{F}_p \oplus \underline{F}_p$.

(0.3) $A = \underline{Z}_{(5)\cup(p)}$, $I = 5p^eA$.

Here the group $K_3(\underline{Z}/p^e)$ injects into $K_2(A,I)$. For $p\geq5$, $K_3(\underline{Z}/p^e)$ is known by Aisbett [Aisbett] to be $\underline{Z}/p^{2(e-1)} \oplus K_3(\underline{Z}/p)$, and I believe that Mennicke-type symbols can be used to give a more elementary proof of this result.

§1. The group $MT_2(R,I)$

In this section we give the basic definitions we will use. 'A' will denote a commutative domain, 'F' its quotient field, and 'I' will denote a radical ideal of A. Recall (from [Stienstra, p. 413]) that the abelian group $K_2(A,I)$ is presented by symbols $\langle a,b\rangle$ with (a,b) in A×I or I×A, subject to the defining relations

(D1) $\langle a,b\rangle + \langle b,a\rangle = 0$ if a is in I,

(D2) $\langle a,c\rangle + \langle b,c\rangle = \langle a+b-abc,c\rangle$ if c is in I or if a,b are in I,

(D3) $\langle ca,b\rangle + \langle ab,c\rangle = \langle a,bc\rangle$ if a is in I.

We will frequently use the following identities as well (a or b in I, m an integer):

$$\langle a,b^m\rangle = m\langle ab^{m-1},b\rangle \quad \text{and} \quad m\langle a,b\rangle = \left\langle \frac{1-(1-ab)^m}{b},b\right\rangle.$$

(The second identity makes sense because b is a nonzero divisor.)

Definition (1.1). A Mennicke-type symbol is a group homomorphism from $MT_2(A,I)$ into an abelian group. $MT_2(A,I)$ is the abelian group on symbols $\left[{f \atop g}\right]$ with f in $F-0 = F^*$ and g in $(1+I)^\times$ (the group 1+I under multiplication), subject to the two relations

(i) $\left[{f \atop g}\right]$ is multiplicative in f and g.

(ii) $\left[{f \atop 1-f}\right] = 0$ for f in I.

Alternatively, $MT_2(A,I)$ is the quotient of $F^*\otimes(1+I)^\times$ by the subgroup $\mathcal{R}$ generated by the elements $f\otimes(1-f)$. It is important to notice the fundamental asymmetry of these symbols, which accounts for use of the asymmetrical symbol $\left[{f \atop g}\right]$. It is perhaps worth remarking that the group $MT_2(A,I)$ depends only on F and I, not on A. In particular, this group satisfies "excision."

Example (1.2). The Steinberg symbol $\left[{f \atop g}\right] \mapsto \{f,g\}$ is a Mennicke-type symbol with values in $K_2(F)$. It is relatively easy to show that we cannot always lift this to a symbol with values in $K_2(A)$. However, if A is a semilocal PID and all the primes of A are tame for I, then a Steinberg symbol with values in $K_2(A)$ exists; see Theorem (3.1) below.

Lemma (1.3)(Bloch). There is a well-defined group homomorphism $\psi: K_2(A,I) \to MT_2(A,I)$ given by $\psi(\langle a,b\rangle) = \begin{bmatrix} b \\ 1-ab \end{bmatrix} = \begin{bmatrix} a^{-1} \\ 1-ab \end{bmatrix}$.

Proof. If either a or b is zero, set $\psi\langle a,b\rangle = 0$. Now checking the (Di) identities is immediate. These checks are performed explicitly on p. 58 of [Bloch]. (Forget the context and read the computations.)

Example (1.4) (to show that ψ is not always into). Let p be an element of A such that p^e is in I.

Then in $MT_2(A,I)$ we have

$$\psi((e+1)\langle p^e,p\rangle) = (e+1)\begin{bmatrix} p \\ 1-p^{e+1} \end{bmatrix} = \begin{bmatrix} p^{e+1} \\ 1-p^{e+1} \end{bmatrix} = 0.$$

However, the symbol $(e+1)\langle p^e,p\rangle$ need not be in zero in $K_2(A,I)$ (see Example (2.7) below).

Example (1.5) (to show that ψ is not always onto). Let $\mathfrak{p}$ be a prime ideal of A such that $A_{\mathfrak{p}}$ is a discrete valuation ring, and let ν denote the $\mathfrak{p}$-adic valuation on F. If $I \subseteq \mathfrak{p}$ then $f \otimes g \mapsto \nu(f)(1-g)$ defines an abelian group homomorphism $v: F^* \otimes (1+I)^\times \to I/I^2$. Define e to be the integer $\min\{\nu(I)\}$, i.e., such that $IA_{\mathfrak{p}} = \mathfrak{p}^e A$.

For a in I and b in A, either ab is in I or else $\nu(a) = e, \nu(b) = 0$. Consequently, $v(ab \otimes (1-ab)) = (\nu(a)+\nu(b))ab$ belongs to $\mathfrak{p}I+eI$. Similarly, $v(b \otimes (1-ab))$ and $v(a^{-1} \otimes (1-ab))$ lie in $\mathfrak{p}I+eI$. Therefore, v induces a surjective Mennicke-type symbol:

$$v: MT_2(A,I) \to I/(\mathfrak{p}I+eI)$$

with $v\psi = 0$. If $\mathrm{char}(A/\mathfrak{p})$ divides e, for example, then $I/\mathfrak{p}I \neq 0$ and the map ψ cannot be onto. The map v is a kind of "co-wild" symbol in this case.

§2. Wild-K_2

In this section we try to describe the group wild-$K_2(A,I)$, and in particular show when it vanishes. For this section we drop the requirement that A be a domain. I am grateful to the referee for the present formulation; my original formulation was entirely in terms of the σ-symbols.

Definition (2.1). Let $a_1, \cdots, a_n$ be elements of A such that $\prod_{i \neq j} a_j$ is in I for $i=1,\cdots,n$. Define

$$((a_1,\cdots,a_n)) = \sum_i \langle a_i, \prod_{i\neq j} a_j \rangle \quad \text{in } K_2(A,I).$$

The subgroup of $K_2(A,I)$ generated by these symbols will be denoted wild-$K_2(A,I)$. If p,q are fixed elements of A such that $p^e q$ is in I, define (e,p,q)-wild-$K_2(A,I)$ to be the subset $\{\sigma(a,e,p,q)\colon a \text{ in } A\}$ of $K_2(A,I)$, where

$$\sigma(a,e,p,q) = ((apq,\underbrace{p,\ldots,p}_{e\ \text{times}},q))$$

$$= e\langle p, ap^e q^2\rangle + \langle q, ap^{e+1}q\rangle + \langle apq, p^e q\rangle.$$

Note that if p^e is in I then $\sigma(1,e,p,1) = (e+1)\langle p,p^e\rangle$; see (1.4).

These σ-symbols are called "wild" because no Mennicke-type symbol detects them:

<u>Lemma</u> (2.2) ψ maps wild-$K_2(A,I)$ to zero.

<u>Proof</u>. Let $u = 1-\Pi a_i$. Then $\psi((a_1,\cdots,a_n))$ equals $\sum\left[\begin{smallmatrix}a_i\\u\end{smallmatrix}\right] = \left[\begin{smallmatrix}\Pi a_i\\u\end{smallmatrix}\right] = 0$.

<u>Exercises</u> (2.3):

(a) (Multilinearity) Show that if $((a_1,\cdots,a_n))$ exists, then

$((a_1u_1,\cdots,a_nu_n)) = ((a_1\Pi u_i,a_2,\cdots,a_n))$.

In particular, $\sigma(a,e,up,vq) = \sigma(au^{e+1}v^2,e,p,q)$.

(b) (Contraction) If Πc_i is in I, show that

$((a,b,c_1,\cdots,c_n)) = ((ab,c_1,\cdots,c_n))$.

In particular, $\sigma(a,e+1,p,q) = \sigma(ap,e,p,q)$.

(c) Show that if a,b are in I then $((a,b)) = 0$. Then show that

$((a_1,\cdots,a_n)) = 0$ if any a_i is in I. Conclude that wild-$K_2(A,I) = 0$ when

I is a prime ideal, and that $(1,p,q)$-$K_2(A,I) = 0$.

(d) Show that $((a,c_1,\cdots,c_n)) + ((b,c_1,\cdots,c_n)) = ((a+b-ab\Pi c_i,c_1,\cdots,c_n))$.

Conclude that (e,p,q)-wild-$K_2(A,I)$ is a group.

<u>Proposition</u> (2.4). Fix $c_1,\cdots,c_n$ so that Πc_i is in I. Set

$$J = \bigcap_i \{a \text{ in } A:\ a \prod_{i\neq j} c_i \text{ is in } I\}.$$

Then there is a well-defined group homomorphism from J/I to wild-$K_2(A,I)$ sending a to $((a,c_1,\cdots,c_n))$. In particular, there is a surjection from A/p^{e-1} onto (e,p,q)-wild-$K_2(A,I)$ which sends a to $\sigma(a,e,p,q)$.

<u>Proof</u>. If b is in I then $((b,c_1,\cdots,c_n)) = 0$ by (2.3.c). By (2.3.d) we then have (for a in J, b in I)

$$\begin{aligned}((a+b,c_1,\cdots,c_n)) &= ((a,c_1,\cdots,c_n)) + ((b(1-a\Pi c_i)^{-1},c_1,\cdots,c_n))\\ &= (a,c_1,\cdots,c_n)).\end{aligned}$$

Hence $a \mapsto ((a,c_1,\cdots,c_n))$ is a well-defined set map. It is a group map by (2.3.d) again. In the special case that $(c_1,\cdots,c_n) = (p,\cdots,p,q)$ the ideal pqA lies in J, and the map $A/p^{e-1}A \to pqA/p^eqA \to J/I \to \text{wild-}K_2(A,I)$ sends a to $\sigma(a,e,p,q)$.

Question (2.4.1). Is wild-$K_2(A,I)$ an A/I-module under the action $a\cdot((c_1,\cdots,c_n)) = ((ac_1,\cdots,ac_n))$?

Remark (2.5). The group wild-$K_2(A,I)$ is in the image of $K_3(A/I) \to K_2(A,I)$, because in $K_2(A)$ we have $((a_1,\cdots,a_n)) = \sum\langle a_i, \prod_{j\neq i} a_j\rangle = \langle 1,\Pi a_j\rangle = 0$

If n=3, Loday [L] has constructed an element $\langle a_1,a_2,a_3\rangle$ of $K_3(A/I)$ which maps to $((a_1,a_2,a_3))$; in particular, $\langle ap,p,p\rangle$ maps to $\sigma(a,2,p,1)$ if p^2 is in I. I do not know canonical lifts if $n \geq 4$, however.

We now restrict our attention to the case that p^eq is in I, and develop a lower bound for (e,p,q)-wild-$K_2(A,I)$. One way to show that elements of $K_2(A,I)$ are nonzero is to study their images under the map $K_2(A,I) \to K_2(A/I^2,I/I^2) \to I/I^2 \otimes \Omega_{A/I}$ (see [W-nil, (1.3)]). In the special case $I=p^eqA$ we can further compose with the map

$$I/I^2 \otimes \Omega_{A/I} \to (p^eA/p^{2e}A) \otimes \Omega_{A/p^e} \cong \Omega_{A/p^e}$$

to obtain the following Kähler differential criterion:

Proposition (2.6). Suppose that $(p,q)A=A$, and choose r in A so that $rq^2 \equiv 1 \pmod{p^e}$. The composite of $(a \mapsto \sigma(ar,e,p,q)): A/p^{e-1} \to$ (e,p,q)-wild-$K_2(A,p^eq)$ with the map $K_2(A,p^eq) \to \Omega_{A/p^e}$ is the map $A/p^{e-1} \to \Omega_{A/p^e}$ sending a to $a(e+1)dp + pda$. In particular, (e,p,q)-wild-$K_2(A,p^eq)$ is nonzero unless we have $(e+1)dp = 0$ in Ω_{A/p^e} as well as $p\Omega_{A/p^e} = 0$.

Proof. The element $\sigma(ar,e,p,q)$ maps into Ω_{A/p^e} as the element

$$\begin{aligned} earq^2dp + arpqdq + qd(arpq) &= (e+1)arq^2dp + 2arpqdq + pq^2d(ar) \\ &= (arq^2)(e+1)dp + pd(arq^2) \\ &= a(e+1)dp + pda. \end{aligned}$$

Taking a=1 yields the result.

Example (2.7). Let $A = k[t]_{(t)}$ and let $e > 2$. Then $A/t^{e-1} \otimes \Omega_{(A/t^e)/k} \cong A/t^{e-1}$.

The composite $A/t^{e-1} \to (e,t,1)\text{-wild-}K_2(A,t^e) \to \Omega_{A/t^e} \to A/t^{e-1}$ sends $\bar{a}$ to $(e+1)a + t(da/dt)$. If $(e+1),\dots,(2e-1)$ are nonzero divisors on k it is easy to see that this map is an injection. This implies that $(e,t,1)\text{-wild-}K_2(k[t]_{(t)},t^e) \cong k[t]/(t^{e-1})$, as claimed in Example (0.2) above.

If e=2 and $k=\underline{F}_3$, this method does not tell us whether or not $3\langle t^2,t\rangle$ vanishes, and hence whether or not $(2,t,1)\text{-wild-}K_2(A,t^2)$ is zero. The element $\langle t^2,t\rangle$ has order 3 or 9 in the (3-torsion) subgroup $K_3(\underline{F}_3[t]/(t^2),t)$ of $K_2(A,t^2)$. Compare with [Stienstra, p.412] and [ALS].

Example (2.8). At the other extreme from Example (2.7) are rings for which $\Omega_A = 0$. The simplest case is $A = \underline{Z}_{(p)}$, $I = p^eA$. In this case the image of the group K_3A in $K_3(A/I)$ has order 24 (if $p \neq 2,3$) and $K_3(\underline{Z}/p^e) \cong \underline{Z}/p^{2(e-1)} \oplus \underline{Z}/(p^2-1)$ by [Aisbett]. From

$$K_3(A) \to K_3(\underline{Z}/p^e) \to K_2(A,p^e) \to K_2(A) \to 0$$

and the fact that $\text{wild-}K_2(A,p^e)$ is a subgroup of $\underline{Z}/p^{2(e-1)}$, it follows that $MT_2(A,p^e)$ detects at least some of the image of $K_3(\underline{Z}/p^e) \to K_2(A,p^e)$.

§3. Semilocal PID's

In this section we will assume that A is a semilocal P.I.D. with maximal ideals $\mathfrak{p}_1,\ldots,\mathfrak{p}_r$, and that $I = \Pi\mathfrak{p}_i^{e_i}$. We have seen that ψ is neither into nor onto in general. If $\mathrm{char}(A/\mathfrak{p}_i)$ divides e_i, we say that $\mathfrak{p}_i$ is <u>wild</u> for I, and that $\mathfrak{p}_i$ is <u>tame</u> for I otherwise.

<u>Theorem</u> (3.1). Let A be a semilocal PID.

(a) If I is unramified (i.e., I is a product of distinct primes), then ψ is an isomorphism: $K_2(A,I) \cong MT_2(A,I)$.

(b) ψ is onto exactly when all primes lying over I are tame for I.

(c) In general, there is an exact sequence:

$$0 \to \text{wild-}K_2(A,I) \to K_2(A,I) \xrightarrow{\psi} MT_2(A,I) \xrightarrow{\Pi v_i} \prod_{\substack{\text{wild} \\ \mathfrak{p}_i}} (I/\mathfrak{p}_i I) \to 0.$$

(d) If $I = \Pi\mathfrak{p}_i^{e_i}$, choose p_i, q_i so that $\mathfrak{p}_i = p_i A$ and $I/\mathfrak{p}_i^{e_i} = q_i A$. Then

$$\text{wild-}K_2(A,I) = \sum_i (e_i,p_i,q_i)\text{-wild-}K_2(A,I)$$

Note that parts (a) and (b) follow from (c). Also note that $(e_i,p_i,q_i)\text{-wild-}K_2(A,I)$ is independent of the choice of p_i, q_i by Exercise (2.2a).

We start the proof by creating a convenient notation. For each i, set

$\mathfrak{q}_i = I/\mathfrak{p}_i^{e_i} = \prod_{j\neq i} \mathfrak{p}_j^{e_j}$ and choose p_i in $\mathfrak{p}_i - \mathfrak{p}_i^2$ congruent to 1 mod $\mathfrak{q}_i$. Then set $q_i = \prod_{i\neq j} p_j^{e_j}$ and $s = \Pi p_i^{e_i}$. Thus the principal ideal I is generated by s, and $s = p_i^{e_i} q_i$ for each i.

(3.2) Proof of (3.1.d): By (2.3.b) the group wild-$K_2(A,I)$ is generated by symbols

$$\sigma(a) = ((a,\underbrace{p_1,\cdots,p_1}_{(e_1+1)\text{times}},\cdots,\underbrace{p_r,\cdots,p_r}_{(e_r+1)\text{times}})) = ((a\Pi p_i,\underbrace{p_1,\cdots,p_1}_{e_1\text{ times}},\cdots,\underbrace{p_r,\cdots,p_r}_{e_r\text{ times}}))$$

Choose a_i in A so that $a = \Sigma a_i q_i$; by (2.4) we have $\sigma(a) = \Sigma\sigma(a_i q_i)$, so we may assume that $a = a_1 q_1$. By repeated use of (2.3.a,b) we then have

$$\sigma(a_1q_1) = \big((a_1q_1\Pi p_i,\underbrace{p_1,\cdots,p_1}_{e_1\text{ times}},q_1)\big)$$

$$= \sigma\big(a_1(\prod_{i\neq 1} p_i),e_1,p_1,q_1\big)$$

$$= \sigma(a_1,e_1,p_1,q_1).$$

Our proof of (3.1.c) is based on the following elementary result.

Lemma (3.3). If $\mathfrak{p}_i$ is wild for I, there is an exact sequence of groups:

$$1 \to (1+\mathfrak{p}_i I)^\times \to (1+I)^\times \to I/\mathfrak{p}_i I \to 0.$$

$$(1-t) \mapsto \bar{t}.$$

If $\mathfrak{p}_i$ is tame for I, there is an exact sequence of groups:

$$1 \to (1+I^2)^e \to (1+\mathfrak{p}^e I)\times(1+\mathfrak{q}I)^e \to (1+I) \to 1$$

$$f \longmapsto (f,f^{-1});\ (f,g) \longmapsto fg.$$

(For clarity, we have written $e,\mathfrak{p},\mathfrak{q}$ for $e_i,\mathfrak{p}_i,\mathfrak{q}_i$.)

Proof. The first sequence is obviously exact. Now suppose that $\mathfrak{p}=\mathfrak{p}_i$ is tame for I. Choose d in A so that $de \equiv 1 \bmod p^e$. Given a in A, set $f=(1-as)(1-daqs)^{-e}$. Since modulo p^eI we have $f \equiv 1-as +(de)aqs \equiv 1$, it follows that f is in $1+p^eI$. Exactness of the second sequence is now straightforward.

(3.4) Proof that Πv_i is onto: For a_i in A we have

$$(\Pi v_i) \sum_{\substack{\text{wild} \\ \mathfrak{p}_i}} \begin{bmatrix} p_i \\ 1-a_i s \end{bmatrix} = (a_1 s, \ldots, a_r s).$$

(3.5) Exactness at $MT_2(A,I)$: The composite $\Pi v_i \psi$ is zero by (1.5). Because A is a UFD, the group F^* is the direct sum of A^* (the units of A) and the free abelian group on the elements $p_1, \ldots, p_r$. The kernel of Πv_i is generated by the subgroups $\begin{bmatrix} A^* \\ 1+I \end{bmatrix}$, $\begin{bmatrix} p_i \\ 1+I \end{bmatrix}$ for tame $\mathfrak{p}_i$, and $\begin{bmatrix} p_i \\ 1+p_i I \end{bmatrix}$ for wild $\mathfrak{p}_i$. For u in A^* and a in A we have $\psi(\langle u^{-1}as, u\rangle) = \begin{bmatrix} u \\ 1-as \end{bmatrix}$. When $\mathfrak{p}_i$ is tame, Lemma (3.3) and the identities

$$\psi(\langle ap_i^{e_i - 1} s, p_i\rangle) = \begin{bmatrix} p_i \\ 1-ap_i^{e_i} s \end{bmatrix}$$

$$\psi(\langle aq_i, s\rangle - \langle as, q_i\rangle) = \begin{bmatrix} p_i^{e_i} \\ 1-aq_i s \end{bmatrix} = \begin{bmatrix} p_i \\ (1-aq_i s)^{e_i} \end{bmatrix}$$

show that $\begin{bmatrix} p_i \\ 1+I \end{bmatrix}$ is also in the image of ψ. Finally, if $\mathfrak{p}_i$ is wild, the equation

$$\psi(\langle as, p_i\rangle) = \begin{bmatrix} p_i \\ 1-ap_i s \end{bmatrix}$$

shows that $\begin{bmatrix} p_i \\ 1+p_i I \end{bmatrix}$ is in the image of ψ, too.

Let U be the kernel of the map $F^* \otimes (1+I)^\times \to MT_2(A,I) \to \prod_{\substack{\text{wild} \\ \mathfrak{p}_i}} (I/p_i I)$. The group U decomposes as a direct sum:

$$U = A^* \otimes (1+I)^\times \oplus \coprod_{\substack{\text{tame} \\ \mathfrak{p}_i}} \{p_i\} \otimes (1+I)^\times \oplus \coprod_{\substack{\text{wild} \\ \mathfrak{p}_i}} \{p_i\} \otimes (1+p_i I)^\times$$

We will construct a map ϕ from U to $K_2(A,I)$, and then show that ϕ induces a splitting of $\overline{\psi}: K_2(A,I)/\text{wild-}K_2(A,I) \to \ker[MT_2(A,I) \to \Pi(I/p_i I)]$.

(3.6) <u>Construction of $\phi:U \to K_2(A,I)$</u>. Define maps:

$$\phi_u: A^*\otimes(1+I)^\times \to K_2(A,I)$$

$$\phi_u(u\otimes(1-as)) = \{1-as,u\} = \langle u^{-1}as,u\rangle \qquad (u \text{ in } A^*)$$

$$\phi_i': (1+\mathfrak{p}_i^{e_i}I)^\times \to K_2(A,I)$$

$$\phi_i'(1-ap^es) = \langle ap^{e-1}s,p\rangle \qquad (a \text{ in } A,\ p=p_i,\ e=e_i,\ q=q_i)$$

$$\phi_i'': (1+\mathfrak{q}_i I)^\times \to K_2(A,I)$$

$$\phi_i''(1-aqs) = \langle aq,s\rangle - \langle as,q\rangle$$

$$\phi_i''': (1+\mathfrak{p}_i I)^\times \to K_2(A,I)$$

$$\phi_i'''(1-aps) = \langle as,p\rangle.$$

It is easy to see that these are group homomorphisms. If $\mathfrak{p}_i$ is wild, define $\phi_i:\{p_i\}\otimes(1+p_iI)^\times \to K_2(A,I)$ by $\phi_i(p_i\otimes f) = \phi_i'''(f)$. If $\mathfrak{p}_i$ isn't wild, we use (3.3) to define $\phi_i:\{p_i\}\otimes(1+I)^\times \to K_2(A,I)$ by the formula

$$\phi_i(p_i\otimes(1-ap^es)(1-bqs)^e) = \phi_i'(1-ap^es) + \phi_i''(1-bqs)$$

$$= \langle ap^{e-1}s,p\rangle + \langle bq,s\rangle - \langle bs,q\rangle.$$

To see that ϕ_i is well-defined, we need to show two things:

(i) if $(1-b_1qs)^e = (1-b_2qs)^e$, then $b_1 = b_2$. This is a consequence of the fact that $\mathfrak{p}$ is tame, i.e., that $e \not\equiv 0 \bmod \mathfrak{p}$.

(ii) that the two possible definitions of ϕ_i agree on $p_i\otimes(1+I^2)^e$.

Choose $f = (1-cs^2)^e$ in $(1+I^2)^e$, and let a,b be such that $f = (1-ap^es) = (1-bqs)^e$. Then we have

$$\phi_i'(1-ap^e s) = \langle ap^{e-1}s,p\rangle = e\langle cp^{e-1}qs,p\rangle$$
$$= \langle cqs,p^e\rangle = \langle cs,p^e q\rangle - \langle cp^e s,q\rangle$$
$$= \langle bq,s\rangle - \langle bs,q\rangle \quad = \quad \phi_i''(1-bqs).$$

The ϕ_i combine to give a map $\phi: U \to K_2(A,I)$.

(3.7) Exactness at $K_2(A,I)$: The computations of (3.5) and Lemma (2.2) show that there is a commutative diagram

$$\begin{array}{ccccccc} & U & \xrightarrow{\overline{\phi}} & K_2(A,I)/\text{wild-}K_2(A,I) & & & \\ & \downarrow & \swarrow_{\Psi} & & & & \\ 0 \longrightarrow & U/\mathcal{R} & \longrightarrow & MT_2(A,I) & \longrightarrow & \prod_{\text{wild } \mathfrak{p}} (I/\mathfrak{p}_i I) & \longrightarrow 0 \end{array}$$

We will show in (3.9) below that $\overline{\phi}$ is onto, and in (3.11) that $\overline{\phi}(\mathcal{R}) = 0$. It follows that $\overline{\phi}/\mathcal{R}$ is the inverse of $\overline{\psi}$, which is what was to be shown.

<u>Lemma</u> (3.8). If f is in I then $\phi_i(p_i \otimes (1-p_i f)) \equiv \langle f,p_i\rangle$ modulo wild-$K_2(A,I)$.

<u>Proof</u>. If $\mathfrak{p}_i$ is wild we have equality. We therefore assume that $\mathfrak{p}_i$ is tame and drop the subscript i. Write $(1-pf) = (1-ap^e s)(1-bqps)^e$; this can be done by (3.3) since $1-pf \in 1+I$. Then

$$\langle f,p\rangle = \langle ap^{e-1}s,p\rangle + e\langle bqs,p\rangle$$
$$\phi(p\otimes(1-pf)) = \langle ap^{e-1}s,p\rangle + \langle bqp,s\rangle - \langle bps,q\rangle$$
$$\langle f,p\rangle - \phi(p\otimes(1-pf)) = e\langle bqs,p\rangle - \langle bqp,s\rangle + \langle bps,q\rangle$$
$$= \sigma(b,e,p,q).$$

<u>Corollary</u> (3.9). The induced map $\overline{\phi}: U \to K_2(A,I)/\text{wild-}K_2(A,I)$ is onto.

<u>Proof</u>. Using (D3), it follows that $K_2(A,I)$ is generated by symbols $\langle f,u\rangle$, $\langle f,p_i\rangle$ where f is in I and u is in A^*.

Corollary (3.10). For all a in A, f in I we have

$$\phi(a\otimes(1-af)) \equiv \langle f,a\rangle \text{ modulo wild-}K_2(A,I).$$

Proof. Write $a = u\Pi p_i^{d_i}$ with u a unit, and set $g = 1-af$.

$$\phi(a\otimes g) \equiv \langle u^{-1}af,u\rangle + \sum d_i\langle p_i^{-1}af,p_i\rangle.$$

By (D3) this is just $\langle f,a\rangle$.

Lemma (3.11). $\phi(\mathcal{R})$ is contained in wild-$K_2(A,I)$.

Proof. We have to show that for all a in A we have $\phi(s\otimes(1-as)) \equiv \langle a,s\rangle$, because with (3.10) this will imply that

$$\begin{aligned}\phi(as\otimes(1-as)) &= \phi(a\otimes(1-as)) + \phi(s\otimes(1-as))\\ &\equiv \langle s,a\rangle + \langle a,s\rangle = 0.\end{aligned}$$

If $a = bs$, this follows from (3.10). If $a = bq_i$, we have

$$\begin{aligned}\phi(s\otimes(1-bq_is)) &= \phi(p_i\otimes(1-bq_is)^{e_i}) + \sum_{i\neq j} e_j\phi(p_j\otimes(1-bq_is))\\ &\equiv (\langle bq_i,s\rangle - \langle bs,q_i\rangle) + \langle bs,\Pi p_j^{e_j}\rangle\\ &= \langle bq_i,s\rangle.\end{aligned}$$

Finally, every element of 1+I can be written in the form $g = (1-as^2)\Pi(1-b_iq_is)$ for a,b_i in A (take $b_i=(1-g)/s$ for example). The additivity of ϕ yields

$$\begin{aligned}\phi(s\otimes g) &= \phi(s\otimes(1-as^2)) + \sum\phi(s\otimes(1-b_iq_is))\\ &\equiv \langle as,s\rangle + \sum\langle b_iq_i,s\rangle.\end{aligned}$$

This is $\langle(1-g)/s,s\rangle$ by relation (D2). Done.

Acknowledgements

I would like to thank S. Bloch for defining the map ψ in the first place. In [Bloch, p. 58], Bloch states: "it seems likely that ψ is an isomorphism" (for his choice of A and I). Second, I would like to thank R. Coleman, who states in [C] that "it is not too difficult to check that Bloch's expectation is correct." The reader can see from this paper that it isn't very difficult, but that it isn't trivial either. Their assertions led to my search for the nature of ψ, directly resulting in this essay. I would like to thank the referee for suggesting that I rewrite section 2 in terms of the symbols $((a_1,\cdots,a_n))$. Finally, I would like to thank Jan Stienstra, who showed me (at the Bielefeld conference) that wild-K_2 need not be zero, and Sue Geller, whose comments greatly improved the exposition of this paper.

References

[Aisbett] J. Aisbett, thesis, University of Western Ontario, 1981.

[ALS] J. Aisbett, E. Lluis-Puebla and V. Snaith, On $K_*(\underline{Z}/n)$ and $K_*(\underline{F}_q[t]/(t^2))$, preprint (1982).

[Bloch] S. Bloch, Higher regulators, algebraic K-theory, and zeta functions of elliptic curves, Lectures notes at U.C. Irvine, 1977.

[C] R. Coleman, Dilogarithms, regulators and p-adic L-functions, preprint (1981).

[DS] J. Dupont and C.-H. Sah, Scissors Congruences II, J.P.A.A. 25 (1982), 159-195.

[EF] L. Evens and E. Friedlander, $K_r(\underline{Z}/p^2)$ and $K_r(\underline{Z}/p[\varepsilon])$ for $p\geq 5$ and $r\leq 4$, Bull.AMS 3 (1980), 440-443.

[L] J.-L. Loday, On the boundary map $K_3(\Lambda/I) \to K_2(\Lambda,I)$, Lecture Notes in Math. No. 854, Springer-Verlag, 1981.

[Stienstra] J. Stienstra, On K_2 and K_3 of truncated polynomial rings, Lecture Notes in Math. No. 854, Springer-Verlag, 1981.

[W-nil] C. Weibel, K_2, K_3 and nilpotent ideals, J.P.A.A. 18 (1980), 333-345.

[W-C] C. Weibel, K_3 of the complex numbers, in preparation.

Vol. 900: P. Deligne, J. S. Milne, A. Ogus, and K.-Y. Shih, Hodge Cycles, Motives, and Shimura Varieties. V, 414 pages. 1982.

Vol. 901: Séminaire Bourbaki vol. 1980/81 Exposés 561–578. III, 299 pages. 1981.

Vol. 902: F. Dumortier, P.R. Rodrigues, and R. Roussarie, Germs of Diffeomorphisms in the Plane. IV, 197 pages. 1981.

Vol. 903: Representations of Algebras. Proceedings, 1980. Edited by M. Auslander and E. Lluis. XV, 371 pages. 1981.

Vol. 904: K. Donner, Extension of Positive Operators and Korovkin Theorems. XII, 182 pages. 1982.

Vol. 905: Differential Geometric Methods in Mathematical Physics. Proceedings, 1980. Edited by H.-D. Doebner, S.J. Andersson, and H.R. Petry. VI, 309 pages. 1982.

Vol. 906: Séminaire de Théorie du Potentiel, Paris, No. 6. Proceedings. Edité par F. Hirsch et G. Mokobodzki. IV, 328 pages. 1982.

Vol. 907: P. Schenzel, Dualisierende Komplexe in der lokalen Algebra und Buchsbaum-Ringe. VII, 161 Seiten. 1982.

Vol. 908: Harmonic Analysis. Proceedings, 1981. Edited by F. Ricci and G. Weiss. V, 325 pages. 1982.

Vol. 909: Numerical Analysis. Proceedings, 1981. Edited by J.P. Hennart. VII, 247 pages. 1982.

Vol. 910: S.S. Abhyankar, Weighted Expansions for Canonical Desingularization. VII, 236 pages. 1982.

Vol. 911: O.G. Jørsboe, L. Mejlbro, The Carleson-Hunt Theorem on Fourier Series. IV, 123 pages. 1982.

Vol. 912: Numerical Analysis. Proceedings, 1981. Edited by G. A. Watson. XIII, 245 pages. 1982.

Vol. 913: O. Tammi, Extremum Problems for Bounded Univalent Functions II. VI, 168 pages. 1982.

Vol. 914: M. L. Warshauer, The Witt Group of Degree k Maps and Asymmetric Inner Product Spaces. IV, 269 pages. 1982.

Vol. 915: Categorical Aspects of Topology and Analysis. Proceedings, 1981. Edited by B. Banaschewski. XI, 385 pages. 1982.

Vol. 916: K.-U. Grusa, Zweidimensionale, interpolierende Lg-Splines und ihre Anwendungen. VIII, 238 Seiten. 1982.

Vol. 917: Brauer Groups in Ring Theory and Algebraic Geometry. Proceedings, 1981. Edited by F. van Oystaeyen and A. Verschoren. VIII, 300 pages. 1982.

Vol. 918: Z. Semadeni, Schauder Bases in Banach Spaces of Continuous Functions. V, 136 pages. 1982.

Vol. 919: Séminaire Pierre Lelong – Henri Skoda (Analyse) Années 1980/81 et Colloque de Wimereux, Mai 1981. Proceedings. Edité par P. Lelong et H. Skoda. VII, 383 pages. 1982.

Vol. 920: Séminaire de Probabilités XVI, 1980/81. Proceedings. Edité par J. Azéma et M. Yor. V, 622 pages. 1982.

Vol. 921: Séminaire de Probabilités XVI, 1980/81. Supplément: Géométrie Différentielle Stochastique. Proceedings. Edité par J. Azéma et M. Yor. III, 285 pages. 1982.

Vol. 922: B. Dacorogna, Weak Continuity and Weak Lower Semicontinuity of Non-Linear Functionals. V, 120 pages. 1982.

Vol. 923: Functional Analysis in Markov Processes. Proceedings, 1981. Edited by M. Fukushima. V, 307 pages. 1982.

Vol. 924: Séminaire d'Algèbre Paul Dubreil et Marie-Paule Malliavin. Proceedings, 1981. Edité par M.-P. Malliavin. V, 461 pages. 1982.

Vol. 925: The Riemann Problem, Complete Integrability and Arithmetic Applications. Proceedings, 1979-1980. Edited by D. Chudnovsky and G. Chudnovsky. VI, 373 pages. 1982.

Vol. 926: Geometric Techniques in Gauge Theories. Proceedings, 1981. Edited by R. Martini and E.M.de Jager. IX, 219 pages. 1982.

Vol. 927: Y. Z. Flicker, The Trace Formula and Base Change for GL (3). XII, 204 pages. 1982.

Vol. 928: Probability Measures on Groups. Proceedings 1981. Edited by H. Heyer. X, 477 pages. 1982.

Vol. 929: Ecole d'Eté de Probabilités de Saint-Flour X – 1980. Proceedings, 1980. Edited by P.L. Hennequin. X, 313 pages. 1982.

Vol. 930: P. Berthelot, L. Breen, et W. Messing, Théorie de Dieudonné Cristalline II. XI, 261 pages. 1982.

Vol. 931: D.M. Arnold, Finite Rank Torsion Free Abelian Groups and Rings. VII, 191 pages. 1982.

Vol. 932: Analytic Theory of Continued Fractions. Proceedings, 1981. Edited by W.B. Jones, W.J. Thron, and H. Waadeland. VI, 240 pages. 1982.

Vol. 933: Lie Algebras and Related Topics. Proceedings, 1981. Edited by D. Winter. VI, 236 pages. 1982.

Vol. 934: M. Sakai, Quadrature Domains. IV, 133 pages. 1982.

Vol. 935: R. Sot, Simple Morphisms in Algebraic Geometry. IV, 146 pages. 1982.

Vol. 936: S.M. Khaleelulla, Counterexamples in Topological Vector Spaces. XXI, 179 pages. 1982.

Vol. 937: E. Combet, Intégrales Exponentielles. VIII, 114 pages. 1982.

Vol. 938: Number Theory. Proceedings, 1981. Edited by K. Alladi. IX, 177 pages. 1982.

Vol. 939: Martingale Theory in Harmonic Analysis and Banach Spaces. Proceedings, 1981. Edited by J.-A. Chao and W.A. Woyczyński. VIII, 225 pages. 1982.

Vol. 940: S. Shelah, Proper Forcing. XXIX, 496 pages. 1982.

Vol. 941: A. Legrand, Homotopie des Espaces de Sections. VII, 132 pages. 1982.

Vol. 942: Theory and Applications of Singular Perturbations. Proceedings, 1981. Edited by W. Eckhaus and E.M. de Jager. V, 363 pages. 1982.

Vol. 943: V. Ancona, G. Tomassini, Modifications Analytiques. IV, 120 pages. 1982.

Vol. 944: Representations of Algebras. Workshop Proceedings, 1980. Edited by M. Auslander and E. Lluis. V, 258 pages. 1982.

Vol. 945: Measure Theory. Oberwolfach 1981, Proceedings. Edited by D. Kölzow and D. Maharam-Stone. XV, 431 pages. 1982.

Vol. 946: N. Spaltenstein, Classes Unipotentes et Sous-groupes de Borel. IX, 259 pages. 1982.

Vol. 947: Algebraic Threefolds. Proceedings, 1981. Edited by A. Conte. VII, 315 pages. 1982.

Vol. 948: Functional Analysis. Proceedings, 1981. Edited by D. Butković, H. Kraljević, and S. Kurepa. X, 239 pages. 1982.

Vol. 949: Harmonic Maps. Proceedings, 1980. Edited by R.J. Knill, M. Kalka and H.C.J. Sealey. V, 158 pages. 1982.

Vol. 950: Complex Analysis. Proceedings, 1980. Edited by J. Eells. IV, 428 pages. 1982.

Vol. 951: Advances in Non-Commutative Ring Theory. Proceedings, 1981. Edited by P.J. Fleury. V, 142 pages. 1982.

Vol. 952: Combinatorial Mathematics IX. Proceedings, 1981. Edited by E. Billington, S. Oates-Williams, and A.P. Street. XI, 443 pages. 1982.

Vol. 953: Iterative Solution of Nonlinear Systems of Equations. Proceedings, 1982. Edited by R. Ansorge, Th. Meis, and W. Törnig. VII, 202 pages. 1982.

Vol. 954: S.G. Pandit, S.G. Deo, Differential Systems Involving Impulses. VII, 102 pages. 1982.

Vol. 955: G. Gierz, Bundles of Topological Vector Spaces and Their Duality. IV, 296 pages. 1982.

Vol. 956: Group Actions and Vector Fields. Proceedings, 1981. Edited by J.B. Carrell. V, 144 pages. 1982.

Vol. 957: Differential Equations. Proceedings, 1981. Edited by D.G. de Figueiredo. VIII, 301 pages. 1982.

Vol. 958: F.R. Beyl, J. Tappe, Group Extensions, Representations, and the Schur Multiplicator. IV, 278 pages. 1982.

Vol. 959: Géométrie Algébrique Réelle et Formes Quadratiques, Proceedings, 1981. Edité par J.-L. Colliot-Thélène, M. Coste, L. Mahé, et M.-F. Roy. X, 458 pages. 1982.

Vol. 960: Multigrid Methods. Proceedings, 1981. Edited by W. Hackbusch and U. Trottenberg. VII, 652 pages. 1982.

Vol. 961: Algebraic Geometry. Proceedings, 1981. Edited by J.M. Aroca, R. Buchweitz, M. Giusti, and M. Merle. X, 500 pages. 1982.

Vol. 962: Category Theory. Proceedings, 1981. Edited by K.H. Kamps, D. Pumplün, and W. Tholen, XV, 322 pages. 1982.

Vol. 963: R. Nottrot, Optimal Processes on Manifolds. VI, 124 pages. 1982.

Vol. 964: Ordinary and Partial Differential Equations. Proceedings, 1982. Edited by W.N. Everitt and B.D. Sleeman. XVIII, 726 pages. 1982.

Vol. 965: Topics in Numerical Analysis. Proceedings, 1981. Edited by P.R. Turner. IX, 202 pages. 1982.

Vol. 966: Algebraic K-Theory. Proceedings, 1980, Part I. Edited by R.K. Dennis. VIII, 407 pages. 1982.

Vol. 967: Algebraic K-Theory. Proceedings, 1980. Part II. VIII, 409 pages. 1982.

Vol. 968: Numerical Integration of Differential Equations and Large Linear Systems. Proceedings, 1980. Edited by J. Hinze. VI, 412 pages. 1982.

Vol. 969: Combinatorial Theory. Proceedings, 1982. Edited by D. Jungnickel and K. Vedder. V, 326 pages. 1982.

Vol. 970: Twistor Geometry and Non-Linear Systems. Proceedings, 1980. Edited by H.-D. Doebner and T.D. Palev. V, 216 pages. 1982.

Vol. 971: Kleinian Groups and Related Topics. Proceedings, 1981. Edited by D.M. Gallo and R.M. Porter. V, 117 pages. 1983.

Vol. 972: Nonlinear Filtering and Stochastic Control. Proceedings, 1981. Edited by S.K. Mitter and A. Moro. VIII, 297 pages. 1983.

Vol. 973: Matrix Pencils. Proceedings, 1982. Edited by B. Kågström and A. Ruhe. XI, 293 pages. 1983.

Vol. 974: A. Draux, Polynômes Orthogonaux Formels – Applications. VI, 625 pages. 1983.

Vol. 975: Radical Banach Algebras and Automatic Continuity. Proceedings, 1981. Edited by J.M. Bachar, W.G. Bade, P.C. Curtis Jr., H.G. Dales and M.P. Thomas. VIII, 470 pages. 1983.

Vol. 976: X. Fernique, P.W. Millar, D.W. Stroock, M. Weber, Ecole d'Eté de Probabilités de Saint-Flour XI – 1981. Edited by P.L. Hennequin. XI, 465 pages. 1983.

Vol. 977: T. Parthasarathy, On Global Univalence Theorems. VIII, 106 pages. 1983.

Vol. 978: J. Ławrynowicz, J. Krzyż, Quasiconformal Mappings in the Plane. VI, 177 pages. 1983.

Vol. 979: Mathematical Theories of Optimization. Proceedings, 1981. Edited by J.P. Cecconi and T. Zolezzi. V, 268 pages. 1983.

Vol. 980: L. Breen. Fonctions thêta et théorème du cube. XIII, 115 pages. 1983.

Vol. 981: Value Distribution Theory. Proceedings, 1981. Edited by I. Laine and S. Rickman. VIII, 245 pages. 1983.

Vol. 982: Stability Problems for Stochastic Models. Proceedings, 1982. Edited by V.V. Kalashnikov and V.M. Zolotarev. XVII, 295 pages. 1983.

Vol. 983: Nonstandard Analysis-Recent Developments. Edited by A.E. Hurd. V, 213 pages. 1983.

Vol. 984: A. Bove, J.E. Lewis, C. Parenti, Propagation of Singularities for Fuchsian Operators. IV, 161 pages. 1983.

Vol. 985: Asymptotic Analysis II. Edited by F. Verhulst. VI, 497 pages. 1983.

Vol. 986: Séminaire de Probabilités XVII 1981/82. Proceedings. Edited by J. Azéma and M. Yor. V, 512 pages. 1983.

Vol. 987: C.J. Bushnell, A. Fröhlich, Gauss Sums and p-adic Division Algebras. XI, 187 pages. 1983.

Vol. 988: J. Schwermer, Kohomologie arithmetisch definierter Gruppen und Eisensteinreihen. III, 170 pages. 1983.

Vol. 989: A.B. Mingarelli, Volterra-Stieltjes Integral Equations and Generalized Ordinary Differential Expressions. XIV, 318 pages. 1983.

Vol. 990: Probability in Banach Spaces IV. Proceedings, 1982. Edited by A. Beck and K. Jacobs. V, 234 pages. 1983.

Vol. 991: Banach Space Theory and its Applications. Proceedings, 1981. Edited by A. Pietsch, N. Popa and I. Singer. X, 302 pages. 1983.

Vol. 992: Harmonic Analysis, Proceedings, 1982. Edited by G. Mauceri, F. Ricci and G. Weiss. X, 449 pages. 1983.

Vol. 993: R.D. Bourgin, Geometric Aspects of Convex Sets with the Radon-Nikodým Property. XII, 474 pages. 1983.

Vol. 994: J.-L. Journé, Calderón-Zygmund Operators, Pseudo-Differential Operators and the Cauchy Integral of Calderón. VI, 129 pages. 1983.

Vol. 995: Banach Spaces, Harmonic Analysis, and Probability Theory. Proceedings, 1980–1981. Edited by R.C. Blei and S.J. Sidney. V, 173 pages. 1983.

Vol. 996: Invariant Theory. Proceedings, 1982. Edited by F. Gherardelli. V, 159 pages. 1983.

Vol. 997: Algebraic Geometry – Open Problems. Edited by C. Ciliberto, F. Ghione and F. Orecchia. VIII, 411 pages. 1983.

Vol. 998: Recent Developments in the Algebraic, Analytical, and Topological Theory of Semigroups. Proceedings, 1981. Edited by K.H. Hofmann, H. Jürgensen and H.J. Weinert. VI, 486 pages. 1983.

Vol. 999: C. Preston, Iterates of Maps on an Interval. VII, 205 pages. 1983.

Vol. 1000: H. Hopf, Differential Geometry in the Large, VII, 184 pages. 1983.

Vol. 1001: D.A. Hejhal, The Selberg Trace Formula for PSL(2, ℝ). Volume 2. VIII, 806 pages. 1983.

Vol. 1002: A. Edrei, E.B. Saff, R.S. Varga, Zeros of Sections of Power Series. VIII, 115 pages. 1983.

Vol. 1003: J. Schmets, Spaces of Vector-Valued Continuous Functions. VI, 117 pages. 1983.

Vol. 1004: Universal Algebra and Lattice Theory. Proceedings, 1982. Edited by R.S. Freese and O.C. Garcia. VI, 308 pages. 1983.

Vol. 1005: Numerical Methods. Proceedings, 1982. Edited by V. Pereyra and A. Reinoza. V, 296 pages. 1983.

Vol. 1006: Abelian Group Theory. Proceedings, 1982/83. Edited by R. Göbel, L. Lady and A. Mader. XVI, 771 pages. 1983.

Vol. 1007: Geometric Dynamics. Proceedings, 1981. Edited by J. Palis Jr. IX, 827 pages. 1983.

Vol. 1008: Algebraic Geometry. Proceedings, 1981. Edited by J. Dolgachev. V, 138 pages. 1983.

Vol. 1009: T.A. Chapman, Controlled Simple Homotopy Theory and Applications. III, 94 pages. 1983.

Vol. 1010: J.-E. Dies, Chaînes de Markov sur les permutations. IX, 226 pages. 1983.